D0931645

FUNDAMENTAL STATISTICS IN PSYCHOLOGY AND EDUCATION

McGRAW-HILL SERIES IN PSYCHOLOGY

Consulting Editors
Norman Garmezy
Lyle V. Jones

Adams Human Memory
Berlyne Conflict, Arousal, and Curiosity
Blum Psychoanalytic Theories of Personality
Bock Multivariate Statistical Methods in Behavioral Research
Brown The Motivation of Behavior
Butcher MMPI: Research Developments and Clinical Applications
Campbell, Dunnette, Lawler, and Weick Managerial Behavior, Performance, and Effectiveness
Crites Vocational Psychology
D'Amato Experimental Psychology: Methodology, Psychophysics, and Learning
Dollard and Miller Personality and Psychotherapy
Edgington Statistical Inference: The Distribution-free Approach
Ferguson Statistical Analysis in Psychology and Education
Fodor, Bever, and Garrett The Psychology of Language: An Introduction to Psycholinguistics and Generative Grammar
Forgus and Melamed Perception: A Cognitive-Stage Approach
Franks Behavior Therapy: Appraisal and Status
Gilmer and Deci Industrial and Organizational Psychology
Guilford Psychometric Methods
Guilford The Nature of Human Intelligence
Guilford and Fruchter Fundamental Statistics in Psychology and Education
Guilford and Hoepfner The Analysis of Intelligence
Guion Personnel Testing
Hetherington and Parke Child Psychology: A Contemporary Viewpoint
Hirsh The Measurement of Hearing
Hjelle and Ziegler Personality Theories: Basic Assumptions, Research, and Applications
Horowitz Elements of Statistics for Psychology and Education
Hulse, Deese, and Egeth The Psychology of Learning
Hurlock Adolescent Development
Hurlock Child Development
Hurlock Developmental Psychology
Krech, Crutchfield, and Ballachey Individual in Society

Lakin Interpersonal Encounter: Theory and Practice in Sensitivity Training
Lawler Pay and Organizational Effectiveness: A Psychological View
Lazarus, A. Behavior Therapy and Beyond
Lazarus, R. Patterns of Adjustment
Lewin A Dynamic Theory of Personality
Maher Principles of Psychopathology
Marascuilo Statistical Methods for Behavioral Science Research
Marx and Hillix Systems and Theories in Psychology
Miller Language and Communication
Morgan Physiological Psychology
Mulaik The Foundations of Factor Analysis
Novick and Jackson Statistical Methods for Educational and Psychological Research
Nunnally Introduction to Statistics for Psychology and Education
Nunnally Psychometric Theory
Overall and Klett Applied Multivariate Analysis
Porter, Lawler, and Hackman Behavior in Organizations
Restle Learning: Animal Behavior and Human Cognition
Robinson and Robinson The Mentally Retarded Child
Rosenthal Genetic Theory and Abnormal Behavior
Ross Psychological Disorders of Children: A Behavioral Approach to Theory, Research, and Therapy
Schwitzgebel and Kolb Changing Human Behavior: Principles of Planned Intervention
Shaw Group Dynamics: The Psychology of Small Group Behavior
Shaw and Costanzo Theories of Social Psychology
Shaw and Wright Scales for the Measurement of Attitudes
Sidowski Experimental Methods and Instrumentation in Psychology
Siegel Nonparametric Statistics for the Behavioral Sciences
Spencer and Kass Perspectives in Child Psychology
Stagner Psychology of Personality
Steers and Porter Motivation and Work Behavior
Vinacke The Psychology of Thinking
Winer Statistical Principles in Experimental Design

McGraw-Hill Book Company
New York St. Louis San Francisco
Auckland Bogotá Düsseldorf Johannesburg
London Madrid Mexico Montreal
New Delhi Panama Paris Sao Paulo
Singapore Sydney Tokyo Toronto

J. P. Guilford
Emeritus Professor of Psychology
University of Southern California

Benjamin Fruchter
Professor of Educational Psychology
University of Texas at Austin

SIXTH EDITION

FUNDAMENTAL STATISTICS IN PSYCHOLOGY AND EDUCATION

FUNDAMENTAL STATISTICS IN PSYCHOLOGY AND EDUCATION

1234567890 FGFG 783210987

This book was set in Optima by Progressive Typographers.
The editors were Richard R. Wright and James R. Belser;
the designer was Nicholas Krenitsky;
the production supervisor was Dominick Petrellese.
Fairfield Graphics was printer and binder.

Library of Congress Cataloging in Publication Data

Guilford, Joy Paul, date
Fundamental statistics in psychology and education.

(McGraw-Hill series in psychology)
Includes bibliographical references and index.
1. Statistics. 2. Educational statistics.
3. Psychometrics. I. Fruchter, Benjamin, joint author.
II. Title.
HA29.G9 1978 519.5 77-5768
ISBN 0-07-025150-9

Contents

Preface

The major objectives of this volume remain the same—to present to the student and the general user of statistics in the behavioral sciences a clear and usable account of an intricate subject. A wide range of content is covered, keeping in mind the variety of potential needs of students and investigators. The presentation of statistical methods aims at much more than a superficial understanding of the principles involved. Many suggestions are offered regarding where and how methods may be appropriately applied as well as their limitations and places where they should not be applied.

This edition expands further the chapter on analysis of variance. For example, a three-way analysis is fully explained and illustrated. Principles are also given for extending such analyses to more complex problems. New, more generalized indices of agreement have been added. A much enlarged table of significant t values is presented, along with separate tables of significant correlation coefficients, including the multiple R.

In order to make room for this new content without lengthening the volume, there have been a few deletions. The major one was the elimination of Appendix A with its many mathematical proofs underlying statistics. The Kolmogorov-Smirnov test and the table that goes with it were dropped, with the belief that it was more dispensable than alternative hypothesis-testing methods. Minor instances of textual condensation occur in several places.

A major change involves statistical symbols. In order to bring this text more in accordance with such consensus as seems to exist in the use of statistical symbols, Greek letters are largely restricted to indicate population parameters and Roman letters to indicate statistics pertaining to samples with a few exceptions, such as Greek letters in some subscripts.

J. P. GUILFORD
BENJAMIN FRUCHTER

1 Introduction for students

WHY THE STUDENT NEEDS STATISTICS

Most seasoned workers in psychology or in education usually take statistical methods for granted as an essential part of their routine, some more so and some less. The initiate may at first react to statistics as a frightful bogey whose mysteries loom forbiddingly before him, and he is likely to ask, "What is the good of them, anyway?" This is particularly true of one who feels he has always had trouble with numbers. Students who enter a first course in statistical methods in psychology or education, and probably in all related social sciences, range all the way from those who find mathematics generally easy and to their liking to those at the other extreme who say they have difficulty adding 2 and 2. Somehow, all these must acquire what they can of a subject for which they are so unequally prepared.

Probably no other subject demonstrates so clearly that there are several kinds of intelligence. No less a person intellectually than Charles Darwin had trouble with statistics, as he is said to have frankly admitted. His almost equally illustrious cousin, Sir Francis Galton, who had so much to do with introducing statistics into psychology, had to turn some of his mathematical problems over to others for aid.

There are different ways of understanding the same things. One student will grasp the new ideas offered by statistics in the way that a mathematician would understand them; another will appreciate the logical rules of thinking and the concepts provided as aids in thinking; still another will master rule-of-thumb operations and be able to carry through computations with a minimum grasp of what they are all about.

Learning without achieving insight into, and an appreciation of, the inner nature of things is learning without full motivation and enthusiasm and is not very satisfying. The average student will necessarily have to be content with levels of insight that fall short of those of the mathematician, remembering that even mathematicians have by no means exhausted the meanings and ramifications of statistical ideas. On the other hand, alert students should strive to inject as much meaning and significance, in their own way, as they can. The proper and optimal use of statistical methods and statistical thinking requires a certain minimal achievement of understanding. Clerks can be taught to carry out many of the computational procedures; it is not the primary purpose of this book or of those who teach with it to develop computational clerks. The purpose is to develop those who could be supervisors of clerks.

To be more specific, there are four simple, undeniable reasons why students who take a required course in statistics must develop some mastery of that subject:

1 *They must be able to read professional literature.* There is no questioning the fact that learning in any field comes largely through reading. Thorough students never finish the extension of their skill in the art of reading. In any specialized field, reading is largely a matter of emlarging vocabulary. One cannot read much of the literature in any specialized field in the social sciences, particularly in the behavioral sciences, without encountering statistical symbols, concepts, and ideas on every hand. One could do as young children do when they tackle reading matter that is somewhat beyond them— "skip over the hard places." But this is hardly excusable in adults who are reading material that should not be beyond them and in which the "hard places" may, in fact, contain the crucial parts of the content. Those who dodge such parts are likely to be dependent upon the conclusions of others for their own conclusions and opinions. This is hardly independent judgment or a symptom of mature scholarship. It is not necessary for every student to be able to sail through the "heavier" mathematical contributions of the specialist in statistics. However, persons who cannot read the average research paper in their field with intelligence and with some appreciation as to whether sound conclusions have been reached are severely limited. The chances are that this appreciation will require familiarity with basic statistical ideas.

2 *They must master techniques needed in advanced courses.* Whether the advanced course is a laboratory course or a practicum, there are usually certain incidental techniques that are commonly used in the operations involved. In the laboratory course, results cannot be treated or reports written without at least minimal statistical operations. A field survey or the checking of a report also involves inevitable statistical steps.

3 *Statistics are an essential part of professional training.* Trained psychologists or educators like to think of themselves as professional persons.

To some extent, statistical logic, statistical thinking, and statistical operations are a necessary part of either profession. To the extent that they use in their practice the common technical instruments, such as tests, psychologists or educators will depend upon statistical background in their administration and in the interpretation of the results. Using tests without knowledge of the statistical reasoning upon which they depend is like the medical diagnostician's using clinical tests without a knowledge of physiology and pathology.

4 *Statistics are everywhere basic to research activities.* To the extent that either psychologists or educators intend to keep alive their research interests and research activities, they will necessarily lean upon their knowledge and skills in statistical methods. The relation of statistics to research will be elaborated upon in the next paragraphs. Here it is merely emphasized that in any professional fields where there are still so many unknowns as in the behavioral sciences, the advancement of those professions and of the competence of their members depends to a high degree upon the continued research attitude and research efforts of those members.

WHY STATISTICS ARE IMPORTANT IN RESEARCH

Briefly, the advantages of statistical thinking and operations in research are as follows:

1 *They permit the most exact kind of description.* When all is said and done, the goal of science is description of phenomena, description so complete and so accurate that it is useful to any who can understand it when they read the symbols in terms of which those phenomena are described. Mathematics and statistics are a part of our descriptive language, an outgrowth of our verbal symbols, peculiarly adapted to the efficient kind of description that the scientist demands.

2 *They force us to be definite and exact in our procedures and in our thinking.* One of the authors once heard a prominent psychologist defend his rather vague conclusions by saying that he would rather be vague and right than definite and wrong. But the alternatives are not to be either "vague and right" or "definite and wrong." One can also be definite and right, and it is the authors' contention that the odds for being right are overwhelmingly on the "definite" side of the matter.

3 *They enable us to summarize our results in a meaningful and convenient form.* Masses of observations taken by themselves are bewildering and almost meaningless. Before we can see the forest as well as the trees, order must be given to the data. Statistics provide an unrivaled device for bringing order out of chaos, for seeing the general picture in one's results.

4 *They enable us to draw general conclusions,* and the process of extracting conclusions is carried out according to accepted rules. Furthermore, by means of statistical steps, we can say about how much

faith should be placed in any conclusion and about how far we may extend our generalization.

5 *They enable us to predict* "how much" of a thing will happen under conditions we know and have measured. For example, we can predict the probable mark a freshman will earn in college algebra if we know the student's score in a general academic-aptitude test, score in a special algebra-aptitude test, average mark in high school mathematics, and perhaps the number of hours per week that the student devotes to studying algebra. Our prediction may be somewhat in error because of other factors that we have not accounted for, but statistical methods will also tell us about how much margin of error to allow in making predictions. Thus not only can we make predictions, but we can also know how much faith to place in them.

6 *They enable us to analyze some of the causal factors underlying complex and otherwise bewildering events.* It is generally true in the social sciences, and in psychology and education in common with them, that any event or outcome is a resultant of numerous causal factors. The reasons why a person fails in a business or in a profession, for example, are varied and many. Causal factors are usually best uncovered and proved by means of experimental method. If it could be shown that, all other factors being held constant, certain businesspeople fail to the extent that they possess some defect in personality "*X*"—a trait—then it is probable that *X* is a cause of failure in this type of business.

Unfortunately, the social scientist cannot manage people and their affairs sufficiently to set up a good experiment of this type. The next best thing is to make a statistical study, taking business people as we find them, working under their normal conditions. Life insurance experts do this when they follow the trail of all possible factors that influence the length of life and determine how important they are. On the basis of these statistical findings, they can predict about how long an individual of a certain type will probably live, and the insurance company can plan a policy accordingly. Statistical methods are therefore often a necessary substitute for experiments. Even where experiments are possible, the experimental data must ordinarily receive appropriate statistical treatment. Statistical methods are hence the constant companions of experiments.

WHAT THIS VOLUME'S TREATMENT OF STATISTICS WILL INCLUDE

In the next few paragraphs we shall take a hasty overview of the things to come. The second chapter will give many more details of a general and preparatory nature. Here we shall try to look at the whole forest before we enter it.

DESCRIPTIVE AND SAMPLING STATISTICS It is common to make a broad distinction between *descriptive* and *sampling* statistics. This distinction refers to two important uses of statistics.

In the first place, statistics are used to describe. For example, averages tell us "how much" of certain quantities we have in a group of individuals or in a group of observations. An average (for example, *arithmetic mean, median,* or *mode*) is a general-level concept. That is, a single number tells how high one group, or sample, stands on a certain scale as compared with another.

Other statistics tell us how much variability, or scatter, the individuals in a group show. A statistic known as the *standard deviation* has been the almost universal indicator of the amount of variability in a set of individuals or observations, though there are other indicators.

A *coefficient of correlation* describes the closeness of relationship between two sets of measures of the same group of individuals or observations. Most of science is concerned with finding out what things go with what, and what things are independent of what. Correlation methods, in the social sciences at least, are the most useful devices for answering these questions of interrelationships. Averages, indices of dispersion, and correlations are the basic descriptive statistics.

Sampling statistics tell us how well the statistics we obtain from measurements of single samples probably represent the larger populations from which the samples were drawn. Almost every descriptive statistic has a *standard error.* A standard error is an index number that leads us to conclusions concerning how far the statistic derived from the sample probably differs from the value we would obtain if we had measured an entire population. A *population* is a well-defined group of individuals or observations. For example, a population could be composed of Wistar Institute albino rats between the ages of 30 and 60 days, or it could consist of all possible reproductions a certain observer makes of a line 10 cm long under the same conditions of rest, time of day, and method of reproduction, for example, by drawing a line with a pencil. A sample in either case would be a limited number of observations out of the entire population. Arriving at conclusions that can be generalized to a population depends upon reducing discrepancies between population values and sample values to as small a size as possible. This is probably best illustrated by public-opinion polling, in which the margin of error of voting outcome can be expressed in terms of percentage of error.

In connection with sampling statistics, there is much in this volume on testing hypotheses. Scientific investigation proceeds from hypothesis to hypothesis. There are numerous hypotheses but relatively few established facts of a general nature. The sooner the research student realizes this point, the better it will be for his clear thinking. Unfortunately, there are some investigators, many of them experienced, who do not make this distinction between a hypothesis and a fact; they mistake hypotheses for facts. For example, there is the hypothesis, stemming from Freudian psychology, that children suffering from asthma are of the "oral-dependent" type and that the breathing spasms are expressions of a cry for aid and love. The plausibility of the hypothesis, and its consistency with other hypotheses, may suffice to lead many a clinical or psychiatric investigator to act as if the problem were solved, as if the hypothesis were a fact. The properly skeptical investigator makes a study of a sample of asthmatic children and of their nonasthmatic siblings to see whether there is any greater incidence of dependency among the one group than among the other. The most fruitful scientific investigations, at least those which lead to dependable answers or which go beyond the exploratory stages, start by setting up a hypothesis, or several alternative hypotheses. Conditions are then arranged in such a way that if the results turn out one way, the hypothesis, or one of its alternatives, is supported and other hypotheses are rendered doubtful. The results must usually be cast in a statistical form, which makes possible a decision between hypotheses.

The simplest example of this is seen where we are studying the effects of one thing on another, say, the effect of Benzedrine on the ability to reason. We restrict our problem to two alternative and mutually exclusive hypotheses: (1) that Benzedrine will affect thinking output or efficiency and (2) that it will not. The first hypothesis can be subdivided into two more: that thinking will be facilitated and that thinking will be hindered. The typical experimental operations would be somewhat as follows, briefly described. We develop or adapt a test of reasoning power. From a pool of individuals of the same sex and of similar age and educational levels, members are assigned at random to two groups for the experiment. We administer the drug to one group and a control dose, or placebo, to the other. Neither group knows which has taken the drug. We administer the reasoning test. We obtain two average scores, and there is some difference in a certain direction. The question is, "Does this difference support hypothesis 1, or could we still tolerate hypothesis 2? Could the difference have occurred by chance?" If not, it must have been due to the drug, for as far as we know, there is no other difference in conditions between the two groups that could account for it. It would require a test of the statistical significance of the difference to permit us to reject hypothesis 2 and accept hypothesis 1. Having rejected the idea that the difference was due to chance, we may accept the idea that it probably was due to the drug. Without the statistical test we would be helpless in reaching a dependable answer.

DISTRIBUTION CURVES Every student is familiar with the normal-distribution curve; mention of it is ubiquitous in psychological and educational literature. There has been much use and abuse of it, and many erroneous things are said about it. The curve itself is a mathematical conception. It does not occur in nature, and it is not a biological or a psychological curve. It is an ideal pattern or model that we can *apply* to useful purpose in many a situation. That there is a distinction between statistics and applied statistics (like that between mathematics and applied mathematics) must be kept in mind. Many fruitful applications of the normal-distribution curve will be described in later chapters. Familiarity with the normal curve and its properties is therefore essential. Other, less familiar distribution curves will be encountered, including those for the statistics t, F, and chi square.

PREDICTION AND STATISTICS Two chapters in this book are organized under the heading of *prediction*. Most elementary psychology textbooks start by saying that it is the purpose of psychology to predict and control human behavior. Dealing with the very complex and intricate set of phenomena that the behavior of living organisms presents, and realizing the limitations to accurate predictions, it is appropriate for us to be modest on the subject. We should not feel guilty, however, about our failures to make predictions comparable to those in the physical sciences. We should make candid and realistic efforts to achieve the predictions that are possible, and we need not disparage results obtained under the limitations inherent in the subject matter.

The operation called *prediction* is actually carried out even when we do not realize it. The vocational counselor who tells a client to consider seriously vocations P, Q, and R and to shy away from vocations U, V, and W is tacitly predicting relative success in the one group and relative failure in the other. The clinician who diagnoses a person as having an anxiety neurosis is saying that certain behavior can be expected of this individual. In prescribing a certain program of therapy, the clinician is predicting improvement under that treatment, as opposed to lack of improvement if it is not applied. The promotion of a child to the next higher grade is a prediction that the child will probably adjust better to that assignment than to reassignment to the same grade. Thus, almost all therapies and administrative decisions imply predictions, whether those who make those prescriptions would be willing to put themselves on record as making predictions or not.

Predictions in psychology and education are often called *actuarial;* that is, they are made on a statistical basis and with the knowledge that only "in the long run" will the practice represented by any prediction be better than other practices, based upon other predictions. Prediction of the single case is recognized as involving many chance elements. For the single case, the prediction is either correct or incorrect. In predicting in large numbers, there are certain probabilities of being right and being wrong which can be determined. Statistical methods provide a basis for choosing what prediction to make and also a basis for knowing what the odds are of being right or wrong. The various ways of making predictions and the ways of determining their degree of accuracy will be treated at length in Chaps. 15 and 16.

TEST PRACTICE AND STATISTICS Because tests play such an important role in psychology and education, considerable attention has been given to them in this volume. Recent investigations by statistical psychologists and educators have drastically changed our former understanding of tests as instruments of measurement. Many of these findings have been reflected in the chapters treating tests, particularly Chaps. 17 and 18. Certain ideas of reliability and validity of tests which have become securely entrenched in the thought and practice of test users are reexamined, and newer experiences have been used to advantage in the applications of statistics to test practice.

THE STUDENT'S AIMS IN THE STUDY OF STATISTICS

With this overview of content and with a general idea of the advantages of statistics, what should the student, particularly the beginner, aim to do? In order to make the task more specific, we may list the beginner's aims as follows:

1 *To master the vocabulary of statistics.* In order to read and understand a foreign language, it is always necessary to build up an adequate vocabulary. The beginner should regard statistics as a foreign language—one that will not for long remain entirely foreign. The vocabulary consists of concepts that are symbolized by words and by letter symbols that are substituted for them. Along with mathematics in general, statistics shares the ordinary symbols for numerical operations. Thus, much of the vocabulary is already known to the student. As for the new concepts, their meanings will continue to grow as the student uses them.

2 *To acquire, or to revive, and to extend skill in computation.* Although it was stated earlier that it is not an important aim for the student to become a statistical clerk, computation is important. For many people, understanding of the concepts themselves comes largely through applying them in computing operations. The mere step-by-step activities with numbers, when certain goals are in mind, provide opportunities for gaining new insights. The average investigator always has a certain amount of computational work to do. Like any skill, computational skill, and this includes application of formulas as well as planning efficient operations, grows with practice.

3 *To learn to interpret statistical results correctly.* Statistical results can be useful only to the extent that they are correctly interpreted. With full and proper interpretations extracted from data, statistical results are a most powerful source of meaning and significance. Inadequately interpreted, they may represent something worse than wasted effort. Erroneously understood, they are worse than useless. It is the latter even-

tuality that leads to the common sour-grapish remark, "Anything can be proved by statistics." In the hands of skilled operators, statistics make data "talk." It is therefore very important that the implications of any statistical result be realized and that their proper meaning be made manifest. The average reader is less able to interpret the result than the investigator should be. Upon the investigator's shoulders rests the responsibility of telling the reader what the conclusions should be and also of including some indication of the limitations of those conclusions.

4 *To grasp the logic of statistics.* Statistics provides a way of thinking as well as a vocabulary and a language. It is a logical system, like all mathematics, which is peculiarly adaptable to the handling of scientific problems. This is hard to explain to the beginner. It is hoped that it may become more apparent as later chapters, particularly those dealing with sampling errors, hypotheses, predictions, and factor analysis, are encountered. The most efficient investigator is the one who masters the logical aspects of his research problem before taking recourse to experiment or to field study. Proper formulation of a research problem is more than half the battle. Too many inexperienced investigators think of a question or a problem and rush to gather data before knowing what it is they really want to observe. Because they realize that data of some kind must be collected, they waste much time and effort collecting them, without thinking through the problem and coming to the proper decision as to just what data are needed, or they collect data in such a manner that no statistical operations now known are adequate to treat them so as to extract an answer. *Well-planned investigations always include in their design clear considerations of the specific statistical operations to be employed.*

5 *To learn where to apply statistics and where not to.* While all statistical devices can illuminate data, each has its limitations. It is in this respect that the average student will probably suffer most from lack of mathematical background, whether the student realizes it or not. Every statistic is developed as a purely mathematical idea. As such, it rests upon certain assumptions. If those assumptions are true of the particular data with which we have to deal, the statistic may be appropriately applied. The student should note that wherever a new statistic is introduced, there are likely to be mentioned certain assumptions or properties of the situation in which that statistic may be utilized. Unfortunately, one can encounter masses of numbers that look as if they are candidates for the use of a certain statistic, for example, a biserial coefficient of correlation (see Chap. 14), when actually to apply that statistic would be meaningless if not misleading. Students without mathematical background will have to learn these exceptions by rote or be satisfied with commonsense reasons. They certainly will prefer to avoid making ridiculous applications, and when in doubt they should seek advice or refrain from making doubtful applications.

6 *To understand the underlying mathematics of statistics.* This objective will not apply to all students, but it should apply to more than those with unusual previous mathematical training. Many an intelligent student who has not been introduced to analytic geometry or calculus can nevertheless grasp many of the mathematical relationships underlying statistics. This will give a more than commonsense understanding of what goes on in the use of formulas.

ONE

BASIC DESCRIPTIVE STATISTICS

2 Counting and measuring

TWO KINDS OF NUMERICAL DATA

Numerical data generally fall into two major categories. Things are counted, and this yields *frequencies;* or things are measured, and this yields *metric values,* or *scale values*. Data of the first kind are often called *enumeration data,* and data of the second kind are called *measurements,* or *metric data*.

Statistical procedures deal with both kinds of data, which is the reason for this chapter. There are certain fundamental ideas about numbers and their use that it is well to have in mind before we go ahead. Perhaps it may seem strange to the readers, who have been counting and measuring as long as they can remember, that we should have to devote an entire chapter to these topics. The experts, who, we must admit, have had a great deal more experience with numbers and their use than most of us, never cease to report new ideas about the properties of the number system and its applications. It is well to keep in mind, incidentally, that there is a real difference between the number system, as such, and its application to counting and measuring. Much confused thinking has resulted from ignoring this fact. The world does not necessarily owe its existence to number and quantity. Numbers were invented by humans as a symbolic system of internally consistent ideas which they can use effectively in describing the world as they know it, thus gaining control over it.

DATA AND STATISTICS

Before we go further, there are two frequently used terms that should be defined: *statistics* and *data*. The word *statistics* itself has several meanings. First, it refers to a branch of mathematics which specializes in enumeration data and their relation to metric data. That is the meaning of the word as it is used in the title of this book.

Another meaning, popular but not used by technical people, is implied in a mother's statement such as "Bobbie, stay out of the street, or you will become a vital statistic." Here the term in the singular refers to a fact of classification, which is a chief source of all statistics. What the mother means is that Bobbie's classification would change from the category "living" to the category "dead." This use of the term *statistics* is more common among those agencies which keep such records. The numerical records *are* the statistics. While this use of the term is recognized by teachers and writers who specialize in statistics, their use of the term and the use of it in this book will usually mean something else. In the textbook and the classroom, we use the word *data* to refer to details in numerical records or reports. The fact that Bobbie is classified among either the living or the nonliving is a *datum*. The word *data* is plural, always referring to more than one fact.

In the textbook and the classroom situation, also, the singular term *statistic* means a derived numerical value such as an average, a coefficient of correlation, or some other single descriptive concept. It may refer either to the *idea* of an average, a median, a standard deviation, etc., or to a particular value computed from a set of data. The reader can usually tell from the context which usage of these terms is meant.

Data in categories

Probably most social data are in the form of categorical frequencies, the numbers of cases in defined classes or categories. The number of births, marriages, and deaths constitutes the bulk of the so-called vital statistics. The number of accidents, fatal or otherwise; the number of arrests for different reasons; and the number of new cases of poliomyelitis constitute other important information by means of which social agencies keep a finger on the pulse of human affairs. Political and economic interests also have their "barometers" for keeping informed of the trend of events, though some of these depend upon measurements of variables as well as upon counting cases.

CLASSIFICATION Before we count, in order to accumulate useful information, we must know what it is we are counting. We do not count indiscriminately. The frequency that we record refers to a particular class of objects or events, and this involves the process of classification. Classification is a basic psychological process which can be seen in rudimentary form even in the simplest conditioned response. Wherever discriminations are made, along with generalizations, classification of a sort occurs. Useful classifications for counting purposes, however, depend upon a high type of logical analysis. Much of science, following Aristotle, has been of the classificatory type. The classification of plant and animal life into species, genus, and order is the best example. Things thus become ordered, and principles emerge.

As science progresses, it is likely to abstract *variables* from its data. Variables are continuous variations in single directions. Continuity provides an opportunity to make refined measurements. In spite of this general trend in a science, however, the classification of phenomena will probably never cease to be useful. Besides, there are some absolute categories that seem not reducible to continuous variables, such as living and dead, married and unmarried, male and female, and voter and nonvoter. Such discrete classes must be recognized and are usefully dealt with in research as well as in public affairs. Classification, then, is a very useful and necessary process in science as well as in practical life. It is the procedure by which objects become categorized for counting.

SOME PSYCHOLOGICAL CATEGORIES Before specifying the way in which categories should be set up and utilized, it may be well to have in mind some examples of the more common kinds from the field of psychology. In experimental psychology, particularly in psychophysical studies, we have categories of judgment. The second of a pair of stimuli is judged as "greater than," "equal to," or "less than" the first. In public-opinion polling, responses are obtained in a small number of categories that are intended to be meaningful for interpretation purposes. In answer to the question, "Are you in favor of an atomic test ban?" the response might be "Yes," "No," "I do not know what an atomic test ban is," or "I know what a test ban is, but I am undecided." In taking a vocational-interest test the examinee may be required to respond in one of three categories—"L" (for like), "I" (for indifferent), or "D" (for dislike)—concerning the thing proposed. In a problem-solving experiment with rats, after preliminary observations, solutions might be categorized as falling among, say, four types. Clinical types in psychopathology are mostly categories having long-standing acceptance: neurotic versus psychotic, schizophrenic, manic-depressive, paranoid, etc. And so one could continue. Many categories used in research are not static; they change as new light is thrown on the field of study. Some categories are invented for temporary duty as provisional scaffolding upon which to arrange data for better inspection.

There is not space here to give detailed instructions on how to choose or construct useful categories. It may suffice to say, and it may seem trite to do so, that categories should be *well defined, mutually exclusive* (if possible), and *exhaustive*. The importance of having good definitions cannot be overestimated. Making proper assignment of cases to classes depends upon it. Being understood by one's colleagues also depends upon it. A prime requirement of scientific findings is that they be communicable to others. Other investigators should be able, if they so desire, to test our results by repeating our operations. The requirement of mutual exclusiveness is perhaps the most difficult to achieve. Lack of it means that something is probably missing in defining the basis of classification; it means that there is some overlapping, interdependence, and loss of power to draw clear-cut conclusions. A set of mutually exclusive categories means that there is one and only one basis of classification. To group schoolchildren into three classes—boys, girls, and adolescents—is to inject two principles or bases: sex difference and age differences. Anything so grossly absurd is easily avoided; it is the more subtle confusion of attributes that causes trouble.

By being exhaustive, a set of categories provides places for all cases. If there are only two classes—say, delinquent and nondelinquent—and if they are well differentiated by objective criteria, two categories can be exhaustive. In many a system, especially where more than two categories are needed, a miscellaneous category is often necessary. Into this category we put all cases that do not fit into any other category. These cases are often ignored, but if they are numerous it probably means that there is biased sampling in other categories. It also probably means that the classificatory system is inadequate as a whole.

QUALITATIVE AND QUANTITATIVE CATEGORIES Most of the examples of categories given thus far have been *qualitative*. The classes of objects are different in kind. There is no reason for saying that one is greater or less, higher or lower, or better or worse than another. There might be some intrinsic or extrinsic basis for thinking of the classes as being ordered on some scale of more or less, but if so, we are unaware of it or regard it as being irrelevant.

There are many classifications, however, in which the groups can be ordered according to quantity or amount. These are *quantitative* classifications. It may be that the cases vary continuously along a continuum which we recognize, such as degree of effort, on which we can group cases, but only in a gross manner. Ratings on a scale of five points are an example. Another example is seen where the variation of experimental conditions is in graded steps, e.g., four amounts of given information in a learning study. To cite another example, in selection of personnel by means of tests, examinees are categorized into the accepted and rejected groups. Later, after the accepted examinees have been trained or have served on the job, there is a further classification between those who are satisfactory and those who are not. Experimental and technological practices are full of such examples. Later chapters will explain methods for dealing with them. The next chapter will show how continuously graded measurements are most conveniently treated by somewhat arbitrary groupings in successive categories.

FREQUENCIES, PERCENTAGES, PROPORTIONS, RATIOS A *frequency* has already been defined as the number of objects or events in a category. There are some other related concepts that, though common in advanced arithmetic, most students do not appreciate fully. They play an important role throughout this volume. We cannot review all the arithmetical features of these concepts here, but there are certain new uses of them that should be stressed and certain pitfalls to be pointed out.

Let us consider an example to illustrate the use of percentages. Table 2.1 gives some original data in the form of frequencies in 12 categories. The categories are in a two-way classification, one qualitative and the other quantitative. The data pertain to the number of students in training and the number of these eliminated in each of four bombardier schools in the Army Air Force during the early part of World War II. In each school the students had been categorized in three levels as to aptitude. The categorization by schools is qualitative, and that by aptitude is quantitative. Such a table would probably be set up to study the relation of elimination rate to aptitude and also to differences between schools. We can make comparisons both ways.

PERCENTAGE AS A RATE INDEX If we wanted to compare schools as to eliminations, the *number* eliminated in each school would be a poor index, particularly if our comparison were made at constant levels of aptitude. For example, at the low level of aptitude, the numbers of eliminations were not very different: 26, 23, 20, and 21. If we gave

TABLE 2.1 Elimination rates for bombardier students of three levels of aptitude in four Army Air Force training schools*

School	*Low*			*Moderate*			*High*			*All levels*		
	No. in training	*No. eliminated*	*% eliminated*	*No. in training*	*No. eliminated*	*% eliminated*	*No. in training*	*No. eliminated*	*% eliminated*	*No. in training*	*No. eliminated*	*% eliminated*
A	**62**	**26**	**41.9**	**340**	**105**	**30.9**	**162**	**29**	**17.9**	**564**	**160**	**28.4**
B	**69**	**23**	**33.3**	**274**	**51**	**18.6**	**125**	**10**	**8.0**	**468**	**84**	**17.9**
C	**69**	**20**	**29.0**	**334**	**43**	**12.9**	**166**	**15**	**9.0**	**569**	**78**	**13.7**
D	**139**	**21**	**15.1**	**274**	**19**	**6.9**	**149**	**9**	**6.0**	**562**	**49**	**8.7**
All schools	**339**	**90**	**26.5**	**1,222**	**218**	**17.8**	**602**	**63**	**10.5**	**2,163**	**371**	**17.2**

* Aptitude was measured in terms of a composite score on psychological tests. The data were selected from results during World War II. (Adapted from unpublished data of the AAF Training Command. This will be true of other AAF data used in this volume unless otherwise specified.)

credence to such small differences, we should place the schools in the rank order *A, B, D,* and *C,* from most to least eliminations. Schools *A, B,* and *C* had comparable numbers in training, but school *D* had about twice as many. To put the schools on a fair basis, we need to find an index of elimination *rate.* We should ask what the elimination "scores" would have been if all schools had had equal numbers in training. If we assume that common number in training to be 100, the number eliminated per 100 is a familiar percentage. The percentages of eliminations for students of low aptitude are 41.9, 33.3, 29.0, and 15.1. Now we see that there are larger differences (due partly to the fact that three of the denominators—62, 69, and 69—are less than 100) between schools, and the rank order is now *A, B, C,* and *D.* The inversion of the order of *C* and *D* is decisive; at least *D*'s position below *C* now seems decisive. The point of this illustration is that percentages are used to compare groups of objects on an equitable basis.

SOME LIMITATIONS TO THE USE OF PERCENTAGES Several precautions should be pointed out concerning the use of percentages. If the total number is less than 100, a change, by chance, of only one case added to, or removed from, a category would mean a change of more than 1 percent. If we ask what percent 15 is of 25, the answer is 60. But if the frequency were to gain one case, the percentage would be 64. If a lower limit must be mentioned as a total below which computation of percentages is unwise, it might be placed at 20. At this number, a change of one case would mean a corresponding change of 5 percent. This is being quite liberal for the sake of applying a very useful index.

In line with the discussion above, it would seem to be not very meaningful to report percentages to any decimal places unless the total number of cases exceeds 100. When we want a percentage for use in further computations, however, it would be wise to retain at least one decimal place. Frequencies are "exact" numbers, and percentages based upon them are accurate to as many decimal places as we wish to use. They thus describe the sample in terms of *per hundred.* It is when we become interested in letting an obtained percentage stand for a population value (see Chap. 8) that we must become conservative about reporting it. In Table 2.1 all percentages were reported to one decimal place because most of them were based upon totals greater than 100 and all were made consistent. Consistency of this sort carries some weight, but should not be pushed too far.

When a percentage turns out to be less than 1.0 (for example, .2 percent), it is not so meaningful as larger ones, and, what is worse, it may be mistaken for a proportion (all proportions are less than, if not equal to, 1.0). In some social statistics a series of percentages may be this small. In this case it is common practice to change the base from 100 to 1,000 or even more, for example, to report 15 deaths per 100,000, 5 cases in 1,000 and the like. As percentages these would read .015 and .5, respectively. To avoid confusion with proportions, these should be written as 0.015 percent and 0.5 percent.

PROPORTIONS Whereas with percentages the base is 100, with proportions the base, or total, is 1.0. A proportion is a part, or fraction, of 1.0. A proportion is 1/100 of a percentage, and a percentage is 100 times a proportion. Careless individuals often call a percentage a proportion, and vice versa. By definition, and in all strictness, the two are different concepts. The symbol used for percentage is a capital P; for proportion the symbol is a lowercase p. This should help to fix the idea of the relative sizes of the two, The *proportion* of eliminees among low-aptitude students at school *A* was .419 (see Table 2.1); for high-aptitude students at school *B* the proportion of eliminees was .080.

Compared with percentages, proportions have advantages as well as disadvantages. They are less familiar to nonmathematical individuals than percentages are. Whenever results are reported to the general reader, then, percentages are almost always to be preferred. Percentages have another advantage in that we can speak of percentage of gain or of loss. Proportions are always parts of something and can never exceed the total, which is 1.0. They have no place in expressing gain or loss, though presumably losses could be expressed in terms of proportions if we chose, for losses cannot exceed the total; but we never use a proportion for this purpose.

The advantages of proportions are best seen in later chapters. They are used more than percentages in connection with the normal-distribution curve, item analysis of tests, certain correlation methods, and so on. It has already been said that percentages may be mistaken for proportions when they are less than 1.0. Since proportions can never be greater than 1.0, they are much less likely to be mistaken for percentages.

PROBABILITIES Another advantage of proportions is their relation to *probabilities*. Every probability can be expressed in the form of a proportion. We say that the probability of getting a head in tossing a coin is 1/2, or 1 chance in 2. This is a more manageable figure if expressed as a probability of .5. We say that in throwing a die, the probability of getting a 6 spot is 1 in 6. Expressed as a proportion, this is .167. In general, for computational purposes, decimal fractions are much preferred to common fractions; they are much more easily manipulated in addition and subtraction and in finding squares and square roots. The interchangeability of proportions and probabilities will be found to be very common in the later chapters.[1]

[1] At this point, it is suggested that readers who are not sure of their grasp of the concepts under discussion, and others who wish to test themselves, do Exercises 1 and 2 at the end of this chapter, and then check their answers with those given following the exercises.

RATIOS A ratio is a fraction. The ratio of *a* to *b* is the fraction *a*/*b*. A proportion is a special ratio, the ratio of a part to a total. We may also have ratios of one part to another. For example, there were 69 low-aptitude students in training school *B* (Table 2.1), of whom 23 were eliminated and 46 were graduated. The ratio of graduates to eliminees was $^{46}/_{23}$, or 2 to 1. This ratio can also be expressed as 2.0. The ratio of eliminees to graduates was $^{23}/_{46}$, or .5. This could also be expressed as .5 to 1, but ordinarily would not be. At any rate, in a ratio the base is 1.0, as it is in a proportion. The chief difference is that a proportion is restricted to the ratio of part to total, whereas a ratio is not.[1]

Ratios are useful as *index numbers,* examples of which follow immediately. They describe rates and relationships. The IQ is an index number of rate of general mental growth—the ratio of mental age to chronological age (multiplied by 100). Comparisons of incomes in different regions are made in per capita terms—the ratio of total income to population. Costs of education are more meaningful if stated in terms of dollars per pupil per day attended rather than in terms of total expenditures. In dealing with index numbers, one should keep in mind the operations by which they were derived. It sometimes makes a difference when they are used in computation, as in averaging them or in correlation problems (see p. 57).

Measurements

SOME EXAMPLES OF PSYCHOLOGICAL MEASUREMENT

In order to make the discussion concrete and specific, let us consider some typical examples of measurements commonly made by psychologists. Perhaps the first examples that come to mind are scores on tests of mental ability. These are usually in terms of the number of correct responses to test items. A similar kind of measurement is seen in scores on a personality questionnaire or a vocational-interest inventory. In these cases the score is not the number of "correct" responses but the number of responses indicating the same interest or trait, often weighted in proportion to their supposed diagnostic value. Also in the area of mental tests we find frequent references to "chronological age," "mental age," and that ratio between the two, the "intelligence quotient."

In the experimental laboratory as well as in the clinic, we frequently measure in terms of the time required to complete a specified test or task. In memory experiments, we measure learning efficiency in terms of the number of trials required to attain a certain standard of performance or in terms of the "goodness" of performance at the end of a certain trial or time. We measure efficiency of retention in terms of the time required for relearning (overcoming the forgetting

[1] See Exercise 3.

that has taken place) and the efficiency of recall in terms of association time or in terms of the number of items correctly recited.

In the sphere of motivation, we gauge the strength of drive in terms of the amount of punishment (electric shock) an organism (for example, a rat) will endure in order to reach its immediate goal or in terms of the number of times it will take a constant punishment in order to attain the same result. The difficulty of a task or test item can now be specified in quantitative terms, as can the affective value (degree of liking or disliking) for a color, a sound, or a pictorial design. In studies of sensory and perceptual powers, the threshold stimulus and the differential limen are given in terms of stimulus magnitudes.[1] The span of perception or of apprehension is given in terms of the average number of items that the observer can report correctly after momentary exposures. The galvanic skin response, the pupillary response, and the amount of salivation also serve as quantitative indicators of amounts of psychological happenings.

SOME EXAMPLES OF EDUCATIONAL MEASUREMENT

Many an educational problem is also a psychological problem, and its mode of measurement has been indicated in the preceding paragraphs. Achievement in any area of learning, like any mental ability, is measurable in terms of test scores. Marks, however obtained, have been the traditional mode of evaluating students in specific units of formal education. Attendance records and data on size of classes, on budgets, on supplies, and on other material aspects of the well-regulated school system constitute another list of measurements in education. Outcomes of educational effort are often expressed quantitatively in terms of promotion statistics, achievement ratios, and estimates of teaching success. Whether for purposes of research in education or for systematic and meaningful record keeping, statistical methods are indispensable.

BASIC KINDS OF MEASUREMENT SCALES

Philosophers and scientists have investigated the operations of measurement, and we shall take advantage of many of the results of their thinking. Conceptions of measurement are now much broader, with a more logical foundation than existed formerly. It is fairly well agreed that measurement should be defined as the *assignment of numerals or numbers to objects and events according to logically acceptable rules*. Beneath this simple-sounding statement lies a wealth of ideas; we cannot go into all of them, by any means.

The number system is highly logical, offering a multiplicity of possibilities for logical manipulations. If we can legitimately assign numbers in the describing of objects and events, we can then operate

1 A "threshold stimulus" is one so weak that it elicits a response a certain proportion of the time, and a "differential limen" is a stimulus increment that is observable a certain proportion of the time.

with those numbers in all permissible ways and emerge with conclusions that we can apply back to the observed phenomena that we measured. We are justified in describing real things with numbers, provided there is a sufficient degree of *isomorphism* (similarity of properties or form) between those things and the number system.

There are certain properties of numbers that must have parallels in the observed phenomena. For example, every number is unique; no other is exactly the same. A number has *identity.* Thus, any object or event to which a number is applied must also have identifty. In the number system, numbers have the property of *order* or rank, one being greater than another. The objects to which they are applied must be orderable along some continuum, if the order of the numbers assigned to them is to lend description of order. Numbers also have the property of *additivity,* which means that the summing of a certain number with a certain other number must invariably yield a unique number. This property is the basis for almost all the more useful operations we can perform with numbers, for if we can add them we can also subtract (add negative numbers), multiply (add the same number to itself a number of times), or divide (produce successive subtractions).

It is not necessary that the phenomena to which we apply numbers have all the properties of identity, order, and additivity in order to be measurable. But the usefulness of the numbers applied in measurement depends upon how many of those properties do apply. Several levels of measurement are dependent upon the number of those properties which do apply. We consider those levels of measurement next.

NOMINAL MEASUREMENT AND NOMINAL SCALES The most limited type of measurement is the distinction of classes or categories—in other words, classification, about which much was said earlier in this chapter. Each group can be assigned a number to act as a distinguishing label, thus taking advantage of the property of identity. The assignment of numbers to qualitative classes is purely arbitrary, since one number would do as well as another—group 1, group 2, group 3, and so on. Having assigned the numerical labels, we must be logically consistent in identifying class members as group 1 cases, group 2 cases, and so on. Statistically, we may count the number of cases in each class, which gives us frequencies. Many statistical methods are designed for dealing with the frequencies in categorical data. Categorical data fall under the heading of *nominal measurement.* But there is nothing we can do meaningfully by way of computations with the numbers used merely as labels for categories. In effect, they are numerals, not numbers. Numerals are merely symbols for numbers, while numbers have unique values and belong to a logical system.

It may seem odd that we should refer to nominal *scales,* when "scale" suggests the idea of a continuum of some kind. A continuum has the property of order, which does not apply to nominal scales. But dictionary definitions sometimes refer to a scale as "that which discriminates," a conception that justifies the use of the expression *nominal scale*. The idea of discrimination or classification extends also to scales of higher types. Thus classification is the very basis of measurement of all kinds. The classification on scales of higher types is merely more refined. While keeping the property of identity, it adds others as well.

ORDINAL MEASUREMENT AND ORDINAL SCALES Ordinal measurement corresponds to what was earlier called "quantitative classification." The classes are ordered on some continuum, and it can be said that one class is higher than another on some defined variable. There may be one member to a class, as when we give a complete rank ordering for observed cases, or each class may have a frequency greater than 1.

Suppose that we place three boys—Charles, Robert, and David—in order of height (Charles is the tallest) and assign them the three numbers 3, 2, and 1, respectively. All we have is information about serial arrangement. We cannot say that Charles is as much taller than Robert as Robert is taller than David, even though the three numbers assigned to them are equally spaced on the scale of measurement. We are not at liberty to operate with these numbers by way of addition or subtraction, and so on; doing so will provide no further information about the boys.

As with nominal measurements, we can find frequencies for the categories and operate with those numbers. But because the class numbers are meaningful ranks, we are provided with a few statistical opportunities such as correlations and tests of statistical significance. These opportunities will be treated in later chapters.

INTERVAL MEASUREMENT AND INTERVAL SCALES If we actually apply a meter scale to the three boys and find their heights to be 195 cm, 180 cm, and 150 cm, respectively, we have values on a scale of equal units. Equality of units is the requirement for an interval scale. We can say that Charles is 15 cm taller than Robert and 45 cm taller than David. We could also say that the difference between Robert and David is twice that between Charles and Robert. Equality of units means that the same numerical distance is associated with the same empirical distance on some real continuum. Some scales in the behavioral sciences are measurements of physical variables, such as temperature, time, or pressure. As far as such physical measurements are concerned, the units are equal, and interval scales have been achieved.

However, one must ask whether the variation in the *psychological* phenomenon that is being measured indirectly is being scaled with equal units.

Most measurements in the behavioral sciences cannot possess the advantages of physical scales. Typical measurements in testing and in laboratory experiments, as in studies of learning, memory, or thinking, are in terms of number of correct answers or responses. Although it is rarely known whether the units on such scales are equal, it is generally assumed that they approach equality. By assuming that the scales are of the interval type, we can take advantage of most of the statistical methods that will be encountered in the following chapters.

RATIO MEASUREMENT AND RATIO SCALES One thing is certain: Scales of the kinds just mentioned have no absolute zero point. They fail to qualify as ratio scales, in which one number can be justifiably stated to be a certain multiple of another. The above-mentioned physical scales (ordinary temperature scales excepted), each of which has an arbitrary zero point, are genuinely ratio scales when applied to physical phenomena. We could say that Charles is 30 percent taller than David. On a test that is scored in terms of the number of correct answers, we could not justifiably say that a score of 50 is twice as high as a score of 25. In most such tests, it is unlikely that a score of zero means that the person being tested has none of the ability of the kind being scaled. In the example just mentioned, if we were arbitrarily to move the zero point 25 points lower, the two scores given above would become 75 and 50, and the ratio would be 75:50, or 3:2, rather than 50:25, or 2:1.

One of the areas in psychology in which some investigators claim to have achieved ratio scales is that of psychophysics, when so-called ratio-judgment methods are used. In such experiments, for example, the subject is instructed to say which stimuli are twice as great as others (in perceived weight, distance, or time). It is often assumed that the subject is able to perform his judgmental task faithfully, and the results seem consistent with that assumption.

Fortunately, there is almost nothing in the way of statistical methods that require ratio scales of measurement of empirical quantities. Interval-scale measurements will do for most purposes. It should be pointed out, however, that the enumeration we do in deriving frequencies does give us values on ratio scales. A zero frequency corresponds to the null situation—no cases at all. It can also be said that in statistical operations we create meaningful zero points, for example, at the mean of a distribution or at a difference of zero. Deviations from these generated zero points can be treated as ratio-scale measurements, permitting the operations of multiplication, division, and the taking of square roots.

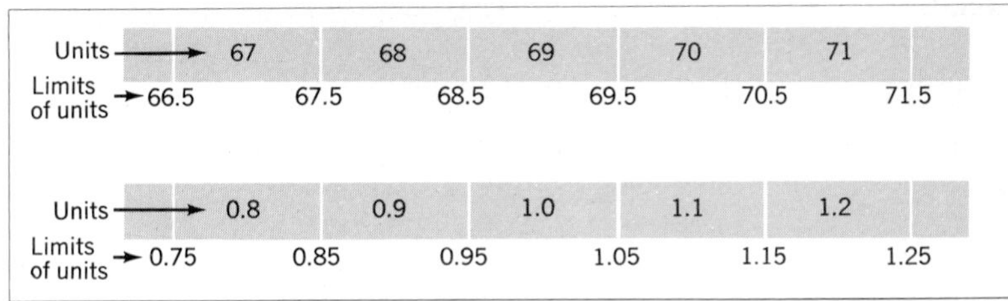

FIGURE 2.1 An illustration of two metric scales, showing selected units and their limits

HOW NUMBERS SHOULD BE REGARDED IN MEASUREMENT

Most interval-scale measurements are taken to the nearest unit—foot, inch, centimeter, or millimeter—depending upon the fineness of the measuring instrument and the accuracy we demand for the purposes at hand. In giving the height of a tree, measurement to the nearest foot—for example, 107 ft—would be adequate. In giving the height of a girl, we should resort to inches or perhaps centimeters as our practical unit. In giving the length of a needle, we should probably report in terms of millimeters, and in giving its diameter as seen under a micrometer, we should resort to some smaller unit. In any case, we may notice that our object does not contain an exact number of our chosen units. Our tree is more than 107 ft, but is closer to 107 than it is to 108; our girl is not exactly 156 cm, but is closer to 156 than to 155; etc. The result is that our report of 107 for the tree means anything between 106.5 and 107.5 ft, and our report for the girl means anything between 155.5 and 156.5 cm. Figure 2.1 shows a graphic illustration of units and their limits.[1]

And so it is with most psychological and educational measurements. A score of 48 means from 47.5 to 48.5, and a score of 70 means from 69.5 to 70.5. We assume that a score is never a point on the scale, but occupies an interval from a half unit below to a half unit above the given number.

A more important practical consideration dictates *the taking of a score as occupying a whole interval on the scale,* as the student will appreciate later. If we did not do this, an average computed from a set of ungrouped measurements would not be consistent with one computed when the same measurements are grouped. Even in dealing with discrete measurements—for example, the number of children in a family—we customarily proceed *as if* 8 children meant anywhere from 7.5 to 8.5. The only notable exception to this general rule is in dealing with chronological age as given to the *last* birthday and the like. Then a twelve-year-old child is anywhere from 12.0 to 13.0. If ages are given *to the nearest birthday,* however, our rule again applies, and a twelve-year-old falls in the interval 11.5 to 12.5.

[1] To test yourself, see Exercise 6.

EXERCISES

1 At a dog show, one room contains 80 contestants, of whom 32 are airdales, 20 are bulldogs, 12 are collies, and 16 are terriers of different varieties. Complete the following table:

Breed	*Frequency*	*Percentage*	*Proportion*
Airedale	32	___	___
Bulldog	___	25.0	___
Collie	___	___	.15
Terrier	___	___	___

2 In selecting a dog at random from this group of 80, what is the probability of getting a bulldog? A collie? A terrier? Either a bulldog or a terrier?

3 What is the ratio of bulldogs to terriers? Of airedales to collies? Of terriers to airedales?

4 At the dog show the year before, the same room contained the following numbers of dogs of each kind: airedales, 47; bulldogs, 27; collies, 11; and terriers, 15. Two years earlier, the numbers were 66, 30, 6, and 18, respectively. Prepare a tabulation of the data for the three years. State some conclusions drawn from the table.

5 For each of the following instances, state the highest level of measurement scale involved:

a. Numbers of men and women in a certain class in statistics
b. Number of pounds that a boy can lift
c. The boy's guesses as to the number of pounds he is lifting when he lifts different weights
d. Degree of embarrassment as indicated by change in skin temperature of the face
e. Numbers assigned consecutively to students as they complete an examination consisting of 100 items
f. Number of items a student answers correctly in an examination containing 10 items
g. Numbers assigned to different groups of students, each of which solved a certain problem by a different method

6 State the exact limits of the following scores or measurements: 57 sec; 150 kg; 65 score points; 0 score points; 14.5 cm; .125 sec; 15 years (to the last birthday).

ANSWERS

1 Frequencies: 32, 20, 12, 16.
Percentages: 40.0, 25.0, 15.0, 20.0.
Proportions: .40, .25, .15, .20.

2 Probabilities: 1/4, 3/20, 1/5, 9/20, (20/80 + 16/80), or, as more commonly stated, .25, .15, .20, .45.

3 5/4, 8/3, 1/2.

5 a. ratio; b. ratio; c. ordinal; d. the temperature scale is an interval one, and the embarrassment scale is an ordinal one; e. ordinal; f. ordinal; g. nominal.

6 56.5 to 57.5; 149.5 to 150.5; 64.5 to 65.5; −0.5 to +0.5; 14.45 to 14.55; .1245 to .1255; 15.0 to 15.99 (or 16.0).

3 Frequency distributions

After we obtain a set of measurements, a common next step is to put them in systematic order by grouping them in classes. A set of individual measurements, taken as they come, as in the list in Table 3.1, does not convey much useful information. We have merely a vague, general conception of about how large they run numerically, but that is about all. The data in Table 3.1 are scores made by 50 students in an inkblot test. Each score is the number of objects the student reported in observing 10 inkblots during a period of 10 min. We usually want to know several things concerning such a set of data. One is what kind of score the average or typical student makes; another is the amount of variability there is in the group or how large the individual differences are; and a third is something about the shape of the distribution of scores, i.e., whether the students tend to bunch up at either end of the range or at the middle or whether they are about equally scattered over the entire range. The first steps in the direction of answering these questions require setting up a frequency distribution.

The class interval and frequencies

THE SIZE OF CLASS INTERVAL

We could begin by asking how many scores of 25, 26, 27, etc., there are but this would not give us an adequate picture because in a group of only 50 individuals whose scores range from 10 to 55, many scores do not occur at all, and others occur only once. We therefore combine the scores into a relatively small number of *class intervals,* each class interval covering the same range of score units on the scale of measurement.

TABLE 3.1 Scores in an inkblot test

25	33	35	37	55	27	40	33	39	28
34	29	44	36	22	51	29	21	28	29
33	42	15	36	41	20	25	38	47	32
15	27	27	33	46	10	16	34	18	14
46	21	19	26	19	17	24	21	27	16

The first thing to be decided is the size of the class interval. How many units shall it contain? This choice is dictated by two general customs to which experience has led. *One is the rule that we should prefer not fewer than 10 or more than 20 class intervals.* More commonly, the number used is 10 to 15. An advantage of a small number is that we have fewer frequencies with which to deal. An advantage of a larger number is higher accuracy of computation. The process of grouping introduces minor errors into the calculations, and the coarser the grouping, the greater this tendency is.

SOME SIZES PREFERRED *The second rule determining the choice of class interval is that certain ranges of units (scores) are preferred.* Those ranges are 2, 3, 5, 10, and 20. These five interval sizes will take care of almost all sets of data. To apply these rules to the data in Table 3.1, we need to know that the lowest score is 10 and that there is a range of 45 points (the highest score minus the lowest). An interval of 3 points would give the best number of classes, according to the first rule. The range (45) divided by the class size (3) gives 15. If we chose an interval of 5 units, we should expect 9 groups. In view of the relatively small number of cases in the sample, and because an interval of 5 will actually give 10 groups, we choose 5 as the size of class interval.

WHERE TO START CLASS INTERVALS

It is natural to start the intervals with their lowest scores at multiples of the size of the interval—when the interval is 3, to start with 9, 12, 15, 18, etc.; when the interval is 5, to start with 10, 15, 20, 25, etc. This is by far the most common practice, though it is arbitrary.

SCORE LIMITS OF CLASS INTERVALS We shall follow the usual practice here. The intervals are therefore labeled 10 to 14, 15 to 19, 20 to 24, and so on. The top and bottom scores for each interval are called the *score limits*. They are useful for labeling the intervals and in tallying scores within the intervals.

EXACT LIMITS OF CLASS INTERVALS In computations, however, it is often necessary to work with exact limits. Remember that a score of 10 actu-

FIGURE 3.1
Exact limits of class intervals with different sizes of interval and of unit of measurement

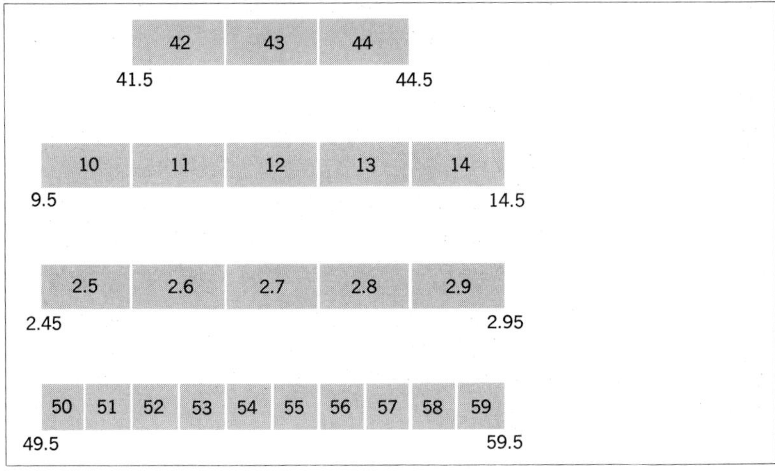

ally means from 9.5 to 10.5 and that a score of 14 means from 13.5 to 14.5. Thus the interval containing scores 10 through 14 actually covers a range from 9.5 to 14.5 on the scale of measurement. The interval labeled 55 to 59 extends from 54.5 to 59.5. The same principle holds no matter what the size of interval or where it begins in terms of a given score. Following this principle, each interval begins exactly where the one just below ends, as it should (see Fig. 3.1).[1]

TALLYING THE FREQUENCIES

Having adopted a set of class intervals, we are ready to list them, as in Table 3.2. Taking each score in Table 3.1 as it comes, we locate it within its proper interval and write a tally mark in the row for that in-

1 Strictly, limits such as 54.5 and 59.5 also stand for small distances rather than points. When measurements are integers, however, they serve as point values. At this point Exercise 1 might be done.

TABLE 3.2
Frequency distribution of the inkblot scores that were listed in Table 3.1

(1) Scores	(2) Tally marks	(3) Frequencies, f
55–59	/	1
50–54	/	1
45–49	///	3
40–44	////	4
35–39	~~////~~ /	6
30–34	~~////~~ //	7
25–29	~~////~~ ~~////~~ //	12
20–24	~~////~~ /	6
15–19	~~////~~ ///	8
10–14	//	2
		$\Sigma f = 50 = N$

terval. Having completed the tallying, we count up the number of cases within each group to find the frequency (f), or total number of cases within the interval. The frequencies are listed in column 3 of Table 3.2.

CHECKING THE TALLYING Next we sum the frequencies, and if the tallying has omitted none and duplicated none, the sum should equal the total number of scores in the sample. At the bottom of column 3 we find the symbol Σf, in which Σ (capital Greek sigma) stands for the "sum of" whatever follows it. Thus, Σf is the sum of the frequencies. The total number of scores in the sample is symbolized by the capital letter N, which stands for "number." If Σf does not equal N, there has been an error in tallying, and the process should be repeated until the check is satisfied. Even if Σf does equal N, there could have been a tally or two placed in the wrong interval. There is no way of checking this kind of error except by repeating the tallying.[1]

Graphic representation of frequency distributions

The frequency distribution in Table 3.2, particularly the array of tally marks, gives a rough picture of the sample of individuals as a whole. For example, we can see that the most frequent scores fall in the interval 25–29, that the very low and very high scores are quite rare, and that the greatest bunching of cases comes in the lower half of

1 See Exercises 2 and 3.

FIGURE 3.2
A frequency polygon for the distribution of scores on the inkblot test

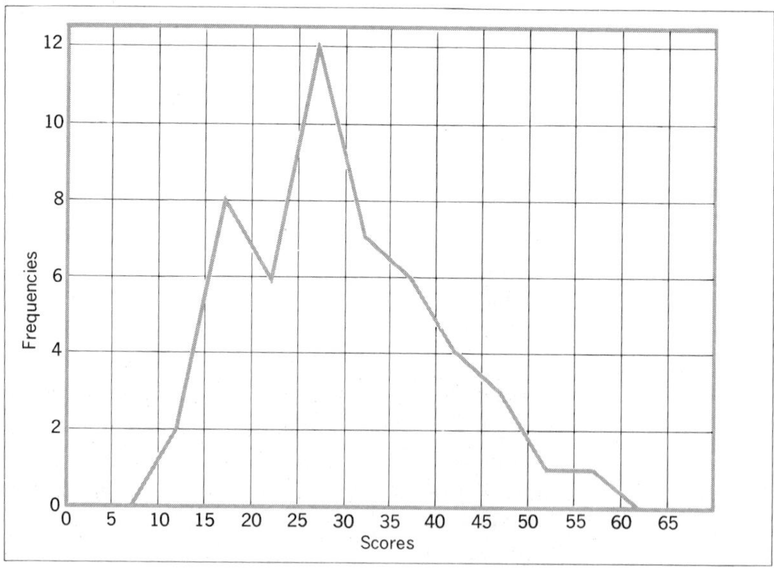

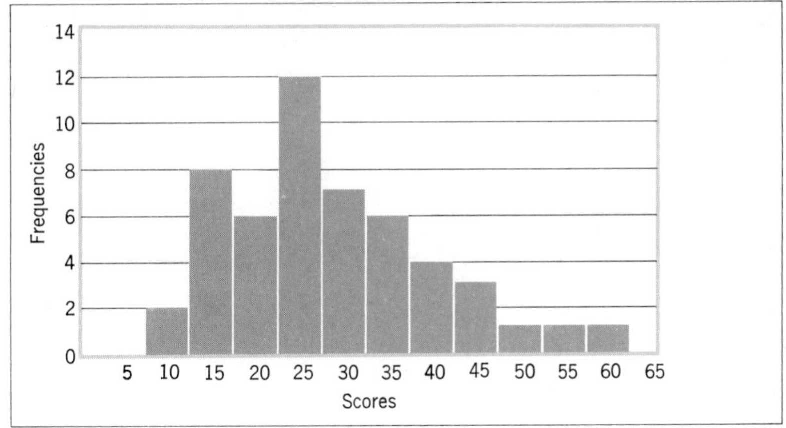

FIGURE 3.3 A histogram for the same distribution as in Fig 3.2

the range. Much better pictures of this distribution are afforded by Figs. 3.2 and 3.3, where the contour of the distribution is more accurately represented and the numbers of cases in the various intervals are more clearly shown. Figure 3.2 is known as a *frequency polygon,* and Fig. 3.3 as a *histogram.*

THE FREQUENCY POLYGON AND HOW TO PLOT IT

A polygon is a many-sided figure—hence the name of the picture in Fig. 3.2. There are a number of considerations to be kept in mind in drawing such a figure.

THE KIND OF GRAPH PAPER In general, the most convenient type of cross-section paper is ruled in heavy lines 1 in. apart, subdivided into tenths of inches.

THE WIDTH OF THE DIAGRAM The question of the width and height of the figure arises. For the sake of readability, the figure should be at least 5 in. wide. In the illustrative data we have 10 intervals in which there are frequencies, but in drawing the diagram, we should allow for at least one additional interval at each end, making 12. This step permits bringing the ends of the polygon down to the base line, as in Fig. 3.2. In that figure, one additional interval has been added to avoid an appearance of crowding.

LABELING THE BASE LINE In deciding how many intervals to allow to the inch, we consider the measurement scale and the range of measurements. In the inkblot data we have an interval of 5, and with the 14 intervals just adopted, we have a total range of 70 units. A width of 7 in. takes care of 70 units nicely, with a ½ in. to each class interval. We label each ½ in. with a multiple of 5.

TABLE 3.3 Class intervals and their midpoints

Score limits	*Exact limits*	*Midpoints*	*Frequencies*
60–64	59.5–64.5	62	0
55–59	54.5–59.5	57	1
50–54	49.5–54.5	52	1
45–49	44.5–49.5	47	3
40–44	39.5–44.5	42	4
35–39	34.5–39.5	37	6
30–34	29.5–34.5	32	7
25–29	24.5–29.5	27	12
20–24	19.5–24.5	22	6
15–19	14.5–19.5	17	8
10–14	9.5–14.5	12	2
5–9	4.5–9.5	7	0

THE HEIGHT OF THE FIGURE A further consideration is the relative height of the figure. For the sake of appearance and also of readability, the total height of the polygon is commonly made roughly 60 percent of its total width. In the inkblot data, this ratio is approached if we allow 3/10 in. to each person, on the kind of graph paper described.

HOW TO LOCATE A MIDPOINT In plotting a dot to represent the frequency in each interval, we must decide directly above what point on the base line the dot will be. It should be at the midpoint of the interval, and the midpoint is exactly midway between the *exact* lower and upper limits of the interval. The midpoint of the interval 55 to 59 is 57.0. When the size of the interval is 5, with the lowest score in each interval a multiple of 5, the midpoints end in 2 and 7 systematically. A complete list of the midpoints for the illustrative data may be seen in Table 3.3. For a more general illustration of midpoints, see Fig. 3.4.

PLOTTING THE POINTS Each frequency is plotted as a point directly above the midpoint of its interval. Frequencies of zero are plotted on the base line. Lines are then drawn to connect neighboring points, as in Fig. 3.2. When all is done, it appears that the points are not midway between multiples of 5, but this is as it should be. The exact limits of the intervals are .5 of a unit below those multiples on the scale of measurement.

THE HISTOGRAM AND HOW TO PLOT IT Many of the steps in plotting the frequency polygon also apply in plotting the histogram. The choices concerning size, proportions, and units per square of graph paper are all the same. The only important differences involve putting a horizontal line through each plotted point

FIGURE 3.4 Midpoints of intervals with differing numbers of units

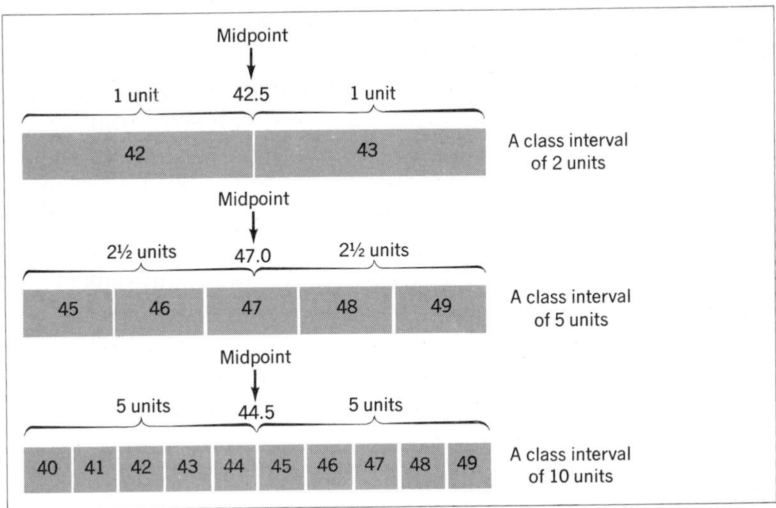

covering the width of the interval and erecting vertical lines at the exact interval limits to complete each rectangle, as in Fig. 3.3.

ADVANTAGES AND DISADVANTAGES OF THE TWO TYPES OF FIGURES

On the whole, the frequency polygon seems generally preferable to the histogram. For one thing, it gives a much better conception of the contour of the distribution; the transition from one interval to another is direct and probably describes the distribution we would have if a very large number of cases were in the sample.

The histogram gives a stepwise change from interval to interval, based upon the assumption that the cases falling within each interval are evenly distributed over the interval. The polygon gives the more correct impression that on both sides of the highest point (directly above the mode),[1] the cases within an interval are more frequent on the side nearer the mode, except where there are inversions in the general trend (as between scores of 15 and 25 in Fig. 3.2).

On the other hand, the histogram gives a more readily grasped representation of the number of cases within each class interval; each measurement or individual occupies exactly the same amount of area. One more advantage of the polygon is that when we wish to plot two distributions overlapping on the same base line—for example, for two different age groups or the two sexes—the histogram gives a very confused picture, whereas the polygon usually provides a clear comparison.[2]

1 As we shall see in the next chapter, the mode is that value on the measurement scale for which the frequency is greatest.

2 See Exercise 4.

PLOTTING TWO OR MORE DISTRIBUTIONS WHEN *N* DIFFERS

The comparison of two distributions graphically raises a new question when the numbers of individuals in the two groups differ. With large differences, naturally, there is the question of scale, or how much space to give the figure. If the smaller distribution is large enough to be clearly legible, the larger one may extend beyond reasonable bounds. Furthermore, if it should be general shapes and general levels on the measuring scale and dispersions or spreads that we wish to compare, the marked difference in size may make such comparisons very unsatisfactory. A common solution to this difficulty is to reduce both distributions to *percentage frequencies* instead of plotting the original frequencies. It is then as if we had two distributions, each of whose *N*'s equals 100. This makes their two total areas approximately equal in the polygon form, and comparisons of shape, level, and dispersion are then quite satisfactory.

HOW TO FIND PERCENTAGE FREQUENCIES The data in Table 3.4 are presented as an example of how to transform frequencies into percentages. In all cases, the frequencies in the distribution are each multiplied by 100 and then divided by *N*. A shorter procedure would be to find the quotient 100/*N* to four or more decimal places and then multiply each frequency in turn by this ratio. In distribution I the ratio is 100/51, which equals 1.9608, and in distribution II it is 100/160, which equals 0.6250. Multiplying each frequency f_1 by 1.9608, we obtain the list of percentages P_1 in column 4, and multiplying each frequency f_2 by 0.625, we obtain the list in column 5. Plotting these percentages above the corresponding midpoints of class intervals, we obtain the

TABLE 3.4 Frequency distributions of scores on a college-aptitude test for freshmen at two different colleges

(1) Scores	(2) f_1	(3) f_2	(4) P_1	(5) P_2
140–149		8		5.0
130–139		32		20.0
120–129		48		30.0
110–119	1	29	2.0	18.1
100–109	0	18	0.0	11.2
90–99	3	14	5.9	8.8
80–89	5	5	9.8	3.1
70–79	6	5	11.8	3.1
60–69	14	0	27.5	0.0
50–59	7	1	13.7	0.6
40–49	11		21.6	
30–39	4		7.8	
Sums	51	160	100.1	99.9

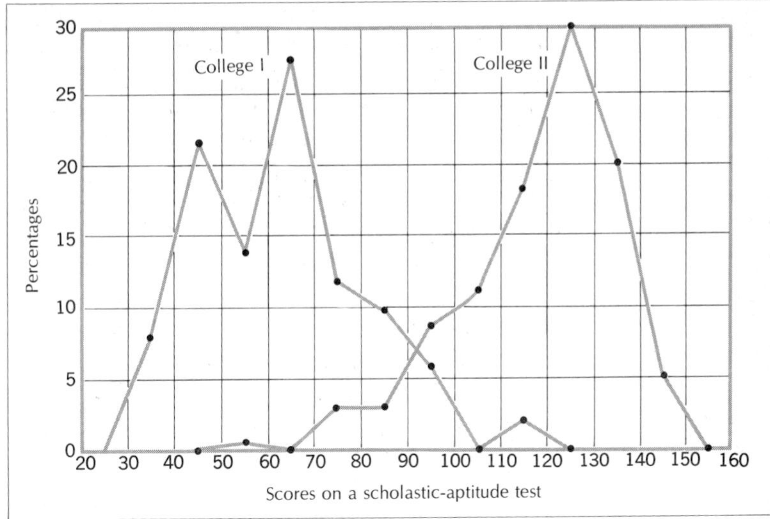

FIGURE 3.5 Distributions of scores on an aptitude test in two colleges. Frequencies have been reduced to a percentage basis.

distribution curves in Fig. 3.5. Although it is apparent in Table 3.4 that the second group was higher on the scale than the first and that there was overlapping of scores between the two, these facts are more clearly brought out in graphic form. Also much clearer is the somewhat narrower dispersion in the second group than in the first.

SKEWED DISTRIBUTIONS In addition, it is more clear that the first group bunches at the left in its own range and has relatively few high scores, whereas the second group bunches at the upper end of its range, with relatively few low scores. We describe the first distribution as being *positively skewed* (pointed end toward the right, or positive direction) and the second distribution as being *negatively skewed* (pointed end toward the left, or negative direction). In noting whether skewing is positive or negative, it is helpful to remember that *skewing* comes from the same source as the word *skewer,* a pin used to hold roasts or small pieces of meat together. We shall see other references to skewed distributions in later chapters.

Cumulative frequencies and cumulative distributions

Many statistical procedures, particularly those applied to test scores and to psychophysical data, are based upon cumulative frequency distributions. In the preceding sections, frequencies have been presented as belonging to certain score ranges or class intervals. In this section we are interested in the numbers of scores or other observations falling

TABLE 3.5
Cumulative frequency distribution for the inkblot-test data

(1) Scores in the intervals	*(2)* Exact upper limit of the interval	*(3)* f Frequencies	*(4)* cf Cumulative frequencies
55–59	59.5	1	50
50–54	54.5	1	49
45–49	49.5	3	48
40–44	44.5	4	45
35–39	39.5	6	41
30–34	34.5	7	35
25–29	29.5	12	28
20–24	24.5	6	16
15–19	19.5	8	10
10–14	14.5	2	2

below certain points on the measuring scale. *The cumulative frequency corresponding to any class interval is the number of cases within that interval plus all those in intervals lower on the scale.*

HOW TO FIND CUMULATIVE FREQUENCIES

The cumulative frequencies *cf* are readily found from the ordinary noncumulative frequencies by a process of successive additions. We use the familiar inkblot-test scores (see Table 3.5). The scores for the intervals are listed in column 1, just as before, with high scores at the top. Next we want a single score value to assign to each interval. Where before we used the midpoint of the interval, we now use the exact upper limit. The reason for this is that the frequency to be given corresponding to it includes all the cases *within* the class and *below it.* All those cases fall below the exact upper limit of the class interval. In column 3 are given the ordinary, noncumulative frequencies, merely because they are to be used in the cumulation process. The cumulation is started at the bottom of the list in column 3. Below the upper limit of the lowest interval (14.5) are two cases. Below the upper limit of the second interval (19.5) are these two plus the eight in the second interval, giving 10 as the cumulative frequency for the second interval. In the third interval we have six cases to add to what we already have, making 16 for the third interval. And so it goes, each cumulative frequency being the sum of the preceding one and the frequency in the class interval itself. This continues until the top interval is reached. The last cumulative frequency should be equal to N, which here is 50; if it is not, some error has been made.

PLOTTING THE CUMULATIVE DISTRIBUTION

In plotting the cumulative frequencies to show the trend in the relation of frequencies to the score scale, an ordinary histogram, which shows the frequency associated with each interval as a rectangle from the base line upward, is rather full of lines. In the cumulative histogram, we represent each class frequency starting upward where the preceding one left off (see Fig. 3.6), to show how each frequency adds to those below. With straight lines we can also connect the points at which the neighboring rectangles touch; this produces a cumulated frequency polygon. These points of contact represent the cumulative frequencies at the exact upper limits of the intervals.

It will be noted that the general trend of the cumulative-distribution curve is progressively rising; there are no inversions or setbacks. This is because all noncumulative frequencies are positive values, except for an occasional zero frequency. The upward rise is not a straight line. When the noncumulative distribution is symmetrical, the cumulative distribution is usually S-shaped. In Fig. 3.6, the upper branch approaches its limit (N) more gradually than the lower branch approaches its limit of zero. This is because the noncumulative distribution is positively skewed. Thus skewing shows up in the cumulative distribution, and it is in the direction of the tail that approaches its limit more gradually.

CUMULATIVE PERCENTAGES AND PROPORTIONS

Previously we had reason to transform frequencies into percentages for the sake of comparing two distributions where N differs (Fig. 3.5). The same reason, plus more important ones, prompts us more frequently to transform *cumulative* frequencies into percentages or into proportions.

FIGURE 3.6 A representation of the cumulative frequencies for the inkblot test score distribution. The rectangles show the frequencies for the intervals, each one beginning where the one below it ends. The diagonal lines connecting their corners constitute the most common way of drafting a cumulative curve.

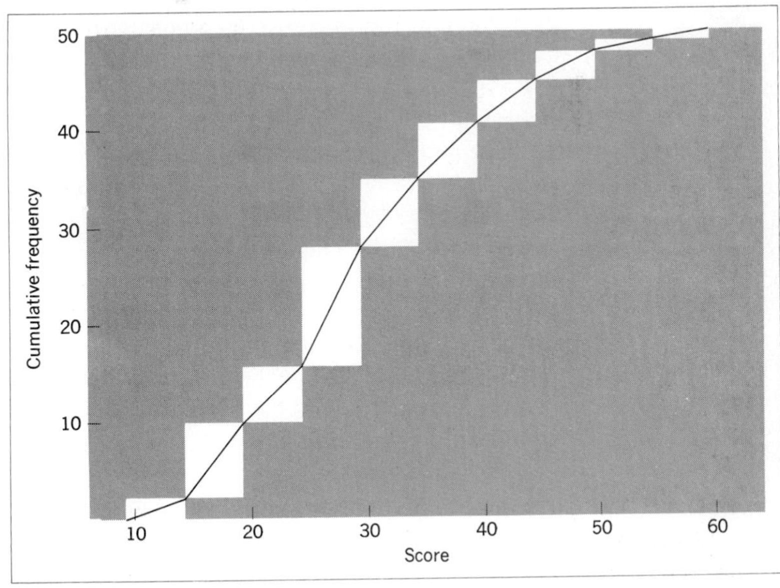

TABLE 3.6 Cumulative frequencies, percentages, and proportions for memory-test scores

(1) Scores	(2) X	(3) f	(4) cf	(5) cP	(6) cp
41–43	43.5	1	86	100.0	1.000
38–40	40.5	4	85	98.8	.988
35–37	37.5	5	81	94.2	.942
32–34	34.5	8	76	88.4	.884
29–31	31.5	14	68	79.1	.791
26–28	28.5	17	54	62.8	.628
23–25	25.5	9	37	43.0	.430
20–22	22.5	13	28	32.6	.326
17–19	19.5	8	15	17.4	.174
14–16	16.5	3	7	8.1	.081
11–13	13.5	4	4	4.7	.047
8–10	10.5	0	0	0.0	.000

This, of course, standardizes the total of the frequencies at 100 or at 1.0, respectively.

In Table 3.6, another example of cumulative frequencies is given. They are obtained here, in column 4, just as before. We now wish to find what percentage of 86 each cumulative frequency is. The arithmetic is simply a matter of multiplying each cumulative frequency by $100/N$. This fraction, $^{100}/_{86}$, is equal to 1.1628. To obtain the cumulative proportions, we find the products of $1/N$ times cf, or .011628 cf, or we divide the cumulative percentages by 100.

In Fig. 3.7, the cumulative percentages are plotted as points against the corresponding score points (exact upper limits of class intervals). Again, an S-shaped curve results, more clearly than in Fig. 3.6 because there is very little skew. In drawing the curve, some smoothing was done by inspection, since the points are so clearly in line. A cumulative percentage or proportion curve is often called an *ogive*. One reason for smoothing the ogive, when this can be reasonably done, is to gloss over some of the irregularities of a distribution that is derived from a limited number of cases in a sample in order to see how the distribution for the population would probably look.

CENTILE RANKS AND CENTILE POINTS

A number of uses for cumulative-distribution curves will be discussed in later chapters, but a special application will be given here. This use is in connection with what is commonly known as a *percentile* scale. The scores obtained from any test, such as the memory test just mentioned, are on an arbitrary scale. Numerical values, such as 5, 17, or 36, from one test are probably not equivalent to the same numerical values from another test. There is a need for a common scale of some

FIGURE 3.7 Smoothed cumulative distribution curve for the memory-test scores. Frequencies are in terms of percentages.

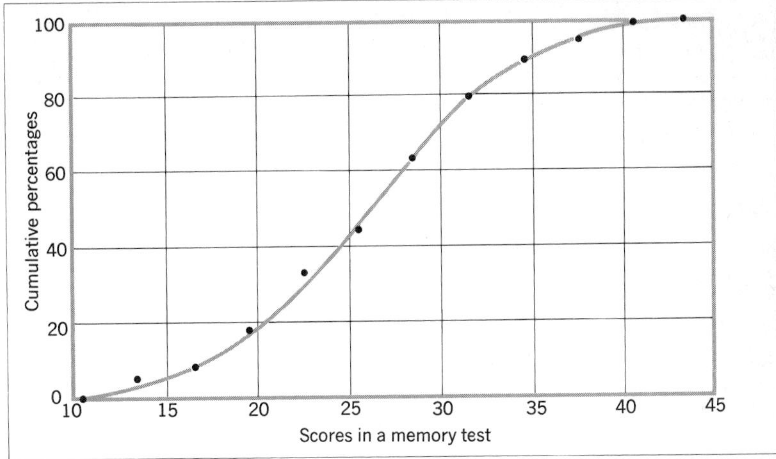

kind so that the same numbers indicate at least roughly the same quantity—so that the obtained-score values from different tests have a common reference frame that is meaningful. One such frame of reference is the percentile scale, although here the term *centile* is preferred.[1] The centile scale is divided into 100 units or centiles. If a person is given a centile value of 95, this means that in a typical sample of 100, the individual would excel 95 others of lower rank. Another person, with a centile value of 44, is higher than the lowest 44 persons in 100, or exceeds 44 percent of the cases in the distribution. The reference to centiles as ranks suggests that a centile scale is an ordinal one, not an interval one, a distinction that was made in the preceding chapter. More will be said on this later.

In setting up centile norms so that we can go readily from an obtained test score to a centile equivalent, it is necessary to derive a "map" of equivalent values all along the two ranges. In one method, the ogive curve is of considerable value. From a curve such as that in Fig. 3.7, particularly if it were magnified on graph paper, we could take any value on the one scale and find the corresponding value on the other scale. This mapping could be done in two ways. If we were concerned with only one test, the most satisfactory solution would be to start with each integral obtained score and read off the vertical the corresponding centile-rank value. But more often, several tests are involved, and reading in the reverse direction is the common practice. In order to find corresponding obtained-score point values on the ob-

1 Distributions are also divided into tenths, in decile ranks, but we do not say "perdeciles," and into fourths, in which case we speak of "quartiles," not "perquartiles." Consistency calls for "centiles," not "percentiles," even though the latter term seems to persist in popular use.

TABLE 3.7 Centile points corresponding to centile ranks, as determined by two different methods, for the memory-test distribution

(1) Centile ranks	(2) Graphic centile points	(3) Interpolated centile points	(4) Integral centile scores
95	38.0	38.0	38
90	35.0	35.3	35
80	32.0	31.8	32
70	29.5	29.8	30
60	28.0	28.1	28
50	26.5	26.6	27
40	24.5	24.6	25
30	22.5	22.0	23
20	20.5	20.0	21
10	17.5	17.1	18
5	14.5	13.8	15

tained-score scales, certain landmark values are selected on the centile-rank scale. A profile chart on which centile landmarks are given may be seen in Fig. 19.5 (p. 488). The centile scale is written only once, and the other scales are made to correspond.

The centile ranks commonly chosen are those in Table 3.7. In the second column are given corresponding centile points from the obtained-score scale as read off the curve in Fig. 3.7. No greater accuracy than half-unit differences was attempted. Perhaps a more common method of finding centile-point values is to interpolate them within the sample distribution, as in Table 3.7. We shall see how interpolations are carried out in the next chapter, where the median value (actually the 50th centile point) is computed, and in Chap. 5, where quartile points, which are also the 25th and 75th centile points, are computed. The values in column 3 of Table 3.7 were obtained by the interpolation method. It may be noted that there is less regularity of differences from one value to the next than in column 2, if we may assume accuracy to one-tenth of a score unit. This is due partly to the smoothing of the distribution in the graphic method. This greater regularity may be taken as justification for using the smoothing operation, but confidence in the results of smoothing should depend upon the closeness of the obtained points to an ogive curve.

Because an interpreter of obtained scores—a counselor, for example—deals with integral scores, not fractional ones, a mapping in terms of whole numbers is called for. Integral obtained scores corresponding to the various centile ranks are given in the fourth column. Because the graphic-derived centile points are more regular, the rounding to whole numbers was done from that list. The rounding is in the upward direction because an integral score must be wholly or in large part above its corresponding centile-rank value. Centile-rank values for other integral obtained scores can be found by interpolation. For example, the centile-rank value corresponding to a score of 26 could be taken as equivalent to the 45th centile rank, and that corresponding to a score of 33 as equivalent to the 83d.

Returning to the question of the kind of scale that is given by centile ranks, it can be said first that reference to Fig. 3.7 shows that if the obtained scores are on an interval scale, the centile ranks cannot be, for the relationship between the two scales is curved. We do not know that an obtained-score scale has equal units and is therefore an interval scale, but at least the shape of frequency distributions on such scales is usually what one should expect from a sample of cases from a homogeneous population when measured on an interval scale. That is, there is a bunching of cases near the center and very few extreme cases in either direction. Obtained distributions are so often found in this form when interval scales are known to be used that we may infer that most obtained test-score scales at least approach the interval type.

A frequency distribution based upon a centile-rank scale, which gives equal rank differences equal scale distances, is rectangular in form, with equal numbers of cases in the scale intervals. With a class interval of 10, a tenth of the cases are in each interval, if the sample is large. The extreme tenths of the cases extend over such a relatively large portion of the *obtained-score* scale that the centile ranks of 5 and 95 have been added to take care of the more extreme cases. Sometimes even the ranks of 1 and 99 are also added where the sample is large enough to enable us to locate those points accurately.

EXERCISES

1 For each of the following ranges of measurements, state your judgment of the best size of class interval, the score limits of the lowest class interval, the exact limits of the same interval, and its midpoint:
a. 83 to 197
b. 4 to 39
c. 17 to 32
d. 35 to 96
e. 0 to 188
f. 0.141 to 0.205

2 Given the list of scores in a "nervousness" test in Data 3*A* and using a class interval of 5, set up a frequency distribution. Begin the lowest class interval with a score of 35. List all exact limits of class intervals and also exact midpoints.

3 Given a list of scores, each of which is the percentage of 400 words judged pleasant by an individual (Data 3*B*), prepare a frequency distribution making the wisest choice of class interval and class limits.

4 Plot a frequency polygon and a histogram for Data 3*C*, group I. State your conclusions concerning these data as revealed by your plotted distributions.

DATA 3*A* Scores in a nervousness inventory

59	48	53	47	57	64	62	62	65	57	57	81	83
48	65	76	53	61	60	37	51	51	63	81	60	77
71	57	82	66	54	47	61	76	50	57	58	52	57
40	53	66	71	61	61	55	73	50	70	59	50	59
69	67	66	47	56	60	43	54	47	81	76	69	

DATA 3*B* Affectivity ratios (all have been rounded to the nearest whole number)

43	62	52	48	46	65	43	48	52	51	57	48	48
38	42	44	46	43	35	42	45	45	44	46	40	40
47	52	38	51	45	38	51	40	46	45	54	55	41
50	59	42	39	56	44	43	47	51	43	50	34	40
53	42	31	44	51	43	48	41	43	48	41	55	

DATA 3C
Distributions of chemistry-aptitude scores in two freshman chemistry courses, I and II

Scores	*Frequencies for group I*	*Frequencies for group II*
90–94	4	2
85–89	10	0
80–84	14	0
75–79	19	0
70–74	32	2
65–69	31	4
60–64	40	5
55–59	28	12
50–54	29	13
45–49	21	21
40–44	18	21
35–39	10	19
30–34	6	20
25–29	1	14
20–24	3	1
Sums	266	134

5 Reduce distributions I and II (Data 3C) to percentage distributions and plot them on the same diagram. Make a descriptive comparison of the two distributions as drawn.

6 Carry through the following steps for the first distribution of chemistry-aptitude scores in Data 3C:

a. Find the cumulative frequencies and tabulate them.
b. Plot a cumulative distribution curve similar to Fig. 3.6.
c. Find the cumulative percentages and proportions and tabulate them.
d. Plot the ogive distribution, smoothing the curve.

7 Using the ogive developed in Exercise 6, derive the kind of centile-norm information that is shown in Table 3.7, omitting interpolated centile-point values. Interpret your findings.

ANSWERS

1

i	*Score limits*	*Exact limits*	*Midpoints*
a. **10**	**80 to 89**	**79.5 to 89.5**	**84.5**
b. **3**	**3 to 5**	**2.5 to 5.5**	**4.0**
c. **1**	**17**	**16.5 to 17.5**	**17.0**
d. **5**	**35 to 39**	**34.5 to 39.5**	**37.0**
e. **20**	**0 to 19**	**–0.5 to 19.5**	**9.5**
f. **.005**	**0.140 to 0.144**	**0.1395 to 0.1445**	**0.142**

2 Frequencies, first solution: 5, 4, 4, 8, 11, 12, 11, 6, 2, 1; second solution: 1, 4, 5, 5, 8, 13, 13, 8, 5, 1, 1.

3 Frequencies ($i = 3$, with lowest interval 30 – 32): 1, 1, 2, 4, 9, 8, 10, 15, 8, 3, 2, 1.

5 Percentages:
I. 1.5, 3.8, 5.3, 7.1, 12.0, 11.6, 15.0, 10.5, 10.9, 7.9, 6.8, 3.8, 2.3, 0.4, 1.1.
II. 1.5, 0.0, 0.0, 0.0, 1.5, 3.0, 3.7, 9.0, 9.7, 15.7, 15.7, 14.2, 14.9, 10.4, 0.8.

6 a. *cf*: 266, 262, 252, 238, 219, 187, 156, 116, 88, 59, 38, 20, 10, 4, 3.
c. *cP*: 100.0, 98.5, 94.7, 89.5, 82.3, 70.3, 58.6, 43.6, 33.1, 22.2, 14.3, 7.5, 3.8, 1.5, 1.1.

7 Graphic centile points (for centile ranks from 99 to 1 inclusive): 94.0, 84.2, 78.8, 72.4, 67.5, 63.7, 60.2, 56.4, 52.4, 47.8, 41.4, 36.0, 26.0. Integral equivalents: 94, 85, 79, 73, 68, 64, 61, 57, 53, 48, 42, 36, 26.

4 Measures of central value

This chapter is about averages, of which there are several kinds. Three of them—the *arithmetic mean* (or *mean*), the *median,* and the *mode*—will be explained.

An *average* is a number indicating the central value of a group of observations or of individuals. To the question, "How good is a sixth-grade class in arithmetic?" the most reliable and meaningful kind of answer would be the mean or median in some acceptable test of arithmetical achievement. To the question, "What is the weakest tone to which this dog will respond?" the best kind of answer would state the average result from a number of trials. In either case, a single score or a single measurement would be highly unreliable, for not all measurements, even from repeated observations of the same thing, have the same value. To answer such questions by reciting the long list of individual measurements would be highly uneconomical in the reporting and not very enlightening to the questioner.

The average, whether it be a mean, median, or mode, serves two important purposes. First, it is a shorthand *description* of a mass of quantitative data obtained from a sample. It is more meaningful and economical to let one number stand for a group than to try to note and remember all the particular numbers. An average is therefore descriptive of a sample obtained at a particular time in a particular way. Second, it also describes indirectly but with some accuracy the *population* from which the sample was drawn. If the sample of sixth-grade children is representative of all the sixth-grade children in the same school, in the same city, or even in the same county, then the average of their scores tells us much about the average that would be made by the population that they represent, be it schoolwide, citywide, or countywide. If we examine the dog's hearing under a set of conditions that is characteristic of its general, day-to-day existence, the sample average will be very close to one that we could actually obtain by testing him day after day on many days.

It is only because sample averages are close estimates of larger population averages that we can generalize beyond particular samples at all and make predictions beyond the limits of a sample. This means considerable economy of effort, but, far more important, it makes possible all scientific investigation. We rarely or never know the average of a population; consequently, we do not know by how much our obtained average has missed it. However, if our sampling has been done in the proper manner, we can estimate our approximate error, as will be shown in Chap. 8. In the present chapter we shall be concerned only with the methods of computing averages from sample data.

The arithmetic mean

THE MEAN OF UNGROUPED DATA Most readers already know that to find the arithmetic mean (popularly called the *average*), we sum the measurements and then divide by the number of measurements or cases. In terms of a formula,

$$\bar{X} = \frac{\Sigma X}{N} \qquad \text{(The arithmetic mean)}^1 \qquad (4.1)$$

where $\bar{X}$ = arithmetic mean
Σ = "the sum of"
X = each of the measurements or scores in turn
N = number of measurements or scores

In an experiment to determine the lowest frequency of vibration of a sound wave that would yield a tone for a human observer, 10 trials were given, with the following results: 13, 17, 15, 11, 13, 11, 17, 13, 11, 11 (cycles per second). The sum of these measurements is 132,

1 To be completely explicit mathematically, we should designate a single observed measurement by the symbol X_i, which can stand for each of a series of measurements X_1, X_2, X_3, . . . X_n, denoting the first, second, third, to the nth measurement. The complete expression for formula (4.1) would read

$$\bar{X} = \frac{\sum_{i=1}^{n} X_i}{N}$$

where the symbols below and above the summation sign indicate that the things summed range from the first X_i to the last, or X_n. In most places in this text the limits of the summed items will not be written, since they will be readily understood from the nature of the formula and from the context. Formulas are generally much easier to read without such additional symbols.

and therefore the mean is 13.2 cycles per second. Note that in reporting a mean, it is given in terms of the unit of measurement, which is specifically stated.

As another example, the scores on the inkblot test found in Table 3.1, when summed, give ΣX equal to 1,480. The mean, with the use of formula (4.1), is

$$\bar{X} = \frac{\Sigma X}{N} = \frac{1{,}480}{50} = 29.60$$

The mean inkblot score is 29.60 score units. In practice, it is customary in reporting a mean to round to one more figure at the right than the original measurements had—in this case, to keep one decimal place, where the original scores were whole numbers. We report the mean as 29.6 score units.

THE MEAN OF GROUPED DATA

When data come to us grouped, when they are too lengthy for comfortable addition without the aid of a calculating machine, or when we are going to group them for other purposes anyway, we find it more convenient to apply another formula for the mean:

$$\bar{X} = \frac{\Sigma f X_c}{N} \qquad \text{(Arithmetic mean from grouped data)} \qquad (4.2)$$

where the symbols N and Σ have the same meanings as before, X_c = midpoint of a class interval, and f = number of cases within the interval. The expression fX_c means the frequency times the X value for the interval. All such products must be found first and then summed.

The solution by way of this formula is illustrated in Table 4.1. Here we have only as many different X values as there are class intervals, instead of as many as there are original measurements. Each class interval has as its X value the midpoint of that interval, which is given the special symbol X_c. This practice assumes that the midpoint of the interval correctly represents all the scores within that interval. This will not be exactly true in some instances, but the discrepancy is small in any case, and in computing the mean, most of the discrepancies tend to counterbalance others, giving a mean that is essentially correct.

In column 2 of Table 4.1, the midpoints of the intervals are given. We must add each midpoint into our total as many times as there are cases within that interval. This means finding for each interval the product of f times X_c, or fX_c. The fX_c products are listed in column 4. The sum of the fX_c products (ΣfX_c) is equal to 1,480. Dividing this by N, we find the mean to be 29.60, as it was for the same data ungrouped. As was indicated before, we should not be surprised to find a minor discrepancy between the means calculated from grouped

TABLE 4.1
Computation of the mean in grouped data

(1) Scores	(2) X_c Midpoint	(3) f	(4) fX_c
55–59	57	1	57
50–54	52	1	52
45–49	47	3	141
40–44	42	4	168
35–39	37	6	222
30–34	32	7	224
25–29	27	12	324
20–24	22	6	132
15–19	17	8	136
10–14	12	2	24
Sums		50	1,480
		N	ΣfX_c

$$\textbf{Mean} = \frac{\Sigma fX_c}{N} = \frac{1{,}480}{50} = 29.60$$

and ungrouped data. It happened here that the discrepancy was zero. We may also expect trivial discrepancies in means when the same data are grouped differently, i.e., with different size of class interval.

The median and centile values

The *median* is defined as that point on the scale of measurement above which are exactly half the cases and below which are the other half. Note that, in general, it is defined as a *point* and not as a score or any particular measurement. If this conception is kept clearly in mind, many difficulties will be forestalled.

THE MEDIAN FROM GROUPED DATA

It is probably easier to grasp the process of computing a median in grouped data. For a first illustration, consider Table 4.2. Here there are 28 cases, and so the median is that point on the measuring scale above which there are 14 cases and below which there are 14. Counting frequencies from the bottom upward, we find that $4+1+1+10=16$ cases, or 2 more than we want. To make 14 cases exactly, we need 8 out of the 10. The median lies somewhere within the interval 15–19, whose *exact* limits are 14.5 and 19.5. We assume for the sake of computation that the 10 cases within this interval are evenly spread over the distance from 14.5 to 19.5 (see Fig. 4.1). We must interpolate

TABLE 4.2 Computation of the median size of class in a certain school, with the use of grouped data

Class size	*f*	
40–44	1	
35–39	0	**12 = number of cases**
30–34	3	**above the interval**
25–29	5	**containing the median**
20–24	3	
15–19	10	
10–14	1	**6 = number of cases**
5–9	1	**below the interval**
0–4	4	**containing the median**
		$Mdn = 14.5 + 8/10 \times 5 = 14.5 + 4.0 = 18.5$
	$N = 28$	$Mdn = 19.5 - 2/10 \times 5 = 19.5 - 1.0 = 18.5$

within this range to find how far above 14.5 we need to go in order to include the 8 cases we need below the median. We must go 8/10 of the way, for 8 is the number we require, and 10 is the total number in the interval. The total distance is 5 units, and so on the scale of measurement we go 8/10 of 5, or exactly 4.0 units. Adding this 4.0 to the lower limit of the class interval, we get 14.5 + 4.0 = 18.5 as the median.

We can check this by counting down from the top of the distribu-

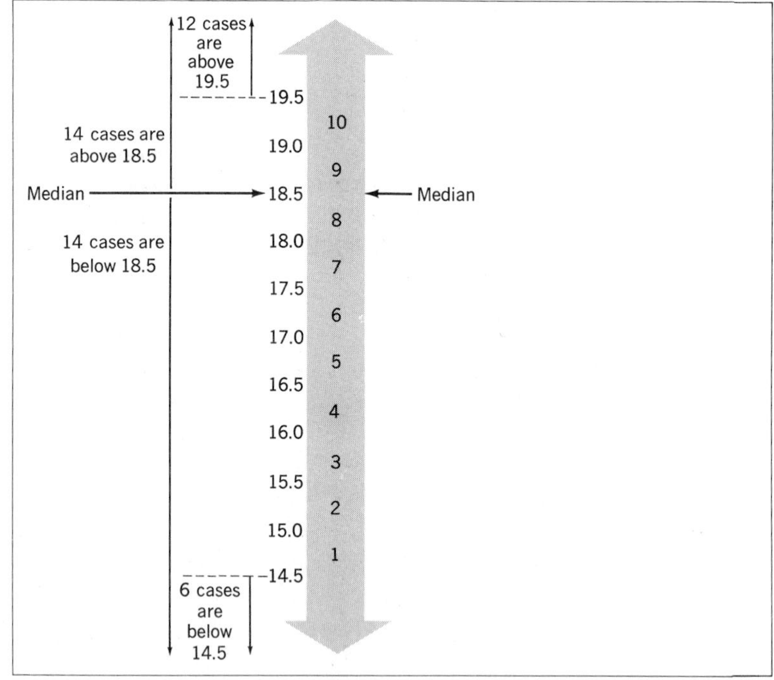

FIGURE 4.1 Showing how the 10 cases in the interval 14.5 to 19.5 are distributed. Each case is assumed to occupy a tenth of the interval, or one-half of a score unit. The eighth case extends up to the point 18.5, which is the median.

tion until we include $N/2$ of the cases, 14 in this problem. Starting at the top, we find that

$$1 + 0 + 3 + 5 + 3 = 12$$

We need 2 more cases out of the next group of 10. We must go 2/10 of the way below the *upper* limit of the interval, i.e., below 19.5. This means 2/10 of 5, or exactly 1.0 unit. The upper limit, 19.5 minus 1.0, gives us 18.5 for the median, which checks with the one obtained by counting up from below. It is well always to check the determination of a median in this manner, and to do so involves very little work. If the two estimates do not agree exactly, something is wrong.

To take another example with grouped data, consider Table 4.3, where N is an odd number. Here $N/2$ is 18.5, but the principle of interpolating within an interval for the exact median is just the same. Counting up from below, we find that $1 + 5 + 8 = 14$, which lacks 4.5 cases of including the lower half. In the next interval, we must go 4.5/8 of the way, or 4.5/8 times 2, which equals 9/8, or 1.125. Adding this many units to the lower limit of the interval (22.5), we have 23.625 as the median; or dropping all but one decimal place, we report the median as 23.6 score units. Checking by counting down from the top, we find 15 cases above the point 24.5. Going 3.5/8 of the way down into the interval of 2 units, we find that we must deduct 0.875 from 24.5 to find the median. When rounded to one decimal place, the

TABLE 4.3 Computation of the median score in a sentence-construction test as given to 37 men

Scores	*f*	
37–38	1	
35–36	2	15 = number of cases above the interval containing the median
33–34	0	
31–32	1	
29–30	0	
27–28	6	
25–26	5	
23–24	8	
21–22	8	14 = number of cases below the interval containing the median
19–20	5	
17–18	1	
	$N = 37$	$\frac{N}{2} = 18.5$

$$Mdn = 22.5 + \frac{4.5}{8} \times 2 = 22.5 + \frac{9}{8} = 22.5 + 1.125 = 23.6$$

$$Mdn = 24.5 - \frac{3.5}{8} \times 2 = 24.5 - \frac{7}{8} = 24.5 - .875 = 23.6$$

median is 23.6, as before. In terms of a formula, the interpolated median is found from below by

$$Mdn = l + \left(\frac{N/2 - F_b}{f_p}\right)i \qquad \text{(Interpolation of a median from below)} \qquad (4.3a)$$

where l = exact lower limit of class interval containing the median, F_b = sum of all frequencies below l, f_p = frequency of the interval containing the *Mdn*, and i is the size of class interval.

In terms of a similar formula, the median is found from above by

$$Mdn = u - \left(\frac{N/2 - F_a}{f_p}\right)i \qquad \text{(Interpolation of a median from above)} \qquad (4.3b)$$

where u = exact upper limit of the interval containing the median and F_a = sum of all frequencies above u.

A SUMMARY OF THE STEPS FOR INTERPOLATING A MEDIAN The steps for computing a median from grouped data may be summarized as follows:

step 1 Find $N/2$, or half the number of cases in the distribution.
step 2 Count up from below until the interval containing the median is located.
step 3 Determine how many cases are needed out of this interval to make $N/2$ cases.
step 4 Divide this number needed by the number of cases within the interval.
step 5 Multiply this by the size of class interval.
step 6 Add this to the exact lower limit of the interval containing the median.
step 7 Check by adding down from the top to find to what point the upper half of the cases extend in a manner analogous to that described in steps 2 to 5 inclusive.
step 8 Deduct the number of score units found in step 7 from the exact upper limit of the interval containing the median.

SOME SPECIAL SITUATIONS

There are some instances in which things do not turn out just as they did in the two illustrative examples.

WHEN THE MEDIAN FALLS BETWEEN INTERVALS If it should happen, in adding up cases from below, that half the cases take in *all* the cases in the last interval, the median is then the exact upper limit of that interval. In counting down from above, it would be found that all the cases in the interval just above this one would also be required to make $N/2$, and so its exact bottom limit would be the median. This coincides with the exact upper limit of the interval below; thus the median checks.

WHEN THERE ARE NO CASES WITHIN THE INTERVAL CONTAINING THE MEDIAN

Another question arises when the median falls within an interval where there are *no* cases. It is even possible that in the region of the median, two or more intervals have frequencies of zero. If the range having no cases is one interval, the median may be taken as the midpoint of that interval, but this gives a crude estimate unless the size of the interval is small—for example, not over three units. If that range covers two or more intervals, no good estimate can be made, but one could use the midpoint of that range.

THE MEDIAN FROM UNGROUPED DATA

The principle for finding a median in grouped distributions applies almost intact to the operation with ungrouped data. The median is a *point* on the measuring scale. In ungrouped data, each score or measurement is assumed to occupy a *range* of one unit. The median falls either within one of those units or between units. The first step is to arrange the measurements in order of their size. The list of 10 measurements of the threshold for pitch given earlier, when placed in rank order, becomes

11, 11, 11, 11, 13, 13, 13, 15, 17, 17

As in the case of grouped data, it is assumed that the four 11s occupy the range from 10.5 to 11.5, the three 13s occupy the range from 12.5 to 13.5, and so on. Counting from below to include five cases brings us to the first 13 that must be included among the five. We must therefore extend one-third of the way into the interval of one unit, or 0.33 unit into the interval that starts at 12.5. The median is $12.5 + 0.33$, which equals 12.83, or, rounded, 12.8. In checking from above, the reader should find the same value.

In the series of measurements

2, 5, 7, 8, 9, 10, 17

the median comes midway in the fourth one, which is 8. The median is exactly 8. In the series of measurements

7, 9, 10, 12, 13, 15, 18, 20

four are 13 or above, and four are 12 or below. The median is 12.5.

OTHER INTERPOLATED VALUES—CENTILE POINTS

In Chap. 3 the concepts of centile rank and centile point were introduced, and a graphic method was presented for deriving particular centile points for a set of measurements to correspond to certain selected centile ranks. In practice, probably the more common method of accomplishing the same results is to interpolate the centile points in the original distribution of scores. That method will now be very briefly illustrated; it involves the same principles as those used for interpolating a median.

The median is at the centile point that corresponds to the centile rank of 50. Just as the median is found by interpolation, so may any centile point be found. For example, if we want to know where the 80th centile point is located in the distribution of scores for the test represented in Table 4.1, we first determine how many of the 50 cases correspond to 80 percent. Eighty percent of 50 is 40 of the cases. Counting frequencies from below, we find that 35 of them are below the interval with exact limits of 34.5 and 39.5. We need 5 more of the 6 cases within this interval to reach exactly 40. The amount to be added to the lower limit is 5/6 of the interval's 5 units, which is 4.2 score points. Adding this quantity to the exact lower limit of the interval (34.5) gives 38.7 for the 80th centile point. This value could be checked by interpolating from the upper end of the distribution. Interpolations of any other centile point would follow the same procedures.

The mode

The mode is defined as the *point on the scale of measurement with maximum frequency in a distribution*. When we have ungrouped data, the mode is the measurement that occurs most frequently. Usually it is somewhere near the center of the distribution, and in a unimodal, symmetrical distribution, it coincides with the mean and the median.

ESTIMATION OF THE MODE

In a distribution of grouped data, the mode is estimated at the midpoint of that class interval having the greatest frequency. In Table 4.1, the highest frequency is 12, for the interval 25–29. The midpoint is 27, which is the estimate of the mode. In Table 4.2 the estimated mode is clearly 17. In Table 4.3, however, the maximum frequency is shared by two neighboring intervals. We do the reasonable thing and estimate the mode at the dividing point between the two, which is 22.5. Should there be two clearly larger frequencies with lower frequency or frequencies between, the distribution has two modes; it is *bimodal*. In distributions with small N, no accurate estimate of a mode can be made.

When to employ the mean, median, and mode

CERTAIN ADVANTAGES OF THE MEAN

The arithmetic mean is to be preferred whenever possible because of several desirable properties. First, it is generally the most reliable or accurate of the three measures of central value. By this we mean that from sample to sample from the same population, the mean ordinarily fluctuates less widely than either the mode or the median. Second, the mean is better suited to further arithmetical computations. Deviations of single cases from the central value give important information regarding any distribution. Much is done with these deviations, as will be seen in the next chapter. It will also be found that we square the deviations, and there is occasion for doing this only when the deviations are taken from the mean. For descriptive purposes, especially when distributions are reasonably symmetrical, we may almost always use the mean and should prefer it to the median and mode. On the other hand, there are instances, particularly when distributions are skewed, where the mean may give less descriptive ideas about distributions.

COMPARISONS OF THE MEAN WITH THE MEDIAN AND MODE

SOME MATHEMATICAL PROPERTIES OF THE MEAN AND MEDIAN A better appreciation of the nature of the mean and median may be gained by noting some of their mathematical peculiarities. To illustrate, let us use the data presented in Table 4.4. For the scores of six individuals, the mean is 6.0 and the median is 4.5.

The first feature to be noted is that the mean is the *center of gravity* of the distribution of scores. In Fig. 4.2 we have the six scores represented on the measurement scale. Imagine that the scale itself is a weightless bar. The six persons may be regarded as exactly the same in all respects except for their scores on this scale. Each "weighs" the same; his effect upon tilting the bar depends only upon his position on it. If we rest the bar on a fulcrum in such a position that the bar will be perfectly balanced, that position must coincide with the mean. The measurements in any sample are perfectly balanced about their arithmetic mean.

Each individual in this small distribution carries an effective weight in proportion to his distance from the mean. In column 3 of

TABLE 4.4 Illustration of certain properties of the arithmetic mean and the median

(1) Person	*(2)* Score	*(3)* Deviations from the mean	*(4)* Deviations from the median	*(5)* Deviations from the mean, squared	*(6)* Deviations from the median, squared
A	2	−4	−2.5	16	6.25
B	3	−3	−1.5	9	2.25
C	4	−2	−0.5	4	0.25
D	5	−1	+0.5	1	0.25
E	9	+3	+4.5	9	20.25
F	13	+7	+8.5	49	72.25
Sums	36	0	+9.0	88	101.50
Means	6.0	0.0	+1.5		
Median	4.5				

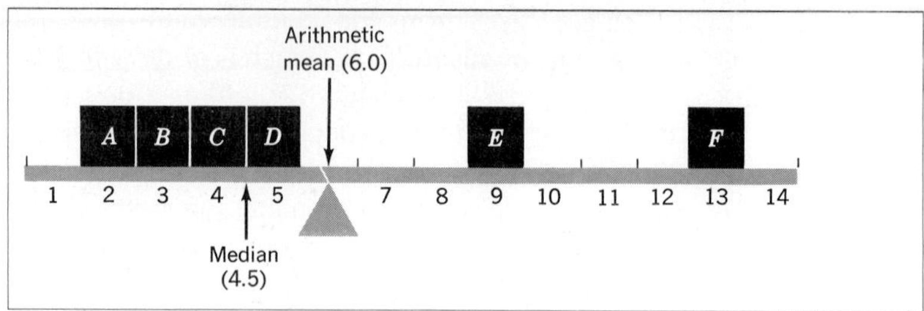

FIGURE 4.2 Illustration of the positions of six cases with respect to the arithmetic mean and the median. With all cases carrying equal intrinsic weight, they are perfectly balanced when the fulcrum is placed at the arithmetic mean.

Table 4.4 these distances are given, and they are called *deviations from the mean,* or simply *deviations.* The size of each deviation indicates how much effective weight each observation carries, and its algebraic sign tells in what direction that weight is applied. The algebraic sum of the deviations is zero, indicating perfect balance. Figure 4.2 illustrates how the mean is the center of gravity.

The arithmetic mean is the only value in a distribution from which the deviations always sum exactly to zero. To show that the median does not qualify in this respect (when it differs from the mean), let us find the deviations of the six scores from the median and sum them (see Table 4.4). The algebraic sum of these deviations is 9.0, with a net balance on the plus side. A fulcrum placed at the point 4.5 on the scale in Fig. 4.2 would cause a serious imbalance. This results from the ignoring of distances of cases from the central value when a median is computed.

Another peculiarity of the mean is that the sum of the squared deviations about it is smaller than that of the squared deviations about any other value. This fact has a double significance. In most of the following chapters we shall very often be concerned with squared deviations from the mean. For the present, it is significant that when squared deviations are considered, the arithmetic mean is closest to the measurements of the sample as a whole. In this respect it represents those measurements best. In Table 4.4 we can see that for this small sample, the sum of the squared deviations is definitely smaller when the reference point is the mean than when it is the median, the two sums being 88 and 101.5. The reader may verify the "least-squares" principle by showing that the sum of 88 is smaller than for deviations from any other value.

CENTRAL VALUES IN SKEWED DISTRIBUTIONS

In skewed distributions, the mean is always pulled toward the skewed (pointed) end of the curve, as Fig. 4.3 shows. As the center of gravity of the distribution, the mean is weighted toward extreme values. The *sum* of the deviations on the one side of it equals the *sum* of the deviations

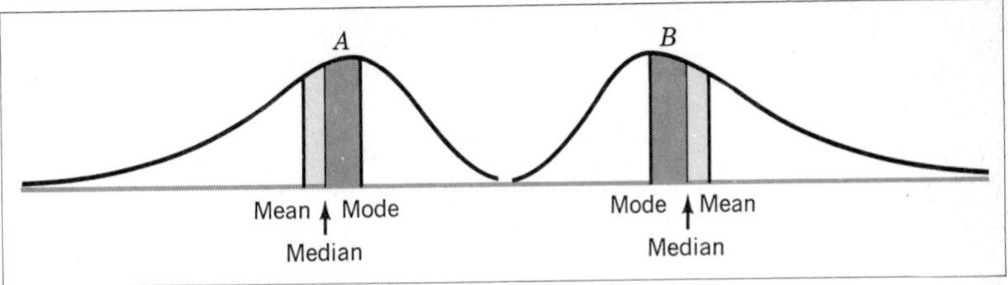

FIGURE 4.3 Two skewed distributions, (*A*) skewed negatively and (*B*) skewed positively, showing the relative positions of modes, medians, and means. Note that the mean is displaced farther from the mode toward the skewed end of the distribution and that the median is displaced two-thirds as far.

on the other. The median comes at a point that divides the *area* under the distribution curve into two equal parts. The *number* of scores on the one side of it equals the *number* of scores on the other. These features can help in interpreting means and medians.

Let us consider an example of a markedly skewed distribution and see how well the three measures of central value serve to describe it. In a study of class size in a certain part of a university, among 62 classes two were found to have 200 students each and two had between 100 and 200; the remaining classes, except two, had less than 60 students each. The average size of the 62 classes was 34, but this value was not at all typical because half the classes had 20 or fewer students (the median was 20.5). The *most typical* class size was 17, which was the mode.

What are the consequences of these unusually large discrepancies among the three indicators of central value? If one objective happened to be to equalize the size of classes, assuming that this were practical, there should be 34 students in each class. If we wanted to decide as a matter of educational policy whether or not there were too many small classes in general and if we had concluded beforehand that most teachers could handle successfully 30 students in a group, then the median would tell us that there are entirely too many small classes for efficient instruction. The mean would not have told us this because it is higher than 30. If we were piloting visitors about the buildings while classes were in session and wished to prepare them for the class size we would be most likely to find at random, we would tell them about the mode.

In reporting central values for a skewed distribution for descriptive purposes, it is sometimes desirable to state both mean and median. Each tells its own story, and from the difference between the two and the direction of the difference we can infer something about the direction and amount of skewing. The mode is rarely reported unless there is some reason for conveying information regarding the most probable value. When a distribution is completely symmetrical about its mode, the three averages coincide, and so only one of them, preferably the mean, would need to be reported.

TRUNCATED DISTRIBUTIONS A truncated distribution is one that has a sharp cutoff of observations at one end. An example would be a test in which an unusually large number of individuals make zero scores or a test in which an unusually large number make perfect scores. Had the test contained more easy items in the first case and more difficult items in the second, the distributions might have had tails rather than cliffs. In both cases, some individuals are simply not measured.

Another example would be a time-scored test, i.e., a test on which the score is the length of time it takes to complete a task or the time each person will keep working voluntarily. For practical purposes in group testing, all examinees may be arbitrarily stopped at some fixed time. In such cases, too, the distributions are truncated.

In truncated distributions, a mean is not a good representative value, since some extreme measurements are missing. A mode or a median can be determined, however, provided that a minority of the cases are in the cutoff tail.

Means of means, percentages, and proportions

The measures of central value described thus far will take care of the great majority of situations in which such statistics must be computed. There are some problems which, though rare, require other treatment. Two of these will be briefly mentioned: means of arithmetic means and means of percentages (and proportions).

FINDING MEANS OF ARITHMETIC MEANS

When one has the means of several samples, presumably from the same population, on the same test or scale, he may wish to know the overall mean for the samples combined. At first thought, it might seem appropriate simply to average the several means, just as one would average single observations. This would be proper procedure provided the samples are all of the same size. If the N's in the samples differ, however, the means are not equally reliable.

In order to extract the best information about the central value of the composite sample, we should weight each mean according to the number of cases in the sample from which it was derived, for a mean's reliability is in proportion to the size of sample. This procedure is equivalent to pooling all the single measurements from the different samples and computing a single overall mean. We can accomplish the same end by computing a weighted mean of the means, which we already know. The general formula for computing a weighted mean is

$$_w\overline{X} = \frac{\Sigma WX}{\Sigma W} \qquad \text{(A weighted arithmetic mean)} \qquad (4.4)$$

where $_w\overline{X}$ = weighted mean
W = weight
ΣWX = sum of the values being averaged, each multiplied by its appropriate weight
ΣW = sum of the weights

THE MEAN OF PERCENTAGES OR OF PROPORTIONS

The weighting procedure just described is even more important in determining the mean of a series of percentages or of proportions. Table 4.5 illustrates this point. The data in that table have to do with the percentage of pilot students eliminated in certain schools during one training period. Had the schools had the same enrollment, or even very nearly the same, the unweighted mean would have sufficed. Since the largest class is nearly four times as great as the smallest, however, and since elimination rates vary from 3.3 to 27.2, there is a marked difference between weighted and unweighted means. If we wished to know the overall elimination rate in order to make decisions for some administrative purpose, the unweighted mean would be misleading. Certainly, when the percentage or the proportion in a composite is wanted for further computations, the weighting procedure is essential, unless the sample N's are equal.

In terms of a formula, the weighted mean of a percentage is

$$_w\overline{X}_p = \frac{\Sigma N_i P_i}{\Sigma N_i} \qquad \text{(Mean of percentages where } N\text{'s di fer)} \qquad (4.5)$$

TABLE 4.5 Computation of an average percentage*

(1) School	(2) No. enrolled N_i	(3) No. eliminated $N_iP_i/100$	(4) % eliminated P_i
G	**243**	**55**	**22.6**
H	**63**	**7**	**11.1**
K	**196**	**43**	**21.9**
L	**61**	**2**	**3.3**
S	**125**	**34**	**27.2**
Sums	**688** $= \Sigma N_i$	**141** $= \Sigma N_iP_i/100$	**86.1** $= \Sigma P_i$
Means	**137.6** $= \overline{X}_N$		**17.2** $= \overline{X}_p$†

* The data represent students enrolled in five AAF pilot schools selected to illustrate this procedure.
† The weighted mean of the percentages equals 14,100/688 = 20.5. The value 17.2 is the unweighted mean.

where N_i = number in each sample
P_i = percentage for each sample
$\Sigma N_i P_i$ = sum of products of each percentage times its corresponding N
ΣN_i = sum of the sample N's

For a weighted mean of proportions, simply substitute p_i for P_i in the formula.

EXERCISES

1 Compute the arithmetic mean for any or all the distributions in Data 4*A* to 4*F* inclusive, using the methods that seem most feasible. In Data 4*E*, you will need to make some assumptions regarding the cases in the two highest intervals. If means are computed for these distributions, state your assumptions.
2 Where defensible, compute medians for any or all distributions in Data 4*A* to 4*F* inclusive.
3 Give the estimated modes for all distributions in Data 4*A* to 4*F*.
4 Compute and list the means and medians for the distributions in Data 4*G*.
5 For each distribution in Data 4*G*, tell to which measure of central value you give first preference. Judge in terms of how well it describes or represents the measurements in the sample. Give reasons.
6 For the distributions in Data 4*A*, 4*B*, and 4*E*, state which measure or measures of central value you would prefer, in terms of descriptive value or because of any particular purposes for which you think the data might be utilized.

DATA 4*A* Scores in an English-usage examination

Scores	*f*
52–53	1
50–51	0
48–49	5
46–47	10
44–45	9
42–43	14
40–41	7
38–39	8
36–37	6
34–35	5
32–33	3
Sum	68

DATA 4*B* Affectivity scores (percent of 400 words marked "pleasant")

Scores	*f*
95–99	6
90–94	11
85–89	16
80–84	7
75–79	9
70–74	8
65–69	2
60–64	3
55–59	2
50–54	1
Sum	65

DATA 4C Scores made by graduates and eliminees in the Complex Coordination Test by student pilots

Scores	*Frequencies*	
	Graduates	*Eliminees*
95–99	1	
90–94	1	
85–89	7	1
80–84	13	2
75–79	37	6
70–74	75	23
65–69	189	34
60–64	297	94
55–59	406	144
50–54	425	208
45–49	341	209
40–44	174	205
35–39	81	105
30–34	16	34
25–29	5	15
20–24	0	2
15–19	1	

DATA 4D Scores in an adjustment inventory obtained from alcoholics and nonalcoholics of both sexes*

Scores	*Frequencies*			
	Males		*Females*	
	Alcoholics	*Nonalcoholics*	*Alcoholics*	*Nonalcoholics*
66–71	1			
60–65	6		3	
54–59	13	1	2	1
48–53	13	1	10	2
42–47	17	3	11	1
36–41	33	3	12	1
30–35	32	2	8	8
24–29	32	9	11	17
18–23	23	16	5	26
12–17	24	36	2	40
6–11	7	43	2	49
0–5	1	25		21

* Manson, M. P. A psychometric differentiation between alcoholics and nonalcoholics. *Quart. J. Stud. Alcohol.*, 1948, **9**, 175–206.

DATA 4E Ages of college freshmen

Age at last birthday	*Men*	*Women*
31–35	1	2
26–30	3	6
25	7	6
24	6	7
23	11	7
22	20	6
21	23	16
20	40	13
19	88	48
18	117	67
17	69	57
16	2	6
Sums	387	241

DATA 4F Aiming-test scores (In terms of average error in millimeters)

Score	*Men*	*Women*
8.0–8.4	1	
7.5–7.9	5	
7.0–7.4	2	
6.5–6.9	7	2
6.0–6.4	6	4
5.5–5.9	11	3
5.0–5.4	10	9
4.5–4.9	16	7
4.0–4.4	18	15
3.5–3.9	19	12
3.0–3.4	17	15
2.5–2.9	17	13
2.0–2.4	14	14
1.5–1.9	13	10
1.0–1.4	8	1
0.5–0.9	1	
Sums	165	105

7 Find the weighted means of the four means 15, 16, 18, and 21. These means were derived from samples in which the *N*'s were 6, 10, 25, and 20, respectively. Compute the unweighted arithmetic mean of the four for comparison. Interpret your result.

8 Find the weighted mean of the proportions .25, .30, .32, and .33. These proportions were based upon samples whose *N*'s were 44, 32, 18, and 25, respectively. Compute an unweighted arithmetic mean of these proportions for comparison. Interpret your results.

9 For the aiming-test scores for men, in Data 4*F*, find the centile-point values for the following centile ranks: 90, 75, 25, and 10.

DATA 4G Some ungrouped data

a. 8, 15, 13, 6, 10, 16, 7, 12, 11, 14, 9
b. 12, 10, 18, 13, 4, 8, 17, 15, 6, 14
c. 9, 8, 9, 15, 3, 9, 11, 9, 13
d. 12, 28, 19, 15, 15, 35, 14, 15
e. 7, 18, 20, 14, 27, 23, 13, 3

ANSWERS *1, 2,* and *3*:

Data	*4A*	*4B*	*4C*		*4D*				*4E*		*4F*	
Mean	41.7	81.8	54.8	49.3	32.8	13.9	37.2	15.3	19.6	19.7	3.91	3.57
Median	42.2	84.7	54.4	48.8	32.1	11.8	38.0	13.4	19.1	18.9	3.78	3.43
Mode	42.5	87	52	47	38.5	8.5	38.5	8.5	18.5	18.5	3.7	3.7

4:

	a	*b*	*c*	*d*	*e*
Mean	11.0	11.7	9.6	19.1	15.6
Median	11.0	12.5	9.1	15.2	
Mode			9	15	

7 18.4; 17.5.

8 .291; .300.

9 Centile points: 6.33; 4.99; 2.60; 1.74.

5 Measures of variability

Knowing the central value of a set of measurements tells us much, but it does not by any means give us the total picture of the sample we have measured. Two groups of six-year-old children may have the same average IQ of 105, from which we would conclude that, taken as a whole, one group is as intelligent as the other, in the sense in which IQ indicates intelligence. We might therefore expect from the two groups the same average level of performance in school, and also out of school in areas of life where IQ is important.

Yet when we are told, in addition, that one group contains no individuals with IQs below 95 or above 115, and that the other includes individuals with IQs ranging from 75 to 135, we recognize immediately that there is a decided difference between the two groups in variability or dispersion. The first group is decidedly more homogeneous than the second with respect to IQ. We should expect the first group to be more teachable in that they will all be likely to grasp new ideas at nearly the same rate. We should expect the second group to show considerable disparity in speed of grasping new ideas. The distributions of two such groups would resemble those in Fig. 5.1.

It is the purpose of this chapter to explain and to illustrate the methods of indicating degree of variability or dispersion by the use of single numbers, just as in the preceding chapter we saw how the central value of a distribution could be indicated by a single number. The three customary values to indicate variability are (1) the total range, (2) the semi-interquartile range Q, and (3) the standard deviation S.

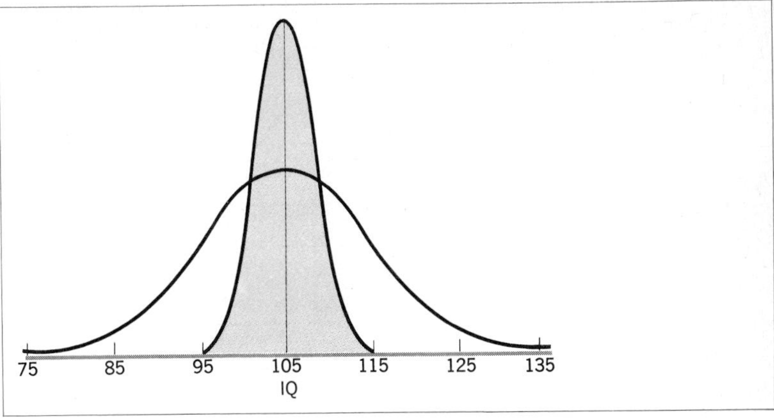

FIGURE 5.1 Two distributions with the same mean (IQ = 105) but with decidedly different dispersions or ranges

The total range

The total range is the easiest and most quickly ascertained value, but it is also the most unreliable. It is thus almost entirely limited to the purpose of preliminary inspection. In the illustration of the preceding paragraphs, the range of the first group (from an IQ of 95 to one of 115) is 20 points. The range of the second group is from 75 to 135, or 60 points. The range is given by the highest score minus the lowest score. The second group has three times the range of the first.

The semi-interquartile range Q

The semi-interquartile range Q is *one-half* the range of the middle 50 percent of the cases. First we find by interpolation the range of the middle 50 percent, or interquartile range, and then divide this range by 2. See Fig. 5.2 for a general picture of the relation of Q to a frequency distribution.

QUARTILES AND QUARTERS When we count up from below to include the lowest, or first, quarter of the cases, we find the point called the *first quartile,* which is given the symbol Q_1. Counting down from above to include the highest, or fourth, quarter of the cases, we locate the third quartile, or Q_3. Incidentally, the median, which separates the second and third quarters of the distribution, is Q_2. Note that the quartiles Q_1, Q_2, and Q_3 are *points* on the measuring scale. They are division points between the *quarters.* We may say of an individual that he is *in* the highest *quarter* (or fourth quarter), and we may say of another that he is *at* the third *quartile.* We should never say of an individual that he is *in* a certain *quartile.*

INTERPOLATION OF Q_1 AND Q_3 In the distribution of inkblot scores again, we locate the third and first quartiles by interpolation (see Table 5.1). One-fourth of the cases ($N/4$) is 12.5. Counting up from the

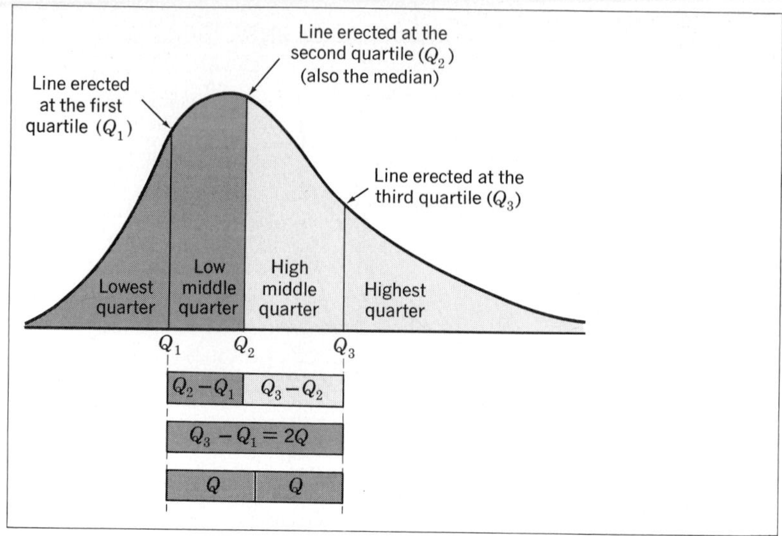

FIGURE 5.2 Illustration of the quartiles Q_1, Q_2, and Q_3, the interquartile ranges, and the quarters in the sample in a slightly skewed distribution

bottom to include 12.5 cases, we find that we need 2.5 out of the 6 cases in the third class interval. As in earlier solutions in connection with computing a median, 2.5/6 times 5 gives 2.08. Added to 19.5, this gives 21.58 as the position of Q_1. Counting down from the top, we find that we need 3.5 cases out of 5 in the fifth class interval. Then 3.5/6 of 5 gives 2.92. Deducted from 39.5, this leaves 36.59 as our estimate of Q_3.

THE INTERQUARTILE RANGE AND Q

The interquartile range, or the distance from Q_1 to Q_3, is given by $Q_3 - Q_1$, or 36.58 − 21.58, which equals 15.00.

The semi-interquartile range is one-half of this, or 7.5. In terms of a formula,

TABLE 5.1 Determination of Q_3, Q_1, and Q (the semi-interquartile range) for the inkblot-test scores

Scores	f		
55–59	1		
50–54	1		
45–49	3		
40–44	4		
35–39	6	← Q_3 lies within	
30–34	7	this interval	
25–29	12		$Q_1 = 19.5 + \frac{2.5}{6} \times 5 = 19.5 + 2.08 = 21.58$
20–24	6	← Q_1 lies within	
15–19	8	this interval	$Q_3 = 39.5 - \frac{3.5}{6} \times 5 = 39.5 - 2.92 = 36.58$
10–14	2		
	$N = 50$		$Q = \frac{36.58 - 21.58}{2} = \frac{15.00}{2} = 7.5$

$$Q = \frac{Q_3 - Q_1}{2} \qquad \text{(Semi-interquartile range)} \tag{5.1}$$

where Q_3 = third quartile and Q_1 = first quartile.

The standard deviation

The standard deviation is by far the most commonly used indicator of degree of dispersion and is the most dependable estimate of the variability in the population from which the sample came. It also enters into numerous other statistical formulas that we shall encounter later.

GENERAL FORMULA FOR THE STANDARD DEVIATION

The standard deviation is a kind of average of all the deviation from the mean, but it is not a simple arithmetic mean. The fundamental formula for computing this index of variability in a sample is

$$S = \sqrt{\frac{\Sigma x^2}{N}} \qquad \text{(A standard deviation of a sample)} \tag{5.2}$$

where x = a deviation from the mean $(X - \bar{X})$ and N = the size of sample.

As a general concept, the standard deviation is often symbolized by SD, but much more often by simply S, as in the formula. Those who want a better estimate of the *population* standard deviation utilize the statistic s, which is computed by a formula very similar to (5.2). The difference is the use of $(N - 1)$ instead of N in the denominator.[1] Only when samples are small (N less then 30) will s differ appreciably from S.

Formula (5.2) calls for several steps in computation, in a particular order:

step 1 Find each deviation from the mean, x, which equals $X - \bar{X}$.
step 2 Square each deviation, finding x^2.
step 3 Sum the squared deviations, finding Σx^2.
step 4 Divide this sum by N, finding $\Sigma x^2/N$.
step 5 Extract the positive square root of the result of step 4, using Table *A* in the Appendix. The result is the standard deviation.

These steps are illustrated by Tables 5.2 and 5.3 and in Fig. 5.3.

1 The expression $(N - 1)$ is called the "number of degrees of freedom," and is explained in Chap. 8 (p. 127).

VARIABILITY, SUM OF SQUARES, AND MEAN SQUARE, OR VARIANCE

Before proceeding to apply the formula, let us consider some important interrelated concepts. In verbal terms, a standard deviation is the square root of the arithmetic mean of the squared deviations of measurements from their means. It has accordingly often been called the *root-mean-square deviation*. In this statement lies considerable statistical meaning. Latent in the few steps enumerated above are three important statistical concepts of which we shall see a great deal, particularly in Chap. 13, on analysis of variance. At the end of step 3, we have the *sum of squares*, symbolized by SS. At the end of step 4, we have the *mean square*, symbolized by S^2. The mean square is also known as the *variance* of a distribution. Let us elaborate upon these concepts by using an illustration.

In Table 5.2 are listed seven fictitious measurements representing a sample of seven individuals. The mean is 10.0, as shown in column 2. Column 3 shows the deviations, with their algebraic sum and their mean equal to zero. In column 4 we find the squared deviations. Their sum, 88, is the sum of squares, SS. Their mean, 12.57, is the mean square, S^2, also the variance in this sample. The square root of the mean square is 3.55, the standard deviation. Let us see what this means in terms of a geometric view of the problem, via another mode of understanding.

A GEOMETRIC PICTURE OF DEVIATIONS, VARIANCE, AND STANDARD DEVIATION

For a geometric representation of these ideas, see Fig. 5.3. In the first diagram the scale of measurement is shown on the base line. Instead of the original measurements X, however, the scale is given in terms of deviations x. The mean is the reference point and is labeled "zero." All the seven individuals retain their relative positions as would be in-

TABLE 5.2
Data illustrating sum of squares, variance, and standard deviation

(1) Person	(2) Score X	(3) Deviation x	(4) Deviation squared x^2
A	15	+5	25
B	14	+4	16
C	11	+1	1
D	10	0	0
E	9	−1	1
F	7	−3	9
G	4	−6	36
Sums	$70 = \Sigma X$	$0 = \Sigma x$	$88 = \Sigma x^2$
Means	10.0	0.0	$12.57 = S^2$
Standard deviation			$3.55 = S$

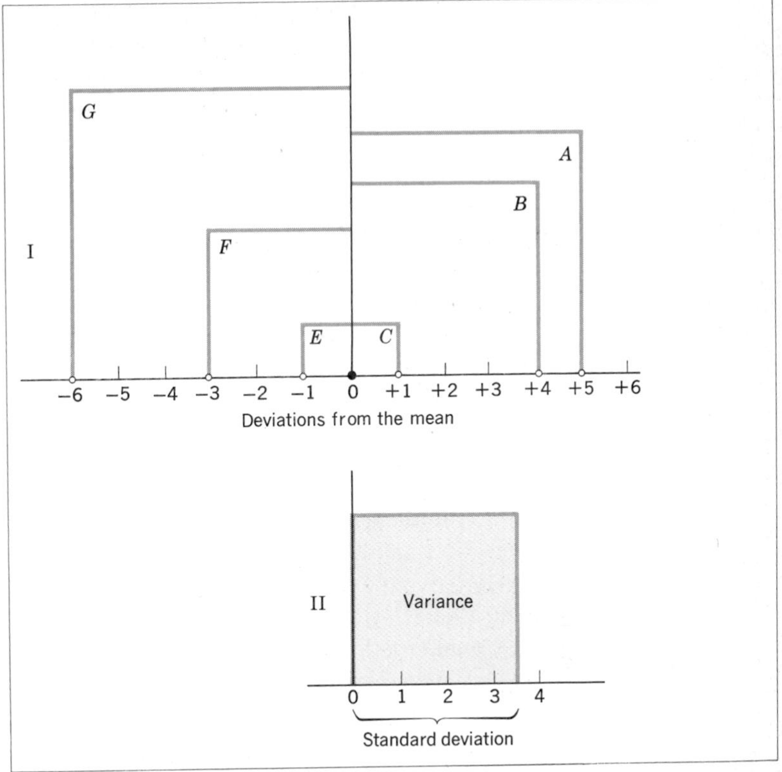

FIGURE 5.3 Illustration of deviations from the arithmetic mean, their squares, the mean of the squared deviations (which is the variance), and the standard deviation (which measures the variability) in a sample of seven cases

dicated by *X* values. We have just moved the zero point up to the mean.

So much for representing the deviations seen in Table 5.2. Consider now the squaring of the deviations. Whereas deviations themselves are represented as linear distances from a common reference point, squared deviations must be represented by areas, namely, squares. The squares belonging to persons *A* to *G* are shown in Fig. 5.3. The areas of the squares correspond to the numerical values given in column 4 of Table 5.2.

The sum of the squares, SS, would be represented geometrically as an area equal to a combination of all the areas in the squares in Fig. 5.3. This surface would contain 88 units, each unit equal in size to those representing persons *C* and *E*. Dividing this total area by 7 gives an area for S^2, represented by the single square labeled variance" in Fig. 5.3. This square is shown on a base line like that in the first diagram. Its length of side represents the standard deviation.

FURTHER INTERPRETATIONS OF VARIABILITY AND VARIANCE One of the most common sources of variance in statistical data is individual differences, where each measurement comes from a different person. Suppose that we have a sample of only one, with only one score. There is no possible basis for individual differences in such a sample, and therefore there is no variance. Bring into the picture a second individual with some other score in the same test or experiment. We now have one difference. Bring in a third case, and we have two additional differences, three altogether. Bring in a fourth, a fifth, and so on. There are as many differences as there are pairs of individuals. We could compute all these interpair differences and average them to obtain a single, representative value. We could square the differences before averaging, if that were desired. It is far more economical, however, to find a mean of all the scores and to use that value as a common reference point. Each difference then becomes a deviation from that reference point, and there are only as many deviations as there are individuals. Either the variance or the standard deviation is a single representative value for all the individual differences when taken from a common reference point.

Let us think of variance from a somewhat different point of view. Consider giving a certain test of n items to a group of persons. Before the first item is given to the group, as far as any information from this test is concerned, the individuals are all alike. There is no variance. This may seem absurd, but it has a real bearing upon what comes next. Now administer the first item to the group. Some pass it and some fail. Some now have scores of 1, and some have scores of zero. There are two groups of individuals. There is this much variation, this much

TABLE 5.3
Calculation of the standard deviation in ungrouped data

(1) Score X	(2) Deviation x	(3) x^2
13	−0.2	.04
17	+3.8	14.44
15	+1.8	3.24
11	−2.2	4.84
13	−0.2	.04
17	+3.8	14.44
13	−0.2	.04
11	−2.2	4.84
11	−2.2	4.84
11	−2.2	4.84
		51.60
		Σx^2

$$S = \sqrt{\frac{51.60}{10}} = \sqrt{5.160} = 2.27, \text{ or } 2.3$$

variance. Give a second item. Of those who passed the first, some will pass the second and some will fail it, unless the two items are perfectly correlated. Of those who failed the first, some will also pass the second and some will fail. There are now three possible scores: 0, 1, and 2. More variance has been introduced. Carry the illustration further, adding item by item. The differences between scores will keep increasing, and also, by computation, the variance and the variability, as indicated by S.

Psychological and educational testing depends almost entirely upon the phenomenon of individual differences and therefore upon variance. A very small percentage of tests commonly used yield scores on absolute scales. The significance of any score is ordinarily its usefulness in the placement of a person somewhere in the range of individual scores. The greater the variance among the scores, other things being equal, the more accurately each person is placed.

COMPUTATION AND INTERPRETATION OF A STANDARD DEVIATION

As an illustrative problem in computing S by formula (5.2), let us use the 10 measurements of the threshold for pitch (see Table 5.3). Their mean is 13.2. The deviations from the mean are given in column 2 and their squares in column 3. The sum of squares is 51.60. The mean of these squares is 5.160, and the standard deviation is its square root, which is 2.27. In terms of the unit of the measuring scale, this is 2.3 cycles per second.[1]

THE INTERPRETATION OF A STANDARD DEVIATION The usual and most accepted interpretation of a standard deviation is in terms of the percentage of cases included within the range from one standard deviation below the mean to one standard deviation above the mean. In a normal distribution, the range from -1σ to $+1\sigma$ contains 68.27 percent of the cases. Since most samples yield distributions that depart to some extent from normality, we say "about two-thirds," which is, of course, a little less than 68.27 percent. Figure 5.4 illustrates the division of the area under a normal curve into regions marked off at -1σ and $+1\sigma$. With two-thirds of the surface within those limits, there is left one-third of the area divided equally between the two "tails" of the distribution—one-sixth below the point at -1σ and one-sixth above the point $+1\sigma$.

In the problem just solved, where we found S equal to 2.3, the distance from $-1S$ to $+1S$ on the scale of measurement is 10.9 to 15.5 cycles; i.e., the mean 13.2 minus 2.3 is 10.9, and the mean plus 2.3 is 15.5 cycles. Within these limits are measurements of 11, 12, 13, 14, and 15. By actual count, there are four 11s, three 13s, and one 15, or 8

[1] The better estimate of the population standard deviation, or s, is 2.39 for this distribution. The difference between S and s is appreciable here because the sample is so small. The symbol for the population standard deviation is the Greek letter sigma, or σ.

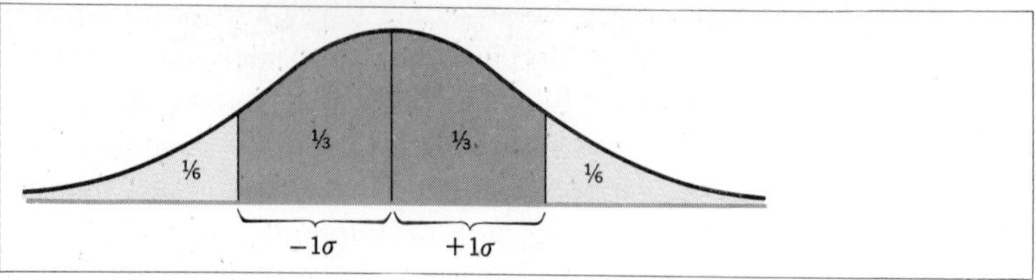

FIGURE 5.4 Approximate fractions of the area under a normal-distribution curve (thus fractions of the cases in a normally distributed sample) that lie within one standard deviation of the mean and also beyond the limits of one standard deviation, in either direction

of the 10 measurements within these limits, whereas we should have expected 7. But, because of the small number of cases and the fact that the distribution is irregular, we should not be surprised at this result. In other problems this comparison serves as a rough check upon the accuracy of computation of S. It will not catch all errors, but will indicate gross errors if the sample is not too small and the distribution is fairly normal.

GROUPING DEVIATIONS AS A SHORTCUT When the measurements are in a frequency distribution with data grouped in class intervals, the following procedure is recommended. The data in Table 4.1 are used as an illustration, with the steps illustrated in Table 5.4. As in calculating a mean from such data, the midpoints of class intervals, X_c, are used to represent the cases within the intervals. These values are listed in column 2.

TABLE 5.4 Computing a standard deviation in grouped data

Scores	X_c	x $X_c - \overline{X}$	x^2	f	fx^2
55–59	57	+27.4	750.76	1	750.76
50–54	52	+22.4	501.76	1	501.76
45–49	47	+17.4	302.76	3	908.28
40–44	42	+12.4	153.76	4	615.04
35–39	37	+7.4	54.76	6	328.56
30–34	32	+2.4	5.76	7	40.32
25–29	27	−2.6	6.76	12	81.12
20–24	22	−7.6	57.76	6	346.56
15–19	17	−12.6	158.76	8	1270.08
10–14	12	−17.6	309.76	2	619.52
					5462.00 Σfx^2
					109.24 S^2
					10.45 S

Further steps are indicated, column by column. The mean, from Table 4.1, is 29.6. $X_c - X$ gives the deviation x for each interval. The deviations are squared, as in column 4. Frequencies for the intervals are listed in column 5, and the fx^2 products in the last column. The last three steps are given at the bottom of column 6—summing the fx^2 products, dividing this sum by N to find S^2, and taking the square root to find S. These last operations are summarized in formula (5.3), which reads

$$S = \sqrt{\frac{\Sigma fx^2}{N}} \quad \text{(A standard deviation calculated from grouped data)} \quad (5.3)$$

We may now say that about two-thirds of the individuals should be expected between the mean minus 10.45 and the mean plus 10.45. Since the mean is 29.6, these limits are 19.2 and 40.0. Fortunately, for the sake of checking on this conclusion, these limits are close to the division points between class intervals (see Table 5.4). The four intervals included within these limits have in them 31 cases, or 62 percent of the whole sample. This figure is a little short of two-thirds but not unreasonably so.

ROUGH CHECKS FOR A COMPUTED STANDARD DEVIATION The kind of comparison just mentioned is a very rough check on the correctness of the solution for the standard deviation. If the actual percentage of cases between $+1S$ and $-1S$ deviates too far from 68 percent, there is probably something wrong with the calculation and a recalculation is in order. This check cannot always be satisfactorily applied with grouped data because the $-1S$ to $+1S$ do not come close to class limits and the needed frequency cannot be acurately determined.

Another rough check is to compare the standard deviation obtained with the total range of measurements. In very large samples ($N = 500$ or more) the standard deviation is about one-sixth of the total range. In other words, the total range is about six standard deviations. In smaller samples the ratio of range to standard deviation can be expected to be smaller, as indicated in Table 5.5.

TABLE 5.5
Ratios of the total range to the standard deviation in a distribution for different values of N*

N	*Range/S*	N	*Range/S*	N	*Range/S*
5	2.3	40	4.3	400	5.9
10	3.1	50	4.5	500	6.1
15	3.5	100	5.0	700	6.3
20	3.7	200	5.5	1,000	6.5

* Adapted from Snedecor, G. W. *Statistical Methods*. Ames, Iowa: Collegiate, 1967. P. 40.

In the inkblot data, since $N = 50$ we should expect the range to be 4.5 times the standard deviation. The standard deviation 10.45 times 4.5 gives an expected range of about 47 points. Actually, the range was 46 points, which checks so closely as to lend confidence that the obtained standard deviation is at least not grossly in error.

It may seem strange that we use a less reliable statistic like range as a criterion of accuracy of a more reliable statistic like the standard deviation. The reasons are that (1) there can hardly be any error in computing a simple thing like a range, whereas (2) there are chances for gross errors in calculating S because of the several steps involved.

THE STANDARD DEVIATION FROM ORIGINAL MEASUREMENTS

If the number of measurements is not large, if the values are small numbers, and when a calculating machine is available, the best procedure for computing a standard deviation is by means of the formula

$$S = \frac{1}{N}\sqrt{N\Sigma X^2 - (\Sigma X)^2} \qquad \text{(Standard deviation computed without knowledge of deviations)} \qquad (5.4)$$

in which the essential steps are:

step 1 Square each measurement.
step 2 Sum the squared measurements to give ΣX^2.
step 3 Multiply ΣX^2 by N to give $N\Sigma X^2$.
step 4 Sum the X's to find ΣX.
step 5 Square the ΣX to obtain $(\Sigma X)^2$.
step 6 Find the difference $N\Sigma X^2 - (\Sigma X)^2$.
step 7 Find the positive square root of the number found in step 6.
step 8 Divide the number found in step 7 by N or multiply it by $1/N$.

A solution by formula (5.4) is illustrated in Table 5.6.

TABLE 5.6 Calculation of the standard deviation from original measurements and ungrouped data

X	X^2
13	169
17	289
15	225
11	121
13	169
17	289
11	121
13	169
11	121
11	121
$\Sigma X = 132$	$1{,}794 = \Sigma X^2$

$$S = \frac{1}{10}\sqrt{10(1{,}794) - 132^2}$$

$$= \frac{1}{10}\sqrt{17{,}940 - 17{,}424}$$

$$= \frac{1}{10}\sqrt{516}$$

$$= \frac{22.7}{10}$$

$$= 2.27, \text{ or } 2.3$$

When an estimate of the population standard deviation is wanted, by a formula that involves the number of degrees of freedom, the following modified equation should be used:[1]

$$s = \sqrt{\frac{\Sigma X^2 - \frac{(\Sigma X)^2}{N}}{N - 1}} \qquad \text{(SD of a population distribution estimated from observed measurements)} \qquad (5.5)$$

Descriptive use of statistics

Thus far, the chief uses indicated for measures of central value and dispersion have been as simple values descriptive of distributions. This use is best appreciated when we compare different samples. As an illustration of this activity see Table 5.7, which gives a few samples of Army General Classification Test data, each derived from a different civilian occupational group. We shall not concern ourselves with the question of how adequate these samples are with respect to either size or representativeness of the populations from which they are purported to have come. These considerations are, of course, important if we want to generalize our conclusions to those populations. Nevertheless, we can still compare samples as such.

Some general conclusions can be drawn from an inspection of Table 5.7. When the means and medians are placed in rank order, the occupational groups are seen to fall into an approximate rank order for socioeconomic level. It is also apparent, as should have been expected, that occupations requiring more "headwork" are highest in the list. The test emphasized verbal, reasoning, and numerical facilities.

The importance of having available both means and medians lies in the information they give concerning skewness. For the lower occupational groups, particularly, the medians are slightly higher than the means. This indicates slight negative skewing. This is a somewhat surprising result, for one should expect that the higher the mean, the greater the negative skewing, and the lower the mean, the greater the positive skewing. When a test of moderate difficulty is administered to a group of low average ability, scores tend to bunch at the lower end of the scale (positive skewing). When the same test is given to a group of high average ability, the bunching is expected near the upper end of the scale (negative skewing). Since in the data of Table 5.7 the skewing seems to be negative for most occupational groups and most marked

[1] A formula suggested by R. A. Charter in a personal communication.

TABLE 5.7
Statistics describing distributions of scores for selected occupational groups who took the Army General Classification Test during World War II*

Occupation	N	$\overline{X}$	*Mdn*	S	*Range*
Accountant	172	128.1	128.1	11.7	94–157
Lawyer	94	127.6	126.8	10.9	96–157
Reporter	45	124.5	125.7	11.7	100–157
Sales clerk	492	109.2	110.4	16.3	42–149
Plumber	128	102.7	104.8	16.0	56–139
Truck driver	817	96.2	97.8	19.7	16–149
Farm hand	817	91.4	94.0	20.7	24–141
Teamster	77	87.7	89.0	19.6	45–145

* From Harrell, T. W., and Harrell, M. S. Army General Classification Test scores for civilian occupations. *Educational and psychological Measurement,* 1945, **5** 229–240. By permission of the publisher.

for those of low average ability, some explanation is demanded. We can only speculate, which means that we can suggest several hypotheses which would need further investigation before their worth could be evaluated. One hypothesis might be that in any occupational group, particularly among those of lower ability in the test, a minority of the more able examinees were very poorly motivated or took the test under adverse conditions, so that they did not do full justice to themselves.

Two indices of dispersion are given: the standard deviation and the total range. Each tells its own story. Standard deviations are more meaningful here if it is remembered that for the *total* range of scores, all occupational groups combined, the standard deviation was approximately 20.0. The scaling that was utilized aimed at a standard deviation of 20.0 and a mean of 100. The mean in some forms of the test turned out to be somewhat above 100. We should expect dispersions within selected occupational groups to be smaller than dispersions for all occupations combined. With three exceptions in Table 5.7, this is true. On the whole, the higher the occupational group and the higher the mean, the smaller the dispersion. The higher groups should not be expected to scatter so far from the mean because the mean score approaches the highest scores made by individuals in *any* group. We might expect a similar curtailment for groups with lowest means. But a study of the ranges will show that this did not occur.

The ranges, as such, are surprisingly large for all groups. It is hard to imagine any individuals in the professional groups with scores below the general average, unless those scores were low because of poor motivation or because of advancing age, which is associated with a slower rate of work. The test was a time-limit test. The lowest scores for the lower occupational groups are in line with expectations, but the maximum scores in those same groups are illuminating. Many a clerk or truck driver could evidently have successfully undertaken training for one of the professional occupations. For some reason, in their prewar assignments they did not take full vocational advantage of their abilities. This fact, and also the fact that men of very low academic abilities can engage successfully in occupations like those of farm hand and teamster, is largely responsible for the unusually wide dispersions of scores in such occupational groups.

In this discussion we are not particularly interested in settling points concerning the relation of mental abilities to occupational level or success. The data are presented here merely as an illustration of the kind of inferences one may draw from a set of statistics and the hypotheses that may be set up for further investigation, possibly of a very fruitful nature. Such inferences and hypotheses would be impossible to make without this kind of inspection, and the inspection is made possible by having the statistical information.

Uses and interrelationships of different measures of dispersion

CHOICE OF THE STATISTIC TO USE Several considerations come into the picture when we decide what measure of variability to employ in a given situation. One is the sampling stability of the statistic—its relative constancy in repeated samples. In this respect, when sampling is random, the statistics come in the order from most reliable to least reliable: standard deviation, semi-interquartile range, and total range. As to ease and quickness of computation, of course, the three are in reverse order to that just given. If further statistical computations are in order, such as estimating significance of differences between means, correlation coefficients, or regression equations, then by all odds the standard deviation is the one to use.

If there should be only a few very extreme measurements, as in badly skewed distributions, because Q is less affected by such deviations it has an advantage over the standard deviation. When the median is the index of central value, Q is a natural companion index of variability, since both are interpolated. When distributions are truncated or have some indeterminate end values, only Q can justifiably be used to indicate variability.

In this discussion the variance statistic has only been mentioned because it is not a value on a linear scale of measurement for the data and is therefore not very descriptive. In later chapters on sampling statistics it will be found to play numerous roles.

EXERCISES

1. Compute the interquartile and semi-interquartile ranges for the distributions in Data 4*A*, 4*B*, and 4*F* (see pp. 58 to 60). Interpret your findings.
2. Compute the standard deviations for any or all of the distributions in Data 4*G* (see p. 60).
3. Compute the standard deviation for any or all of the distributions in Data 4*A* to 4*F*, inclusive.
4. Decide which index of variability it is wisest to employ with each of the distributions in Data 4*A*, 4*B*, and 4*F*. Give reasons.

ANSWERS

1. Q: 3.5, 7.7, 1.19.
2. S: 3.2, 4.4, 3.2, 7.6, 7.5.
3. 4.58, 10.86, 9.78, 8.75, 13.92, 10.42, 12.64, 9.97, 2.12, 2.77, 1.69. 1.30.

6 Correlation

No single statistical procedure has opened up so many new avenues of discovery in psychology, and possibly in the behavioral sciences in general, as that of correlation. This is understandable when we remember that scientific progress depends upon finding out what things are co-related and what things are not. A *coefficient of correlation* is a single number that tells us to what extent two things are related, to what extent variations in the one go with variations in the other. Without the knowledge of how one thing varies with another, it would be impossible to make predictions. And wherever causal relationships are involved, without knowledge of covariation we should be unable to control one thing by manipulating another.

For example, when we know that the higher a person's score in a clerical-aptitude test, the higher the average performance that person is likely to exhibit after training, we can thereafter use scores on this test to predict level of proficiency. If the predictions are very accurate, we say that there is a high positive correlation between aptitude-test score and clerical success. We discover this fact by finding a coefficient of correlation between scores made by a number of people and measures of clerical performance obtained later for the same people. We cannot compute a coefficient of correlation from just two such measurements on one person alone, nor can we compute it without having made two sets of measurements on the same individuals or on matched pairs of individuals.

In the same clerical-aptitude example, if we consider that the aptitude test has measured individual differences in some quality or qualities that lead to success, i.e., in the sense of a "cause" of clerical success, then we can not only predict future success for individuals but also promote high general efficiency in any group of clerks by selecting those with high scores. Thus studies leading to prediction and control of human affairs are conducted because correlation techniques are available. Without some device like this for checking up on a test, we can have only very rough notions concerning its effectiveness, unless, indeed, its effectiveness is so obvious from direct observation as to require no inspection by correlation methods, a state of affairs that is highly unlikely.

The meaning of correlation

SOME EXAMPLES OF CORRELATION BETWEEN TWO VARIABLES

The coefficient of correlation is one of those summarizing numbers, like a mean or a standard deviation, which, though they are single numbers, tell a story. In different situations it can vary from a value of +1.00, which means perfect positive correlation; through zero, which means complete independence, or no correlation whatever; on down to −1.00, which means perfect negative correlation.

CASES OF PERFECT POSITIVE CORRELATION Figure 6.1 illustrates an instance of perfect positive correlation. It is a fictitious case, for such exact agreement between two things is rarely, if ever, experienced, certainly not in psychology and other behavioral sciences. Here we have assumed two tests, X and Y. Ten individuals have received scores in the two tests. The pairs of scores are as follows:

Individual	*A*	*B*	*C*	*D*	*E*	*F*	*G*	*H*	*I*	*J*
Score in test X	2	4	5	6	7	8	9	10	12	13
Score in test Y	4	6	7	8	9	10	11	12	14	15

Looking down the rows of scores, each pair made by one individual, we readily conclude that each person's score in Y is two points higher than his score in X. In terms of a simple equation, $Y = X + 2$. There are *no exceptions*, which makes the correlation perfect.

To take another instance:

Individual	*A*	*B*	*C*	*D*	*E*	*F*	*G*	*H*	*I*	*J*
Score in test P	1	3	4	5	7	8	9	11	12	15
Score in test Q	2	6	8	10	14	16	18	22	24	30

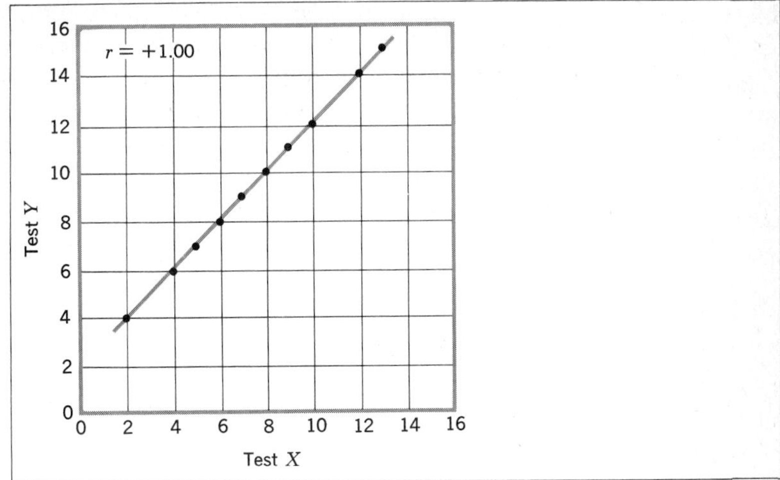

FIGURE 6.1
Simple correlation chart showing the kind of relationship between X and Y scores when the correlation is +1.00

In this situation, each person's score in *Q* is two times that in *P*, again without exception; there is perfect agreement, and the coefficient of correlation would be +1.00. The equation for predicting *Q* from *P* is $Q = 2P$.

A CASE OF HIGH POSITIVE CORRELATION In Fig. 6.2, we have illustrated a case of correlation that is positive but less than +1.00. The graphic picture of the individuals shows that, *in general,* a person who is high in test *X* is also likely to be high in test *Y* and that one who is low in *X* is also likely to be low in *Y*. The actual scores for these 10 people are listed in the first two columns of Table 6.1. Although the individuals are arranged in rank order for scores in *X*, when we inspect their scores

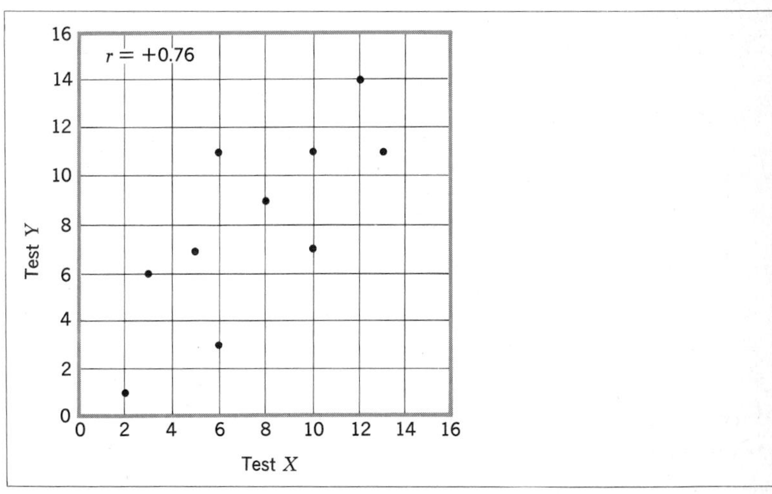

FIGURE 6.2
Correlation chart illustrating the kind of situation where the correlation is +.76

in Y, we see that there are some deviations from this rank order. The coefficient of correlation by computation is equal to +.76. We shall soon see how this was obtained, but we first simply note by comparison of Figs. 6.1 and 6.2 how the individuals are scattered in the two diagrams. In Fig. 6.1 they line up in perfect file from lowest to highest. In Fig. 6.2 they tend to fan out or to diverge from a strict lineup, but a definite *trend* of relationship can be observed. The amount of spreading in Fig. 6.2 as compared with that in Fig. 6.1 (in which there is, of course, none) illustrates the difference between correlations of +1.00 and +.76.

A CASE OF LOW POSITIVE CORRELATION A third instance is shown in Fig. 6.3, in which the spreading effect is even greater. The coefficient of correlation here is +.14, in other words, close to zero. This being true, a person with a high score in X is likely to be almost anywhere within the total range in terms of his Y score. The three highest people in X, with scores of 10, 12, and 13, scatter all the way from 3 to 11 in test Y. The three lowest people in test X, with scores of 1, 3, and 4, scatter all the way from 2 to 9 in test Y. Although there is some relationship between X scores and Y scores, it is very weak. The actual scores may be compared in Table 6.3 (p. 85).

A CASE OF HIGH NEGATIVE CORRELATION The situation that obtains when there is a negative correlation is shown in Fig. 6.4. Here the coefficient is −.69. Compare this diagram with that in Fig. 6.2, and it will be apparent that the trend of the points is along the other diagonal now, from upper left to lower right. This illustrates the fact that persons making high scores in X are likely to make low scores in Y and that persons making low scores in X are likely to make high scores in Y.

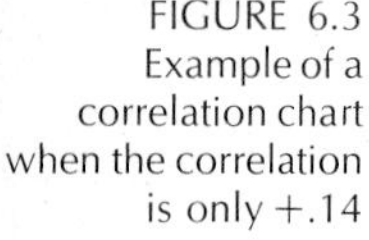
FIGURE 6.3 Example of a correlation chart when the correlation is only +.14

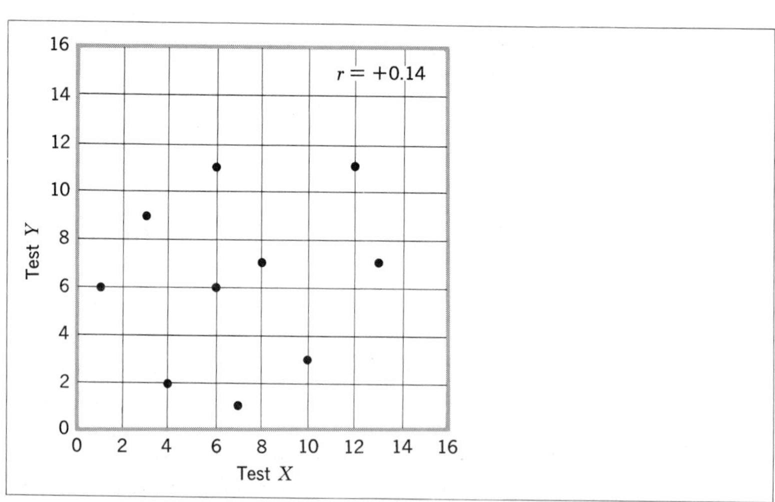

FIGURE 6.4 Example of a correlation chart when the correlation is −.69

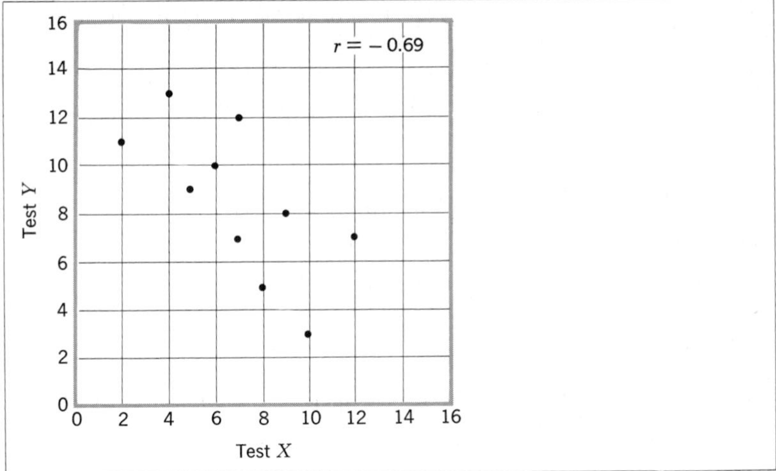

This inverse *order* of relationship is also apparent in the actual scores in the first two columns of Table 6.2. The numerical *size* of the coefficient (.69) is nearly the same as for the correlation in Fig. 6.2 (.76). It will be seen that the width of scatter of the points is about the same in the two cases. A perfect negative correlation would be pictured as a line of dots like that in Fig. 6.1, but it would slant downward instead of upward from left to right. The algebraic sign of the coefficient of correlation therefore has to do merely with the *direction* of the relationship between two things, whether direct or inverse, and the size of the coefficient (distance from zero) has to do with the *strength*, or *closeness*, of the relationship. The standard kind of coefficient of correlation and the one most commonly used is Pearson's product-moment coefficient. The formula for the correlation of two variables in a population ρ (rho) (where the population SDs are known) is

$$\rho_{xy} = \frac{\Sigma xy}{N\sigma_x\sigma_y} \quad \text{(The Pearson product-moment coefficient of correlation in a population)} \tag{6.1}$$

How to compute a coefficient of correlation

THE PRODUCT-MOMENT COEFFICIENT IN A SAMPLE

The basic formula for estimating ρ in a sample or computing the correlation in a sample is

$$r_{xy} = \frac{\Sigma xy}{NS_xS_y} \quad \text{(Computing formula for a Pearson product-moment coefficient of correlation)} \tag{6.2}$$

where r_{xy} = correlation between X and Y

x = deviation of an X score from the mean of X scores

y = deviation of a corresponding Y from the mean of the Y scores

Σxy = sum of all the products of deviations, each x deviation times its corresponding y deviation

S_x and S_y = standard deviations of the sample distributions of X and Y scores

The steps necessary are illustrated in Table 6.1. They are as follows:

step 1 List in parallel columns the paired X and Y scores or other measurements, making sure that corresponding scores are together.

step 2 Determine the two means, $\bar{X}$ and $\bar{Y}$. In Table 6.1 these values are 7.5 and 8.0, respectively.

step 3 Determine for every pair of scores the two deviations, x and y. Check them by finding their algebraic sums, which should be zero.

TABLE 6.1 Correlation between two sets of measurements of the same individuals; ungrouped data; product-moment coefficient of correlation

	X	Y	x	y	x^2	y^2	xy
	13	11	+5.5	+3	30.25	9	+16.5
	12	14	+4.5	+6	20.25	36	+27.0
	10	11	+2.5	+3	6.25	9	+ 7.5
	10	7	+2.5	−1	6.25	1	− 2.5
	8	9	+0.5	+1	0.25	1	+ 0.5
	6	11	−1.5	+3	2.25	9	− 4.5
	6	3	−1.5	−5	2.25	25	+ 7.5
	5	7	−2.5	−1	6.25	1	+ 2.5
	3	6	−4.5	−2	20.25	4	+ 9.0
	2	1	−5.5	−7	30.25	49	+38.5
Sums	75	80	0.0	0	124.50	144	102.0
Means	7.5	8.0			Σx^2	Σy^2	Σxy

$$S_x = \sqrt{\frac{124.50}{10}} = \sqrt{12.450} = 3.528$$

$$S_y = \sqrt{\frac{144}{10}} = \sqrt{14.4} = 3.795$$

$$r_{xy} = \frac{\Sigma xy}{NS_xS_y} = \frac{102.0}{(10)(3.53)(3.79)} = \frac{102.0}{133.90} = +.76$$

An alternative solution without computing the S's:

$$r_{xy} = \frac{\Sigma xy}{\sqrt{(\Sigma x^2)(\Sigma y^2)}} = \frac{102.0}{\sqrt{(124.5)(144)}} = \frac{102.0}{\sqrt{17{,}928.0}} = \frac{102.0}{133.90} = +.76$$

step 4 Square all the deviations and list in two columns. This is for the purpose of computing S_x and S_y.

step 5 Sum the squares of the two sets of deviations to obtain Σx^2 and Σy^2.

step 6 From these values compute S_x and S_y.

step 7 Find each person's *xy* product (as in the last column in Table 6.1) and sum them for Σxy.

step 8 Finally, apply formula (6.2). For the illustrative problem, the arithmetical calculations are given below Table 6.1.

A SHORTER SOLUTION There is an alternative and shorter route that omits the computation of S_x and S_y, should they not be needed for any other purpose. The formula is

$$r_{xy} = \frac{\Sigma xy}{\sqrt{(\Sigma x^2)\,(\Sigma y^2)}} \qquad \text{(Alternative formula for a Pearson } r\text{)} \qquad (6.3)$$

The solution with this formula is also given below Table 6.1, and it leads to the same coefficient. In both cases, only two digits have been saved in *r* because for so small a number of cases, the sampling error in *r* is so relatively large that more than two digits would be very deceptive as to assumed accuracy.

COMPUTING A NEGATIVE COEFFICIENT As another example of the computation of *r*, this time when the correlation is *negative*, Table 6.2 is presented. The operations are just the same, step by step. The only thing new is the care that must be taken with algebraic signs.

COMPUTING *r* FROM ORIGINAL MEASUREMENTS In both examples thus far, we have been dealing with a small number of observations and with ungrouped data. When the data are more numerous, we commonly resort to grouping into class intervals. But first let us see another procedure with ungrouped data, which does not require the use of deviations. It deals entirely with original scores. When raw scores are small numbers or when a good calculating machine is available, this is the best procedure. The formula is

$$r_{xy} = \frac{N\Sigma XY - (\Sigma X)(\Sigma Y)}{\sqrt{[N\Sigma X^2 - (\Sigma X)^2][N\Sigma Y^2 - (\Sigma Y)^2]}} \qquad \text{(A Pearson } r \text{ computed from original data)} \qquad (6.4)$$

where *X* and *Y* are original scores in variables *X* and *Y*. Other symbols tell what is done with them. We follow the steps that are illustrated in Table 6.3.

The authors have found it more convenient, particularly when machine work can be done, to compute r_{xy}^2 first by the formula

$$r_{xy}^2 = \frac{[N\Sigma XY - (\Sigma X)(\Sigma Y)]^2}{[N\Sigma X^2 - (\Sigma X)^2][N\Sigma Y^2 - (\Sigma Y)^2]} \qquad (6.5)$$

TABLE 6.2 A negative correlation in ungrouped data by the product-moment method

	X	Y	x	y	x^2	y^2	xy
	12	7	+5	−1.5	25	2.25	− 7.5
	10	3	+3	−5.5	9	30.25	−16.5
	9	8	+2	−0.5	4	.25	− 1.0
	8	5	+1	−3.5	1	12.25	− 3.5
	7	7	0	−1.5	0	2.25	0.0
	7	12	0	+3.5	0	12.25	0.0
	6	10	−1	+1.5	1	2.25	− 1.5
	5	9	−2	+0.5	4	.25	− 1.0
	4	13	−3	+4.5	9	20.25	−13.5
	2	11	−5	+2.5	25	6.25	−12.5
Sums	70	85	0	0.0	78	88.50	−57.0
Means	7.0	8.5			Σx^2	Σy^2	Σxy

$$S_x = \sqrt{\frac{78}{10}} = \sqrt{7.8} = 2.79$$

$$S_y = \sqrt{\frac{88.5}{10}} = \sqrt{8.85} = 2.97$$

$$r_{xy} = \frac{-57.0}{(10)(2.79)(2.97)} = \frac{-57.0}{82.863}$$

$$= -.69$$

and then finally to extract the square root to find r_{xy}, as shown just below Table 6.3. The algebraic sign of r is the same as that for the number in the numerator before it is squared.

The scatter diagram

Before calculators, and even computers, were as available as they are now, when samples to be correlated were large, or even moderate in size, the common procedure was to group data in both X and Y and to prepare a scatter diagram or correlation diagram to provide some shortcuts in calculation. There are still other reasons for wanting to see how the X and Y values distribute themselves in a bivariate plot, as will be pointed out late in this chapter. Whether or not one is justified in computing a Pearson r may be indicated by the nature of that plot. Some sketchy scatter plots may be seen in Figs. 6.2, 6.3, and 6.4. From them one can also gain a very rough idea of the general level of the correlation coefficient and of its algebraic sign.

PREPARING A SCATTER DIAGRAM The choice of size of class interval and the limits of intervals is made on the same basis of much the same rules that were given in Chap. 3. For the sake of a clearer illustration of the procedure, however, a

TABLE 6.3
Correlation of ungrouped data computed from the original measurements

	X	Y	X^2	Y^2	XY
	13	7	169	49	91
	12	11	144	121	132
	10	3	100	9	30
	8	7	64	49	56
	7	2	49	4	14
	6	12	36	144	72
	6	6	36	36	36
	4	2	16	4	8
	3	9	9	81	27
	1	6	1	36	6
Sums	70	65	624	533	472
	ΣX	ΣY	ΣX^2	ΣY^2	ΣXY

$$r_{xy}^2 = \frac{[N\Sigma XY - (\Sigma X)(\Sigma Y)]^2}{[N\Sigma X^2 - (\Sigma X)^2][N\Sigma Y^2 - (\Sigma Y)^2]}$$

$$= \frac{(4{,}720 - 4{,}550)^2}{(6{,}240 - 4{,}900)(5{,}330 - 4{,}225)}$$

$$= \frac{(170)^2}{(1{,}340)(1{,}105)}$$

$$= \frac{28{,}900}{1{,}480{,}700}$$

$$= .019518$$

$$r_{xy} = \sqrt{.019518}$$

$$= +.14$$

smaller number of classes will be employed in the problem now to be described. The data were scores earned by a class in educational measurements in objectively scored examinations, one of which stressed statistical methods and the other of which stressed tests and measurements.

In setting up a two-way grouping of data, a table is prepared in columns and rows—columns for the dispersions of Y scores within each class interval for the X scale, and rows for the dispersions of X scores within each of the class intervals for the Y scale. Along the top of the table (see Table 6.4) are listed the score limits for the class intervals for X scores. Along the left-hand margin are listed the score limits for the intervals of Y scores. One tally mark is made for each individual's pair of X and Y scores. For example, if one individual has a score of 83 in test X and a score of 121 in test Y, we place a tally mark for that student in the *cell* of the diagram at the intersection of

TABLE 6.4 A scatter diagram of the scores in two achievement tests

Y: Scores in second achievement test	X: Scores in first achievement test 60–64	65–69	70–74	75–79	80–84	85–89	90–94	95–99	f_y	
135–139								1	1	
130–134				1	1		1		3	
125–129				1		2	1		4	
120–124			1	4	4	6	2		17	
115–119			7	5	7	2	1		22	
110–114	1	4	2	9	4	2			22	
105–109	1	1	2	5	1				10	
100–104	1	3		1	1				6	
95–99		2							2	
f_x	3	10	12	26	18	12	5	1	87	N

the column for interval 80–84 in *X* and the row for interval 120–124 in *Y*. All other individuals are similarly located in their proper cells.

When the tallying is completed, we write the number of cases, or the *cell frequency*, in each of the cells. Next we sum the cell frequencies in the rows separately, recording each frequency in the last column under the heading f_y. When this column is filled, we have the total frequency distribution for test *Y*. We also sum the cell frequencies in all the columns, writing sums in the bottom row, headed f_x. When completed, this row gives us the total frequency distribution for test *X*. We can check the summing of the cell frequencies by adding up the last row and last column. Their sums should, of course, both equal *N*—in this case, 87. The check does not, however, guarantee correct tallying. This can be checked partly when we correlate either test with another one and compare total frequency distributions or when we have knowledge of the correct frequency distribution of *Y* or of *X* from any other source.

Interpretations of a coefficient of correlation

HOW HIGH IS ANY GIVEN COEFFICIENT OF CORRELATION?

Any coefficient of correlation that is not zero and that is also statistically significant denotes some degree of relationship between two variables.[1] But we need further discussion of the matter, for the strength of relationship can be regarded from a number of points of

1 For a treatment of the topic of statistical significance of a coefficient of correlation, see Chap. 8. "Statistical significance" of *r*, in this context, means that there is a very small probability that the coefficient obtained in the sample is a chance deviation from a population correlation of zero.

view. The coefficient of correlation does *not* give directly anything like a percentage of relationship. We cannot say that an r of .50 indicates two times the relationship that is indicated by an r of .25. Nor can we say that an increase in correlation from $r = .40$ to $r = .60$ is equivalent to an increase from .70 to .90. The coefficient of correlation is an index number, not a measurement on an interval (equal-unit) scale. It will be seen in Chap. 14 that r^2, however, is not only on an interval scale but a ratio scale. It should be noted that a correlation of $-.60$ indicates just a close a relationship as a correlation of $+.60$. The relation differs only in direction.

PARTICULAR USES HAVE A BEARING ON INTERPRETATION OF *r*

The question about the size of r cannot be answered without making reference to particular uses of r. One common use is to indicate the amount of agreement between scores on an aptitude test and measures of academic or vocational success. Such a correlation gives what is called a validity coefficient. It is an index of the predictive value of a test. Chapter 18 deals extensively with the subject of validity. Common experience shows that the validity coefficient of a single test may be expected in the range .00 to .60, with most indices within the lower part of this range. It is common experience that a college-aptitude test correlates somewhere near .50 with grades in courses, for example. By combining tests of different abilities and by weighting them properly in obtaining a composite score, the validity coefficient for that score for predicting grades in a certain course such as algebra has been known to reach .80.

Another common application of r is in the form of a reliability coefficient, which could be obtained by correlating scores from two alternate, parallel forms of the same test. There has been some consensus that to be a very accurate measure of individual differences in some characteristic, the reliability should be above .90. The truth is, however, that many standard tests with reliabilities as low as .70 prove to be very useful. And tests with reliabilities lower than that can be useful in research.

When one is investigating a purely theoretical problem, even very small correlations, if statistically significant (i.e., ρ is most probably not zero), are often very indicative of a psychological law. Whenever a relationship between two variables is established beyond reasonable doubt, the fact that the correlation coefficient is small may mean merely that the measurement situation is contaminated by some factor or factors uncontrolled or not held constant. One can readily conceive of an experimental situation in which, if all irrelevant factors had been held constant, the r might have been 1.00 rather than .20. For example, the correlation between an ability score and academic achievement is .50, since both are measured in a population whose academic achievement is also allowed to be determined by effort, attitudes, marking peculiarities of the instructors, etc. Were all the other determinants of achievement held constant and were both aptitude and marks perfectly measured, the r would be 1.00 rather than .50. This line of reasoning indicates that where any correlation between two things is established at all, and particularly where a causal relationship is involved, the fundamental law implies a perfect relationship. Thus, in nature, correlations of zero or 1.00 are expected to be the rule between variables when their effects are experimentally completely isolated. The fact that we obtain anything else is due to the inextricable interplay of variables that we cannot measure in isolation.

The practical conclusion to be drawn from this is that *a correlation is always relative to the situation under which it is obtained, and its size does not represent any absolute natural fact.* To speak glibly of *the* correlation between intelligence and achievement is absurd. One needs to say *which* intelligence, measured under *what* circumstances, and in *what* population, and to say *what kind* of achievement, measured by *what* instruments, or judged by *what* standards. *Always, the coefficient of correlation is purely relative to the circumstances under which it was obtained and should be interpreted in the light of those circumstances, very rarely, certainly, in any absolute sense.*

How much faith one should place in any relationship shown by a coefficient of correlation also depends upon the urgency of the outcome. There are probably some medical treatments, such as some surgeries, inoculations, vaccinations, and the like, concerning which the knowledge is rather incomplete but which are administered even though the correlation between the treatment and survival (or between nontreatment and death) is of the order of .10 to .20. Although the probabilities of survival may be increased by only 1 percent by the treatment, the saving of 1 life in 100 is regarded as worth the effort. If a procedure in education promised only 1 percent improvement over guesswork, we should probably pay little attention to it because the low seriousness of outcome would not justify the means.

Graphic representations of correlations

In presenting the facts of correlation to the layman, who is probably not accustomed to thinking in terms of numerical indices in any case and who has probably never heard of the coefficient of correlation, it is better to convey the idea of a relationship in other ways, preferably in the form of a diagram of some kind. Figures 6.5 and 6.6 are two examples of how this might be done. Figure 6.5 is a bar diagram showing for each level of aptitude score, on a nine-point (stanine) scale,

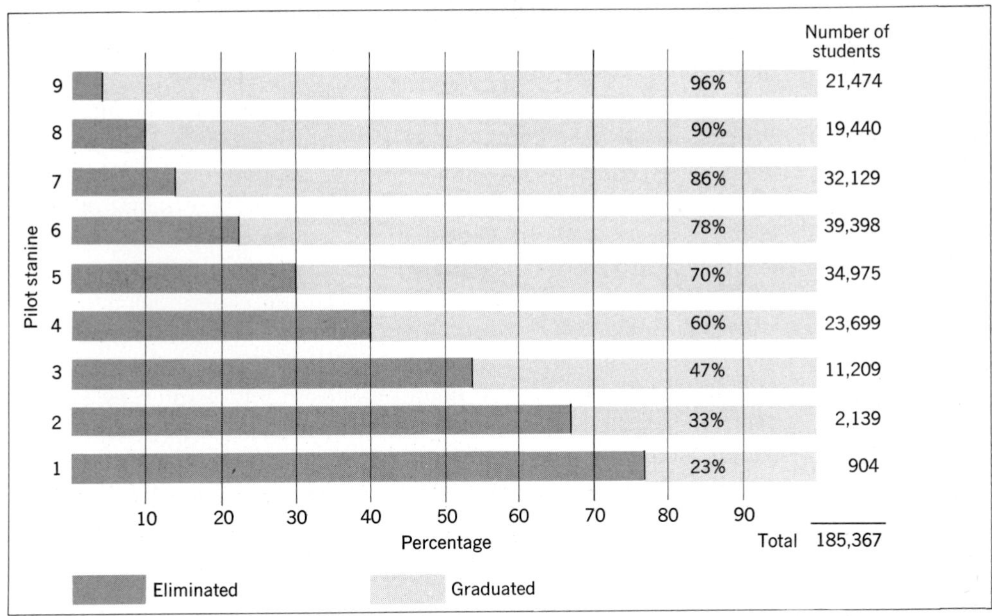

FIGURE 6.5 Correlation between the pilot stanine (pilot-aptitude score) and the criterion of graduation versus elimination from flying training. (From Stanines, in *Selection and classification for air crew duty*. Washington: Army Air Force, 1946.)

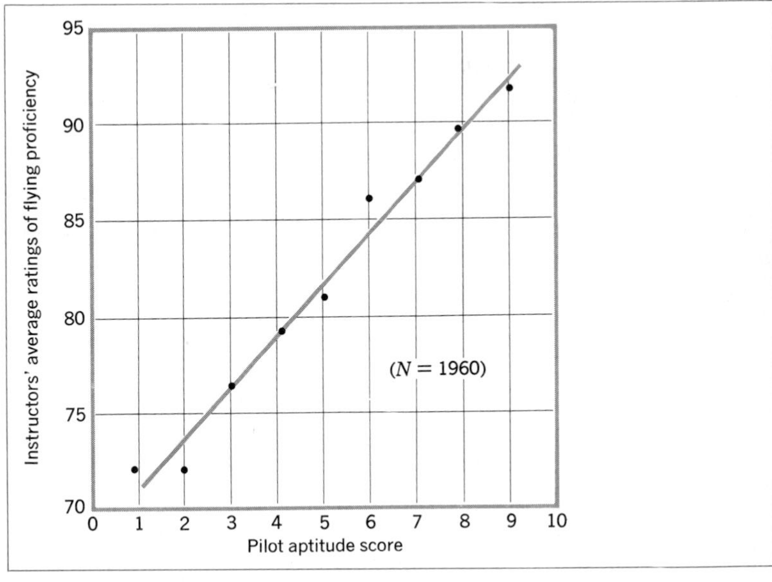

FIGURE 6.6 Correlation between the pilot stanine and instructors' ratings of flying proficiency illustrated by a regression line based upon averages of ratings for different aptitude-score levels

the percentage of pilot students who were graduated from flying-training schools. The actual percentages are given for those who are interested in simple numbers. In spite of the unusually large samples, the percentages are given to two significant digits only. The number of students in each stanine group is given for the benefit of those who have some appreciation of the stability offered by large samples.

The other diagram, Fig. 6.6, shows the average rating of flying proficiency received by cadets at each stanine level, and only the average. A straight line has been fitted to these averages by inspection, which is possible when the fit is as good as it is here. Such a line is known as the *line of regression,* and there are ways of fitting it mathematically, as will be shown in Chap. 15. We do not have the plot of individual cases here, as in figures earlier in this chapter, to indicate very roughly the degree of correlation. Here the extent of correlation is indicated by the slope of the line; the greater the slope, the higher the correlation. If the mean rating at every stanine level were nearly the same, the line would be horizontal, its slope would be zero, and the correlation would be zero. The steeper the slope, the higher the correlation. This slope conception of correlation was the way Sir Francis Galton saw it first. The symbol r originally stood for regression.[1]

Assumptions underlying the product-moment correlation

Before finishing this chapter, the student should be informed about some restrictions that should be observed in the use of the Pearson product-moment coefficient of correlation. The derivation of the formula for the Pearson r assumes that (1) the scores have been obtained in independent pairs, each pair being unconnected with other pairs; (2) the two variables correlated are continuous; and (3) the relationship between the two variables is rectilinear (as in Fig. 6.6).

The most important requirement is the third, rectilinearity, a straight-line regression. This can often be determined by inspection of the scatter diagram. If the distribution of cases within the diagram appears to be elliptical, without any indications of a clear bending of the ellipse, the chances are that the relationship is rectilinear. Even if it is slightly bent, the departure from a straight-line relationship may be so

1 One of the requirements originally set up by Galton for computing a coefficient of correlation was that it should be independent of the size of unit and location of the zero point of either X or Y. The basic equation is $r = \Sigma z_x z_y / N$, where z_x is the standard score for X [that is, $(X - \overline{X})/S_x$] and z_y is the standard score for Y. Thus, r may also be interpreted as the mean of the products of the paired standard scores, also as the slope of the regression line for standard scores. The subject of regression and of regression equations will be treated in Chap. 15.

small that r is still a good index of correlation. A better check would be to use a diagram like that in Fig. 6.6 to see whether the means of the columns approach linearity. When there is definite bending in the trend of cases, a correlation ratio, or some other special coefficient, should be applied, as explained in Chap. 14. For an example of a nonlinear regression, see Fig. 14.1, p. 297.

In psychological and educational research there are certain conditions that can produce artificially curved scatters in the correlation diagram. This could happen when one or both variables have markedly skewed distributions and when the skewing is produced by a faulty measuring scale, with a shifting unit of measurement. If there is good reason to suspect this to be the case, a solution would be to normalize the skewed distribution by methods described in Chap. 19. After a distribution is corrected for skewing, the curvature in the regression is largely eliminated. If any distribution is truncated (scores piled up at one end) it should be dichotomized and a method described in Chap. 14 should be applied.

Nothing in what has just been said demands that the Pearson r be computed only with normal distributions. The forms of distribution may vary, as long as they are fairly symmetrical and unimodal. Even rectangular distributions, with frequencies nearly equal along the range of measurements, would be acceptable. In the case of strongly bimodal distributions and others that tend toward discontinuity, other methods of correlation described in Chap. 14 may be used.

Generality of use of correlation methods

It should be pointed out that correlation methods are by no means limited to scores on tests or other measures of individual differences. It is true that this is where they have found their most extensive applications, and it is easiest to find illustrations from such data.

But correlations may also be usefully obtained relating variables in data from a single individual in a laboratory experiment. Some examples of related variables are ease of memorizing words and the frequencies of use of those words in the population; reaction time and loudness of sound stimuli; rate of work and the number of observers present; and amount of saliva produced as names of different foods are mentioned and degree of palatability of those foods. Wherever observations can be paired off on a one-to-one basis and there is interest in the amount of agreement in the two series of observations, there are correlation problems.

EXERCISES

1 Using the first 10 scores in the lists in Data 6*A*, compute a Pearson r between parts I and II. Use formulas (6.2) and (6.3). Find a similar

DATA 6A Scores earned by 40 high school students in seven parts of the Guilford-Zimmerman Aptitude Survey* (Data 6A continued on next page)

Part I Verbal Comprehension	*Part II Reasoning*	*Part III Numerical Operations*	*Part IV Perceptual Speed*	*Part V Spatial Orientation*	*Part VI Spatial Visualization*	*Part VII Mechanical Knowledge*
22	11	24	29	27	39	30
8	5	22	40	16	23	21
19	6	44	36	14	12	21
32	8	72	32	21	20	33
13	2	25	46	25	20	29
24	5	30	47	2	6	8
22	4	38	49	15	37	35
35	1	54	53	34	28	16
18	7	37	51	37	46	30
13	10	61	50	38	46	35
53	23	56	45	22	41	38
15	9	42	48	18	5	18
34	18	30	25	40	58	46
15	2	42	48	12	21	17
27	4	28	28	31	26	24

* Part I is a vocabulary test; part II is composed of arithmetic-reasoning problems; part III is composed of simple number operations; part IV is on matching visual objects differing very little; part V involves awareness of spatial relationships; part VI requires imagination of an object turned in space; and part VII is on common knowledge of tools and their use, automobile parts and functions, and common trade knowledge. The intercorrelations in this particular sample will be found to be generally low except between parts I and II and between parts V and VI.

coefficient using the last 10 pairs of scores in the same two variables. State a conclusion regarding the consistency of correlation coefficients obtained from different small samples.

2 Compute the correlation between parts II and III for the first 10 pairs of scores, using formulas (6.4) and (6.5). Correlate the same two parts using the last 10 pairs of scores. State your conclusions.

3 Prepare a scatter diagram for the correlation of parts III and IV, including all 40 cases. Inspect the scatter plot for appropriateness of computing a Pearson *r*. Are required conditions apparently satisfied? What is your guess concerning the direction and degree of correlation?

4 Do the same as in exercise 3 for parts V and VI. Using all 40 cases, how many different coefficients of correlation are possible with Data 6A? State a general rule for the number of intercorrelations when there are *n* variables.

5 Find five Pearson *r* coefficients reported in the literature. Tell what

DATA 6A (Continued)

Part I *Verbal Comprehension*	*Part II* *Reasoning*	*Part III* *Numerical Operations*	*Part IV* *Perceptual Speed*	*Part V* *Spatial Orientation*	*Part VI* *Spatial Visualization*	*Part VII* *Mechanical Knowledge*
19	9	32	40	11	13	19
29	4	24	37	26	0	27
24	9	42	58	21	21	23
27	9	54	54	23	20	30
16	5	42	44	29	24	34
56	12	67	48	20	40	26
22	5	58	48	28	41	20
32	4	57	33	20	4	16
18	8	49	47	19	36	42
24	15	87	52	36	34	26
22	12	14	48	25	16	27
22	10	38	46	21	0	20
21	21	32	33	11	43	37
13	10	52	40	29	35	11
23	3	60	49	43	13	37
2	10	29	49	10	21	27
20	4	50	55	22	8	27
25	11	76	43	26	20	26
14	6	40	38	35	8	46
11	2	32	56	37	4	26
2	9	61	45	20	10	20
38	17	56	67	25	20	35
16	6	61	42	29	23	21
14	4	17	44	26	7	21
23	25	61	48	23	29	16

variables are being correlated in each case. Interpret the results. Are the coefficients about the size one should expect for the variables correlated? Were there any special conditions that could have biased the correlation in one way or another?

6 If $r_{xy} = +1.00$, can it be said that X causes Y or that Y causes X? Explain.

7 Give an example or two of pairs of variables that you should expect to correlate positively and other pairs that should correlate negatively.

8 Sketch roughly some scatter plots that you would estimate to represent correlations of +.20, −.50, and +.90.

9 Prove that

$$\frac{\Sigma xy}{NS_xS_y} = \frac{N\Sigma XY - (\Sigma X)(\Sigma Y)}{\sqrt{[N\Sigma X^2 - (\Sigma X)^2][N\Sigma Y^2 - (\Sigma Y)^2]}}$$

ANSWERS

1 The seven parts of the Aptitude Survey were designed to measure different abilities that are relatively independent and hence to correlate low with one another. The correlation r_{12} (between parts I and II) is found to be $-.16$ and $+.47$ in the first and last 10 pairs of scores, respectively. Incidentally, this somewhat large discrepancy shows how widely the correlation between the same two variables can fluctuate from sample to sample when samples are very small. The correlation between the same two tests for all 40 pairs of scores is $+.37$. Typical correlations in larger samples have been .25, .57, and .40, for college men, high-school boys, and high-school girls, respectively.[1]

2 r_{23} is .18 and .49 for the two small samples. In larger samples (the same as for the answer to Exercise 1) r_{23} was .18, .37, and .33.

3 $r_{34} = .25$. In larger samples it was .20, .07, and .31.

4 $r_{56} = .27$. In larger samples it was .61, .34, and .46.
The number of pairs of variables among n equals $n(n - 1)/2$.

5 Decisions regarding causality must be based on other information.

1 For additional information on intercorrelations of these tests, see Michael, W. B., Zimmerman, W. S. and Guilford, J. P. An investigation of the nature of the spatial-relations and visualization factors in two high-school samples. *Educational and Psychological Measurement,* 1951, **11,** 561–577.

TWO

STATISTICAL TESTS AND DECISIONS

7 Probability and mathematical distributions

Thus far, we have been dealing exclusively with descriptive statistics, statistics that enable us to summarize masses of data and to arrive at single descriptive numbers, either for single distributions or for degrees of relationships between variables. In the chapters immediately following we shall be concerned almost exclusively with the estimation of corresponding population values and with the question of how much confidence we should have in those estimates.[1] Most of the developments in the field of statistics during the past 40 years have been in the area of sampling statistics, statistical inference, and tests of significance. This has been a great boon to experimental investigators, for it has enabled them to plan their experiments more wisely and to know whether or not their conclusions are sound.

The need for mathematical models

THE ROLE OF MATHEMATICS IN SCIENCE

Concerning the great value of mathematics in general in science there can be no argument, if we view the development of science as a whole, culminating in modern theoretical physics. Whether or not we believe that the universe, including man and his behavior, is constructed along mathematical lines, the application of mathematical ideas and forms in describing it is an undeniably profitable practice. Think, for example, of all the consequences of $e = mc^2$, Einstein's equation stating the equivalence of mass and energy. This equation is a mathematical model, an expression in symbolic form of a structural idea that describes a whole range of physical phenomena.

Mathematics exists entirely in the realm of ideas. It is a logic-based system of elements and relationships, all of which are precisely

[1] For definitions of "population" and "sample" see pp. 119–120.

defined. It is a completely logical language that can be applied to the description of nature because the events and objects of nature have properties that provide a sufficient parallel to mathematical ideas. There is *isomorphism* (similarity of form) between mathematical ideas and phenomena of nature. Even if the description of nature in mathematical terms is never completely exact, there is enough agreement between the forms of nature and the forms of mathematical expression to make the description acceptable. The approximation is often so close that once we have applied the mathematical description, we can follow where the mathematical logic leads and come out with deductions that also apply to nature.

Take, for example, the normal-distribution curve, which we shall investigate further in this chapter. The normal, or Gaussian, curve is entirely a mathematical idea. It is incorrect to refer to it as either a biological or a psychological curve. It is a particular mathematical model that happens to describe groups of natural objects so well that we can often use its properties to make inferences and predictions about those objects or groups. We have already done this in interpreting the standard deviation, in Chap. 5. Now we need to become better acquainted with the normal-distribution curve in order to capitalize further upon its properties. We shall also meet other statistical models that we can use.

A TYPICAL EXPERIMENT

As an illustration of the foregoing statements, let us consider a typical experiment in which a statistical model is needed. The experiment is in the area of extrasensory perception (ESP): the general problem of whether or not one who perceives objects (the sender) can transmit directly to another person (the receiver) any information about the objects without the use of the receiver's senses.

Suppose that an experiment with the Duke University ESP cards is properly designed to prevent the receiver from being influenced by any cues except possible telepathic stimulation. The materials are five different symbols, each on a different card; each symbol is repeated 5 times in a set of 25 cards. The deck is thoroughly shuffled so that the cards should come up at random. As each card comes up, the sender reads it silently, and the receiver records a judgment as to which symbol is being perceived. The card is returned to the deck, which is reshuffled, and the next card to be transmitted is selected.

Over a run of 25 trials, suppose the receiver gives 8 correct judgments. Does this result show evidence of ESP, or does it not? Without the aid of statistical inferences we should never be able to answer this question. There are two alternative hypotheses to be tolerated before the experiment and before a decision is made as to the meaning of the outcome. One is the hypothesis that some genuine ESP has been at work; the other is that only the laws of chance have produced the result. We know that even by guessing, the receiver could be correct part of the time. Now we cannot find a mathematical model that would enable us to predict how many correct responses there should be if there has been some ESP operating. But we can find a model that will tell us how many correct responses to expect if the outcome is a purely random one, due to the "laws of chance," as we sometimes say.

Without very much mathematical sophistication, it is easy to see that with the receiver merely guessing at random which of 5 symbols is being "transmitted," there is 1 chance in 5 of being correct. In this 1-out-of-5 situation, over many runs of 25 trials each, on the average the receiver should be correct 5 times by chance. The statistical model that we apply should therefore include an expected score of 5.

But that is not all. We can also readily see that in a relatively short run of 25 trials, by what is popularly called "lucky guessing," the receiver might make a score of 6, 7, or more correct responses. If the receiver is "unlucky," the score might just as well go below 5 to about the same extent. The model therefore also involves a family of chance outcomes, which has to be described. This can be done mathematically, as we shall see. To state the question of statistical inference more precisely, we ask whether the obtained score of 8 could be one of the chance-generated outcomes and, if it could be, how likely it is that it could so occur. If a score of 8 or some even higher score could very rarely happen by chance, we can reject the hypothesis that it is chance-generated, which leaves us with some degree of confidence in the other hypothesis, that ESP is responsible.

Principles of probability

There have been several mentions thus far in this chapter of random events, chance, and probability. We therefore need to look into some basic mathematical ideas involving probability. They are basic to the models used in testing hypotheses and drawing statistical inferences. It is well known that the mathematics of probability arose out of interest in gambling and games of chance. Wagering on outcomes of events is a popular pastime. Usually, the wagerer bases odds on an appreciation of subjective probability regarding outcomes of various kinds. Mathematical descriptions of the known chances in a situation permit much better-informed bases for making bets. Such information takes much of the guesswork out of gambling, just as it takes much of the guesswork out of a scientist's evaluation of experimental results. With application of an appropriate probability model, the scientist can state accurately the odds for drawing right or wrong conclusions.

EXAMPLES FROM GAMES OF CHANCE

As is customary in introducing the mathematics of probability, let us begin with some common games of chance—tossed coins, tossed dice, and the drawing of playing cards. In each case there is a specified event, which can turn out in two or more ways. A tossed coin can land with a head or a tail showing. A die can land with one of six numbers of dots on top. A card drawn from a thoroughly shuffled deck of 52 ordinary playing cards will be from one of four suits (club, diamond, heart, or spade) and from one of the 13 denominations within one of those suits. The chance that the coin will land with a head showing is 1 in 2; the chance that a tail will be showing is also 1 in 2. The chance that a die will come up with a 2 spot is 1 in 6; the same is true for any other number of spots. In each case, the probability that the event will occur in the way in which we are interested (sometimes called the *favored way*) is the *ratio of the number of ways in which that favored way can occur to the total number of ways the event can occur*. This is the logical definition of probability. There is one way in which a coin can give a head and a total of two ways in which the coin can fall (excluding the extremely unlikely case of standing on edge). The probability of a head is $1/2$ (also expressible as .5). The probability of a tail is also $1/2$, or .5. The number of ways a single die can yield a 6 is 1, and the total number of ways the die can fall is 6, and so the probability of a 6 is $1/6$, or .167.

Let us apply the definition to the drawing of playing cards. What is the probability that a card drawn at random will be a spade? The number of favorable ways is 13, and the total number of ways is 52; therefore, the probability of drawing a spade (any spade) is $13/52$, which equals $1/4$, or .25. What is the probability of drawing a queen? There are only 4 queens and a total of 52 cards, and so the probability is $4/52$, or $1/13$. What is the probability of drawing the queen of hearts? There is only one way in which this could happen, and so we have $1/52$ as the probability.

SOME MATHEMATICAL THEOREMS ABOUT PROBABILITY

Mathematically, a probability is symbolized by p, which may range from zero, when there is no chance whatever of the favored event, to 1.0, when there is absolute certainty: nothing else could happen. The probabilities we have just been considering are purely theoretical ones, based upon the a priori knowledge of the frequencies of the favored event and of all events, including the favored one and all its alternatives in the situation. It is known that if we actually toss a coin a finite number of times, the ratio of the number of heads to the number of tosses would probably not come out exactly $1/2$. For a definition of probability that is more fully in accord with the outcomes of events, therefore, we need the following modification, although it is not always explicitly stated:

$$p = \lim_{n \to \infty} \frac{n_f}{n}$$

This expression states that as the number of times (n) the event occurs becomes indefinitely large, the ratio of the number of favored outcomes (n_f) to n approaches the probability p as a limit. It is also necessary to stress two requirements: *that the ways of occurrence of the event must be equally likely and that they must be mutually independent.* "Mutual independence" means that one event has no effect whatever on any other event. Granting unbiased coins, dice, and cards, the occurrences of heads and tails are equally likely; the occurrences of different sides of a die coming up are equally likely; and in drawing a card from a well-shuffled deck, any card has as much chance of being drawn as any other.

THE ADDITION THEOREM We can ask other kinds of questions regarding events in games of chance. In tossing a die, what is the probability of *either* a 1 *or* a 2 coming up? The probability of a 1 is $1/6$, and the probability of a 2 is $1/6$. There are *two* ways in which the favored event can occur, out of a total of 6 ways; therefore, by definition, the probability is $2/6$. Note that this probability is the sum of the two separate probabilities. Thus, probabilities are additive.

What is the probability of obtaining four or more spots? The specification "four or more" includes the outcomes 4, 5, and 6. Adding the three probabilities, we have $3/6$ or $1/2$ as the probability of the alternative outcomes. In tossing a coin, what is the probability of getting a head *or* a tail? The addition $1/2 + 1/2$ gives us 1.00, which means that we are certain to obtain either a head or a tail.

Applying this principle to drawing cards, we may ask the probability of drawing either a spade or a club. The two separate probabilities are $1/4$ and $1/4$ which, summed, give $1/2$. What is the probability of drawing a queen or a king? The answer is $1/13 + 1/13$, or $2/13$. In general, *the probability of alternative outcomes is the sum of the probabilities of the outcomes taken separately.*

THE MULTIPLICATION THEOREM We have just treated the either-or kind of case. Here we are concerned with the "this *and* that" kind of case—the *probability of combined outcomes.* In two independent throws of a coin, what is the probability that the outcome will be two heads (a head *and* a head)?

It will help to go back to the basic definition of probability, the relative frequency of an outcome: the ratio of the number of favored ways to the total number of ways. In tossing two coins (either two different coins simultaneously or one coin in succession two times), what is the total number of ways in which to get the outcome? If we wrote them all out, we should have the following: HH HT TH and TT. That is, when the first is a head, the second can be either a head or a tail, and when the first is a tail, the second can be either a head or a tail. There are four total ways and only one favored way (HH), and so the probability is $1/4$. Here the probability that the first event will come out H is $1/2$, and the probability that the second event will come out H is also $1/2$. The probability of the combination of the two outcomes is $1/2 \times 1/2$, or $1/4$. The probability of each of the other outcomes is also $1/4$, and the four add up to 1.0, as they should. We are certain to have one of the four.

We can state another kind of question: What is the probability of getting one (and only one) head? This specification describes the HT and TH cases, of which there are two. The addition theorem applies. The probability for this is $2/4$, or $1/2$. What is the probability of obtaining at least one head? This statement includes also the case HH, and so the probability is $3/4$.

Problems of the following sort are of special interest to those who deal with dice. What is the probability of throwing 12 spots (a pair of 6s)? Here we have two events, the throw of two dice, and a specified combination of two outcomes (six spots in each case). According to the principle of multiplication of probabilities, we have $1/6 \times 1/6 = 1/36$ as the probability of obtaining a pair of 6s. The same probability would apply to a pair of 1 spots. What is the probability of obtaining a total of 11 spots in throwing 2 dice? The total number of ways in which 2 dice can fall is 36 because for every way the one falls, there are six independent ways in which the other one can fall. Eleven spots can come from a 6 on the first and a 5 on the second, or vice versa: two ways. Two over 36 is $1/18$ as the probability for exactly 11 spots. There is special interest in a total of seven spots in two throws. A 7 could occur by combinations of 1-6, 2-5, 3-4, 4-3, 5-2, or 6-1: altogether, six ways. The probability is $6/36$, or $1/6$—the highest for any combination, as any experienced dice thrower knows. A total of seven spots is therefore six times as likely as any specified pair with equal numbers.

Binomial distributions

We should be ready now to consider mathematical models that rest upon the principles of probability, the first of which is the binomial type of distribution. It can be readily explained by reference to coin tossing.

THE BINOMIAL EXPANSION We saw that the independent tossing of two coins yields four possible outcomes, which could be classified in three categories: cases with 2 heads, 1 head, and 0 head, with frequencies of 1, 2, and 1, respectively. In this list of frequencies we have a frequency distribution, which could be plotted with numbers of heads on the abscissa and frequencies on the ordinate, respectively.

Now let us take the case with three coins tossed independently. We could write out all the possible combinations of outcomes for the three successive coins as follows:

	Score	*f*
HHH	**3 heads**	**1**
HHT		
HTH	**2 heads**	**3**
THH		
HTT		
THT	**1 head**	**3**
TTH		
TTT	**0 head**	**1**
	Total	**8**

From this distribution we can say that the probability of obtaining exactly 2 heads is $3/8$ and that the probability of obtaining exactly 3 heads is $1/8$. The latter outcome agrees with our multiplicative theorem, in that $1/2 \times 1/2 \times 1/2 = 1/8$. What is the probability of obtaining *at least* 2 heads? Here we would apply the additive theorem: $1/8 + 3/8 = 4/8 = 1/2$. Half the outcomes are 2 or more heads, and half are 1 or fewer heads. The distribution is symmetrical about the middle. Figure 7.1 shows how we may regard the results as forming a frequency distribution. Each frequency is represented by a vertical bar because of the lack of continuity on the base line.

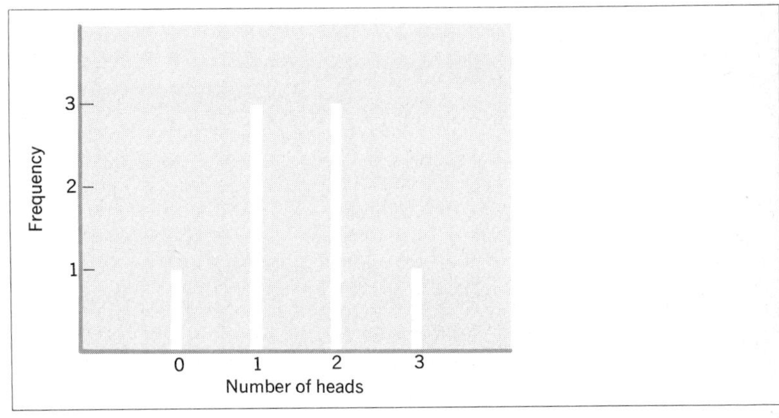

FIGURE 7.1 A simple binomial frequency distribution, showing the number of times out of eight trials one should expect 0, 1, 2, and 3 heads, tossing three coins on each trial

THE BINOMIAL EQUATION We could go on, considering problems with four, five, six, and more coins, generating all the possible combinations and their frequencies. We should expect symmetrical distributions, with the number of categories equal to $n + 1$, where n is the number of coins. A mathematical way of deriving the probabilities for such problems makes use of a well-known kind of equation, the *bionomial expansion*. For the case of two coins, the equation is

$$\left(\frac{1}{2}+\frac{1}{2}\right)^2 = \left(\frac{1}{2}\right)^2 + 2\left(\frac{1}{2}\times\frac{1}{2}\right) + \left(\frac{1}{2}\right)^2$$
$$= \frac{1}{4} + \frac{1}{2} + \frac{1}{4}$$

This is just as in basic algebra, where the student learns that $(a + b)^2 = a^2 + 2ab + b^2$. It will be seen that the three terms in the expansion of $(\frac{1}{2} + \frac{1}{2})^2$ represent the probabilities of obtaining 2, 1, and 0 heads, respectively (or 0, 1, and 2 heads: the distribution is symmetrical). For the case of three coins,

$$\left(\frac{1}{2}+\frac{1}{2}\right)^3 = \left(\frac{1}{2}\right)^3 + 3\left(\frac{1}{2}\right)^2\left(\frac{1}{2}\right) + 3\left(\frac{1}{2}\right)\left(\frac{1}{2}\right)^2 + \left(\frac{1}{2}\right)^3$$
$$= \frac{1}{8} + \frac{3}{8} + \frac{3}{8} + \frac{1}{8}$$

Besides the four terms expressing the probabilities of occurrence of the four outcomes, 0 to 3 heads, the numerators express the frequencies with which each kind of outcome occurs. Since the denominator term expresses the total number of outcomes, each ratio expresses a probability.

There is, of course, a general equation for the expansion of a binomial with any complementary pair of values raised to any power n, but because the student would have rare occasions for using it, it is not given here. The more common uses are covered by the information given in Table 7.1.

TABLE 7.1 Pascal's triangle for cases up to $n = 10$

n												*Sum*
1	1	1										2
2	1	2	1									4
3	1	3	3	1								8
4	1	4	6	4	1							16
5	1	5	10	10	5	1						32
6	1	6	15	20	15	6	1					64
7	1	7	21	35	35	21	7	1				128
8	1	8	28	56	70	56	28	8	1			256
9	1	9	36	84	126	126	84	36	9	1		512
10	1	10	45	120	210	252	210	120	45	10	1	1,024

THE PASCAL TRIANGLE For the common case of $p = 1/2$, the easiest way of determining the frequencies for expansions of binomials when n is small is to utilize Pascal's triangle, illustrated in Table 7.1. Starting with the case n (number of coins) $= 1$, we have two frequencies, 1 and 1, with a total of 2. The frequencies for $n = 2$ can be written by putting 1 at either extreme and a 2 in the middle category. Note that the 2 is the sum of the two frequencies on either side of it in the row above. The third row is generated in the same fashion from the second, and so on down the triangle. Except for the 1s at the extremes, each frequency is the sum of the 2 just above it. The sums at the extreme right of the rows are 2 to the power n.

THE MEAN AND VARIANCE OF A BINOMIAL DISTRIBUTION Just as we describe distributions of samples of measurements by giving their means and variances or standard deviations, we can also describe particular binomial distributions. The mean and variance of a binomial distribution are readily computed from knowledge of p and n. The mean is given by the equation

$$\mu_b = np \qquad \text{(Mean of a binomial distribution)} \tag{7.1}$$

The variance is given by

$$\sigma_b^2 = npq \qquad \text{(Variance of a binomial distribution)} \tag{7.2}$$

where $p =$ probability of the favored outcome
$q = 1 - p$
$n =$ number of replications (occurrences) of the event

The Greek letters in the symbols μ and σ indicate that we are dealing with *parameters* rather than statistics. Parameters are mathematically exact constants, not estimates from samples. For the distribution describing the frequencies to be expected in tossing 10 coins, with $p = 1/2$ and $n = 10$, the mean equals 5.0, and the variance equals 2.5. The student might check this by actually computing the mean number of heads and the variance from the last distribution in Table 7.1.

SOME SIMPLE APPLICATIONS OF THE BINOMIAL MODEL In research on behavior the binomial model is often found to apply quite well. Assume that we have a true-false test of 10 items or, a parallel case in animal research, a sequence of 10 two-choice trials for a rat. We suspect that the behavior in either case will not be entirely random, that in the long run there will be a bias toward right responses. However, to exclude the possibility that the behavior *is* random, we need to set up the random-behavior hypothesis in the form of a binomial expressed by $(1/2 + 1/2)^{10}$. Let us say that the student taking the test does flip a coin for every item and that the rat does something equivalent in its trials. The binomial model, expanded, would have the frequencies seen in the bottom row of the Pascal triangle in Table 7.1, representing the various numbers of right answers given by the subject. The expected, chance-generated number of right responses is 5. Do any such students or rats do enough better than 5 so that we may reject the chance hypothesis and tolerate in its place the hypothesis that they know something about what they are doing? We shall encounter many such problems in later chapters, and we shall find that the question about "enough better" can be answered mathematically.

Let us return to the ESP problem with which we started this whole discussion. In that experiment, the subject has one chance in five of being correct merely by guessing: the probability of a correct response is $1/5$. With 25 trials, the expected chance-generated score would be 5. The binomial would be $(1/5 + 4/5)^{25}$. Expansion of this binomial in order to determine the probabilities for all the possible scores from 0 to 25 would be a formidable task, except on a high-speed computer. It would not be necessary to derive all those probabilities, for our interest would be in the scores as high as, or higher than, the one made by a particular person. The two examples are probably sufficient to show the kinds of problems to which the binomial distribution can be applied. The opportunities for its use are numerous, but we shall leave actual applications to later chapters, particularly Chap. 10.

The normal distribution

RELATION OF THE NORMAL DISTRIBUTION TO THE BINOMIAL DISTRIBUTION

It is easy to show the relation between the normal distribution and the binomial distribution. If we plot the frequencies given in the bottom row of Table 7.1, for the case of $n = 10$, we have a bar diagram as in Fig. 7.2, with 11 score categories, where the score is the number of heads expected. Imagine similar graphs where n equals 20, then 50, and then larger and larger numbers, keeping the total width of the distribution the same. The stair steps of the bar diagram become smaller and smaller until they merge into a smooth contour. *The normal distribution is the limit of the binomial distribution where $p = .5$ and n becomes indefinitely large.* This has been proved mathematically, but the proof will not be given here.

Superimposed upon the bar diagram in Fig. 7.2 is a normal distribution whose mean and standard deviation are the same as those for the binomial distribution: 5.0 for the mean and 1.58 for the standard deviation, from the use of formulas (7.1) and (7.2). The agreement of the two distributions would be even more apparent if the binomial distribution were drawn in the form of a frequency polygon. The bar diagram is preferred for a binomial distribution because the scores are discrete numbers: there is a lack of continuity over the range of possible score values. The comparison of the two distributions can be done analytically by comparing category frequencies, as in Table 7.2. The discrepancies are mostly small, but the largest is 6.5 for the modal cat-

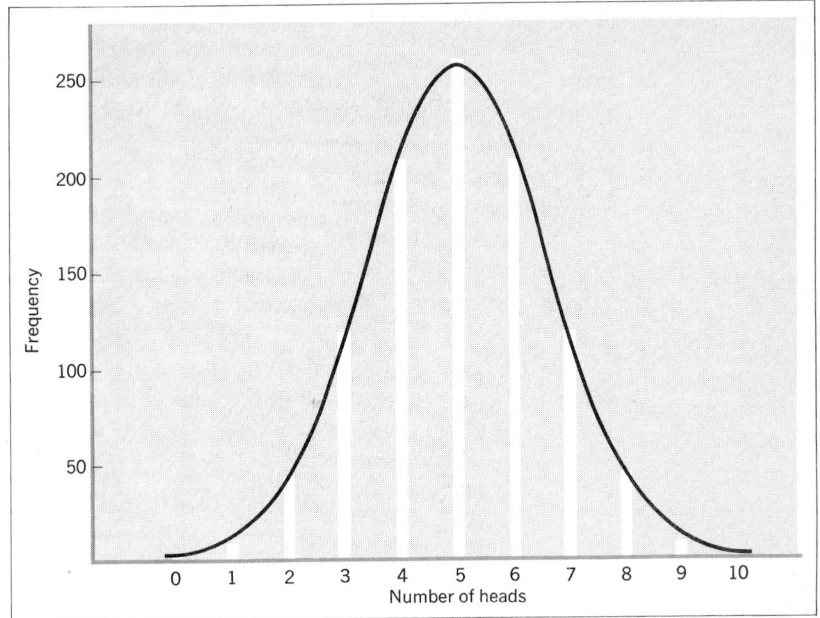

FIGURE 7.2 A binomial frequency distribution of expected numbers of heads in tossing 10 coins 1,024 times. Superimposed is a normal-distribution curve that has the same mean and standard deviation as the binomial distribution.

egory. The normal-distribution frequencies add up to 1,023.5, or half a case short. This is due partly to the fact that some of the cases in the normal distribution extend beyond the terminal values of 0 and 10 for the binomial distribution. If we find the cumulative distributions of proportions for the two sets of frequencies (see the last two columns of Table 7.2), we see that the majority of the discrepancies are .001 and that the largest is only .005. The discrepancies are greatest near the center of the distributions. Since it is the frequencies near the tails that are crucial in hypothesis testing that we do with either kind of distribution, we see why the normal distribution may often be substituted for the binomial as a model for chance-generated results.

THE EQUATION FOR THE NORMAL-DISTRIBUTION CURVE

The general equation that describes mathematically the normal-distribution curve is often written

$$Y = \frac{N}{\sigma\sqrt{2\pi}} e^{\frac{-x^2}{2\sigma^2}} \qquad \text{(General equation for the normal-distribution curve)} \qquad (7.3)$$

where Y = frequency
N = number of observations
σ = standard deviation of the distribution
π = 3.1416 (approximately)
e = 2.718 (approximately), the base of the Napierian system of logarithms
x = deviation of a measurement from the mean $(X - \mu)$

Examination of the equation suggests some features of the normal curve. With fixed N and σ, all elements are constants except x, the independent variable. Since x is squared, either negative or positive values of it have the same effect and yield the same value of Y; hence the curve is bilaterally symmetrical. As x increases, since x^2 has a negative sign, the exponent of e decreases, and Y also decreases. Y is at a maximum when $x=0$, for that is when the exponent of e is a maximum. The mean of the distribution is 0, and the mode and median are also at 0. These features have been mentioned before, but now we see the mathematical reasons.

THE CUMULATIVE NORMAL DISTRIBUTION

Much of the use of the normal distribution in statistical operations is concerned with proportions of the area or surface under the normal curve in relation to certain distances on the base line. As the first approach to this subject, we consider the cumulative normal distribution. We have already seen, first in Chap. 3 and then again in Table 7.2, how ordinary obtained frequency distributions, normal or otherwise, can be cumulated. We start with the frequency in the lowest category and successively add on the frequencies in higher categories in order.

The last column in Table 7.2 is a cumulation of actual frequencies in a normal distribution, with the frequencies in 11 categories. In approaching this problem mathematically in a more general manner, an indefinitely large number of categories is assumed (with correspondingly smaller numbers of cases in the categories, where N is con-

TABLE 7.2 Comparison of similar binomial and normal distributions (with same means and standard deviations) with respect to frequencies and cumulative proportions

Score (heads)	*Frequencies*		*Cumulative proportions*	
	Binomial	*Normal*	*Binomial*	*Normal*
10	1	1.8	1.000	1.000
9	10	10.5	.999	.998
8	45	42.5	.989	.988
7	120	116.1	.945	.946
6	210	211.6	.828	.833
5	252	258.5	.623	.626
4	210	211.6	.377	.374
3	120	116.1	.172	.167
2	45	42.5	.055	.054
1	10	10.5	.011	.012
0	1	1.8	.001	.002
Sum	1,024	1,023.5		

stant). The process is known as *integration,* in integral calculus. The equation for the cumulative normal distribution reads

$$Y = \int_{-\infty}^{x} \frac{N}{\sigma\sqrt{2\pi}} e^{\frac{-x^2}{2\sigma^2}} dx \qquad \text{(Equation for the cumulative normal distribution)} \qquad (7.4)$$

where the symbols are exactly the same as in equation (7.3), except for two additions. The symbol $\int$ is the integration sign, which can be interpreted as "the sum of" what follows it. The limits of integration are $-\infty$ (an infinitely large negative value) at the lower end and x at the upper end. The lower tail of the normal distribution extends to the left indefinitely. The category areas (frequencies) summed are all those below any particular chosen value of x. The expression dx at the right in the equation indicates simply that areas corresponding to very small increments of x are summated.

THE UNIT NORMAL DISTRIBUTION Tables have been prepared that help in doing computations where either the noncumulative or the cumulative normal distribution is concerned. To save space, the tables do not provide for all possible cases of N and all cases of σ. Instead, they adopt a particular curve with $N = 1$ and $\sigma = 1$. The mean, of course, remains at zero. Under these conditions, the total area under the curve is equal to 1, and all frequencies are in the form of proportions. Such are Tables B and C in the Appendix.

With both N and σ equal to 1.0, equation (7.3) becomes

$$y = \frac{1}{\sqrt{2\pi}} e^{-\frac{z^2}{2}}$$

where $z = x/\sigma$ [see formula (7.5)].

Areas under the normal-distribution curve

Since the normal-curve tables are limited to the standard case in which $N = 1$ and $\sigma = 1$, in applying the tables to situations in which N and σ have other values, it is necessary to convert measurements into standard form, which means finding *standard scores.* A standard score z is a deviation from the mean in terms of the standard deviation as the unit. The transformation of an obtained score into a standard score is performed by using the equation

$$z = \frac{x}{S_x} = \frac{X - \bar{X}}{S_x} \qquad \text{(A standard score derived from a deviation score or an obtained score)} \qquad (7.5)$$

The normal distribution involved in Table 7.2 has a mean of 5 and a standard deviation of 1.58. The z corresponding to an X of 5 would, of course, be zero. The z corresponding to a score of 6 would be

$$z = \frac{6-5}{1.58} = \frac{1}{1.58} = +0.63$$

The standard score corresponding to an X of 2 would be

$$z = \frac{2-5}{1.58} = \frac{-3}{1.58} = -1.90$$

It is with such z values that we enter Table B in the Appendix. Next, we shall try a number of different kinds of problems pertaining to relations between areas under the curve and z values.

PROPORTION OF THE AREA BETWEEN THE MEAN AND SOME MEASUREMENT OR SCORE

In connection with interpreting a standard deviation, we have already had occasion to say that the interval extending one standard deviation on either side of the mean includes about two-thirds of the cases. To say the same thing in another way, from the mean to $+1S$ are to be expected about one-third of the cases, and from the mean to $-1S$, another one-third of the cases. We can verify this by referring to Table B in the Appendix and looking up the proportion of the area between the mean and 1σ (i.e., a z equal to 1.00). The area given to four decimal places is .3413, or 3,413 ten-thousandths of the area. If there were a normal distribution with 10,000 cases, 3,413 of them would be expected between the mean and 1σ. In terms of percentage, it would be 34.13 percent, or 34.13 cases in 100. The total interval from $+1\sigma$ to -1σ contains twice this area, or .6826, or 68.26 percent. Figure 7.3 illustrates these facts graphically.

From Table B, we can also see that between the mean and a point 2σ distant (either above or below, i.e., either $+2\sigma$ or -2σ), we should expect .4772 of the total surface, or 47.72 percent of the cases. Included in the range from -2σ to $+2\sigma$ we should find twice this proportion, or .9544 of the area, or 95.44 percent of the cases. From the mean to 3σ is .4987 of the area, and in both directions from the mean to 3σ we find twice this, or .9974 of the area. Only 26 cases in 10,000 (10,000 − 9,974), therefore, should be expected *beyond* the range from -3σ to $+3\sigma$ in a large sample.

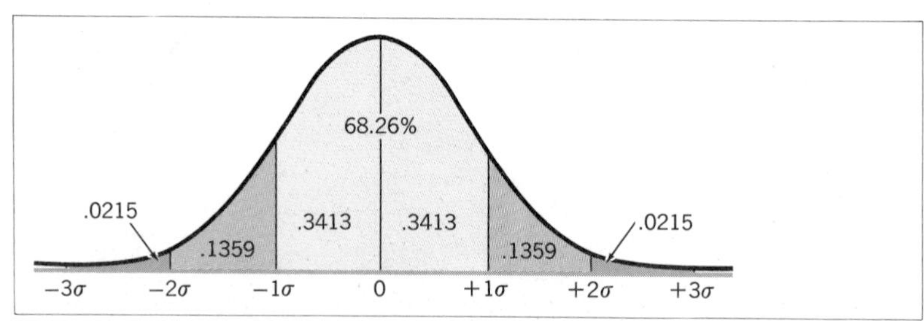

FIGURE 7.3 Different proportions of area under the normal curve within the limits of the various σ units on the base line

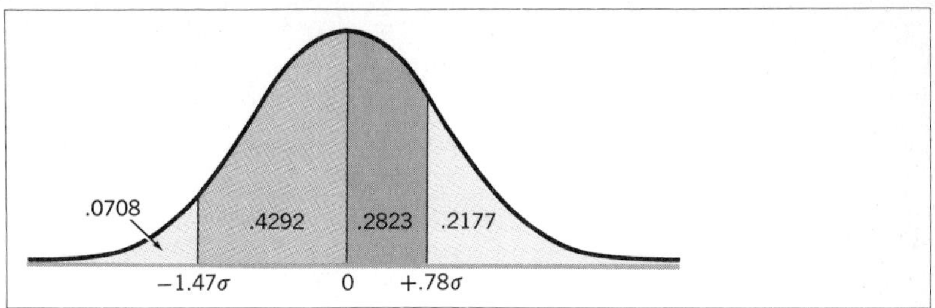

FIGURE 7.4 Proportions of the area under the normal curve within certain standard-score limits on the base line

To take another example of a less special nature, how much of the area under the normal curve will be found between the mean and $+0.78\sigma$? From the table, we find this to be .2823. In still another problem, what proportion of the cases lie between the mean and -1.47σ? From the table, we find this to be .4292. Figure 7.4 illustrates these two cases. It will be seen that the positive or negative sign of z tells us merely whether the area extends above the mean or below. The numerical *size* of z, whether positive or negative, determines the *amount* of area between the mean and the point.

Thus far we have begun each problem of this type with some particular z or standard measurement. Let us now start the problem a step or two further back and begin with some raw score or measurement. In the more practical case, we begin with X, not z. In the memory-test data, we may inquire what proportion of the cases come between the mean (26.1) and a point of 35 on the scale of measurement. This point deviates 8.9 X units from the mean ($X - \bar{X} = +8.9$). This is the deviation x. The standard score z is x/S, which equals $8.9/6.45 = +1.38$. *Everything must be transformed into standard measure before the probability table may be utilized.* Entering the table with a z of 1.38, we find the corresponding area to be .4162 (see Fig. 7.5). In other words, 41.62 percent of the cases in a normal distribution would be

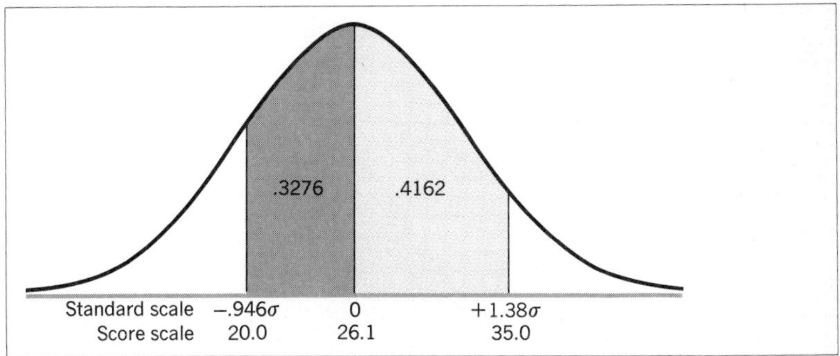

FIGURE 7.5 Proportions of the area under the normal curve between the mean and selected scores on the scale of scores for the memory test

found between the mean and 35 points on the scale. In the memory-test data, 41.62 percent of 86 is 35.8, or, in whole numbers, 36 cases. In a similar manner, which the student should verify, between the mean and a score of 20, .3276 of the cases are to be expected, or approximately 28. Between the mean and a score of 15, about 39 cases of the 86 are to be expected, and if we go on down to a score point of 5, we find 49.95 percent of the cases.

THE AREA ABOVE OR BELOW A CERTAIN POINT ON THE SCALE

For a given deviate or standard score, Table B in the Appendix also gives the proportion of the area above a certain point on the scale or below it. Above a point at $+1\sigma$ will be found .1587 of the area. When a vertical line is erected at $+1\sigma$ (see Fig. 7.6), it divides the total area under the curve into two portions, the one to the right of the line being the smaller of the two. Below the point $+1\sigma$ is the remainder of the area, or the larger portion (found in column *B* of the table), including .8413, or 84.13 percent of the area. If we were interested in the point -1σ, we would note that the larger portion under the curve is now to the right of the point of division and is found in column *B*, whereas the portion to the left, being the smaller of the two, is found in column *C*. The situation is just reversed in the case where the division comes at $+1\sigma$. In this kind of problem it is necessary to keep in mind whether the area we wish to know is under the smaller end of the curve, all on one side of the mean, or whether it is under the larger side of the curve extending across the mean.

The proportion of the area above the point at $+0.78\sigma$ is in the smaller portion, and is found in column *C*; it is .2177. The area below -1.47σ is also under the smaller portion of the curve, and from column *C* we find that it is .0708 (see Fig. 7.4). The area *above* the point -1.47σ would be equal to $1.0 - .0708$, which is .9292. Or it can be found from column *B*, since it occupies the larger portion under the curve, and this also gives us .9292. Or, from Fig. 7.4, we can see that it is the sum of the area from the point to the mean (.4292) plus .5000, which gives the same result.

In the memory-test data, where the mean is 26.1 and *S* is 6.45, we may ask for the percentage of the cases to be expected below a score

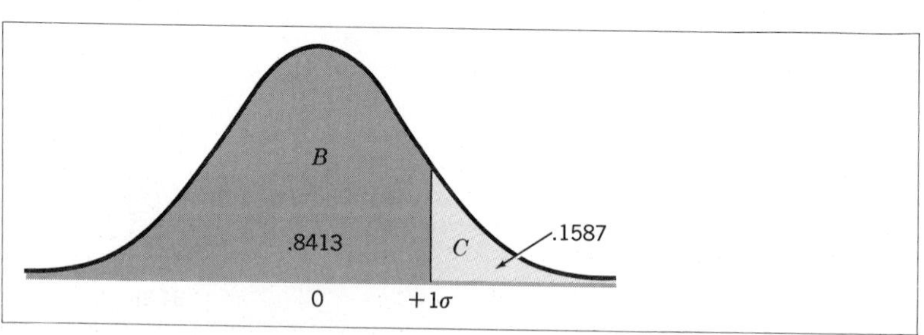

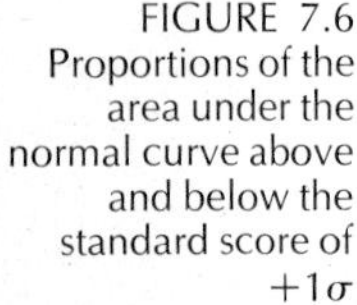
FIGURE 7.6 Proportions of the area under the normal curve above and below the standard score of $+1\sigma$

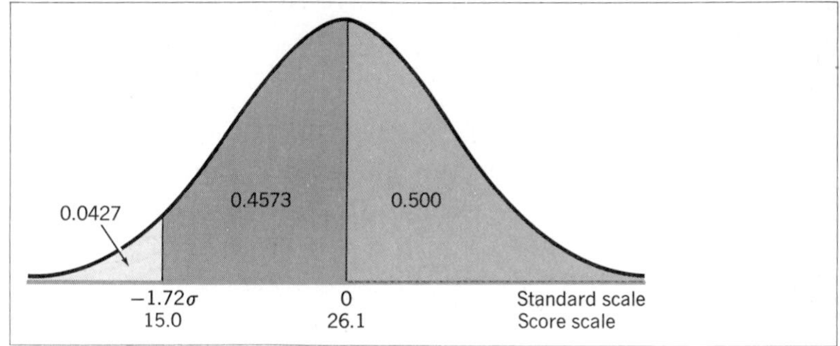

FIGURE 7.7 Proportions of the area under the normal curve above and below a selected score on the scale for the memory test

of 15. The deviation from the mean is 11.1. When this is divided by 6.45, we find that the z score is −1.72. Corresponding to a z of −1.72 is an area of .0427 in the tail of the normal curve (see Fig. 7.7). We may expect 4.27 percent of the cases below a score of 15, or, out of 86, 3.7 cases. Above a score of 15, we should naturally expect the remainder of the cases, i.e., a proportion of .9573, a percentage of 95.73, and in number of cases 82.3.

POINTS ABOVE OR BELOW WHICH CERTAIN PROPORTIONS OF THE CASES FALL

The next problems reverse the processes that have just been described. Before, we had given points on the scale of measurement to determine areas; now we have given areas from which to determine points on the scale. For example, above what point in the normal curve do the highest 10 percent of the cases come? Ten percent is a proportion of .10. We could now use Table B in the Appendix in reverse, but it is much more convenient to utilize Table C, which gives the proportions in even steps. We are faced with a problem that gives the proportion in the tail of the curve, and so we look in the last column for C, the smaller area. We find the z score corresponding to it to be 1.2816. This should be with a plus sign, since we are talking about the highest 10 percent (see Fig. 7.8). Had we asked below what point the *lowest* 10 percent fall, the answer would have been -1.2816σ. If the question is, "Above what score do the highest 80 percent of the cases lie?" we are dealing with the larger proportion under the curve; accordingly, we

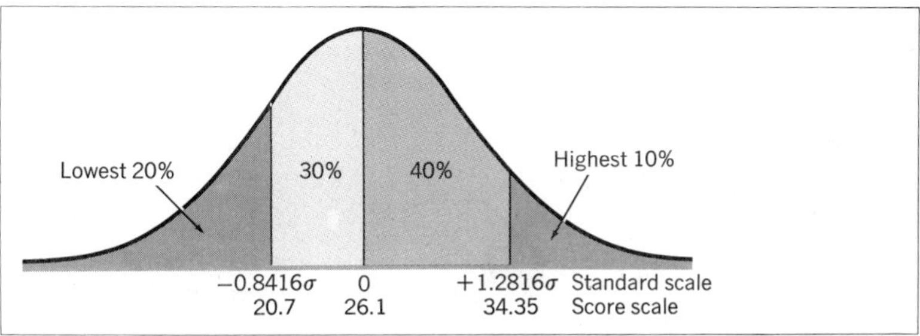

FIGURE 7.8 Score points above or below which certain percentages of the cases are expected in the memory-test distribution, assuming normality of distribution

look for the proportion of .80 in the first column of Table C. The corresponding z score is -0.8416 (see Fig. 7.8). Had we asked for the point below which the *lowest* 80 percent are, the answer would have been $+0.8416\sigma$.

To apply these same questions to the memory-test data, we must go a step further and transform the z scores into terms of the raw-score scale. The highest 10 percent come above a z of $+1.2816$. Multiplying this by S (which is 6.45), we obtain the deviation (x) of $+8.27$. The mean (or 26.1) plus 8.27 gives us a score of 34.37 points. The highest 10 percent in a near-normal curve with mean of 26.1 and S of 6.45 would therefore be expected above the point 34.37. It happens that this point comes close to the division point between two class intervals, or 34.5. In the actual distribution (see Table 3.6), 10 cases, or close to 12 percent, were scores of 35 or above, which is good agreement. Ten percent would have called for 8.6 cases, or 9 in whole numbers.

Above what raw score may the highest 80 percent of the cases, which we found to come above a z score of -0.841, be expected? The deviation of this point from the mean is -5.43 points, or a score of 20.67. This comes close to another division point between class intervals, namely, 20.5. In the actual distribution, 71, or 82.5 percent, of the cases are above a score of 20.5. Again the agreement between obtained proportion and expected proportion is quite close. To take one more case, which yields a point exactly between class intervals, we ask, "Above what point are 93.2 percent of the cases?" The point turns out to be a score of 16.5 units (the student should verify this). The actual percentage of cases above this score point is 92—again a very close agreement.

Other tests and other mathematical models

In this chapter we have dealt with some fundamental matters which are basic to the concepts of the several chapters to follow. We shall see how the principles of probability and the binomial and normal distributions are used in drawing conclusions about "true" means, standard deviations, correlation coefficients, and the like, and we shall learn how much confidence we should have in those conclusions. We shall see how we can decide whether obtained differences between such statistics indicate any "real" differences in the phenomena under investigation. We shall also find that additional models are needed to make certain tests of hypotheses, including Student's t distribution, the F distribution, and the chi-square distribution, which open up a whole new set of possibilities of procedures for use by the scientific investigator and the technical worker.

EXERCISES

1. In drawing one card from a shuffled deck, what is the probability of drawing:
   a. A jack?
   b. A club?
   c. A red card?
   d. A black heart?
   e. A spade that is not a face card (jack, queen, or king)?
2. In tossing four coins, what is the probability of obtaining:
   a. The sequence H T H T?
   b. Four heads?
   c. Three tails?
   d. Four heads *or* three tails?
   e. No heads in two successive tosses?
3. In tossing two dice, what is the probability of obtaining:
   a. A 2 and a 3?
   b. A 2 and then a 3?
   c. A total of five spots?
   d. On repeating a throw of two dice (i.e., throwing two dice twice in succession), what is the probability of repeating exactly seven spots?
4. A clinical psychologist is given Rorschach (inkblot-test) protocols for five patients and told that two are schizophrenics and three are neurotics.
   a. In how many different ways could the psychologist assign the two schizophrenics to the five protocols by random selection?
   b. What is the probability of the psychologist being completely correct by guessing?
5. For the ESP experiment, with 25 trials and a probability of $1/5$ of being correct in each trial, write the appropriate binomial (without expanding it).
6. What is the mean and the standard deviation of the binomial distribution for the preceding problem?
7. Toss a set of six pennies 64 times. After each throw, note and record the number of heads. Compare your obtained frequencies with the expected frequencies (seen in Table 7.1). Plot a bar diagram for the obtained binomial distribution and a frequency polygon for the expected distribution. Compute the mean and standard deviation of your obtained distribution and of the expected distribution.
8. Determine the standard scores (z) for the following scores in the distribution for Data 7*A*: 40, 55, 72, 85.
9. Find the proportions of the areas under the unit normal-distribution curve between the mean and the following z scores: −2.15, −1.85, −0.19, +0.375, +1.10.
10. Assuming a normal-distribution model, find the proportions and numbers of cases to be expected between the mean and the following scores in Data 7*A*: 45, 65, 75, 58.35.
11. Find the proportions of the area in the unit normal distribution *above* the following scores: +2.15, +1.62, +0.175, −0.36, −1.90; also *below* the following scores: −1.225, −0.6745, +0.05, +1.75, +2.30.

12 Find the proportions and numbers of cases to be expected in distribution 7A *above* the following score points: 55, 65, 69.5, 41.5; also *below* the following score points: 45, 56, 77.5, 61.5. Where possible, compare these frequencies with the obtained frequencies.

13 Give in terms of standard measurements z the points *above* which the following percentages of the cases fall in the unit normal distribution: 85, 55, 35, 42.3, 9.4.

14 Give the z scores *below* which the following proportions of the cases fall: .14, .62, .375, .418, .729.

15 *Below* what score points in distribution 7A should we expect the following numbers of cases: 11, 63, 123, 162? Compare with actual cumulative frequencies.

DATA 7A
Distribution of spelling-test scores in a superior group of freshmen*

Scores	*f*
82–85	1
78–81	8
74–77	8
70–73	5
66–69	34
62–65	21
58–61	39
54–57	32
50–53	20
46–49	7
42–45	3
38–41	0
34–37	1
Sum	179
Mean	61.1
S	8.4

* The test was one of the Cooperative series, and the scores are *T* scores (see Chap. 19).

ANSWERS

1 a. 1/13; b. 1/4; c. 1/2; d. 0; e. 5/26.
2 a. 1/16; b. 1/16; c. 1/4; d. 5/16; e. 1/256.
3 a. 1/18; b. 1/36; c. 1/9; d. 1/36.
4 a. 10; b. 1/10.
5 $(1/5 + 4/5)^{25}$.
6 Mean: 5.0; SD: 2.0.
8 −2.51, −0.73, +1.30, +2.84.
9 p: .4842, .4678, .0753, .1461, .3643.
10 p: .4726, .1786, .4510, .1282.
f: 84.4, 32.0, 80.7, 22.9.

11 p above: .0158, .0526, .4306, .6405, .9713.
p below: .1104, .2500, .5199, .9599, .9893.

12 p above: .7660, .3214, .1587, .9902.
f above: 137.1, 57.5, 28.4, 177.2.
p below: .0276, .2720, .9745, .5191.
f below: 4.9, 48.7, 174.4, 92.9.

13 z: −1.0364, −0.1257, +0.3853, +0.1942, +1.3165.

14 z:−1.0803, +0.3055, −0.3186, −0.2070, +0.6098.

15 X_e: 48.1, 57.9, 65.2, 72.0.
f_e: 11, 63, 123, 163.
f_o: 9, 67, 121, 160.

8 Statistical estimations and inferences

In this chapter we raise the very important question, "How near the 'truth' are statistical answers such as means, standard deviations, proportions, and the like?" As was said in Chap. 1, measured samples are often employed to represent larger populations. From the statistical point of view, a population is any arbitrarily defined group. The term will be more fully explained in later paragraphs.

Sampling has to be limited for practical reasons; ordinarily we cannot measure total populations, or at least it is generally inefficient and unnecessary to do so. Yet we usually wish to generalize beyond our sample, arriving at scientific decisions that transcend the observations made at a particular time and in a particular place, or reaching administrative decisions that apply to larger groups of individuals. In earlier chapters we have been concerned primarily with *descriptive statistics*. The computed values were used to describe the properties of particular samples. If we want to apply those same statistics beyond the limits of samples, we should like to know how much risk we take of being wrong. In general terms, the statistics stressed in this chapter are designed to do that very thing. They are known as *sampling statistics*.

To be more specific, when we obtain the mean of a sample that is measured in some respect, before we say that this obtained mean also describes the mean of the population sampled, we need to find some basis for believing that it does not deviate very far from the population mean. Fortunately, there is a statistical procedure that will inform us about how far our obtained mean can deviate from the population mean, provided certain conditions, to be explained later, have been satisfied. The statistic that will do this is known as the *standard error of the mean*. In a similar manner, there are standard errors of other sample statistics—medians, standard deviations, proportions, correlation coefficients, and the like—which inform us of the accuracy of our obtained figures as estimates of the corresponding population values. Examples of sampling distributions of means and correlation coefficients may be seen in Figs. 8.2 and 8.5.

Some principles of sampling

Before going into the treatment of sampling statistics, it is necessary to understand clearly the essential facts about the process of sampling. The application of sampling statistics depends upon certain *conditions* of sampling. If these are not satisfied, standard errors, no matter how accurately computed, may give wrong impressions. At best, they give us only estimates on the basis of which we can make decisions and draw conclusions, never with complete conviction but with various degrees of assurance.

POPULATIONS AND SAMPLES

It is time that we had a better definition of *population*. Some statisticians call it *universe*. In any case, the statistician's idea of population is quite different from the popular idea. Rarely would any statistical study regard the entire population of a nation, a city, or some geographic region as its *universe*.

The population in a statistical investigation is always arbitrarily defined by naming its unique properties. It might be the entering freshman class in a certain university or a part of the freshman class entering a certain college or even a certain course. It might be the male sixteen-year-olds in a given school district, the children of Mexican parentage in a certain city, or the registered Democratic voters in the New England states. All these examples are of groups of human individuals. Populations could, of course, be defined as families, species, or orders of animals or of plants.

There are also populations of observations or of reactions of a certain kind—simple reactions to sound stimuli, word-association reactions, judgments of pleasantness of colors, and the like—from the psychological laboratory. It is probably the nonhuman groups that have seemed to require the more general term *universe* as an alternative to the more restricted term *population*. In this volume we shall use the term *population* in the broad sense to include all sets of individuals, objects, or reactions that can be described as having a unique combination of qualities.

PARAMETERS AND STATISTICS If we were to measure all the individuals of a population and actually compute the indices of central value, dispersion, and correlation, as we ordinarily do for samples, we should obtain what the statistician calls *parameters*.

FIGURE 8.1
A comparison of a population distribution and a sample distribution, also of population parameters and sample statistics

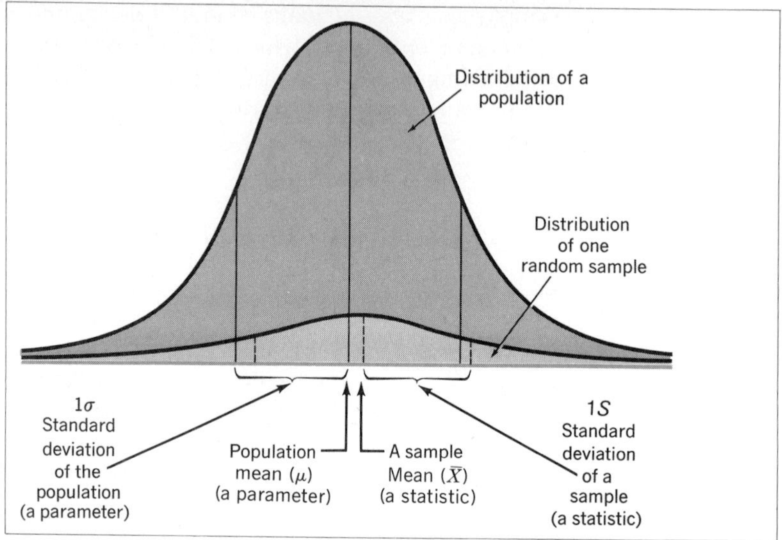

Figure 8.1 illustrates the distinction between population parameters and sample statistics. The larger distribution is that of the entire population. The smaller distribution is a sample drawn at random from that population. The population parameters, mean and standard deviation, are symbolized by μ and σ. It will be noted that in this particular sample the mean X and standard deviation S do not coincide exactly in size with their corresponding parameters μ and σ. This is characteristic. A second sample should be expected to have still different X and S, but also similar to μ and σ in size.

RANDOM SAMPLING It should be kept in mind that the use of sampling statistics (standard errors and the like) rests upon the assumption that the sampling has been random. Random sampling is the selection of cases from the population in such a way that every individual in the population has an equal opportunity to be chosen. In addition, the selection of any one individual is in no relevant way tied to the selection of any other. Selections are independent of one another. This definition calls to mind a well-conducted lottery, picking selective-service numbers, coin tossing, throwing dice, and other operations that allow the "laws of chance" to operate freely.

For the most sophisticated approach, one uses tables of random numbers that have been published as aids to random sampling.[1] The numbers in such tables have been placed in sequence by a kind of lottery procedure. If individuals in a population have been numbered in sequence and are thus identified by numbers, selections can be

1 See Meredith, W. B. *Basic mathematical and statistical tables for psychology and education*. New York: McGraw-Hill, 1967.

made for a sample by following the random numbers in any systematic way. A random sample should be fairly representative of the population. If a sample is small, however, it may not be as representative as we should like.

BIASED SAMPLING In a biased sample there is a systematic error. Certain types of cases have advantages over others in being selected. The likelihood of individuals' being chosen differs from one to another. A common example of this in research is the voluntary return of questionnaires. The names of those who are to receive the questionnaire may have been randomly selected from a larger group, but of the 60 percent who return the questionnaires, a large number may be atypical or not very representative of the population. The personal features that determine whether or not individuals return questionnaires may have no relation to the content of these instruments, but, on the other hand, they may be biasing forces. For example, if the questionnaire asks for information that reflects favorably or unfavorably on the respondent, his associates, or his work, it is quite natural to expect that those with a "good" showing will be more inclined to reply than those with a "bad" showing. If the trait of cooperativeness, responsibility, or dependability of the respondent is involved in the data or even correlated with something wanted in the data, there is also a strong likelihood of bias.

A colossal example of biased sampling is that of the *Literary Digest* public-opinion poll during the 1936 presidential campaign. Several million postcard ballots were said to have been circulated, certainly anticipating a sample of most generous size. But the mailing lists were made up from telephone directories and automobile registration lists. It so happened that in the poll, a majority of the telephone subscribers and car owners voted for the candidate who lost, while the non-telephone subscribers and non-car owners voted at the polls in a more decisive way for the successful candidate. Among those who received postcard ballots there was also probably a selection as to which ones would be most likely to take the trouble to return the card. Those who were most discontented with things as they were and wanted a change would take the trouble to register a protest straw vote. Those who were contented or who felt somewhat secure as to the outcome would be less likely to return the card. This would also tend to make the vote appear to favor the losing candidate, who was running against an incumbent.

The scientific investigator must be eternally vigilant to the possibility of biased sampling. A good, systematic control of experimental conditions is designed to prevent biased samples or to make known their effects. Where there is less than customary experimental control of the observations, every possible effort should be made to find out the conditions under which the data were obtained. Thorough knowledge of the conditions should be a basis for deciding whether selection of cases has been biased. Knowledge of conditions is also essential for the sake of accurate definition of the population sampled.

STRATIFICATION IN SAMPLING One common procedure that is introduced in sampling to help to prevent biases and also to ensure a more representative sample is known as *stratification*. Stratification is a step in the direction of experimental control. It operates with subgroups of more homogeneous composition within the larger population.

A very common example is to be found in public-opinion-polling practices. Suppose the issue to be investigated is public attitude toward a certain piece of labor legislation. It is quite likely that people in the two major political parties would tend to lean in opposite directions on such an issue. It is probable that people in different socioeconomic categories—professional, business, office worker, semiskilled laborer, and unskilled laborer—would react with some systematic differences on the issue. It is possible, though not so likely, that individuals of the two sexes would tend to respond somewhat differently. Other divisions of the population, such as rural versus urban, regional, and educational groups, might also show systematic differences on the issue. In other words, subgroups of the population are considered with respect to any variable that is suspected of correlating appreciably with the variable being studied. It does not matter that some of the variables are themselves intercorrelated unless such an intercorrelation is very high, in which case it would be superfluous to control selection of samples on both of two variables so closely related.

Having decided which variables are important in sampling, the entire population is studied to see what proportions fall into each category, e.g., what proportions are Democrat or Republican, male or female, urban or rural, in each socioeconomic group, and so on. Any sample to be obtained, then, should have proportional representations from all subgroups. Within each defined subpopulation—for example, a male, professional, Republican, New England group—random sampling may then be carried out. Random selection of cases would also be made within each of the other defined subpopulations in appropriate numbers. The total sampling procedure here described is called *stratified-random sampling*.

The importance of the proportional-representation principle and its advantage over a purely random sampling can be readily demonstrated. Suppose that 55 percent of the Republicans and 45 percent of the Democrats are in favor of a certain labor bill. In the general population let us assume that 60 percent are registered Democrats and 40 percent are registered Republicans. In a random sample of 100 voters one should expect in the long run to draw the two party representatives in about the same ratio, 60:40. This would vary from sample to sample, however, even to the extent that the majority could be reversed; for example, it could even be 45:55. In the typical polling sample we should expect a majority of voters to be against the bill. If the sample should by chance contain a majority of Republicans, however, the majority might favor the bill. If stratification were applied, we should make sure that the ratio is 60:40, and with this restriction imposed upon the random sampling we should expect the general population sentiment to be more accurately reflected. Thus it can be seen that a stratified-random sample is likely to be more representative of a total population than a purely random sample.

PURPOSIVE SAMPLES A *purposive sample* is one arbitrarily selected because there is good evidence that it is very representative of the total population. Experience has shown in public-opinion polling that there are certain states or regions that come close to reflecting national opinion time after time. If one is willing to depend upon this experience, one may use the limited population as the source of the sample to employ as a "barometer" for the total population. This is a convenient procedure, but it has the disadvantage that much prior information must have been obtained. There is also a risk that conditions may change to the extent that the particular segment of the population no longer represents the total or does not represent it on some new issue.

INCIDENTAL SAMPLES The term *incidental sample* is applied to those samples which are taken because they are the most available. Many psychological studies have been made with utilization of students of beginning psychology as the samples merely because they are most convenient. Results thus obtained can be generalized beyond such groups with some risk.

Generalizations beyond any sample can be made safely only when we have defined the population that the sample represents in every significant respect. If we know the significant properties of the incidental sample well enough and can show that those properties apply to new individuals, those new individuals may be said to belong to the same population as the members of the sample. By "significant properties" is meant those variables which correlate with the experimental variables involved. They are the kind of properties considered above in connection with stratification of samples. It is unlikely that membership in a political party would have much bearing upon the results of certain experiments performed upon sophomores in a beginning psychology course, but such variables as age, education, social background, and the like may definitely be pertinent.[1]

1 Extensive discussions of problems of sampling can be found in Edgington, E. E. *Statistical inference: The distribution-free approach.* New York: McGraw-Hill, 1969.

Inferences regarding averages

THE DISTRIBUTION OF MEANS OF SAMPLES Suppose that we are dealing with a population whose mean (μ) is 50.0 and whose standard deviation (σ) is 10.0 on the measuring scale we are using. Such a distribution is illustrated by the top diagram in Fig. 8.2.

SAMPLING DISTRIBUTIONS Suppose, next, that we proceed to draw random samples, all of equal size, from this population one at a time. To satisfy the conditions of random sampling in a strictly mathematical sense, we should replace each member drawn, after noting its value, before drawing the next member. Each individual should have an equal opportunity of being selected in *every* drawing. Having lost one member, the population is different from what it was originally. When

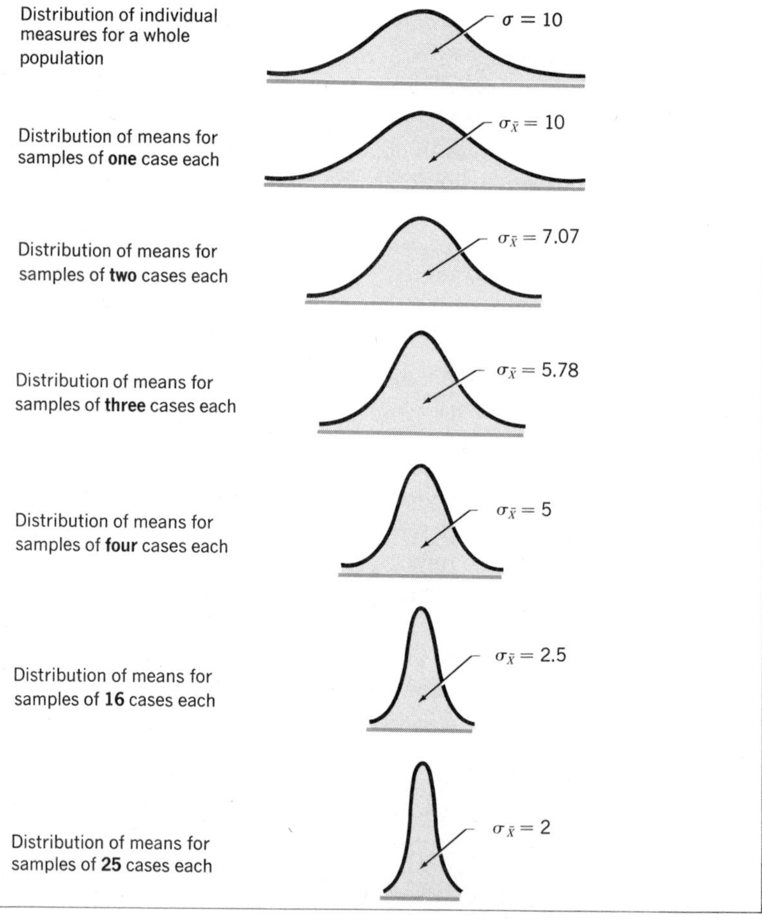

FIGURE 8.2 The hypothetical decrease in variability or fluctuation of the means of samples as we increase the size of the sample drawn from a large population. (Modified from Lindquist, E. F. *A first course in statistics.* Boston: Houghton Mifflin. By permission.)

the population is very large, as compared with the size of sample, however, we can forget about this *replacement* requirement for practical purposes. In this case, one member of the population would hardly be missed; that is, its loss would change the chance conditions only to an inconsequential degree.

To take a specific example of random sampling, with the same population described above in mind, let the size of each sample be 25. The sample mean not only will differ from sample to sample but also will usually deviate from the population parameter (in this example, the mean of 50.0). If we have a number of such sample means, we may treat them just as if each were a single observation and set up a frequency distribution of them. This is known as a *sampling distribution*. Such a frequency distribution will be close to the normal form when the population distribution is not seriously skewed and when N is not small (i.e., not less than about 30).

Normality of distribution of single cases in the total population favors normal distribution of means and of certain other statistics computed from samples drawn from that population. Even when the population distribution departs from normality, however, the distribution of means of samples drawn from it tends to be normal, unless the samples are too small. The smaller the sample, the more the form of distribution of the population affects the form of distribution of the means.

A knowledge of the form of sampling distribution of a statistic is very important. Our ability to draw conclusions known technically as *statistical inferences* depends upon our knowledge of the form of distribution of sample statistics. Without knowledge of the form of sampling distribution, many a scientific result would remain inconclusive. This is where the use of mathematical models comes in. The reasons for this will become clearer as we go into the subject of interpretation of standard errors.

THE STANDARD ERROR OF A MEAN

At this stage of getting acquainted with sampling distributions, we are most interested in the dispersion of statistics, in this case, the dispersion of sample means. The reason is that the amount of this dispersion gives us the clue as to how far such sample means may be expected to depart from the population mean. If we are to use a sample mean as an estimate of the population mean, any deviation of such a sample mean from the population mean may be regarded as an error of estimation. The standard error of a mean tells us how large these errors of estimation are in any particular sampling situation. *The standard error of a mean is a standard deviation of the distribution of sample means.* To distinguish such a standard deviation from the more familiar one that applies to dispersions of individual observations, we call it a *standard error*. In later discussions it will often be referred to by the abbreviation SE.

In order actually to *compute* the standard error of a mean, we need two items of information: the population parameter σ and the size of sample N. Since we do not ordinarily know σ, it would seem that we could very rarely compute the standard error. There are satisfactory ways of *estimating* it, however, as we shall see later. The formula for *computing* the standard error of a mean is

$$\sigma_{\bar{X}} = \frac{\sigma}{\sqrt{N}} \quad \text{(Standard error of an arithmetic mean computed from a known population parameter)} \tag{8.1}$$

where σ = standard deviation of the population and N = number of cases in the sample (not the number of means in a distribution of means).[1]

SAMPLE SIZE AND THE STANDARD ERROR OF A MEAN The standard error of a mean is therefore *directly* proportional to the standard deviation of the population and *inversely* proportional to the size of the sample. More precisely stated, $\sigma_{\bar{X}}$ is inversely proportional to the square root of the size of sample.

In Fig. 8.2 are shown graphically several instances of samples when N varies. The smallest possible sample occurs when $N = 1$. The mean of each sample is then identical with the measurement in that sample. The dispersion of the means is then as great as the dispersion of the whole population; $\sigma_{\bar{X}}$ then equals σ, which we have assumed to be 10. When each sample contains two cases, $\sigma_{\bar{X}} = 10/\sqrt{2} = 7.07$; when each sample contains four cases, $\sigma_{\bar{X}} = 10/\sqrt{4} = 5$; and so on. The remaining cases in Fig. 8.2 should speak for themselves.

ESTIMATING THE STANDARD ERROR OF A MEAN FROM KNOWN STATISTICS

Formula (8.1) requires knowledge of the parameter σ in order to *compute* the standard error of the mean. In ordinary practice we must estimate it.

ESTIMATION OF $s_{\bar{x}}$ FROM S In describing a sample, we usually compute S as well as the mean. When S is thus known, we may estimate the parameter $\sigma_{\bar{X}}$ by the formula

$$s_{\bar{X}} = \frac{S}{\sqrt{N-1}} \quad \text{(Standard error of a mean estimated from } S\text{)} \tag{8.2}$$

The reason for the expression $N - 1$ in this formula can be better understood after we consider the next estimation method. It could be recommended that, for large samples (N of 30 or more), we simply substitute S for σ in formula (8.1), in which case we should have the ratio $S/\sqrt{N}$ instead of the ratio $S/\sqrt{N-1}$. To do this would be to overlook

[1] When necessary to divide by $\sqrt{N}$, it is simpler to look up $1/\sqrt{N}$ Table *A* in the Appendix and multiply.

the fact that S is actually a biased estimate of σ for samples of any size—the smaller the sample the greater the bias. There is no sudden change in this condition at an N of 30. The result of using formula (8.2) is identical with the result obtained with the next procedure, which is favored by many statisticians.

ESTIMATION OF σ FROM A SAMPLE The standard deviation computed for a sample is likely to be smaller than that for the population from which the sample came. Recall from the discussion in Chap. 5 that as samples become smaller, the total range of measurements is more and more curtailed. This is because extreme deviations in the population are rare and are likely to be missed in sampling. A similar effect occurs in the standard deviation although to a smaller extent. A mathematical reason is that the deviations used in computing S are departures from a sample mean rather than from the population mean. If they were taken as departures from the population mean they would be larger. In smaller samples particularly, S gives an estimate of the population σ that is biased downward.

A less biased estimate of σ is given by the formula

$$s = \sqrt{\frac{\Sigma x^2}{N-1}} \quad \text{(Best estimate of population standard deviation)} \qquad (8.3)$$

where Σx^2 = sum of squares of deviations in the sample and N = number of cases in the sample.

DEGREES OF FREEDOM Formula (8.3) involves an important concept that will be liberally utilized hereafter when sampling errors (deviations of statistics from parameters) are mentioned, particularly in connection with small samples.

Compare formula (8.3) with the one given earlier for the standard deviation of a sample [formula (5.4)], and it will be found that they differ only in the denominator terms, $(N - 1)$ and N, respectively. The difference between the two may seem very slight, and it is slight when N is reasonably large, but there is a very important difference in meaning. In this particular formula, $(N - 1)$ is known as the number of *degrees of freedom,* which is symbolized by df. This is a key concept in what is known as *small-sample statistics.* The number of degrees of freedom will not always be $(N - 1)$ but will vary from one statistic to another, as will be pointed out in various places later. Let us see why the df is $(N - 1)$ here.

The "freedom" part of the concept means *freedom to vary.* The standard deviation is computed from the variance, and the variance is computed from deviations from the mean. Statisticians often express the matter by saying that 1 degree of freedom is "used up" when we employ the sample mean as an estimate of a parameter mean. This leaves $(N - 1)$ degrees of freedom for estimating the population variance and standard deviation.

A numerical example will make this clearer. Let us assume five measurements: 5, 7, 10, 12, and 16, the mean of which is 10.0. We now use this value as an estimate of the population mean. A mathematical requirement or property of the arithmetic mean is that the sum of the deviations from it shall equal zero. The five deviations in this sample are −5, −3, 0, +2, and +6, the sum of which is zero. With this condition satisfied, i.e., the sum equal to zero, how many of these deviations could be simultaneously altered (as if by taking new samplings) and still leave the sum equal to zero? With a little thought or trial and error it will be seen that if any four are arbitrarily changed, the fifth is thereby fixed. We could make the first four −8, −4, +1, and −2, which would mean that for the sum to equal zero, the fifth has to be +13. Try any other changes, and if the sum is to remain zero, one of the five deviations is automatically determined. Thus only four $(N - 1)$ are "free to vary" within the restriction imposed.

There were N degrees of freedom in computing the mean because the cases were presumably sampled entirely independently. If they were not independently sampled, there were less than N degrees of freedom in computing the mean. Freedom means independence, and only when there is independence of observations can the "laws of chance" operate freely and the mathematics based upon them be applied.[1]

THE SE OF A MEAN ESTIMATED DIRECTLY FROM A SUM OF SQUARES Whether we precede the estimation of the standard error of a mean, $s_{\bar{X}}$, by computing S or s from the sample, we find ourselves performing the same steps, but in a different order. These steps are dividing by $(N - 1)$ and by N. If we should happen to have no interest in knowing the value of either S or s, we can combine these two operations in a single equation, and we have the formula

$$s_{\bar{X}} = \sqrt{\frac{\Sigma x^2}{N(N-1)}} \quad \text{(Standard error of a mean estimated directly from a sum of squares)} \qquad (8.4)$$

A RECAPITULATION In order to emphasize the distinctions that have just been made, let us recapitulate, with definitions of the various symbols.

σ = a standard deviation of a population; a parameter.

S = a standard deviation of a sample from the population, computed by the expression $\sqrt{\Sigma x^2/N}$; a statistic.

1 For an excellent discussion of the general subject of degrees of freedom, see Walker, H. M. Degrees of freedom *Journal of Educational Psychology*, 1940, **31**, 253–260. See also Johnson, P. O. *Statistical methods in research*. Englewood Cliffs, N.J.: Prentice-Hall, 1949.

s = estimated standard deviation of the population, derived from a sample by means of the expression $\sqrt{\Sigma x^2/(N-1)}$; a statistic.

$s_{\bar{X}}$ = standard error of a mean, estimated by the expression $\sqrt{\Sigma x^2/N(N-1)}$
or, equivalently by $S/\sqrt{N-1}$ or by $s/\sqrt{N}$.

Three kinds of distributions are involved in connection with these values. The quantity σ describes the dispersion of the distribution of the entire population of measurements (of persons or of observations). The quantity S describes the dispersion of the distribution of a particular sample drawn from the population. The quantity s pertains to the same sample distribution and is computed from it, but it is used to estimate σ, the dispersion in the population distribution. The quantity $s_{\bar{X}}$ pertains to an entirely different distribution—a *sampling* distribution of all the means of the samples of a certain size that would be drawn randomly from the population. Figure 8.1 shows a sample distribution and a population distribution, each composed of single measurements. Figure 8.2 shows some *sampling* distributions, all (except the first one) composed of means of samples.

INTERPRETATION OF A STANDARD ERROR OF A MEAN

We are now ready to apply the standard-error formula to a concrete instance and to consider the interpretation of the obtained SE. To revive an example already employed, the inkblot data, we find that S is 10.45 and N is 50. From formula (8.2), $s_{\bar{X}} = 10.45/\sqrt{49} = 1.49$. For simplicity in discussion, let us round this figure to 1.5.

When we estimate this standard error, we are essentially asking, "How far from the population mean are the sample means, like this one we obtained, likely to vary in random sampling?" We do not know what the population mean is, but from the value 1.5 we conclude that means of samples of 50 cases each would not deviate from it in either direction more than 1.5 units about two-thirds of the time. We may conclude this because in a sample as large as 50, we may assume that the sample means are normally distributed. This assumption makes possible a number of inferences that we could not make without it. Remember that even when the population distribution is not normal, the means derived from sampling from the population are likely to be almost normally distributed.

Since, as we have already seen, in these inkblot data we may conclude that two-thirds of the sample means (when N is 50) will lie within 1.5 units, plus or minus, of the population mean, we can also say that there is only 1 chance in 3 for a sample mean to be further than 1.5 units from the population mean in either direction. Or we can say that the odds are 2 to 1 that sample means will be within a range of 3 units, the middle of which is the population mean. The standard error thus brackets a range within which to expect sample means. We shall expand this idea in the discussion to follow.

HYPOTHESES CONCERNING THE POPULATION MEAN The kind of conclusion that we should most like to make is slightly different from the one just given. We are attempting to estimate the population mean, knowing the sample mean. We should therefore like to know how far away from the sample mean the population mean is likely to be.

It might seem that if we can say that two-thirds of the sample means are within one SE of the population mean, we could also say that the odds are 2 to 1 that the population mean is within one SE of the sample mean. But note that the last statement implies a normal distribution about the *sample* mean, whereas, actually, the sampling distribution is about the *population* mean. Logically, we cannot reverse the roles of μ and $\overline{X}$ in this manner. But through some mathematical reasoning, which we can examine only briefly here, we can do something equivalent. The process results in setting up *confidence limits* and *confidence intervals* for the population mean.

Since we do not know the population mean, we are free to make some guesses, or hypotheses, about its value. No matter what reasonable hypothesis we choose, the estimated standard error still applies to the distribution of expected sample means about this hypothetical value.

In the inkblot problem, the sample mean was 29.6. Let us select in turn a number of hypothetical population means. They should, of course, be somewhere in the neighborhood of the sample mean. Figure 8.3 shows five normal sampling distributions, each about a different hypothesized μ and each with a standard error of 1.5. The hypothesized means are all above 29.6; they could just as well have been chosen below that value. They are at the values 30.0, 31.0, 32.0, 33.0, and 34.0.

Consider first the hypothesis that is furthest from the sample mean, namely, a hypothetical μ of 34.0. A sample mean of 29.6 deviates 4.4 score points from this hypothetical μ. This deviation corresponds to a z (standard-score value) of 4.4/1.49, or 2.95. We may enter the normal-curve tables with such a value.

We next ask what the probability is of a deviation *as large as this* occurring by random sampling. This probability is twice the proportion of the area under the tail of the normal curve beyond the point at

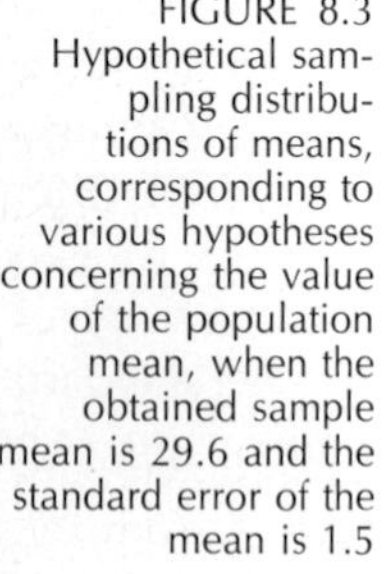
FIGURE 8.3 Hypothetical sampling distributions of means, corresponding to various hypotheses concerning the value of the population mean, when the obtained sample mean is 29.6 and the standard error of the mean is 1.5

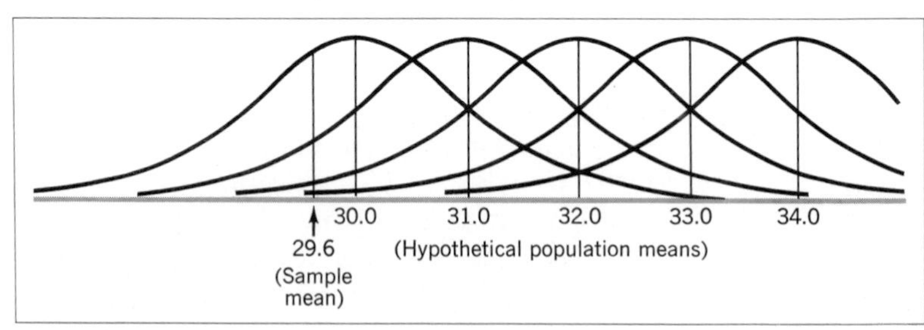

$z = 2.95$. When we say "a deviation as large as this," we actually mean a deviation as large *or larger*. For a deviation of 4.4 we could just as well have chosen a hypothetical μ that far *below* the obtained mean, in other words, at $29.6 - 4.4$, or a score of 25.2. The obtained mean is $2.95z$ above this hypothetical mean, just as it is $2.95z$ below the hypothetical mean of 34.0. Thus we have two hypothetical normal distributions to consider. The area under the tail of the normal distribution beyond a z of 2.96 is .0016, as found from Table B in the Appendix. In the two tails the area is twice .0016, or .0032. We can conclude that if the population mean is as far removed from the obtained mean as 34.0, there is only the slim chance of 32 in 10,000 that a mean of 29.6 will occur by random sampling. Since these odds are so small, we can reject with confidence the hypothesis that the population mean is 34.0 or that it is 25.2.

The next hypothesis is that the μ equals 33.0, which gives a deviation of 3.4 and a z of 2.28. The area under the normal curve beyond this point is .0113. Twice this area is .0226. If the population mean were actually as far removed as 33.0, there are only about 2 chances in 100 for a departure such as a sample mean of 29.6 to occur. If we reject this hypothesis there would be only 2 chances in 100 that we would be wrong. Although we cannot reject this hypothesis with as much confidence as we could the previous one, we can still do so with a high level of assurance.

If we hypothesize a μ of 32.0, the deviation is 2.4, and the tail area (doubled) is .1074. The chances for a random deviation that large is more than 10 in 100. If we hypothesize that $\mu = 31.0$, the deviation is 1.4, z is 0.94, and the probability for so large a deviation is .348. We cannot very well reject the hypothesis that the population mean is 31.0. There would be considerable risk in doing so. In fact, we could say that this hypothesis is rather plausible.

But other hypotheses are even more plausible. If we choose the hypothesis that μ is 30.0, the deviation is 0.4, z is 0.267, and the area beyond this deviation is .788 of the total. Thus, as we approach the sample mean with our hypothetical population mean, the odds in favor of greater deviations than the obtained one keep increasing, and the hypothesis brcomes more and more plausible. The maximum plausibility would be reached when the hypothesis is 29.6, in other words, when it coincides with the sample mean. From this point of view, we can say that the sample mean (when other information is lacking) is the most probable and the most defensible estimation of the population mean. It is an unbiased estimate, since the deviations are as likely to be positive as negative.

CONFIDENCE LIMITS AND CONFIDENCE INTERVALS From this discussion it is clear that we are confronted with a sliding scale of confidence with respect to location of the population mean. Values remote from the

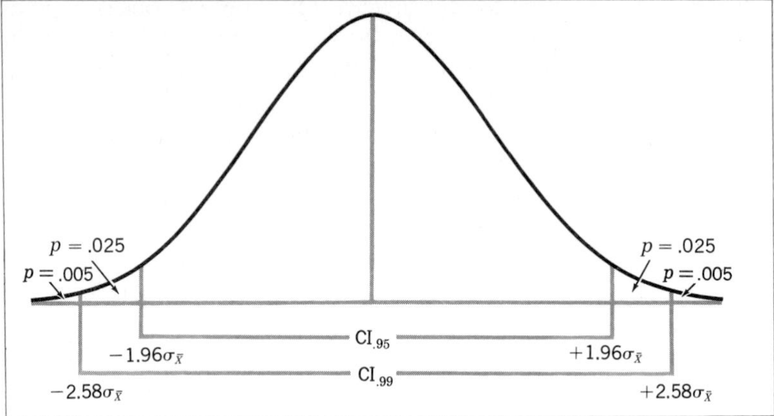

FIGURE 8.4 A normal distribution showing the ranges of the confidence intervals at the .95 and .99 probability levels, with limiting points in terms of standard-error distances from the obtained mean

sample mean can be rejected with much confidence; as we approach the sample mean, hypothetical values can be rejected with less and less confidence. It is not customary to do all the calculations that we have just done merely to interpret the value of an obtained mean and its standard error. By common consent an arbitrary choice has been made to adopt two particular *levels of confidence*. One is known as the *5 percent level* or .05 level and the other as the *1 percent* or .01 level.

At the .05 level there is a deviation that leaves 5 percent of the area in the two tails of the normal distribution; 2.5 percent in each tail. In large samples (N as large as 30), this area at either end is marked off at a z value of 1.96. See Fig. 8.4, where the intervals and the tail probabilities (p) are marked off, as are the confidence intervals $\mathrm{CI}_{.95}$ and $\mathrm{CI}_{.99}$. The .01 level leaves 1 percent of the area in the two tails, .5 of 1 percent in either tail. The z that marks off this much at either end is 2.58. These percentages and these z values are applied regardless of the size of the mean and of its standard error. It must be remembered, however, that they apply only to large samples.[1]

For the inkblot data a z of 1.96 corresponds to a score deviation of 2.9, which is 1.96 times $s_{\bar{X}}$. All hypotheses of population means differing more than 2.9 from the sample mean can be rejected at the .05 level. Only once in 20 times would we be in error by making this decision. (This once would be when the deviation is *really* due to chance.) Since these *confidence limits* are 2.9 units from the sample mean, they come at score values of $29.6 - 2.9$ and $29.6 + 2.9$, or at 26.7 and 32.5, respectively. The score limits of 26.7 and 32.5 mark

[1] For testing such hypotheses, and for determining confidence intervals with small samples, Student's t and the t distribution should be used in place of z and the normal distribution (see Chap. 9 for a discussion of t and its distribution).

off a *confidence interval* within which the population mean probably lies. The probability to be associated with this interval is .95 (i.e., 1.00 − .05). See Figs. 8.3 and 8.4.

We can make a similar interpretation in connection with the .01 level. All hypothetical means differing more than 3.9 (3.9 is 2.58 times $\sigma_{\bar{X}}$) from the sample mean can be rejected, with only 1 chance in 100 of being wrong in doing so. The confidence interval is from 25.7 to 33.5, and the probability to be associated with it is .99. We have a high degree of assurance that the population mean is between 25 and 34. The odds are 99 to 1 in favor of this conclusion. Whether we wish to stake our case on the .05 limits or the .01 limits depends upon our inclinations. The next chapter includes a more extended discussion of the choice of standards of confidence.[1]

From the arithmetic that has been done in recent paragraphs, it can be seen that the confidence intervals for a mean can be set up by applying the following formulas:

$$\text{CI}_{.95} = \bar{X} - 1.96s_{\bar{X}} \text{ to } \bar{X} + 1.96s_{\bar{X}} \qquad (8.5a)$$
$$\text{CI}_{.99} = \bar{X} - 2.58s_{\bar{X}} \text{ to } \bar{X} + 2.58s_{\bar{X}} \qquad (8.5b)$$

(Confidence intervals for a mean)

where $\text{CI}_{.95}$ stands for the confidence interval at the .95 level and $\text{CI}_{.99}$ stands for the corresponding interval at the .99 level.

COMPARISONS OF SOME OBTAINED MEANS AND STANDARD ERRORS Let us apply the interpretation of $s_{\bar{X}}$ to some other data. The practical usefulness of an obtained statistic is often more apparent when we compare it with its counterpart derived from different data. Table 8.1 gives means of Army General Classification Test scores for samples derived from different civilian occupational groups. For the sake of illustration, we shall assume that each occupational group represents a different population, as designated, and that sampling of scores was random within each population. What do the standard errors and confidence intervals in this table tell us?

The mean in which we would have the greatest confidence, as representing the status of the general occupational population, is that for the truck driver. The odds are about 2 to 1 that this sample mean of 96.2 does not deviate more than .7 from the mean of all truck drivers that this sample represents. There is confidence at the .95 level that the interval 94.8–97.6 contains that mean, and there is confidence at the .99 level that the interval 94.4–98.0 contains it: a total range of about 4 units. The mean in which we have least confidence is that for teamsters because of its $s_{\bar{X}}$ of 2.23. For teamsters, our confidence at the .99 level should tolerate the hypothesis of the population mean in the interval of 81.9 to 93.5, a range of nearly 12 units.

[1] Other confidence levels sometimes used are the 10 percent, or .10, level (when z is 1.65); the .02 level (when z is 2.33); and the .005 level (when z is 2.81).

TABLE 8.1 Comparisons of means and their confidence intervals on the Army General Classification Test as applied to men from different civilian occupational categories*

Occupation	N	$\overline{X}$	S	$s_{\overline{X}}$	$CI_{.95}$	$CI_{.99}$
Accountant	172	128.1	11.7	0.88	126.4 – 129.8	125.8 – 130.4
Lawyer	94	127.6	10.9	1.13	125.4 – 129.8	124.7 – 130.5
Reporter	45	124.5	11.7	1.76	121.1 – 127.9	120.0 – 129.0
Sales clerk	492	109.2	16.3	0.74	107.7 – 110.7	107.3 – 111.1
Plumber	128	102.7	16.0	1.42	99.9 – 105.5	99.0 – 106.4
Truck driver	817	96.2	19.7	0.69	94.8 – 97.6	94.4 – 98.0
Farm hand	817	91.4	20.7	0.72	90.0 – 92.8	89.5 – 93.3
Teamster	77	87.7	19.6	2.23	83.3 – 92.1	81.9 – 93.5

* From Harrell, T. W., and Harrell, M. S. Army General Classification Test scores for civilian occupations. *Educational and Psychological Measurement,* 1945, **5,** 229–240. By permission.

Incidentally, the relation of $s_{\overline{X}}$ to both S and N can be seen roughly by comparison of the data for the occupational groups. On the whole, the largest standard errors come for samples where N is smallest—for lawyer, reporter, plumber, and teamster—though the rank orders are not perfect within this list of four. Where sample sizes are comparable, as for lawyer and teamster, and for accountant and plumber, the value for $s_{\overline{X}}$ is more apparently in proportion to the standard deviation of the sample. It can be seen that if the sample is sufficiently large, the standard error can be brought below one scale unit.

SOME SPECIAL PROBLEMS CONCERNING THE SE OF A MEAN

We shall now consider several conditions that have a bearing upon the standard error and the steps that may be taken to deal with them.

WHEN SAMPLING IS NOT RANDOM It has been repeatedly stressed that sampling statistics, including standard errors, apply only when sampling has been random. The reason is that the mathematics of the situation apply only when sampling has been random. It is then that we may use a chance-generated mathematical model, such as the normal distribution. Any condition that tends to interfere with randomness of selection of observations will therefore make the estimation of standard errors and their application in drawing conclusions inaccurate, if not misleading.

There are several noteworthy situations that depart from the random requirement. Some would lead to standard errors that are too small to describe the actual distributions of means, and others would lead to standard errors that are too large. In the former case, we should have too much confidence in the accuracy of the mean, and in the latter we should have too little. Certain variations in the standard-error formulas have been developed to take care of some of these special situations.

SAMPLES WITH BIAS The effect of biased sampling upon the distribution of means can be strikingly illustrated by reference to some data on the training of pilots in the AAF during World War II. All pilot students were given a battery of classification tests from which was derived for each man a "pilot stanine," or composite pilot-aptitude score. At the completion of preflight training, students were formed into class groups, and each was sent to a different primary flying school. In one study which covered a 5-month period, 269 such classes had been sent to 58 training schools divided among three AAF flying training commands. The mean stanine for approximately 52,000 students was 5.56. This value may be taken as the population mean in this situation. The standard deviation of the population was assumed to be 1.96, which should have been ensured by the use of the stanine scale. The average size of sample (each class group in a single school) was 195.[1] From this information, using formula (8.1), we compute a standard error of 0.14. From this we should expect two-thirds of the 269 mean stanines to deviate not more than 0.14 from 5.56, if the sampling had been random. What are the facts?

When the 269 means were actually compiled in a frequency distribution and their standard deviation computed, the dispersion of means was actually found to be very much larger than expected (see Table 8.2). Where one should expect a range of means within the limits 5.2 to 6.0, the actual range was from 4.6 to 6.9. Where the expected standard deviation of the distribution of means was 0.14, the actual standard deviation was 0.37. A comparison of the expected and obtained distribution of means is shown in Fig. 8.5.

The obvious conclusion is that the sampling of aviation students in pilot classes was most probably not random. One can surmise some of the causes after looking into the procedures by which class groups were formed. In each preflight class (i.e., each month) a small percentage of students would fail to pass the course and would be held over, probably to qualify for flight training in the next class. There was a tendency for the "holdovers" to be sent together to the same flight schools. They tended to be of low pilot aptitude. There may also have been some geographic differences in pilot aptitude which would tend to make the averages of stanines differ systematically somewhat from one command to another. This hypothesis could be subjected to experimental check by comparing command averages. There were probably other reasons for students of similar aptitudes to gravitate together, with a consequent biasing of samples.

1 Actually, some classes deviated from 195 in number. For the sake of illustration, however, we may treat the samples as if they were of constant size.

TABLE 8.2 Sampling statistics concerning 269 class groups of pilots in primary training during a period of 5 months in three training commands of the AAF during World War II*

Variable	*Expected results*		*Obtained results*		
	SE	*Range*	*Mean*	*Mean SE*	*Range*
Pilot stanine	0.14	5.2 – 6.0	5.56	0.37	4.6 – 6.9
Graduation rate	3.4	56 – 75	65.3	9.5	40 – 90
Validity coefficient	.073	0.32 – 0.74	0.53	.088	0.21 – 0.71

* Including the pilot stanine, or composite pilot-aptitude score; the graduation rate, or percentage of a class graduating; and validity coefficient, a biserial coefficient of correlation between stanine and graduation versus elimination.

Another study was made of the graduation rates (percentage of a class group graduating) in different samples. The pertinent data are given in Table 8.2. From the overall graduation rate of 65.3 and the size of sample, we should expect [by formula (8.11)] a standard deviation of the distribution of the 269 rates to be 3.4. Actually it was 9.5. Since the probability of graduation for any cadet was strongly correlated with his aptitude score, we should expect the bias in sampling on aptitude to be reflected in biased samples as to graduation rate. This is probably not the whole story, however. There were many other conditions which could contribute to marked variations in graduation rate besides variation in aptitude. Weather conditions varied from school to school and from month to month. Training practices and policies may have varied, in spite of close regulations. Instructor and test-pilot judgments were not standardized and may have varied from school to school.

A third study is mentioned now for comparison, although it involves the sampling errors of coefficients of correlation, which are treated later. This study is concerned with the variation in validity coefficients in the same 269 class groups. The validity of the pilot stanine for predicting the training success of pilots was indicated by what is known as the *biserial coefficient of correlation* (see Chap. 14). This has approximately the same value as a Pearson product-moment r, but is computed when one of the variables, assumed to be normally distributed, is forced into two categories. The two categories for the training criterion were those who graduated and those eliminated. The standard error for a biserial correlation equal to .53, when the size of sample is 195, amounts to .073. The expected and obtained statistics are given in Table 8.2 and illustrated in Fig. 8.5. In drawing the distribution curve, a normal distribution of the coefficients was assumed, whereas the expected distribution should be slightly negatively skewed. The obtained distribution of the 269 coefficients was actually so skewed. At any rate, since the obtained standard deviation (.088)

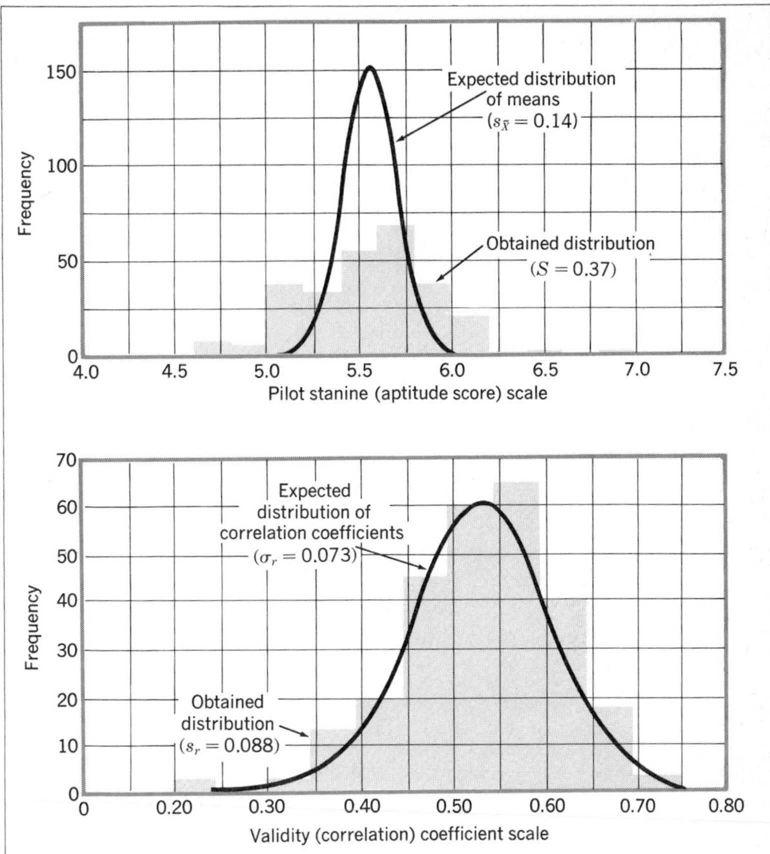

FIGURE 8.5 Distribution of expected and obtained sample means, and of expected and obtained validity coefficients, in connection with 269 samples (class groups) of AAF pilots in primary training during a five-month period in about 60 different schools. Especially to be noted is that the obtained distribution of means was much wider than expected, indicating nonrandom sampling, while the distribution of validity coefficients was about as expected, suggesting random sampling. This difference is possible because different kinds of sampling are involved for the two statistics.

was not so very different from the expected one (.073), we may conclude that if there was biased sampling with respect to the validity of pilot stanines, it was of minor importance.

Although there were seemingly enormous variations in validity from school to school and from time to time, amounting to a total range of .21 to .71, such variations may be regarded as due mostly to sampling errors. Incidentally, this example shows by just how much obtained correlation coefficients may deviate from the population parameter even with samples as large as 195. Any single obtained coefficient may be anywhere in the range of such a distribution, but the saving feature is that extreme deviations are highly improbable, and small ones most probable. These illustrations should demonstrate more clearly some of the practical uses of standard errors, as well as the importance of random sampling, if we are to make accurate and useful interpretations.

WHEN OBSERVATIONS ARE NOT INDEPENDENT Random sampling also implies independence of observations. In the preceding examples, observations were not independent because certain restricting conditions tied cases together: if one student was chosen to go to a certain school at a certain time, one or more others like him were also chosen with him. There are other situations where this occurs, often without the investigator's being aware of it. It is most likely to occur when sampling is obtained from subgroups of the population.

Suppose that we have an experiment in which there are 10 subjects and each has 10 trials in each experimental session. For each session we do not have 100 independent observations, nor do we have merely 10 observations. Because there are individual differences, the 10 observations in each set (from each person) will be somewhat homogeneous, having been derived from a single source. In the larger setting of the 100 observations, they are not independent. In computing $s_{\bar{X}}$ for these 100 observations, the number of degrees of freedom is not 99. It is difficult to say just what the number of df should be. The most conservative approach would be to assume 10 observations, each being the mean derived from one individual, and 9 df. But this would lead to an overestimate of the standard error. In the situation described, we have what is called *cluster sampling*. For special treatments of this subject that include formulas for estimating $s_{\bar{X}}$, the reader is referred to discussions by Marks and by Jarrett and Henry.[1]

Inferences regarding other statistics

THE STANDARD ERROR OF A MEDIAN AND OTHER CENTILES

The variability of sample medians is about 25 percent greater than the variability of means when the population is normally distributed. Under this condition the standard error of a median can be estimated by the formula

$$s_{Mdn} = \frac{1.253S}{\sqrt{N}} \quad \text{(Standard error of a median estimated from } S\text{)} \tag{8.6}$$

As applied to the inkblot-test data,

$$s_{Mdn} = \frac{(1.253)(10.45)}{\sqrt{50}} = 1.85$$

In large samples drawn at random from the population, two-thirds of the sample medians of inkblot scores would be expected within 1.85

[1] Marks, E. S. Sampling in the revision of the Stanford-Binet Scale *Psychological Bulletin,* 1947, **44,** 413–434. Jarrett, R. F., and Henry, F. M. The relative influence on error of replicating measurements or individuals. *Journal of Psychology,* 1951, **31,** 175–180.

units of the population median. Since the population is assumed to be normally distributed, we may also say that the sample medians would not deviate from the population *mean* more than 1.85 units, two-thirds of the time. The median may thus be used as an estimate of the population mean, but with less confidence than we have in the use of the sample mean for the same purpose.

We have seen before that the median is a centile point. It can be symbolized as $C_{.50}$, to indicate that it is the X value below which 50 percent of the cases occur, or a proportion of .50. There are standard errors for other centile points, each different, depending upon which centile value we are talking about: $C_{.25}$, $C_{.75}$, the two quartile points, or some other, perhaps $C_{.90}$, for example. The general equation is

$$s_C = \frac{S_x}{y_p}\sqrt{\frac{pq}{N}} \qquad \text{(Standard error of a centile-point value)} \qquad (8.7)$$

where the subscript C stands for centile, p is the proportion of cases below the centile point, $q = (1 - p)$, and y_p is the ordinate in the unit normal distribution corresponding to p.

From this general formula we can see that the standard error of a median is just a special case, and we can also see whence the constant 1.253 came. In a unit normal distribution the ordinate at the median is equal to $1/\sqrt{2\pi}$. Substituting this term for y_p in formula (8.7), and with $\sqrt{pq}$ equal to .5, we have

$$\frac{.5S\sqrt{2\pi}}{\sqrt{N}} = \frac{.5S\sqrt{6.283}}{\sqrt{N}} = \frac{1.253S}{\sqrt{N}}$$

THE STANDARD ERROR OF A STANDARD DEVIATION

The standard deviation also fluctuates from sample to sample. The sampling distribution is somewhat skewed for small samples (N less than 100), but approaches the normal form so closely with large samples drawn from a normally distributed population that we can draw inferences about a population σ, knowing a sample S and its standard error. The SE of S is estimated by the formula

$$s_S = \frac{S}{\sqrt{2N}} \qquad \text{(Standard error of a standard deviation)} \qquad (8.8)$$

Applying this formula to the standard deviation for the sample of accountants mentioned in Table 8.1, where S is 11.7,

$$s_S = \frac{11.7}{\sqrt{2(172)}} = 0.63$$

Comparing formula (8.8) with formula (8.2) for the SE of a mean, we see that a population standard deviation is more accurately estimated than a population mean, when we compare them as to their respective sampling errors. The denominators of these two formulas contain the values $2N$ and $(N - 1)$, respectively, which means that $s_{\bar{X}}$ is more than 40 percent greater than s_S. In one sense, it is fortunate that the standard deviation is more stable than the mean, because both $s_{\bar{X}}$ and s_S are estimated from it.

THE STANDARD ERRORS OF FREQUENCIES, PROPORTIONS, AND PERCENTAGES

Data in terms of frequencies, percentages, and proportions are so common in the social sciences that the problem of their stability in sampling is very important. Out of 30 students who attempted a certain test item, 18 succeeded and 12 failed. How much confidence can we have that the 18 successes represent the actual success rate for the larger population these 30 students represent?

This particular problem should be recognized as calling for a binomial sampling distribution, with mean Np, which is 18, and variance Npq, which is $30 \times .6 \times .4$, or 7.20. The square root of this value gives the standard deviation of this particular binomial distribution, which is 2.7. Since we are dealing with a sampling distribution, the 2.7 is the standard error of the mean or the frequency 18. The general formula is

$$s_f = \sqrt{Npq} \qquad \text{(Standard error of a frequency)} \qquad (8.9)$$

where N = number of cases in the sample
p = proportion in the category of interest
$q = 1 - p$

First, we need to make an important qualification. Comparison of the binomial data given here and in Chap. 7 will show that in the latter place we were dealing with purely hypothetical frequencies and proportions, whereas here we are dealing with observed or empirical data. The value of p that should actually be utilized in formula (8.9) is the population parameter. That value is unknown here. We use in its place our best estimate of the population parameter $\bar{p}$, the observed statistic p. The latter is a fairly good estimate when samples are reasonably large. It can be pointed out that the outcome of formula (8.9) depends relatively more upon the size of N than upon that of p or q because the product pq remains fairly uniform (between .20 and .25) for a considerable range of values for p (namely, for p between .27 and .73). If we have better information concerning the population p, from some previously obtained data or from some a priori reasoning, we should use the better estimates of $\bar{p}$.

To return to the test-item problem, finding the SE of 2.7, we interpret it as we have other SE's. For the most exact interpretation, we should utilize a binomial distribution, the expansion of the binomial $(.6 + .4)^{30}$, but the calculation would be prohibitive without the use of an electronic computer. As pointed out in Chap. 7, we may often use the normal distribution as a good approximation when the minimum frequency (Np or Nq) is 10 or more. Here the smaller frequency is 12, and so the normal approximation may be applied. On this basis, finding 1.96σ to be 5.4, we may conclude with confidence at the .95 level that the obtained frequency of 18 does not deviate more than 5.4 units from the population value.[1]

THE STANDARD ERROR OF A PROPORTION Since a proportion p is equal to a frequency divided by N, the standard error of a proportion is equal to

$$s_p = \frac{1}{N}\sqrt{Npq} = \sqrt{\frac{pq}{N}} \qquad \text{(SE of a proportion)} \qquad (8.10)$$

Out of 100 students who were quizzed, 65 said that they regularly read a morning newspaper. Does this proportion of .65 represent the larger population of students very closely? By formula (8.10), the standard error of this proportion is .048, or roughly .05. The .95 confidence interval has limits of approximately .55 and .75; the .99 confidence interval has limits of .52 and .78. The whole of the latter range is above .50. Hence we may make the further inference that the majority of students represented by this sample (randomly drawn) read morning newspapers, and we should have a high degree of confidence in this inference.

THE STANDARD ERROR OF A PERCENTAGE If we wish to work in terms of percentages, we should remember that a percentage P is simply 100 times a proportion and that $Q = 100 - P$. The formula is

$$s_P = 100\sqrt{\frac{pq}{N}} = \sqrt{\frac{PQ}{N}} \qquad \text{(SE of a percentage)} \qquad (8.11)$$

HYPOTHESES CONCERNING A POPULATION CORRELATION

The mathematical basis for the standard error of a coefficient of correlation is rather complicated, and the same is true of confidence intervals. The difficulty is in the form of sampling distribution for r. Only when the population ρ is zero or near zero and when N is rather large will the sampling distribution approach the normal form. Reference to the upper part of Fig. 8.6 will show that the upper limit of $\rho = 1.0$ puts a real restriction upon the variation around a population ρ of .8, with a marked negative skew. A similar situation would occur for a ρ of $-.8$, except that there would be a positive skew. Another difficulty is that the form of the distribution also differs with the size of sample, particularly among small samples.

With small obtained r's the chief interest is in whether or not

[1] In all these instances in which the actual distribution is binomial, it should be noted that we are dealing with discrete data: frequencies can change only a full unit at a time. On the other hand, we are interpreting the SE's on continuous scales. Where samples are large, such approximations are good; with small samples, as we shall see in Chaps. 9 and 10, a "correction for continuity" should often be applied.

FIGURE 8.6 Distributions of sample coefficients of correlation when N is very small and when the population correlations are .00 in the first distribution and .80 in the second. Corresponding to them, below, are distributions of Fisher's Z coefficients. Conversion of ρ to Z brings about symmetrical sampling distributions, regardless of the size of ρ.

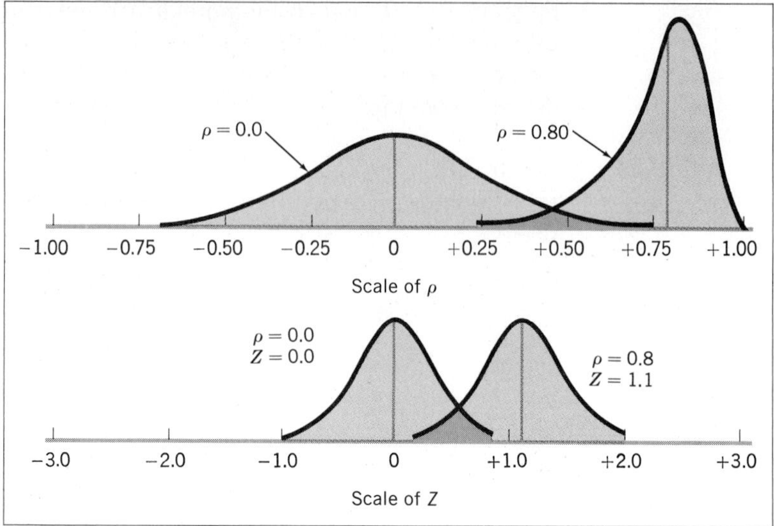

they represent significant departures from population ρ's of zero. With an N as large as 50, we may test the hypothesis that the population ρ is zero by using the standard error obtained from formula (8.12).

$$s_r = \frac{1}{\sqrt{N}} \quad \text{(Approximate standard error of } r \text{ for testing the hypothesis that } \rho = 0\text{)} \tag{8.12}$$

when $\rho = 0$.[1]

With an obtained correlation of .30 and with a sample size of 52 pairs of observations, for $\rho = 0$ $s_r = 1/\sqrt{52} = .14$. A deviation of .30 from the hypothesized zero is .30/.14, or 2.1 standard-error units, which is a z ratio. This z is slightly above that required for rejection of the null hypothesis at the .05 level of confidence. There are fewer than 5 chances in 100 that so large an r could occur in either direction if the population correlation were zero, when $N = 52$.

R. A. Fisher developed a method of testing the hypothesis that $\rho = 0$, for an obtained r under much more general circumstances, in normal bivariate distributions (distributions normal in both X and Y). When the population correlation ρ is zero, a parameter t can be estimated by the formula

$$t = \frac{r\sqrt{N-2}}{\sqrt{1-r^2}} \quad \text{(} t \text{ ratio for testing the significance of a coefficient of correlation)} \tag{8.13}$$

Student's t statistic, of which we shall see a great deal in the

[1] This is an example of what is known as a *null hypothesis*. Many more cases of null hypotheses will be encountered later.

chapters that follow, was actually developed by W. S. Gossett, who published under the name of "Student." It is a ratio, and like z, it can be computed by dividing a deviation from a mean by a standard error. For small samples t has a frequency distribution somewhat different from that of z. Another difference is that N is used in connection with finding the s_r and z with formula (8.12), but the degrees of freedom $(N - 2)$ are used in formula (8.13). Two degrees of freedom are lost in using the two means of X and Y in finding the deviations x and y that are used in the Pearson-r formula.

Applying formula (8.13) to the same illustrative problem, t proves to be 2.2, or about the same as z, in this case. Referring to the table for t (Table D in the Appendix), we do not find exactly 50 df in the first column, but for either 40 or 60 df, a t of 2.0 is significant at the .05 level. A more direct test of significance, with no t computation needed, involves the use of Table Q, which provides the minimum r's for significance at different probability levels, for different degrees of freedom. For 50 df we find that it takes an r of .273 to be significant at the .05 level in a two-tail test. The obtained r of .30 meets that standard but fails to reach the .02 level, which requires an r of .322.

FISHER'S Z COEFFICIENT In order to set up confidence limits and intervals and to gain in general any inferences concerning the accuracy of sample r's not near zero, it is necessary to resort to another innovation of Fisher's—his Z-transformation procedure. His Z coefficient, into which any r can be transformed mathematically, does have a normal distribution regardless of the size of N and the size of population ρ. Nor is an estimate of the population ρ needed in order to determine the standard error.

The range of Z is from $-\infty$ to $+\infty$, but when r reaches .995, Z is still short of 3.0. Up to an r of .25, r and Z are approximately equal. Even when r is .50, Z is only a little larger, being .56. The transformation equation is

$$Z = \frac{1}{2} \log_e(1 + r) - \log_e(1 - r) \qquad \text{(Transformation of } r \text{ into Fisher's } Z) \qquad (8.14a)$$

or

$$Z = 1.1513 \log_{10} \frac{1 + r}{1 - r} \qquad (8.14b)$$

where $\log_e$ stands for a logarithm to the base e of the Naperian system of logarithms.[1] For general practice Table H in the Appendix may be used for the transformation of r into Z or of Z into r. One should not report results in terms of Z but convert back to the more familiar r.

[1] Mathematically, Z is the hyperbolic are tangent of r, or $Z = tanh_r^{-1}$.

Some special problems of estimation

In some special sampling situations, there are a number of refinements that can be applied in estimates of standard errors. These cases include stratified-random sampling, matched samples, and samples from finite populations (populations not much larger than the sample). Methods described in Chap. 13 make such procedures less needed, and so those special standard errors will not be treated here. They are presented in the fourth edition of this volume.

EXERCISES

1 Compute the standard errors of the means for Data 8*A* and interpret your results.
2 Determine the confidence limits and confidence intervals at both the .95 and .99 levels.
3 Compute the standard errors of the two medians and the two 90th centile points for Data 8*A*. Interpret your results.
4 Compute the standard errors of the two standard deviations in Data 8*A* and interpret your results.
5 Compute the standard errors of the frequencies, proportions, and percentages for passing students in Data 8*B* and interpret your results.
6 The correlation between an interest-inventory score and the estimated degree of satisfaction in a certain vocational assignment was .43 in a sample of 101.
   a. Find the SE of r on the assumption of a population correlation of zero.
   b. Find the standard score z for the obtained coefficient of correlation and draw a conclusion regarding its significance.
   c. Use Table Q in the Appendix and interpret the obtained r with reference to relevant information you find in the table.
   d. Compute Fisher's t by using his formula. Compare inferences from this source with those from the other solutions.

DATA 8*A* Results from a test of the ability to name facial expressions in the Ruckmick Photographs

Statistic	*Men*	*Women*
N	95	164
$\bar{X}$	21.1	22.0
S	3.62	3.15
Mdn	21.5	22.2
C_{90}	25.7	26.0

DATA 8*B* Numbers of students in two groups who passed each of three items in a psychological examination

Item	*Group I*	*Group II*
A	24	26
B	33	32
C	30	40
N	37	65

7 Compute Fisher's *t* for the following combinations of *r* and *N*.

r	.25	.25	.50	.50
N	25	50	25	50

8 Transform the *r* of .43 mentioned in Exercise 6 into Fisher's *Z*, determine the standard error of this *Z*, and find the confidence limits for *r*.

ANSWERS

1 $\sigma_{\bar{X}}$: .373, .247.
2 .95 limits: 20.4 and 21.8, 21.5 and 22.5: .99 limits: 20.1 and 22.1, 21.4 and 22.6.
3 s_{Mdn}: .47, .31; $s_{C_{90}}$: .634, .419.
4 s_S: .26, .17.
5 s_f (group I): 2.90, 1.89, 2.38; (group II): 3.95, 4.03, 3.92; s_p (group I): .077, .051, .064; (group II): .060, .062, .060; ($s_P = 100s_p$)
6 $s_r = .10$; $z = 4.30$; *r* of .254 is needed for significance at the .01 level, with 100 df; $t = 4.74$.
7 *t*: 1.24, 1.79, 2.77, 4.00.
8 $Z = .46$; $s_Z = .101$; .95 limits: .25 and .58; .99 limits: .20 and .62.

9 Significance of differences

In the preceding chapter our attention was centered on the estimation of population parameters from experimentally observed sample values and on making inferences regarding the accuracy of those estimates. We were thus concerned with one statistic at a time. In this chapter the emphasis is to be on deviations of one population parameter from another, or, more accurately stated, on whether two observed statistics—such as two means, two proportions, or two correlation coefficients—indicate differences in a corresponding pair of parameters. We observed something of the process of hypothesis testing in the discussions of the logic underlying confidence limits and confidence intervals for means and also of inferring whether a coefficient of correlation differs significantly from zero. A tested hypothesis is often a "null hypothesis." In this chapter we shall continue to investigate null hypotheses and learn more about the use of Student's *t* distribution.

SOME FEATURES OF A NULL HYPOTHESIS

The expression is "*a* null hypothesis" rather than "*the* null hypothesis" because such a hypothesis takes different forms, depending upon which statistic is involved in the process of hypothesis testing. The application of a null hypothesis to the correlation coefficient in Chap. 8 is a good example. In this connection, the null hypothesis is that the population correlation is zero. A correlation of .30 was obtained in a sample of 52 pairs of observations. Under these conditions (N greater than 30 and r not greater than .50) we could adopt as the chance model a population correlation of .00 and to go with it a sampling distribution with a mean of .00 and a standard deviation of approximately $1/\sqrt{52}$, or .14. With a normal distribution of this description, a correlation deviating 1.96σ or more from the mean could occur with a probability of less than .05. The obtained correlation of .30 is more than the deviation required for rejection of the null hypothesis at the .05 level of confidence. The obtained r is not sufficiently large, however, for rejection of the null hypothesis at the .01 level of confidence, for then r would have to be 2.58σ or larger (.36 or larger) in order to meet the .01-level criterion. Remember that .01 of the area under the tails of the normal distribution lies beyond 2.58σ from the mean, at the two extremes.

Some statisticians recommend that the investigator adopt a significance level in advance of knowledge of the results. Levels most commonly adopted are .05, .01, and .001. Whichever it is, it is known as α (Greek alpha). Yet in practice, investigators often just state which level was achieved after the result is known.

In the correlation problem just used as an example, if we reject the null hypothesis, we infer that there is some degree of correlation; the population parameter is probably not zero. If we do not reject the null hypothesis, we are saying that for all we know, the real correlation in the population could be zero. We cannot say that it *is* zero.

Taking into account the variables that were correlated, we could go further. Suppose that variable X is the score on a personality-trait scale for the trait of activeness and that variable Y is the order in which examinees completed answering questions on a personality inventory, with the fastest worker ranked N and the slowest ranked 1. The null hypothesis in this experiment is that there is no correlation between the score for activeness and the rate of work on the inventory. Rejection of the null hypothesis, with a positive r, would lead to the conclusion that there is some relation between activeness score and rate of work of the kind in question.

THE NULL HYPOTHESIS FOR A DIFFERENCE BETWEEN MEANS As another example, let us suppose that we have done an experiment to see whether a certain condition or treatment will make a difference in how much food rats will consume. The rats are deprived of food for a certain period of time. An experimental group is then fed pellets of food that have been soaked with alcohol, and the control group is fed pellets without alcohol. The experimental group consumes pellets with an average of 24.8, whereas the control group consumes 21.3 on the average. Are the two sample means far enough apart to give confidence that there is a real difference in food consumption? The question about "real difference" pertains to population parameters. With sample means fluctuating as they do in sampling, it is possible that the obtained difference of 3.5 pellets could have been due to random sampling from the same population—same, that is, with respect to food consumption under the conditions of the experiment. The size of samples is only 20. The null hypothesis is that there is no difference in amount of food consumption.

The usual investigator of this problem might approach it with the thought that the addition of alcohol would make the food more attractive, and would expect a greater consumption by the experimental group. It may seem rather indirect to test a positive hypothesis by testing its alternative, a null hypothesis. But this is precisely what we have to do. The reason is that we can apply a mathematical model in the case of the null hypothesis, but there is no easy way for testing alternative hypotheses. When the null hypothesis is not true, there is a multitude of other possible hypotheses, each of which would have to be tested in turn. The null hypothesis is a particular, well-defined, testable case.

Differences between means

In considering the significance of differences between means, it is necessary to examine two variations. Different procedures are commonly used when samples are large (30 or more) and when samples are small. When the means are independent, the procedures used are different from those employed when they are correlated. We shall begin with the case of large samples and independently sampled means.

THE STANDARD ERROR OF A DIFFERENCE BETWEEN MEANS Means are independently sampled when derived from samples drawn at random from totally different and unrelated groups. In order to make a test of significance of a difference between means, as usual, we need a special mathematical model describing what would happen in the case of purely random sampling of such pairs of means. With large samples, the model used is a normal distribution of differences $\bar{X}_1 - \bar{X}_2$, with a mean of zero and a standard deviation that is called the SE of a difference between means. Such a standard error can be estimated from the standard errors of the two means $s_{\bar{X}_1}$ and $s_{\bar{X}_2}$. The formula is

$$s_{d_{\bar{X}}} = \sqrt{s_{\bar{X}_1}{}^2 + s_{\bar{X}_2}{}^2} \qquad \text{(SE of a difference between uncorrelated means)} \tag{9.1}$$

VARIANCES OF SUMS AND DIFFERENCES When two variables X_1 and X_2 are summed, if each is normally distributed, the distribution of the set of sums, $X_1 + X_2$, is normal, as well as the distribution of their differences $X_1 - X_2$. In theoretical terms, the variance of the sum $\sigma_s{}^2$ is equal to $\sigma_1{}^2 + \sigma_2{}^2 + 2r_{12}\sigma_1\sigma_2$, where r_{12} is the correlation between X_1 and X_2. Similarly, the distribution of the difference $X_1 - X_2$ is normal

TABLE 9.1
Means and other statistics in the comparison of men and women in a word-building test

Statistic	*Men*	*Women*
N	114	175
$\bar{X}$	19.7	21.0
S	6.08	4.89
$s_{\bar{X}}$	.572	.371
$s_{d_{\bar{x}}}$	.682	
$D_{\bar{x}}$	1.3	
z	1.91	

with a variance σ_d^2, which is equal to $\sigma_1^2 + \sigma_2^2 - 2r_{12}\sigma_1\sigma_2$. When r_{12} equals zero, the variance of the difference is reduced to the first two terms, the square root of which is the standard deviation. The same mathematics applies to the distributions of means in sampling distributions and hence the basis for formula (9.1). This is why means must be independent or uncorrelated when formula (9.1) is used.

A TEST OF SEX DIFFERENCES Let us apply formula (9.1) to a typical problem. Two groups, one made up of 114 men and the other of 175 women, were given the same word-building test. In this test the score is the number of words one can build out of six given letters in 5 minutes. The results are summarized in Table 9.1. The women's mean of 21.0 is 1.3 points higher than that for the men. Although this difference is very small numerically, in view of the relatively large samples there is a possibility that it could be significantly different from zero. The stabilities of the two means are indicated by an SE of .572 for men and .371 for women.

Just as sample means are distributed normally about the population mean when N is large, the sample differences between means are also distributed normally. The central value about which the differences between means fluctuate is also a population value. We do not know what that population value is. We are most concerned, first, about determining whether there is any difference at all, and second, if it is significant at all, about determining its approximate size.

A DISTRIBUTION OF z RATIOS In accordance with the usual null hypothesis, we assume a sampling distribution of differences, with the mean at zero or at $\mu_1 - \mu_2 = 0$.[1] The deviation of each sample difference $\bar{X}_1 - \bar{X}_2$ from this central value is equal to $(\bar{X}_1 -$

1 Actually, one could arbitrarily adopt any other difference as the hypothesis to be tested, testing it as described in Chap. 8 (see pp. 130–131).

$\bar{X}_2) - (\mu_1 - \mu_2)$, or $\bar{X}_1 - \bar{X}_2 - 0$. The division of each sampled difference given in terms of a standard measure would be the difference divided by the SE of the difference, which gives a z value. This operation is symbolized by the formula

$$z = \frac{\bar{X}_1 - \bar{X}_2}{s_{d_{\bar{X}}}} \qquad \text{(A z ratio for a difference between means)} \qquad (9.2)$$

To be quite complete, the numerator should read $\bar{X}_1 - \bar{X}_2 - 0$, as was stated before. But since the zero makes no contribution to the computation, it is dropped from the formula. It will help the user of the formula to think more clearly if the population mean of a zero difference is kept in mind.

Figure 9.1 shows a sampling distribution of z ratios. Such a distribution can exist, although it is rarely or never derived by using actual data. It would be made up of all the sampled differences that might be obtained by taking pairs of samples of the same size from the same population, or from two populations between which there is no mean difference. It would be composed of all z's, each derived by application of formula (9.2) to an obtained difference between sample means.

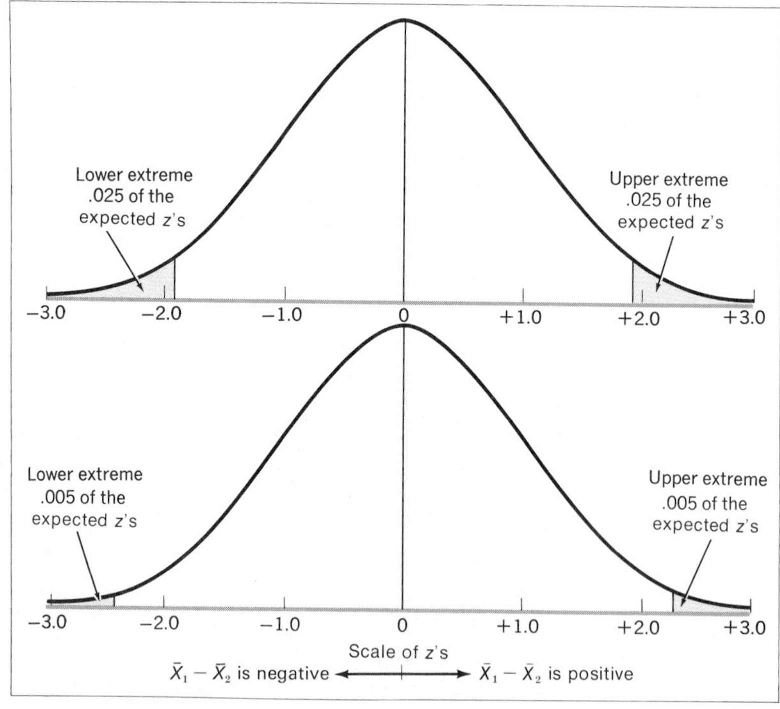

FIGURE 9.1 Two sampling distributions of z, each with a mean of 0, which corresponds to a hypothetical difference between means equal to zero. The shaded areas in the tails represent the critical regions of extreme zs, which lead to rejection of the null hypothesis at the .05 level of confidence in the first curve and at the .01 level of confidence in the second

TESTING THE NULL HYPOTHESIS For the word-building test we have the information (see Table 9.1) that the difference in the obtained sample is 1.3. The algebraic sign of the difference does not concern us at this point; we are interested in the amount. The standard error $s_{d_{\bar{X}}} = .682$. From this,

$$z = \frac{1.3}{.682} = 1.91$$

The value 1.91 tells us how many $s_{d_{\bar{X}}}$'s the obtained difference extends from the mean of the distribution, which is zero. Since the sample is large, we may assume a normal distribution of z's. The obtained z just fails to reach the .05 level of significance, which for large samples is 1.96. Consequently we would not reject the null hypothesis and we would say that the obtained difference is not significant. There may actually be some difference, but we do not have sufficient assurance of it. There are more than 5 chances in 100 that a difference as large as this one could have happened by random sampling from the same population—same with respect to word-building ability. A more practical conclusion would be that we have insufficient evidence of any sex difference in word-building ability, at least in the kind of population sampled. Note that the conclusion was *not* stated to the effect that we have demonstrated no sex difference in word-building ability. *We cannot prove the truth of the null hypothesis; we can only demonstrate its improbability.*

THE STANDARD ERROR OF A MEAN DIFFERENCE IN CORRELATED DATA

When the data are so sampled that there is a correlation between the means in the two sets of measurements (i.e., so that the means in pairs of samples tend to rise and fall together, in a positive correlation, or tend to be contrasting, in a negative correlation), the SE of a difference is estimated by the formula

$$s_{d_{\bar{X}}} = \sqrt{s_{\bar{X}_1}{}^2 + s_{\bar{X}_2}{}^2 - 2r_{12}s_{\bar{X}_1}s_{\bar{X}_2}} \quad \text{(SE of a difference between correlated means)} \qquad (9.3)$$

which is like formula (9.1) except for the last term, in which r_{12} is the correlation between *the two sets of means.*

THE CORRELATION BETWEEN MEANS Fortunately, under the usual circumstances of random sampling, the correlation between two sets of means is approximately equal to the correlation between two sets of single measurements in a sample. Since we ordinarily have only two samples with two means, from which we could not compute the needed r_{12} of the formula, this fact is a great convenience.

But in order to compute the correlation between single measurements, of course, we must have the individual measurements in the two samples paired off in some manner. For example, if the same group of students takes the same word-building test twice we would have first-trial scores to pair off with second-trial scores. Or, if in comparing males with females in the test, we want to standardize our two groups better by taking a brother and a sister from each family, or if we pair boy with girl with respect to age, IQ, social status, or all such conditions, then, if these variables have any relation to word-building score they automatically introduce correlation between pairs and also means in word-building scores. With pairing of individuals we can compute a coefficient of correlation to use in formula (9.3).

A case with correlated samples is shown in Table 9.2. The data are knee-jerk measurements in two sets, both from the same 26 men, each obtained under a different condition. In the first case (*T*) the subjects were squeezing a hand dynamometer just before the stimulus struck the knee, and in the second case (*R*) the subjects were relaxed. Will the subjects show a real difference in height of the knee-jerk response under the two experimental conditions? The two means, with a difference of 3.39 degrees, would suggest that the knee jerk tends to be really higher under the tense condition. But we want to be sure that this difference could not have happened by random sampling from a population difference of zero.

SIGNIFICANCE OF DIFFERENCE WITH AND WITHOUT CONSIDERING CORRELATION If we were to assume no correlation between the tensed and relaxed measurements of the knee jerks, we should apply formula (9.1), or apply formula (9.3) with the assumption that the correlation is zero, an equivalent operation. Such an SE turns out to be 2.37 degrees of arc. The z ratio is then 3.39/2.37, or 1.43, which falls short of significance at the .05 level of confidence. We should conclude, erroneously, that although there is some difference it is not significant. So far as these indications go, we should not be called upon to reject the null hypothesis; the difference of 3.39 could represent merely a result of random sampling.

The correlation between the two sets of knee-jerk measurements proved to be +.82. The subjects were in about the same rank order under the two conditions, *T* and *R*. If the sampling was random we may also conclude that two sets of means of samples would also correlate near .82. When means tend to rise and fall together, the differences between them tend to keep more nearly the same. The dispersion of means therefore remains small. In the extreme case, where $r_{12} = 1.00$, the distribution of differences would have no dispersion and $s_{d_{\bar{X}}}$ would be zero. We could then be certain of a real difference between means. A correlation of +.82 is less than 1.00, however; thus there is still some room for variability among the differences. But from the line of reasoning above, we can see that the $s_{d_{\bar{X}}}$ is going to be smaller than it was when we assumed an r equal to zero.

TABLE 9.2
Strength of the patellar reflex under two conditions, tensed and relaxed, for 26 men, and differences between them (measurements are in terms of degrees of arc)

	T *Tensed*	*R* *Relaxed*	*T* − *R* *Difference*
	31	35	− 4
	19	14	+ 5
	22	19	+ 3
	26	29	− 3
	36	34	+ 2
	30	26	+ 4
	29	19	+10
	36	37	− 1
	33	27	+ 6
	34	24	+10
	19	14	+ 5
	19	19	0
	26	30	− 4
	15	7	+ 8
	18	13	+ 5
	30	20	+10
	18	1	+17
	30	29	+ 1
	26	18	+ 8
	28	21	+ 7
	22	29	− 7
	8	4	+ 4
	16	11	+ 5
	21	23	− 2
	35	31	+ 4
	26	31	− 5
Σ	653	565	+88
$\bar{X}$	25.12	21.73	3.39
S	7.17	9.45	5.50
$s_{\bar{X}}$	1.43	1.89	1.10

By the use of formula (9.3) we find the $s_{d_{\bar{X}}}$ to be 1.10, which is less than half the previous estimate of 2.37. The z ratio is now $3.39/1.10 = 3.06$. A z above 3 is obviously in the "very significant" category.[1] We therefore feel very confident that the difference is genu-

[1] A sample of 26 pairs of observations would be regarded as a small sample by most investigators. A small-sample t test would lead to the same conclusion in this instance, however.

ine. This is not to say that we feel confident that the actual difference is exactly 3.39. It might be more or less than that.

A ONE-TAIL TEST The tests of differences made thus far have been two-tail tests. Some statisticians, but not all, agree that if investigators predict in advance (not after the difference is known) that a difference in a certain direction should be expected, they may make a one-tail test instead of a two-tail test. They would then test the hypothesis that the difference is *zero* or *negative,* for example, rather than the usual hypothesis that the difference is zero. The subject of one-tail tests will be treated more fully in Chap. 10.

TESTING DIFFERENCES BY PAIRING OBSERVATIONS In setting up an experiment with two matched groups of subjects or two corresponding groups of measurements for statistical comparison, it is well to pair off cases, if possible, so that a correlation can be computed.

Often when such pairing is not carried out, there is a correlation between the means of the samples anyway; the full formula for the SE of a difference could not then be applied, and the $s_{d_{\bar{X}}}$ by formula (9.1) is overestimated. It is true that under these circumstances, if the correlation is positive, we can say that the correct $s_{d_{\bar{X}}}$ is smaller and that the correct z ratio is larger than the one we estimated. When we have a significant z under these circumstances, we can be sure that the z we would obtain by taking into account the positive correlation would be even larger.

One difficulty is that when the z obtained under these circumstances is too small to be significant, we cannot come to any definite conclusions, and least of all can we conclude that the actual difference is probably zero. For, had we considered the correlation, we might have found a significantly large z. The process of matching and the inclusion of the correlation factor in the $s_{d_{\bar{X}}}$ formula are said to increase the *power of the test.* By this is meant that the test is more sensitive to a difference when it is genuine. As a result, we are more likely to avoid the error of accepting the null hypothesis when it is incorrect.

It is important that pairing individuals or observations be done on some meaningful basis. Pairing is not worth doing except on the basis of some variable that correlates with the measurements on which the two groups are to be compared. For example, if we were to compare two groups of boys as to ability to do a high jump, one group after training of a certain kind and the control group without such training, it would be important that the two groups be equated as to age, among other factors. Ability in the high jump, regardless of training, is dependent upon age and hence is correlated with it, but the ability is probably not correlated significantly with a grade earned in arithmetic; therefore, there would be no point in matching the groups on this variable.

AN SE OF A DIFFERENCE OBTAINED DIRECTLY FROM DIFFERENCES When individuals have been paired off, we can find the desired statistics directly from differences between pairs. In Table 9.2 we find the difference in knee-jerk measurements ($T - R$), given with algebraic signs, for every individual. If we sum the differences and divide by N, we obtain the mean of the differences, which is equal to the difference between the means. If we calculate the SE of the mean of these differences, we have $s_{d_{\bar{X}}}$. The $s_{d_{\bar{X}}}$ is thus obtained in the most direct manner. We need not even know the SE's of the two means or the amount of correlation present; yet the direct procedure has taken these things into account.

The $s_{d_{\bar{X}}}$ for the knee-jerk data obtained in this manner is identical to the value we found previously, taking r_{tr} into account, as it should be. The interpretations and conclusions concerning the mean difference are as usual. This more direct method is strongly recommended whenever it can conveniently be applied.

DIFFERENCE IN MEANS FROM SMALL SAMPLES

The distinction between large-sample and small-sample statistics is not an absolute one, the one realm merging into and overlapping the other. If one asks, "How small is N before we have a small sample?" the answers from different sources vary. There is general agreement that the division is in the range of 20 to 30. The truth of the matter is that the needs for small-sample treatment of data increase as N decreases, and they may become critical very quickly below an N of 30. Small-sample methods apply regardless of the size of N, but they become imperative for N much below 30.

THE SAMPLING DISTRIBUTION OF t We have already seen that for small samples, some statistics exhibit sampling distributions that depart from normality in various ways; for example, distributions of correlation coefficients, proportions, and standard deviations are often skewed. Another kind of change in sampling distributions is in *kurtosis,* the degree of steepness of the middle part of the distribution. A normal distribution is *mesokurtic,* which means neither very peaked nor very flat across the top. Curves tending toward rectangular form are called *platykurtic*. Those more peaked than normal are called *leptokurtic* (see Fig. 9.2).

Many of the small-sample statistical tests are based upon the statistic known as Student's t. Actually, t is defined as we have defined z. It is the ratio of a deviation from the mean or other parameter, in a distribution of sample statistics, to the standard error of that distribution. In the case of either z or t we have a sampling distribution. Imagine that we computed the ratio for every single sample drawn from the same population with N constant. A frequency distribution of these ratios would be a t (or z) distribution.

A difference between z and t is in degree of generality. Statistic z

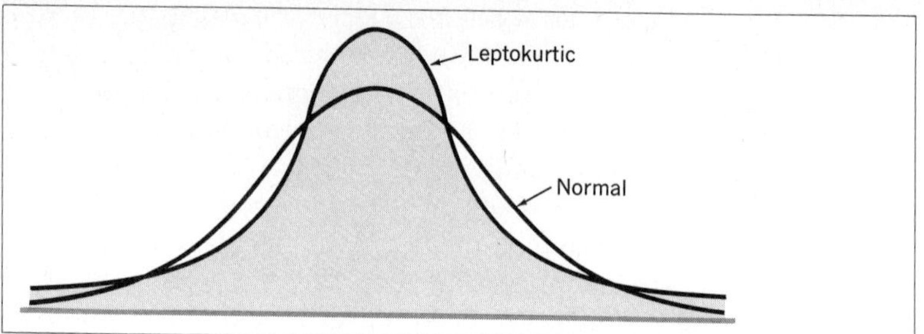

FIGURE 9.2 Comparison of a normal distribution with a leptokurtic distribution when their means and standard deviations are approximately equal

is normally distributed and is so interpreted. It applies when samples are large and sometimes under other restrictions, as when derived from samples of *p* or *r*. Statistic *t*, on the other hand, applies regardless of the size of sample. Where the sampling distribution of *z* is restricted to only one degree of kurtosis (that applying to the normal distribution, a kurtosis of 3.0), the sampling distribution of *t* may vary in kurtosis. Student's *t* distribution becomes increasingly leptokurtic as the number of degrees of freedom decreases (see Fig. 9.3). As the df becomes very large, the distribution of *t* approaches the normal distribution.

Figure 9.3 shows *t* distributions with differing df involved. The most important feature of a leptokurtic distribution, as compared with the normal distribution, for the purposes of hypothesis testing, is the difference at the tails. The tails are higher for the leptokurtic distribution. The effect of this is that we have to go out to greater deviations in order to find the points that set off the regions significant at the .05, .01, and other standard levels. From Fig. 9.3 it can be seen that the *t* dis-

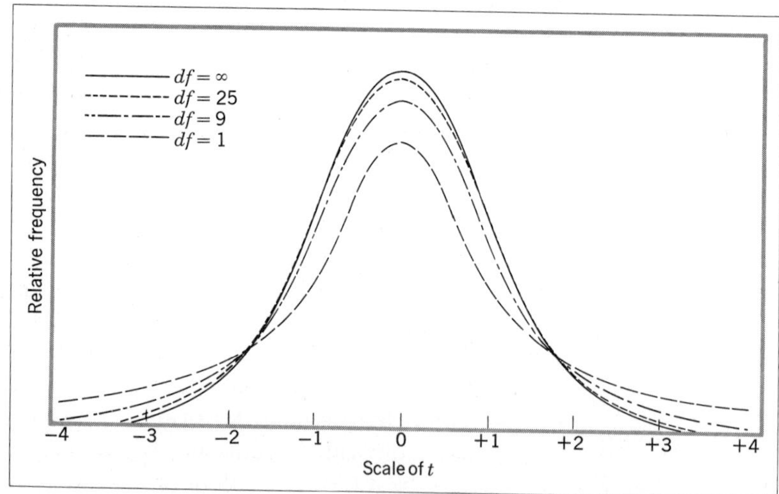

FIGURE 9.3 Sampling distributions of Student's *t* for various numbers of degrees of freedom. When the df becomes very large, the distribution of *t* approaches the normal distribution as its limit

tribution for 25 df comes very close to the normal distribution but that the one for 9 df definitely does not. Before we decide to accept a normal-curve approximation to the t distribution when there are 25 df, however, let us consider what difference it would make in the significance limits.

SIGNIFICANCE LIMITS IN t DISTRIBUTIONS In distributions of t, significant values have been determined at the .05 and .01 levels for different degrees of freedom. These are listed in the columns of Table D in the Appendix. Reference to that table will show that when the number of df is infinite, the two t values are 1.960 and 2.576, the same as for the normal distribution. With 1,000 df the critical values are different from those figures only in the third decimal place. For 100 df there is a small change in the second decimal place. The limits with 100 df are 1.984 and 2.626. Rough limits, by rounding, of 2.0 and 2.7 would do very well even down to about 30 df. With only 10 df, however, t's of 2.23 and 3.17 would be required for the .05 and .01 significance levels. With small samples, then, it becomes imperative to consider the changing t values needed for significance. Even when the number of df is greater than 30, if t turns out to be near the critical limits for rejection of hypotheses, it would be well to refer to Table D to find the exact values.

FISHER'S t FORMULAS Fisher has provided several formulas designed for the computation of t. We have already noted his formula for use in connection with a coefficient of correlation in the preceding chapter.

THE t TEST OF A DIFFERENCE BETWEEN MEANS When means are uncorrelated, the t formula for testing their difference is

$$t = \frac{\bar{X}_1 - \bar{X}_2}{\sqrt{\left(\frac{\Sigma x_1^2 + \Sigma x_2^2}{N_1 + N_2 - 2}\right)\left(\frac{N_1 + N_2}{N_1 N_2}\right)}} \qquad (9.4)$$

(Fisher's t for testing a difference between uncorrelated means)

where $\bar{X}_1$ and $\bar{X}_2$ = means of the two samples
Σx_1^2 and Σx_2^2 = sums of squares in the two samples
N_1 and N_2 = numbers of cases in the two samples

The complete numerator should read $\bar{X}_1 - \bar{X}_2 - 0$, to indicate that it represents a deviation of a difference from the mean of the differences. The denominator as a whole is the SE of the difference between means, as the t ratio requires.

In writing the $s_{d\bar{X}}$ in this form, we take the null hypothesis quite seriously. That is, if there is but *one* population, there should be but one estimate of the population variance. In the first term under the radical we have combined the sums of squares from the two samples (in the numerator) and the degrees of freedom (in the denominator) that come from the two samples. The expression $N_1 + N_2 - 2 = (N_1 - 1) + (N_2 - 1)$. The effect of the second expression under the radical is to give the SE of the mean difference.

When two samples are of equal size, i.e., $N_1 = N_2$, formula (9.4) simplifies to

$$t = \frac{\bar{X}_1 - \bar{X}_2}{\sqrt{\dfrac{\Sigma x_1^2 + \Sigma x_2^2}{N_i(N_i - 1)}}} \qquad \text{(t for difference between uncorrelated means in two samples of equal size)} \qquad (9.5)$$

where N_i = size of either sample.

When means of paired samples are not independent but correlated, the best formula to use for deriving t directly from sums of squares is

$$t = \frac{\bar{X}_d}{\sqrt{\dfrac{\Sigma x_d^2}{N(N - 1)}}} \qquad \text{(t for differences between correlated pairs of means)} \qquad (9.6)$$

where $\bar{X}_d$ = mean of the N differences of paired observations and $\bar{x}_d$ = deviation of a difference from the mean of the differences.

The procedure implied by this formula was actually applied in connection with the knee-jerk data under two experimental conditions (see Table 9.2). The number of df to use with t in this case is $N - 1$, where N is the number of *pairs* of observations. For the knee-jerk problem $N = 26$ and there are 25 df, which indicates (from Table D in the Appendix) that t's of 2.06 and 2.79 are significant at the customary levels. The obtained t is 3.06.

WHEN t TESTS DO NOT APPLY If there is a good reason to believe that the population distribution is not normal but is seriously skewed, and especially if the samples are small, t tests do not apply. For skewed distributions, Festinger and others have suggested substitute tests.[1] There are also available, as substitutes for the t test, a number of distribution-free tests, some of which are described in Chap. 12.

The reader should also be warned that if the two samples have markedly differing variances, the t test is questionable. Whether or not two sample variances are significantly different can be determined by making an F test, which is described later in this chapter. Cochran and Cox[2] have provided a method for meeting the case of unequal variances. One should also have some hesitation in using t formulas

1 Festinger, L. The significance of difference between means without reference to the frequency distribution function. *Psychometrika*, 1946, **11**, 97–105.
2 Cochran, W. G. and Cox, G. M. *Experimental design*. New York: Wiley, 1950.

if the N's in the two samples differ markedly. Differing N's do not seem to affect the use of formulas (9.1) and (9.4) in the same way. Bonneau[1] has investigated by means of sampling from nonnormal distributions, with unequal variances of samples and with differing N's, the effects of conditions such as these, which violate the assumptions of the t test, upon rejections of hypotheses at the .05 and .01 levels. On the whole, t is not markedly affected by rather strong violations unless N is very small.

Differences between proportions and frequencies

DIFFERENCES BETWEEN UNCORRELATED PROPORTIONS

When a null hypothesis is assumed with regard to two observed proportions, Fisher recommends that we use just one estimate of the population variance and not two estimates, one from each sample p. This suggestion involves finding a weighted mean of the two sample proportions, after the manner of formula (4.5). Repeating that formula here, with minor variation, the estimated population proportion is given by the formula

$$\bar{p}_e = \frac{N_1p_1 + N_2p_2}{N_1 + N_2} \qquad (9.7)$$

(Weighted mean of two sample proportions to estimate a population proportion)

The test of significance of a difference between two proportions is not suited to small-sample techniques. The sampling distribution of $p_1 - p_2$ approaches the normal form when both the Np's and Nq's are as large as 10. The test of significance can then be made using a z ratio. The formula for such a z is

$$z = \frac{p_1 - p_2}{\sqrt{\bar{p}_e\bar{q}_e\left(\dfrac{N_1 + N_2}{N_1N_2}\right)}} \qquad (9.8)$$

(A z for a difference between uncorrelated proportions)

where $\bar{q}_e = 1 - \bar{p}_e$.

To apply the last two formulas, let us take the case in which two groups of students were sampled. In the first group, 60 out of 100 members said that they approved of the way in which the president was doing his job. In the second group, 35 of the 50 members said that they approved. Thus, p_1 is .60 and p_2 is .70; N_1 is 100 and N_2 is 50. The estimated population proportion, given by a weighted mean, is

$$\bar{p}_e = \frac{60 + 35}{100 + 50} = \frac{95}{150} = .6333$$

1 Bonneau, C. A. The effects of violations of assumptions underlying the t test. *Psychological Bulletin,* 1960, **57,** 49–64.

The variance of the estimated population proportion is .6333 times .3667, which equals .2322. Applying formula (9.8), we find

$$z = \frac{.70 - .60}{\sqrt{.2322\left[\frac{100+50}{(100)(50)}\right]}} = \frac{.10}{\sqrt{(.2322)(.0300)}} = \frac{.10}{.0835} = 1.20$$

A z of 1.20 falls short of significance at the .05 level. Therefore, we do not reject the hypothesis that the two sample proportions arose from the same population. The sentiment with regard to approval of the president could have been equally strong in the two populations, or it was really just one population with respect to approval of the president.

When the two samples are of the same size, with $N_1 = N_2 = N_i$, formula (9.8) simplifies to

$$z = \frac{p_1 - p_2}{\sqrt{\frac{2\bar{p}_e\bar{q}_e}{N_i}}} \qquad (9.9)$$

in which $\bar{p}_e$ is the simple mean of p_1 and p_2, and in which N_i is the number of cases in either sample.

This formula applies in a study in which 400 college men and 400 college women were asked whether or not they found the word "symphony" pleasant. The proportions responding affirmatively were .6850 for the men and .8875 for the women. Thus we have

$$z = \frac{.8875 - .6850}{\sqrt{\frac{2(.1681)}{400}}} = \frac{2025}{\frac{.3362}{400}} = \frac{.2025}{.0029} = 6.99$$

where $\bar{p}_e = .78635$ and $\bar{p}_e\bar{q}_e = .1681$. The hypothesis of no sex difference should be rejected well beyond the .01 level of confidence.

It may have been noted that in both problems p_1 and p_2 were assigned so as to yield a positive difference and a positive z. In making a two-tailed test the sign of z is immaterial, since the region for making the rejection of the null hypothesis is in both positive and negative extremes of the sampling distribution.

A ONE-TAIL TEST OF A DIFFERENCE[1] In the first of the two illustrative problems, let us suppose that the first group, in which the p_1 was .60, was composed of Republicans and that the second group, in which p_2 was .70, was composed of Democrats. Let us also suppose that the president was a Democrat. We should then have expected that the

[1] One-tail tests are treated much more fully in Chap. 10.

difference $p_2 - p_1$ would be positive. We should then make a one-tail test. The hypothesis then tested is that the difference is zero or it is *negative*. The alternative hypothesis is that $p_2 - p_1$ is positive. If alpha, the probability of an extreme z deviation, remains at .05, the region for rejection is above a z of 1.64 rather than 1.96. It is much easier to find a significant z in a one-tail test than in a two-tail test, with the same alpha. The danger of making an error in rejecting the tested hypothesis is the same so long as alpha is the same.

We can make a one-tail test, also, in the case of reactions to the word *symphony*. Since the word pertains to art and since in our culture it is more accepted for females than males to show pleasure in response to things artistic, we should have expected an outcome in the obtained direction. In neither problem, as it turned out, would a one-tail test have changed the decisions: to accept the tested hypothesis in the first case and to reject it in the second. In the second case the tested hypothesis was, as usual, that there would be a zero or negative z when the difference was $p_2 - p_1$ (women's proportion minus the men's). In other instances the one-tail test may lead to rejection of the tested hypothesis where the two-tail test would not.

CORRECTION FOR CONTINUITY Even when the Np or Nq product (whichever is smaller) is between 5 and 10, it is possible to make a z test if we make a correction for continuity. This will be more fully explained in the next chapter. Here we shall only remind the reader that frequencies, from which p's are derived, change by discrete jumps, whereas the z parameter is a continuous variable. When we use the normal curve as an approximation to the binomial, the smaller the samples are, the more important it is to take this discrepancy into account. An allowance for continuity can be made by introducing an adjustment in the numerators of formulas (9.8) and (9.9). The numerator, $p_1 - p_2$, should be reduced in absolute size (whether it is positive or negative) to the extent of the value

$$\frac{1}{2}\left(\frac{1}{N_1} + \frac{1}{N_2}\right) \quad \text{or} \quad \frac{1}{2}\left(\frac{N_1 + N_2}{N_1 N_2}\right)$$

If the smaller Np or Nq is less than 5, we can still possibly resort to the use of a chi-square test, which is described in Chap. 11.

DIFFERENCES BETWEEN CORRELATED PROPORTIONS

For the case in which the sampled proportions are correlated, an economical procedure has been proposed by McNemar.[1] The formula avoids the necessity for computing the standard error of the estimated population proportion as well as the correlation coefficient. As with

1 McNemar, Q. Note on the sampling error of the difference between correlated proportions or percentages. *Psychometrika,* 1947, **12,** 153–157.

correlated means, we expect to find correlated proportions in two samples when the same individuals are involved in both or when we are dealing with matched pairs, such as twins, siblings, littermates, and the like.

Suppose that we have administered two test items to a sample of 100 students. Item I is answered correctly by 60 of the group, and item II by 70. Is item II actually easier than item I? In making the z test to answer this question, we must definitely face the possibility of correlation between the two items and consequently between the two proportions. To handle this problem properly, we need to arrange the data in the form of a four-cell contingency table, as in Table 9.3. At the left in the table are the four cell frequencies, including those who pass item I and either pass or fail item II, and those who fail item I and either pass or fail item II. In the section at the right are letter symbols standing for frequencies in the four categories. With these symbols, the formula reads

$$z = \frac{b - c}{\sqrt{b + c}} \quad \text{(A } z \text{ ratio for difference between correlated proportions)} \tag{9.10}$$

It will help to ensure the proper application of this formula if we note that the symbols b and c stand for the discordant cases in the four-cell table. In this problem, b and c stand for individuals who pass one item and fail the other. It will help to know that the difference $b - c$ divided by N equals the difference between p_1 and p_2. It is therefore the difference between two *frequencies*, i.e., $b - c = Np_1 - Np_2$. Finding the difference that is being tested in the numerator of the z ratio is not a new experience. The denominator must therefore somehow represent the SE of a difference between frequencies, *with the correlation between them being taken into account*. In this formula, also, is implied but one estimate of the population variance, and it is derived from an average of the sample proportions. What we are actually testing with formula (9.10) is whether the *change* in the frequencies is significant.

TABLE 9.3 A four-cell contingency table of frequencies of students who passed or failed each of two test items

Frequency table

	Item II Fail	Pass	Both
Item I Pass	5	55	60
Fail	25	15	40
Both	30	70	100

Symbolic table

	Item II Fail	Pass	Both
Item I Pass	b	a	$a + b$
Fail	d	c	$c + d$
Both	$b + d$	$a + c$	N

Applying formula (9.10) to the test-item data, we have

$$z = \frac{5 - 15}{\sqrt{5 + 15}} = \frac{-10}{\sqrt{20}} = -2.24$$

We infer that the difference is significant between the .05 and .01 levels—item II is very probably easier than item I.

Had we not taken into account the correlation between the two items, by using formula (9.9) we should have found a z ratio of only 1.48, which would have led to a wrong decision. The correlation between the two items, a phi coefficient (see Chap. 14), was found to be +.58. It should be added that one restriction in the use of formula (9.10) is that $b + c$ should be 10 or greater.

Differences between coefficients of correlation

In the preceding chapter it was noted that the sampling distribution of the Pearson r is so variable, depending as it does upon the sizes of both N and r, that only under highly restricted conditions could a z distribution be used. No test of differences between population ρ's based upon standard errors of r's is very satisfactory. The best general recourse we have is to use Fisher's transformation to Z, whose standard error is related only to N.

WHEN COEFFICIENTS OF CORRELATION ARE UNCORRELATED

With uncorrelated r's, as when we have two correlations between the same two variables X_1 and X_2 derived from two different, unmatched samples, the standard error of a difference between Fisher's Z's can be computed by the formula

$$s_{d_Z} = \sqrt{\frac{1}{N_1 - 3} + \frac{1}{N_2 - 3}} \qquad \text{(SE of a difference between two independent } Z \text{ coefficients)} \qquad (9.11)$$

Consider two r's from two different samples, .82 and .92. The corresponding Z coefficients from Table H in the Appendix are 1.16 and 1.59, respectively. Corresponding N's are 50 and 60. From these data,

$$s_{d_Z} = \sqrt{1/47 + 1/57} = .197$$

and

$$z = \frac{1.59 - 1.16}{.197} = 2.18$$

The sampling distribution of Fisher's Z is normal, therefore the sampling distribution of $Z_1 - Z_2$ is also normal, so z may be interpreted as a standard score. The difference in Z's deviates from a difference of 0.0 to the extent of 2.18SE, which means that it is significant at the .05

level. We reject the null hypothesis of no difference in the Z's, a decision that can also be made with respect to the difference between the two r's.

WHEN COEFFICIENTS OF CORRELATION ARE CORRELATED

There is one kind of comparison of r's that comes up sufficiently often to call for treatment when the r's are correlated. It is the case in which the correlations are r_{12} and r_{13}, in other words, the correlations of two variables X_2 and X_3 with the same third variable X_1. An example of this is the comparison of two validity coefficients; we have correlated two predictor-test variables with the same criterion of successful behavior of some kind. Perhaps one of the two predictors is a new one that someone claims to be superior to an older one. Let us say that the older one, X_2, correlates .45 with the criterion of academic achievement, a score on an algebra-achievement test X_1. The new test correlates .55 with the same criterion. The sample is composed of the same 200 students for the two validity coefficients. Is the second test genuinely superior to the old one in validity?

For this particular kind of problem, Hotelling[1] has developed a t test that takes into account the correlation between r_{12} and r_{13}. The Hotelling formula is

$$t_{d_r} = (r_{12} - r_{13}) \sqrt{\frac{(N-3)(1+r_{23})}{2(1 - r_{23}{}^2 - r_{12}{}^2 - r_{13}{}^2 + 2r_{23}r_{12}r_{13})}} \tag{9.12}$$

The number of df may be taken as $N - 3$.

In order to apply formula (9.12) to the two validity coefficients r_{12} and r_{13}, we need to know the intercorrelation r_{23}, which is .60. Then, with $N = 200$,

$$t_{d_r} = (.55 - .45) \sqrt{\frac{197(1+.60)}{2[1-.3600-.3025-.2025+2(.60)(.45)(.55)]}}$$

$$= .10 \sqrt{\frac{315.20}{.864}}$$

$$= 1.91$$

The alternative hypothesis is that the r of .55 represents a genuinely higher correlation than the r of .45; hence a one-tail test is called for. If we adopt the .05 level and a one-tail test, the t needed is 1.64. This can be determined from the normal-curve table, for with large N's, z is interchangeable with t for hypothesis-testing purposes. We would accordingly reject the hypothesis being tested: that the true difference in correlation is either zero or negative.

[1] Hotelling, H. The selection of variates for use in prediction, with some comments on the general problem of nuisance parameters. *Annals of Mathematical Statistics,* 1940, **11,** 271–283.

Differences between variances

We saw in Chap. 8 that the use of standard errors of standard deviations is precluded except for very large samples. The same limitation applies to testing the significance of differences between standard deviations by the method used in the case of other statistics. Fortunately, procedures have been developed that apply to measures of variations in large and small samples alike. One of them involves a variance ratio, signified by the symbol F, which has known sampling distributions. By this approach, we can test whether or not two variances could probably have arisen by random sampling from the same population of observations or from two populations with the same variance.

WHEN TWO VARIANCES ARE INDEPENDENT When estimates of population variance are obtained from two independent samples (with no matching of samples in any way involved), their difference is tested, not by the usual operation of subtraction, or $s_1^2 - s_2^2$, but by forming their ratio, s_1^2/s_2^2. The ratio that satisfies the null hypothesis completely is equal to 1.00. As the ratio departs from 1.00, the differences are greater. In comparing two sample variance estimates (there are other uses of the F ratio, some of which will be seen in Chap. 13), it is customary to put the larger s^2 in the numerator, giving a ratio greater than 1.00. The equation for an F ratio is [1]

$$F = \frac{s_1^2}{s_2^2} \qquad (s_1^2 \text{ being greater than or equal to } s_2^2) \qquad (9.13)$$

It should be noted that the variances being compared are the estimates of population variance s^2, not the sample variance S^2.

A small set of data will illustrate the operation of this procedure. Assume that two sets of scores, in one of which $N_1 = 8$ and in the other of which $N_2 = 5$, have sums of squares $\Sigma x_1^2 = 132$ and $\Sigma x_2^2 = 26$. The numbers of degrees of freedom are 7 and 4, respectively, and so the estimated population variances, independently derived, are $132/(N_1 - 1) = 18.86$ and $26/(N_2 - 1) = 6.5$. The variance ratio F is 18.86/6.5, which equals 2.90.

THE SAMPLING DISTRIBUTION OF F In random sampling, the distribution of F can be predicted from mathematical relationships. The shape of the distribution depends upon the two different numbers of degrees of freedom involved, but the general shape is that of marked positive skew, with a mean of 1.0, as we might expect. Figure 9.4 shows three distribution curves of F for three different pairs of numbers of degrees

1 It is of interest that the F ratio was proposed by G. W. Snedecor, who based it upon earlier work by R. A. Fisher, in honor of whom the ratio was symbolized by F.

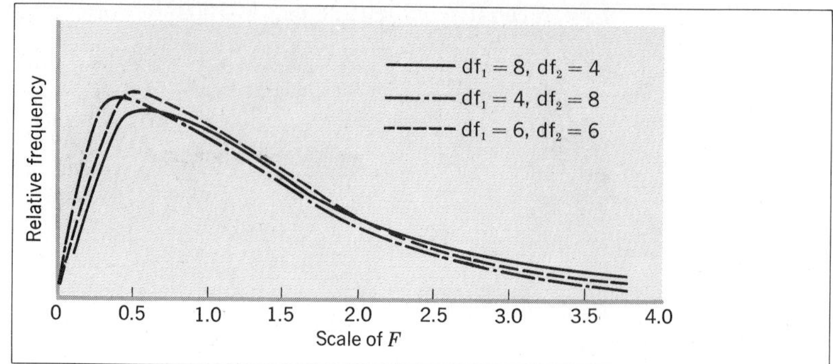

FIGURE 9.4 Sampling distributions of Snedecor's *F* ratio for various combinations of degrees of freedom

of freedom, df_1 and df_2. These curves are probability distributions, the total area under each one having a value of 1.00 (the sum of all probabilities for different values of *F*).

Table F in the Appendix gives the *F* limits that are significant at the .05 and .01 levels for different combinations of df's. For the illustrative problem above, the df's are 7 and 4, for the numerator and denominator of the ratio, respectively. We look for the numerator df at the heading of the appropriate column and for the denominator df at the heading of the appropriate row of the table. We find the *F* values significant at the .05 and .01 levels at the intersection of the appropriate column and row. For the combination of df's 7 and 4, we find the two significance levels to be *F*'s of 6.09 and 14.98, respectively. The obtained *F* of 2.90 does not come close to the smaller of these two values. We therefore do not reject the null hypothesis and decide that as far as variances are concerned, the two samples could well have come from the same population or from two populations that have equal variances. The same decision applies to variabilities.

In this particular use of the variance ratio *F*, we have to consider an important modification in interpreting an obtained *F* by the use of Table F in the Appendix. This is because Table F is set up in terms of one-tail tests. The arbitrary placing of the larger variance in the numerator doubles the probability of obtaining deviations above the mean, in other words, on the right side of the distribution. Consequently, for a two-tail test we have to double the probabilities of the .05 and .01 regions, making them .10 and .02 respectively. In other uses of the *F* ratio, where the larger variance is not arbitrarily placed in the numerator, the .05 and .01 probabilities still hold. Such cases will be discussed in Chap. 13.

WHEN VARIANCES ARE CORRELATED

When the two variances to be compared arise from samples that are matched in some way, there is likely to be some positive correlation between the variances, and the *F* test does not apply. A *t* test has been developed, however, to take care of such cases.

Suppose we are interested in determining whether a group of subjects changes in variability in performance of the same task as a result of intervening practice between an initial and a final test. There will naturally be some positive correlation between initial and final test scores. To take a specific case, suppose the initial variance estimate is 16.0 and the final one is 36.0. Other needed information is an N of 54 and a correlation of .65. The formula for t is

$$t = \frac{(s_2^2 - s_1^2)\sqrt{N-2}}{2s_1s_2\sqrt{1-r_{12}^2}} \qquad \text{(t ratio for testing difference between correlated variances)} \qquad (9.14)$$

where s_1 and s_2 are the two estimates of population variance derived from two matched samples, N is the number of cases in the sample, and r_{12} is the correlation between observations in samples 1 and 2. Inserting the illustrative values,

$$t = \frac{(6.0^2 - 4.0^2)\sqrt{54-2}}{2(6.0)(4.0)\sqrt{1-.65^2}}$$

$$= \frac{20\sqrt{52}}{48\sqrt{.5775}}$$

$$= 3.95$$

Table D in the Appendix shows that it takes a t of 2.68 to be significant at the .01 level when there are 50 df. We may conclude that the obtained t well exceeds that level of significance, and so in a two-tail test we quickly reject the hypothesis of no difference in variances.

A more convenient formula for computing purposes, one that does not require the computation of the variances or the coefficient of correlation, is given by Walker and Lev:[1]

$$t = \frac{(\Sigma x_2^2 - \Sigma x_1^2)\sqrt{N-2}}{2\sqrt{\Sigma x_1^2 \Sigma x_2^2 - (\Sigma x_1 x_2)^2}} \qquad \text{(t ratio for testing difference between correlated variances)} \qquad (9.15)$$

EXERCISES

1. Estimate the standard error of the difference between means for Data 8A (see p. 144) and make a z test. Interpret your results.
2. Estimate the SE of the difference between means for Data 9A and make a z test. Interpret your results.
3. Compute a t for the difference between means in the following data: $N_1 = 11$; $N_2 = 26$; $\bar{X}_1 = 17.5$; $\bar{X}_2 = 14.8$; $\Sigma x_1^2 = 44$; $\Sigma x_2^2 = 65$. The means and variances are uncorrelated.
4. Apply a t test to the difference between means for the first 10 pairs of observations of knee-jerk data in Table 9.2. Interpret your results.
5. Make z tests for differences between groups for the proportions

[1] Walker, H. M., and Lev, J. *Statistical inference*. New York: Holt, 1953.

derived from Data 8*B* (see p. 144) for each of the three items. Interpret your results.

6 In a certain precinct, 200 voters cast votes in both the 1970 and 1974 elections. Of these, 20 switched from the Democratic candidate for president to the Republican candidate, whereas 10 switched in the reverse direction. Was there a significant trend? Give statistics.

7 In a group of 145 boys the correlation between tests A and B was .65, and in a group of 135 girls the correlation between the same two tests was .75. Was there a genuine difference in the size of correlation in the two groups? Give statistical evidence.

8 The predictive validity of a composite score with a pass-fail criterion in flying training was indicated by a correlation of .55 for 150 students, whereas the validity coefficient for another composite score made up of different tests was .45. The correlation between the two composites was .60. Was one composite more valid than the other? Give evidence.

9 Apply an *F* test to the two variances represented in Exercise 3. Interpret your results. Is the application of a *t* test for differences between means in exercise 3 justifiable? Explain.

10 Is there a significant difference between the two standard deviations given in Data 9*A*? Answer the question by making a *t* test.

ANSWERS

1 $s_{d\bar{X}} = .448$; $z = 2.01$. Null hypothesis is rejected at the .05 level.

2 $s_{d\bar{X}} = .658$; $z = 2.89$. Null hypothesis rejected at the .01 level.

3 $t = 4.25$ ($s_{d\bar{X}} = .635$). With 35 df, null hypothesis is rejected at the .01 level.

4 $t = 2.07$. With 9 df, t is not significant.

5 $s_{d_p} = .103, .099, .096$; $z = 2.42, 4.04, 2.04$. Differences for items A and C are significant at the .05 level and for item B at the .01 level.

6 $z = 1.923$. No significant trend is indicated.

7 $z = 1.64$. Probably no real difference in a two-tail test.

8 $t = 1.65$. Probably no difference in a two-tail test.

9 $F = 1.69$. Difference in either variances or in S's not significant.

10 $t = 1.518$; not significant.

10 Hypothesis testing

Although we have examined the process of testing hypotheses in the preceding chapters, we did so without going very deeply into the logic of statistical decision making. We shall now look further into the matter, for a deeper understanding of the problems and principles involved is especially necessary before considering a greater variety of applications. There are qualifications and elaborations to be made in connection with what has already been presented in Chaps. 8 and 9. In addition to the general considerations of statistical decision making and the errors involved in that process, we shall also direct our attention to the sizes of samples needed to achieve certain levels of confidence that our decisions are correct, and we shall also discuss some statistical tests utilizing binomial distributions.

Some rules for statistical decisions

Let us begin with a simple psychophysical test situation. A student asserts that he can distinguish between two tones whose stimuli differ by only 2 cycles per second. That is his hypothesis—that he possesses genuine power to discriminate so small a difference in pitch. We doubt him, which means that we adopt a null hypothesis. Out of six presentations of stimuli, how many should we require him to judge correctly before we give up our hypothesis and yield to his? Our hypothesis implies that when he judges the pair of stimuli, he might just as well flip a coin and report "second higher" for "heads" and "second lower" for "tails." By such guessing, we should expect him to be correct half the time, or three times out of six. But how many more than three correct judgments will it take to convince us that he is not merely guessing?

In a set of six trials, there are seven possible outcomes—all the way from 6 to 0 correct judgments. Table 10.1 lists the seven possibilities and the probability that each event will occur by random sampling. According to the probabilities, we should expect only *one* "score" of 6 in 64 samples of 6 judgments each; we should expect 6 scores of 5, 15 scores of 4, and so on. These expectations are according to the expansion of the binomial $(1/2 + 1/2)^6$, as we saw in Chap. 7. The model we apply here to describe the null hypothesis is this binomial distribution.

TESTING DEVIATIONS FROM EXPECTED VALUES

In determining whether the student's hypothesis about his acuity for pitch discrimination should be accepted, we are interested in how far the score he obtains deviates from the one most to be expected by chance. The most probable chance-generated score in this situation is three correct judgments: the mean of the binomial distribution, given by $np = 6 \times .5 = 3.0$. How much deviation from a score of 3 does the student need in order to lead us to reject the null hypothesis and to tolerate, if not to accept, his alternative hypothesis?

A score of 6 would be expected $1/64$ of the time. One chance in 64 lies between the familiar .05 and .01 levels commonly applied as standards of rejection of the null hypothesis. But before we conclude that we should reject the null hypothesis, we have to consider whether we are making a one-tail or a two-tail test.

ONE-TAIL VERSUS TWO-TAIL TESTS If we begin the experiment with the belief that either the student's judgments are determined by chance or they are not, we must use a two-tail test. For the alternative hy-

TABLE 10.1 Expected occurrences and probabilities of specified numbers of correct judgments in making six judgments at random

Number of correct judgments	*Times expected in 64 sets of judgments*	*Probability of this number occurring in random sampling*	*Probability of as many or more occurring*	*Probability of as few or less occurring*
6	1	1/64	1/64	64/64
5	6	6/64	7/64	63/64
4	15	15/64	22/64	57/64
3	20	20/64	42/64	42/64
2	15	15/64	57/64	22/64
1	6	6/64	63/64	7/64
0	1	1/64	64/64	1/64

pothesis—that his judgments are not directed by chance—there are *two* possible outcomes: an extreme positive deviation or an extreme negative deviation. Either outcome falls into a single *logical* region, called a *critical region,* in spite of the fact that the two extremes occur at opposite tails in a frequency distribution. If this is our line of thought as we start the experiment, we must remember that a deviation provided by a score of 0 is just as probable by chance as a score of 6. The confidence level attached to the occurrence of a score of 0 is $\frac{1}{64}$, so that the probability of *either* a 0 or a 6 (by the addition theorem of probability) is $\frac{2}{64}$, or $\frac{1}{32}$. The two-tail test thus also leads to a rejection of the null hypothesis beyond the .05 level (with a probability less than .05 but not less than .01). The statistical decision would be the same in this particular example, whether we make a one-tail or a two-tail test.

In this psychophysical problem, a score of 0 would be interesting to interpret. Of course, if we had adopted in advance a critical region for rejection at the .01 level of confidence, no further conjectures would be called for. But if we had adopted the .05 level and the obtained score were 0, we might naturally ask what this could mean, psychologically. It could indicate a bias of some kind, one that operates toward making a discrimination in the wrong direction. The source of the bias might be inferred, perhaps from additional information from another source. Any such hypothesis might furnish the starting point for a new experiment, with new conditions.

If we were to adopt the one-tail test in the illustrative experiment, scores below 3 would be regarded differently. First, we have less interest in them. Since the difference of opinion which brought the experiment about involved two alternatives—the student either *can* or *cannot* sense a difference between a pair of stimuli differing by 2 cycles per second—a one-tail test seems more logical than a two-tail test. If he *surpasses* our adopted standard, statistically defined, we will give credence to his belief. *Any* other score, then, whether it is on the positive or the negative side of the mean, is in the noncritical region and has the same meaning: no significant positive deviation, no ability. All outcomes not in the critical region are regarded as generated by chance. The student's hypothesis is supported (some would even say "accepted") if the result comes out in the critical region, which contains only the score of 6. The null hypothesis is supported or "accepted" if any score in the noncritical region, including scores 0 through 5, is obtained. The region of rejection of the null hypothesis in the two-tail test includes scores 6 and 0; the noncritical region includes scores 1 through 5.

Figure 10.1 illustrates the difference between one- and two-tail tests applied with the normal-curve sampling-distribution model and with the same confidence level of .05; the critical regions have been

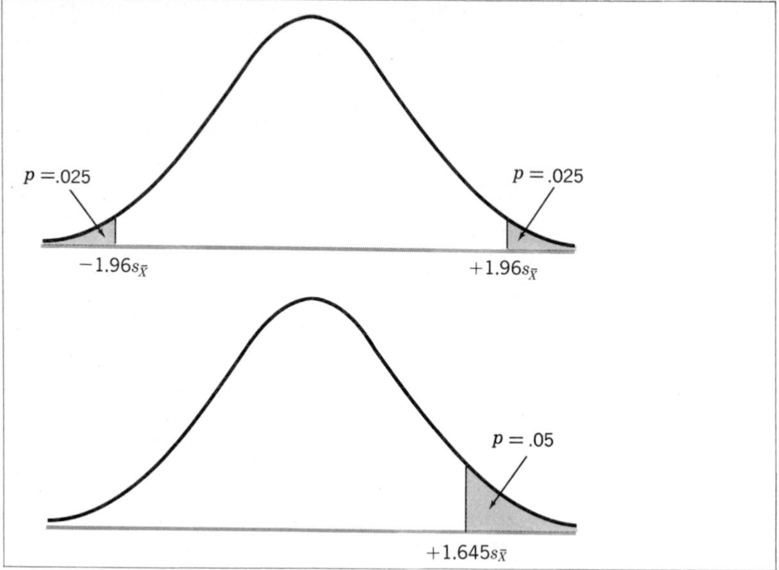

FIGURE 10.1 Two sampling distributions of means, showing the extreme (shaded) portions of area lying in the critical zone when the level of significance (alpha) is .05, for the two-tail test (where the critical zone is in two equal parts, in the two tails) and for the one-tail test (where the critical zone is all at one end of the distribution)

shaded. For the two-tail test, the critical region is evenly divided in the two tails of the normal distribution at distances of $1.96s_{\bar{X}}$ from the mean. It is all at one end of the distribution, beginning at a distance of $1.645s_{\bar{X}}$ from the mean, for the one-tail test.

COMBINING PROBABILITIES IN CRITICAL REGIONS In the psychophysical problem we concluded that a score of 6 is significant between the .05 and .01 levels, whether we apply a one- or a two-tail test. Let us ask whether a score of 5 would be significantly different from the mean in either case.

This question does not ask whether a score of precisely 5 is significant, even though it makes sense to state the probability of obtaining a score of exactly 5 and we see such a probability given in Table 10.1. What we are really asking is whether a score of 5 *or higher* deviates far enough from the mean to be sufficiently rare to lead to rejection of the hypothesis of a chance-generated mean of 3. This question makes sense because if a score of 5 is far enough removed to be significant, all scores further removed must also be significant. In the one-tail binomial-model test in this psychophysical experiment, the critical region under discussion combines scores of 5 and 6. The probabilities of these two scores individually are $^6/_{64}$ and $^1/_{64}$, respectively. The probability of obtaining a 5 *or* a 6 is the sum of the two probabilities, by application of the additive theorem regarding probabilities (see Chap. 7). The probability of obtaining a score of 5 or higher is $^7/_{64}$, which is decidedly in the noncritical region, and therefore we do not reject the null hypothesis. Just for comparison, if we were applying a two-tail test, we should include scores of 1 and 0 also, giving a total probability of $^{14}/_{64}$, or almost $^{1}/_{4}$.

SOME MORE GENERAL CONCEPTIONS OF HYPOTHESIS TESTING

From the preceding discussion it can be seen that a sampling distribution provides not only a model for what should happen in a particular chance situation but also a basis for a clear-cut division of the total outcome space[1] into two mutually exclusive regions for rejection and acceptance of alternative hypotheses, once we have adopted either a one-tail test or a two-tail test along with a confidence level. We shall now put these ideas in more general form.

Let us say that the hypotheses with which we are concerned have to do with a sex difference in verbal-comprehension ability. Logically, there are *three* alternatives: (1) males have more ability than females; (2) females have more ability than males; or (3) there is no sex difference connected with this ability. In terms of symbols, these three alternatives may be expressed as follows: $\mu_m > \mu_f$, $\mu_m < \mu_f$, and $\mu_m = \mu_f$, where the μ's stand for population means and the subscripts denote male and female.

Investigators who approach this problem with open minds simply ask, "Is there a genuine sex difference here?" They would make a two-tail test. They would combine the first two alternatives into one, stated as follows: $\mu_m \neq \mu_f$ (the mean for males is not equal to that for females). Their alternatives in their statistical test would be this hypothesis against the third, $\mu_m = \mu_f$. They are prepared to find a significant deviation from a zero difference in either direction and, if a significant difference is found, to accept it. If they reject the null hypothesis, in accepting the alternative, $\mu_m \neq \mu_f$, they also accept the algebraic sign as being meaningful. For if they were to mark off confidence limits for the mean difference, the confidence interval would be entirely or predominantly on one side of the point of zero difference (see Fig. 8.3, p. 130). Using the information provided by the algebraic sign, the investigators could not only reject the null hypothesis but also make a decision between the two possibilities included in the hypothesis $\mu_m \neq \mu_f$.

Investigators who think they have reason to favor the hypothesis $\mu_m > \mu_f$ or the hypothesis $\mu_m < \mu_f$—on logical grounds, on the basis of previous experience, or both—would make a one-tail test. If they believe that females are superior to males in verbal-comprehension ability, they will reduce the situation to two alternatives by making another kind of combination. That is, the hypothesis to be tested would be $\mu_m \geqq \mu_f$, with the alternative hypothesis (which they expect to be true) $\mu_m < \mu_f$. They would expect a significant deviation in the negative direction in the distribution of the quantities $\mu_m - \mu_f$

1 Such as the area under the unit normal distribution or the total of the frequencies of cases in a binomial distribution.

about the hypothetical mean (of the distribution of differences between pairs of sample means) of zero.

In either a one- or a two-tail test, the reduction of three alternative hypotheses to two is an important simplifying step that facilitates decision making. The two alternatives are often symbolized by H_0 and H_a. H_0 represents the hypothesis that is tested, and H_a is its alternative.

Errors in statistical decisions

THE CHOICE OF SIGNIFICANCE LEVEL Thus far, we have not considered very seriously the question of what significance level or levels to adopt. This topic might well have been discussed under "rules for statistical decisions," but it is so intimately connected with errors of decision that it is better discussed here.

Since the investigator controls the adoption of the significance level that is to set a boundary between the critical and noncritical regions, some guidance is required on this subject. Most statisticians insist that the investigator adopt a single standard of significance before the study or experiment starts. When the time for making a decision comes, it is easy to make because one follows the rule adopted in advance. The reasons for statisticians' insistence on this procedure will soon become apparent. Despite this urging, many investigators prefer not to adopt in advance any rigid standard of rejection or acceptance of hypotheses. They appear to be content to observe the level of significance achieved (in accordance with conventional limits such as the .05 and .01 levels) and to report their findings. We cannot discuss this matter adequately without considering errors in decision and their consequences.

TWO KINDS OF ERRORS OF STATISTICAL DECISION The choice of a standard of significance depends very much upon the amount of risk we are willing to take of being wrong in making the statistical decision to accept or reject the tested hypothesis. Two distinct types of error are possible:

Type I: rejecting hypothesis H_0 when in fact it is true
Type II: accepting hypothesis H_0 when in fact it is false

Figure 10.2 clearly displays these two kinds of errors in relation to two decisions and two alternatives for H_0 (true or false). With two categories of decision and two of veracity, four combinations are possible, two being correct decisions and two incorrect decisions.

PROBABILITIES OF ERRORS OF TYPES I AND II The probability of making a type I error is very simply and directly indicated by α, the probability level the investigators choose for rejecting H_0. Whether they make a

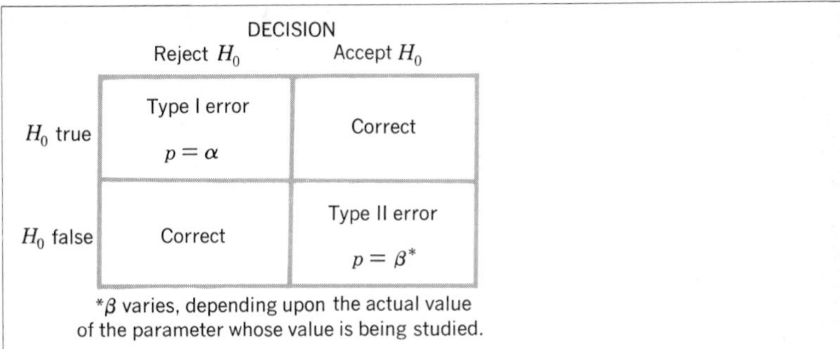

	DECISION	
	Reject H_0	Accept H_0
H_0 true	Type I error $p = \alpha$	Correct
H_0 false	Correct	Type II error $p = \beta$*

*β varies, depending upon the actual value of the parameter whose value is being studied.

FIGURE 10.2 The four cases generated by combinations of two decisions (reject or accept the hypothesis) with two *actual* situations (tested hypothesis true or false). Errors of types I and II are identified in two cells of the table.

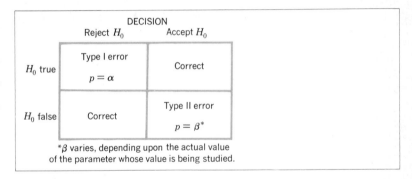

one- or a two-tail test, an alpha of .05 means that there are five chances in a hundred of their being wrong in rejecting H_0 when it is true. If they adopt the .01 level of significance, there is only one chance in a hundred of being wrong. With alpha equal to .001 there is only one chance in a thousand of being wrong. Thus, not only does alpha indicate the probability of making an error of type I, but also its relationship to that probability tells us that the smaller we make alpha, the less likely we are to make a type I error. Why, then, not adopt an extremely small alpha and practically never make such a mistake?

The trouble is that as we decrease α, we automatically increase the chances of making an error of the second type—of accepting the tested hypothesis when it is false. The probability of making an error of type II is symbolized by β (beta). Alpha and beta are inversely related: as the one decreases, the other increases. Alpha is under our direct control; beta is only indirectly under our control through its inverse relation to alpha.

But it should be said immediately that the relation between α and β, although inverse, is by no means simple. (The ways of estimating β will soon be explained.) For a concrete view of the inverse relationship, see Fig. 10.1. The shaded portions are in the critical regions: the regions for rejection of H_0. The clear areas under the normal curves represent noncritical regions: regions of acceptance of H_0. It is easy to see that as the shaded areas are made smaller, the noncritical regions grow larger, and if the true parameter is not exactly at the mean, the decision to accept H_0 is in error. Thus, increasing the amount of clear space increases the chances of errors of type II. Remember, however, that the amount of clear space is by no means a simple indicator of the size of beta. Beta plus alpha does not equal 1 except under a very special condition, which will be brought out later.

RELATIVE IMPORTANCE OF ERRORS OF TYPES I AND II The crux of the dilemma is how much weight we want to give to errors of the two kinds. Overly cautious scientists abhor the type I error more than they do the type II error. They want to be very sure that their finding is not due to chance. The conventional choice of alpha as small as .05 and .01 is evidence of the caution exercised by most investigators against making a type I error. Such decisions on choice of alpha are almost always made without consideration for beta. The result of too much caution, and very small alphas, is that relatively few nonchance conclusions are drawn, and few differences and relationships are accepted as "established."

Some kind of balance is called for. Considerations external to the data themselves should be noted and given weight. There may be serious theoretical or practical reasons why it would be costly to make one kind of error or the other. Thus, this question ultimately cannot be decided on purely statistical grounds. If certain commonsense decisions can be reached regarding the relative seriousness of the consequences of making each type of error, statistical statements can then often be introduced which will further guide the choice of alpha.

In research on important theoretical issues, such as whether or not telepathy and clairvoyance exist or whether there is inheritance of acquired traits, a higher-than-usual level of confidence (lower α) may well be demanded. The potential social impact of conclusions about these questions justifies such practice. If the investigation is on the selection of the best of several insecticides when one is sorely needed and none does any harm to noninsects, a larger α might well be tolerated. If it is a matter of the use of a new anesthetic, in a concentration needed for effectiveness but where there is a danger of death if given in overdose, a much smaller α might well be demanded.

In general scientific practice, where externally determined risks are of little or no consequence, there is another possibility. Instead of confining ourselves to a two-choice decision—rejection or acceptance—we might allow a third possibility, that of suspended judgment, which usually calls for a replication of the experiment. For example, if the deviation is significant at the .01 level or better, we might reject H_0; if the deviation is smaller than the boundary of the critical region at the .10 level, we might accept H_0. Between the two levels, .10 and .01, we might suspend judgment.

THE PROBABILITY OF AN ERROR OF TYPE II

When we wrongly reject hypothesis H_0, we are rejecting a specific value—for example, a difference of zero or a correlation of zero—and we have an available estimate of the probability of being wrong, namely, α. When we wrongly *accept* hypothesis H_0, however, there are many other values that may be correct. In this fact lies an important difficulty of estimating the probability of being wrong in making a type II error. We can make such estimates only for *specific* alternative hypotheses. Lacking a good reason for choosing any other hypothetical value as a specific alternative hypothesis H_a, the best we can do is to select arbitrarily a number of reasonable alternatives, each in the vicinity of the value for H_0. We shall next apply this approach, computing a probability β for a number of specific alternative hypotheses.

DETERMINING BETA PROBABILITIES FOR A ONE-TAIL TEST The simpler case is that with a one-tail test. Consider the hypothetical problem in which an aptitude test has been given to a certain group of 50 students in the ninth grade randomly chosen from a school system. The national norm for the test is a mean of 50 with a standard deviation of 8.63. The teacher's impression is that his group averages a little below the national norm. A statistical one-tail test is made to determine whether the mean score for the group is significantly below the norm. The mean score obtained for the group was 48.2. With an SE of the mean of 1.22, the obtained mean failed to reach a deviation significant at the .05 level.

This experimental situation is shown graphically in the first distribution in Fig. 10.3. With alpha equal to .05, it takes a deviation of $1.64s_{\bar{X}}$ to be significant, or a deviation of 2.0 score points. The critical region marked off by an alpha of .05 has its upper boundary at a score of exactly 48. This boundary is denoted as C, for critical limit. The noncritical region is above C. It figures very importantly in determining the beta probabilities because it is the region representing all acceptances of hypothesis H_0.

Although hypothesis H_0 can be accepted, the truth may be that the population mean is located somewhere other than at exactly H_0. As a first choice for a specific alternative H_a, let us consider the value 49.

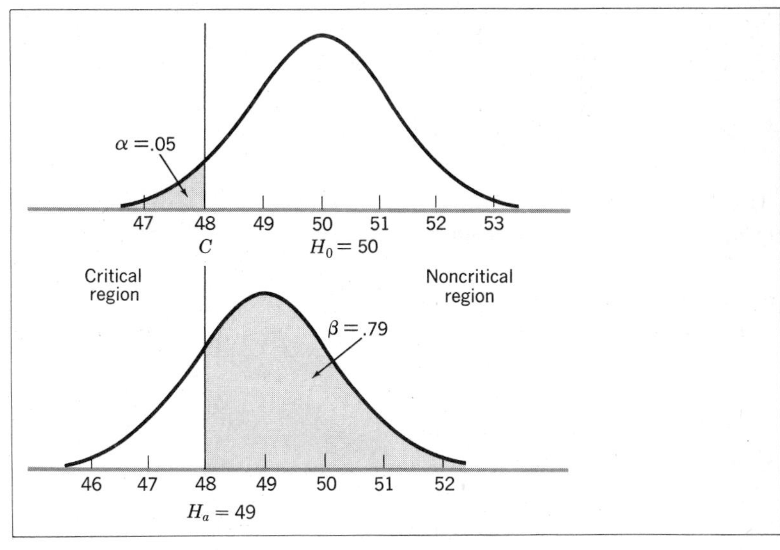

FIGURE 10.3 Illustration of the probability of an error of type II as the area under the normal curve lying in the noncritical zone, when the tested hypothesis is a mean of 50. The test is one-tail with an alpha of .05, and the special alternative hypothesis is a mean of 49

Now we have to think in terms of a new hypothetical sampling distribution, normal in form, with an SE of 1.22 but with a mean of 49. The second distribution in Fig. 10.3 shows this new hypothetical picture. The critical limit is still at 48, with critical and noncritical regions below and above it, respectively, as before. All the score values above 48 still represent those which, if obtained as sample means, would lead to acceptance of H_0, but now not so much of the sampling distribution is above the critical point. Therefore, an error of type II would not be made so often.

The probability of a type II error is given by the shaded area under the second normal-distribution curve in Fig. 10.3. This proportion of the area is evaluated in the usual way from a knowledge of the z value of a score at the cutoff point, 48. The z for a score of 48 is given by the ratio $(48 - \bar{X})/s_{\bar{X}}$, which is $(48 - 49)/1.22$, or -0.82. From the normal-curve table we find the shaded area to be .79, which is the beta probability for the H_a of 49.

We go through similar procedures for other H_a values in the region of 48, with results shown in Table 10.2. It will be noted that the beta proportions vary from a high of .993 when H_a is 51 down to .007 when H_a is 45. The higher the special hypothetical mean, the more of the sampling distribution is above the critical point, *which remains at 48 throughout these operations*. We include H_a equals 50 in spite of the fact that 50 is also H_o. When H_a equals H_0, we find that β is the complement of α; that is, $\alpha + \beta = 1.0$. This is the only case in which that simple relationship holds. The noncritical part of the first distribution in Fig. 10.3 is obviously .95 of the total area.

THE POWER OF A STATISTICAL TEST; POWER FUNCTIONS Whereas beta gives us the probability of making a type II error, its complement, $1 - \beta$, generally indicates the probability of *not* making a type II error. The way not to make a type II error is to reject hypothesis H_0. The probability of rejecting H_o when it is not true is known as the *power of a statistical test*. The last column of Table 10.2 gives the power indices under the conditions of the various hypothesized true

TABLE 10.2 Determination of probabilities of errors of type II and power values of a one-tail test of the hypothesis that $\mu = 50.0$ with $s_{\bar{x}} = 1.22$ and $\alpha = .05$

H_a	z	β	$1 - \beta$
51	−2.46	.993	.007
50	−1.64	.95	.05
49	−0.82	.79	.21
48	0.00	.50	.50
47	+0.82	.21	.79
46	+1.64	.05	.95
45	+2.46	.007	.993

population values in the illustrative problem. As the true mean decreases, the power increases, since the lower the actual mean of the population the less likely it is that the sample means will go into the noncritical region.

The relationship of the power index $1 - \beta$ to hypothetical parameters (μ) is a continuous function. In Table 10.2 only selected parameter values are represented for H_a. A plot of the relation of the power index to H_a is shown in Fig. 10.4. Such a function can be constructed to fit any case of hypothesis testing.

The curve will vary under different conditions. Had we chosen a smaller alpha, the betas would have been larger, as one can see from considering what would happen in Fig. 10.3 if the critical limit were at 47.5 instead of 48, for example. If the betas are larger, the power indices become smaller. In the illustrative problem, the one-tail test was at the lower end of the distribution. When it is at the upper end, the power function is found to be a rising curve on the upper side of H_0. With a two-tail test the power function is composed of two branches, as we shall see next.

DETERMINING THE BETA PROBABILITIES IN A TWO-TAIL TEST Finding the beta probabilities of type II errors in the instance of a two-tail test is a bit more complicated, but follows the same principles as for a one-tail test. This procedure will be illustrated by another hypothetical problem. This time, the hypothesis to be tested, H_0, is that there is no difference between two population means μ_1 and μ_2. The size of sample was 50, with an SE of 2.55 and an alpha of .05, divided in two tails of a normal distribution. The first distribution in Fig. 10.5 illustrates the sampling of differences $\mu_1 - \mu_2$, with two critical limits of

FIGURE 10.4
A power function for a one-tail test, with the probability $1 - \beta$ shown as a function of various special alternative hypothetical means H_a

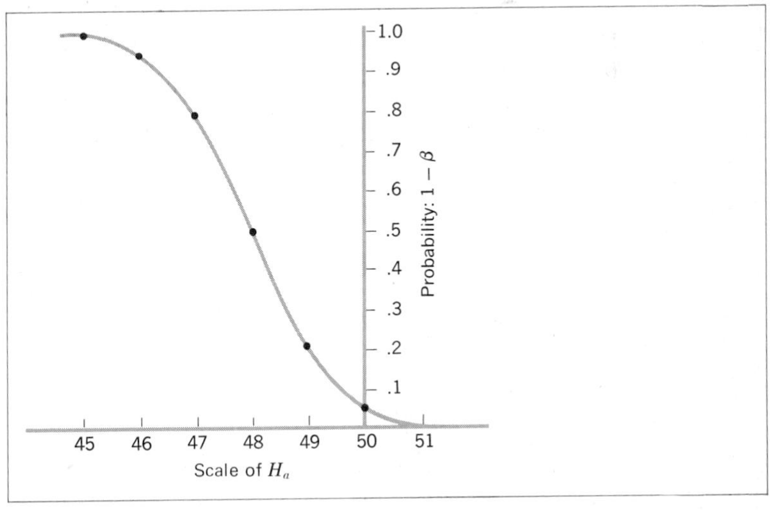

FIGURE 10.5 Illustration of the probability of committing an error of type II with a two-tail test, with alpha equal to .05 and a special alternative hypothesis of a mean difference of +6

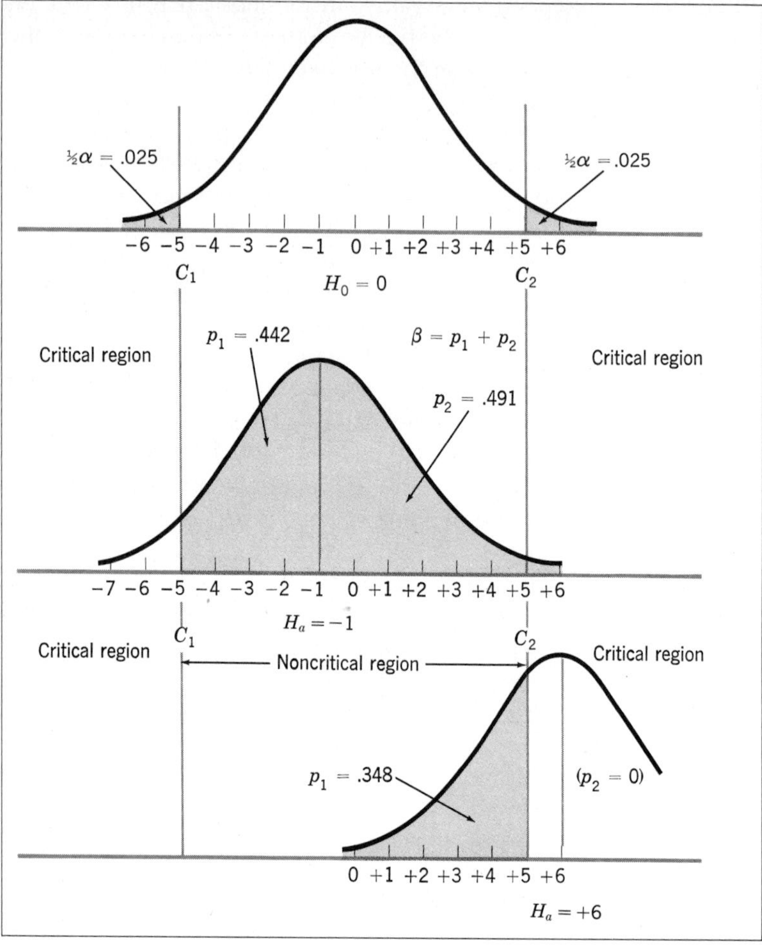

C_1 and C_2 dividing the total range of differences into a central, noncritical region and two critical regions on either side.

Again, we must successively adopt different H_a values and determine a beta for each. The second distribution in Fig. 10.5 illustrates the case of H_a equal to −1 (the supposition that $\mu_1 - \mu_2 = -1$). The sampling distribution of mean differences is about a mean of −1, with an SE of 2.55. As before, the beta probability is represented by the total area under the sampling distribution falling within the noncritical region. It is most feasible to determine this area in two portions: that below the mean of the sampling distribution for H_a and that above, indicated by p_1 and p_2, respectively. We note the z deviation on the base of this distribution at the point C_1, which is at a value of −5 on the scale of differences. The z is therefore found by the usual ratio, $(X - \bar{X})/s$, which in this case is $(-5 - \bar{X})/s_{d\bar{X}}$ or $(-5 - -1)/2.55$, or

-1.57. From the normal-curve table, the area between -1.57 and the mean is .442. This is the p_1 component contributing toward beta. In a similar manner, we find that C_2 cuts the sampling distribution at a z of $+2.35$, which marks off .491 of the area from the mean. The sum of the two components p_1 and p_2 is .933. This is beta, the probability of making a type II error under the conditions of the chosen critical limits and the chosen H_a of -1.

Table 10.3 presents other betas, found in a similar manner, of H_a values from -4 to $+4$. Within this range, both critical limits are involved. Beyond this range, only one critical limit is appreciably concerned, and the procedure must be somewhat different. Note the third distribution in Fig. 10.5, which illustrates this kind of case. Here the H_a is $+6$, which is *above* the limit C_2. The p_1 region is now still in the lower part of the sampling distribution, but it is the *tail* area rather than the area between z and the mean. Except for this difference, the operations are just the same. The value of p_2 is zero for all H_a hypotheses at and above C_2, and p_1 is zero for all H_a hypotheses below C_1.

THE POWER FUNCTION IN A TWO-TAIL TEST The procedure for obtaining a power function is just the same as in a one-tail test. It is composed of relations of the power values $1-\beta$ to H_a. Table 10.3 shows that the power value is at a minimum equal to alpha when $H_a = H_0$, as before. Figure 10.6 presents the graphic picture of the power function.

TABLE 10.3 Determination of probabilities β that errors of type II will be made, assuming various hypothetical differences between means, and also probabilities by which a power function is determined (see Fig. 10.5)

H_a*	z_1	z_2	p_1	p_2	β	$1-\beta$
+12		−2.75	.003	.000	.003	.997
+10		−1.96	.025	.000	.025	.975
+ 8		−1.18	.119	.000	.119	.881
+ 6		−0.39	.348	.000	.348	.652
+ 4	−3.53	+0.39	.500	.152	.652	.348
+ 2	−2.75	+1.18	.497	.381	.878	.122
+ 0	−1.96	+1.96	.475	.475	.950	.050
− 2	−1.18	+2.75	.381	.497	.878	.122
− 4	−0.39	+3.53	.152	.500	.652	.348
− 6	+0.39		.000	.348	.348	.652
− 8	+1.18		.000	.119	.119	.881
−10	+1.96		.000	.025	.025	.975
−12	+2.75		.000	.003	.003	.997

* H_a = the hypothetical difference being tested; z_1 = the ratio $(-5 - H_a)/2.55$; z_2 = the ratio $(5 - H_a)/2.55$; $\beta = p_1 + p_2$.

FIGURE 10.6
A power function for a two-tail test of differences between means

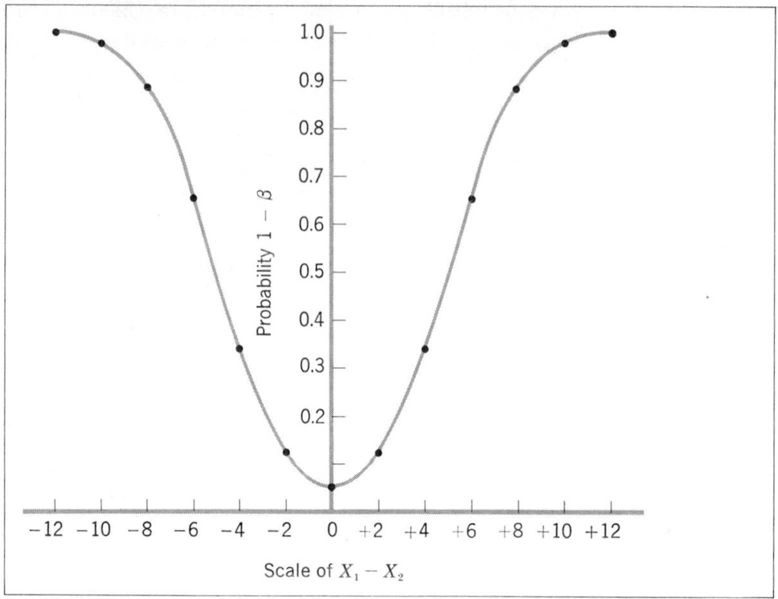

If a one-tail test had been applied to the differences between means in this illustrative problem, with the same critical point at the end of the distribution (alpha equal to .025), the betas would have been much smaller, and the power of the test correspondingly greater. For the same size of deviation from the null hypothesis and the same obtained deviation found in the sample, then, a one-tail test is more powerful than a two-tail test.

WAYS OF INCREASING POWER OF STATISTICAL TESTS There are other ways of increasing the power of a statistical test. Some kinds of tests are more powerful than others, and in the following chapters references to power will occasionally be made. But the most important factor in increasing the power of statistical tests is reduction in the size of the SE. The most obvious way in which to reduce the size of the SE is to use a larger N, since SE $= s/\sqrt{N}$. Other devices for reducing the SE include using matched samples and stratified sampling. We shall next consider the question of what size samples would be needed to bring the probability of errors of decision down to desired levels.

Needed sample sizes

In starting an investigation, the question naturally arises, "How large a sample do I need?" Statistical thinking enables us to develop satisfactory answers to this question, provided we have sufficient information of certain kinds. Heretofore, we have been concerned with a quite different question, namely, how much deviation from a null hypothesis we need for rejection of the tested hypothesis.

The question of size of sample is often of practical significance, as in public-opinion polls, where the collection of data is expensive. An answer to the question can be reached taking into account the probability alpha only or both alpha and beta.

ESTIMATION OF SAMPLE SIZE NEEDED ON THE BASIS OF ALPHA ONLY

The question of sample size should ordinarily be answered with due consideration given to errors of decision of both types, I and II, and this can be done. Sometimes, however, not enough information is available to take type II errors into account. In spite of this lack, a great deal can be done simply in terms of ensuring rejection of a null hypothesis at a specified level of significance. We shall consider this kind of case first.

Let us assume a public issue where a majority vote is decisive. Let us also assume that a sample of opinion has been properly obtained, with good representation of the voting population, assuming random sampling.[1] The null hypothesis implies a mean proportion of .50. We ask first how large a sample will be needed to give us confidence that an obtained poll result of .55 in favor of the proposition means a majority sentiment in that direction and is not merely a chance-generated result from a population in which the split is exactly 50:50. The formula we need is

$$N = \frac{z^2\sigma^2}{d^2} \qquad \text{(}N\text{ needed to achieve a significant deviation of a specified amount, alpha being known)} \qquad (10.1)$$

where z = normal-curve deviate corresponding to alpha
σ = standard deviation of the population
d = specified deviation

In the polling problem under discussion, in a one-tail test, let alpha be .05, from which z (from the normal-curve table) is 1.64. The deviation d is .05 (.55 − .50). The population variance σ^2 is pq.[2]

$$\begin{aligned} N &= \frac{1.64^2(.5 \times .5)}{.05^2} \\ &= \frac{2.6896(.25)}{.0025} \\ &= 268.96 \qquad \textit{or 269 as an integer} \end{aligned}$$

[1] In much public-opinion polling, of course, sampling is stratified-random, in which case the SE would be smaller.

[2] The expression pq stands for the variance in a point distribution (where observed values are 1 and 0 only).

A similar application of the same formula to the two-tail case, where, with an alpha of .05 z is 1.96, gives a needed N of 384. It requires an N of 384 to lead to the rejection of the hypothesis that the population is evenly divided on the issue when the obtained deviation is .05 in either direction.

But where much is at stake, we should not be satisfied with these odds against the null hypothesis. We might reduce the probability of a type I error to .01 by lowering alpha to that quantity. In this case, the obtained p must be at $2.58s_p$, and formula (10.1) estimates a required N of 666. On very critical issues, we might demand a much smaller d_p than .05; it might be .01. With d_p equal to .01 and an alpha of .05, N should be 9,604. With the same d_p and an alpha of .01, N would need to be about 16,640. Thus, for the detection of very small differences with high assurance against errors of type I, samples must be of considerable size.

SIZE OF SAMPLE WHEN ERRORS OF TYPE II ARE SPECIFIED The procedure just described is satisfactory when the results are sufficiently decisive to reject H_0. Then we must be concerned with a type I error, and the alpha level tells the probability of having made that kind of error. But if the outcome is in the noncritical region and we do not reject the tested hypothesis, what is the possibility that we have overlooked some genuine departure from H_0, that we have made an error of type II? The following procedures allow for the likelihood of such an error.

THE CASE OF A ONE-TAIL TEST The procedures for estimating the N needed to fit a specified test situation are not different for one- and two-tail problems. But there are different logical problems, and therefore the two cases will be discussed separately.

In order to solve any specific problem, we must have certain information, and certain choices must be made. We must choose not only alpha but beta as well, which means that we must weigh carefully the alternatives of risking each kind of error. We must know the value of H_0 for the hypothesis we want to test, and we must be concerned with a specific alternative hypothesis H_a, a requirement that we encountered before in consideration of beta probabilities. Another piece of needed information is a good estimate of the population standard deviation.

As an illustration, let us return to the problem of aptitude testing, in which a one-tail test was applied. The H_0 value was an assumed national-norm value of 50.0. The sample size was 50 and the best estimate of the population σ was 8.63, giving an SE of 1.22. Let us say that the specific alternative hypothesis (H_a) adopted is a mean score of 48. Keeping the alpha level at .05, let us say that we are willing to take the risk of .10 of an error of type II; that is, beta equals .10. The alpha of .05 and the beta of .10 indicate that we are more willing to risk an error

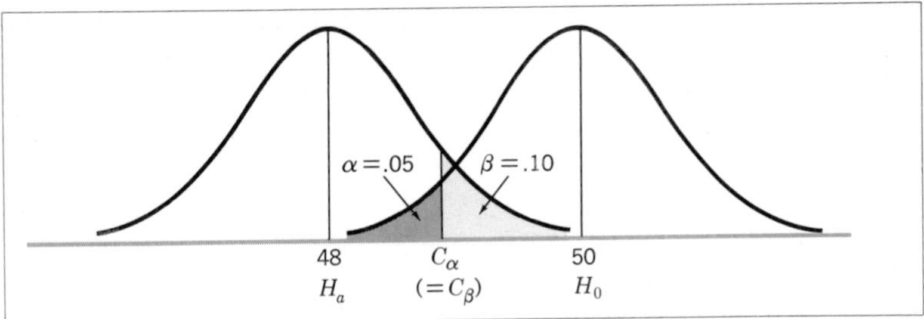

FIGURE 10.7 Two sampling distributions illustrating the arrangement whereby a certain deviation from a mean H_0 (the tested hypothesis) with an alpha probability of .05 is identical with the deviation from the mean H_a (the alternative hypothesis) with a beta probability of .10. The illustration applies to the problem of estimating the size of sample needed to reject hypothesis H_0 under the conditions illustrated, for a one-tail test.

of type II than to risk one of type I: we are more concerned about wrongly rejecting hypothesis H_0 than we are about wrongly accepting it when the obtained mean deviates -2 from H_0.

Figure 10.7 presents the situation graphically. The sampling distribution about the mean H_0 is given at the right, with the critical limit C_α being determined at $-1.64s_{\bar{X}}$ by the choice of .05 for alpha. The critical region for rejection is left of C_α. The sampling distribution on the assumption that 48 is the true mean is at the left. The region for errors of the second kind is to the right of the critical limit C_β, which now also becomes a critical limit for deciding to accept H_0 when H_a is actually true. The curves overlap in such a way that *the minus deviation of $z = -1.64$ from mean H_0 coincides with a plus deviation of $+1.28$ from mean H_a*. The fact that C_α and C_β are at identical values on the scale of means is the key to the solution of the problem of sample size.

C_α and C_β can be at the same point on the scale only when there is a certain SE of the mean for both distributions, and this also means only when N is of a certain value. It is that value of N that we seek. Fortunately, we can express C_α and C_β in terms of information that we know, with only N unknown. The formula is

$$N = \frac{\sigma^2(z_\beta - z_\alpha)^2}{D_{0a}{}^2} \qquad \text{(Size of sample needed for given } \alpha \text{ and } \beta \text{ with a one-tail test)} \qquad (10.2)$$

where σ = population standard deviation
z_α = distance of C_α from H_a in σ units (with appropriate sign)
z_β = distance of C_β from H_o in σ units (with appropriate sign)
D_{0a} = scale distance between H_a and H_0

Using the data illustrated in Fig. 10.7,

$$N = \frac{8.63^2[1.28 - (-1.64)]^2}{2^2}$$

$$= 158.75 \qquad \text{or, in whole number, 159}$$

The validity of the solution can be checked by computing a new SE of the mean, using the new N, and then determining whether the two hypothetical means, 50 and 48, with a new critical limit, would yield a beta probability of .10.

THE CASE OF A TWO-TAIL TEST For an illustration with a two-tail test, let us use the same data from which beta probabilities were derived in testing the difference between two means. H_0 was a mean difference of zero. Alpha for the two-tail test was .05, with .025 being given to the two critical regions. Retaining the same alpha, let us tolerate an error of type II with a probability of .10. Let us assume a deviation D_{0a} of 4.

It turns out that formula (10.2) also applies to this situation because we are concerned with overlapping of only the two tails, as in Fig. 10.7. The only information yet lacking for the use of formula (10.2) is the population standard deviation. We know that the standard error of the mean difference is 2.55. From this and other known information, the population standard deviation σ_x is estimated to be 12.7. The critical limit C_α or C_β is at a z_α of -1.96 and a z_β of $+1.28$. With D_{0a} of four units, we have

$$N = \frac{(12.7)^2[1.28 - (-1.96)]^2}{4^2}$$

$$= 105.8 \qquad \text{or, rounded upward, 106}$$

It is interesting to note that if we specify a beta probability of .50, z_β becomes zero for use in formula (10.2). In this case, this formula reduces essentially to formula (10.1), in connection with which the probability of an error of type II is not explicitly stated. But that probability is, in fact, .50. In applying formula (10.1), therefore, we are implicitly specifying a situation in which we are as likely as not to make an error of type II.

Testing hypotheses with the binomial model

Something more must be said regarding the use of the binomial model in hypothesis testing. We saw a simple case of this kind of application early in this chapter in connection with the problem of the student with the psychophysical-judgment claim. The particular model used was the distribution of probabilities given by expansion of the binomial $(1/2 + 1/2)^6$. Exact probabilities could be stated for departures from the mean of the distribution. Wherever the probabilities of two alternative, mutually exclusive events add up to 1.0, use of such an approach is possible. So many experimental situations can be brought under such a pattern that the binomial test has very wide application.

As we saw in Chap. 7, when the mean Np is sufficiently large, a binomial distribution approaches a normal distribution with the same mean and variance so closely that a z test can be applied. This was the approach taken in Chap. 8, where large samples were supposed. The polling problem used earlier in this chapter to illustrate the estimation of the size of sample required to reject a null hypothesis is another example employing the normal-curve approximation. The normal-curve substitution is a much more efficient way of testing a hypothesis where the binomial model applies, but with the smaller samples in which this approach can be taken, there is a need for a minor *correction for continuity*. This kind of correction will be explained, and some varied problems in which the binomial model can be applied will be mentioned.

CORRECTION FOR CONTINUITY

In a certain elementary-psychology laboratory experiment, there is the problem of determining whether students can perceive from photographs whether or not a man has been convicted of a crime. Pictures of 20 pairs of men matched for certain qualities are exhibited, and the students are to judge which one of a pair is probably the criminal. The null hypothesis calls for 10 correct responses, provided only random guessing has accounted for the score. How much higher must the score be before it may be taken to indicate the operation of something other than chance?

Now we could expand the binomial $(\frac{1}{2} + \frac{1}{2})^{20}$, but that would be quite a task without the aid of a high-speed computer. For help in such problems, where N does not exceed 25, Table N in the Appendix is provided. It gives exact probabilities to three decimal places of outcomes in one tail of a binomial distribution, when $p = q = \frac{1}{2}$. For example, with $N = 20$, as in the criminal-judgment problem, we find the first entry of .001 under the heading of either category 3 or category $N - 3$ (i.e., 17: both are the same distance from the mean). This means that a score as high as 17 or higher would occur by chance only once in 1,000 times. The same statement can be made about a score as low as 3 or lower. The next entry at the right in the row for N of 20 is .006, which means that a score of 16 or higher would be expected 6 times in 1,000. This is beyond the .01 level. A score of 14 would not achieve significance at the .05 level (one-tail) because the probability given for the category of 14, $(N - 6)$, or higher is .058.

For the experimental problem at hand, however, Np equals 10, which is large enough to justify a normal-curve approximation. Yet N is small enough to require a correction for continuity. The appropriate normal-curve model is given in Fig. 10.8. The SE of this distribution is $\sqrt{Npq} = 2.236$. We can ask the same question concerning this distribution that we just asked concerning the binomial distribution: What are the probabilities of scores as high as 16, 15, and 14?

At this point, the correction for continuity must be considered.

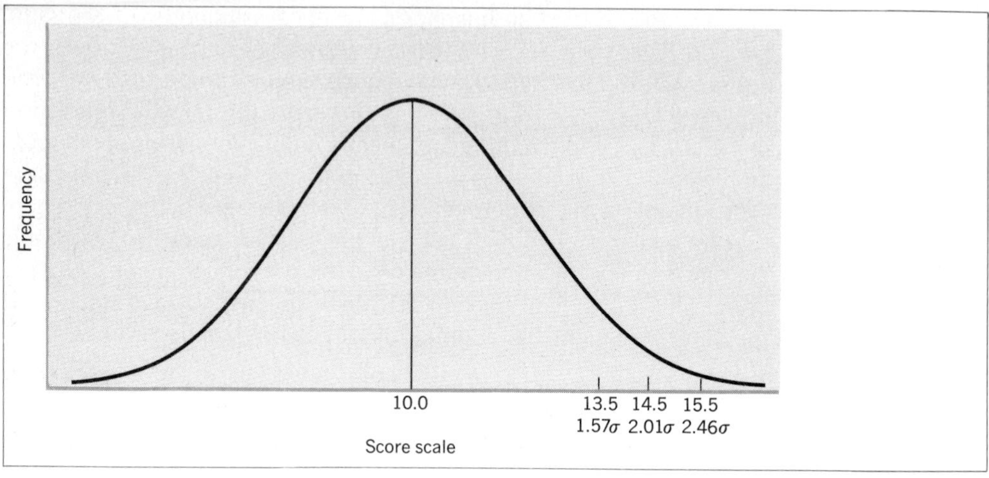

FIGURE 10.8 Illustration of the need for correction for continuity when a normal-curve model is substituted for a binomial-distribution model having the same mean and standard deviation. Integral values are points on a binomial scale, but represent ranges of values on a normal-curve scale.

The binomial distribution is on a scale of integral values, where 16 means exactly 16.0 and 15 means exactly 15.0. The probabilities found in Table N in the Appendix pertain to observations at exact, integral scores. The normal curve, however, is on a base line with continuity, where fractional values have real meaning. A score of 16 extends from 15.5 to 16.5. When we ask the probability for a score of 16 or higher in connection with the normal-curve model, the critical point is not at 16.0 but at exactly 15.5.

A score point of 15.5 deviates 5.5 units from the mean of 10. A deviation of 5.5 divided by the SE, 2.236, gives $z = 2.47$. Reference to the normal-curve table shows that there are about 7 chances in 1,000 of reaching that deviation. Table N gives .006, which shows that with correction for continuity, we have an excellent approximation. A score of 15 begins at 14.5, a deviation of 4.5 units from the mean. The corresponding z is 2.01. The corresponding probability is .022, which is very close to the .021 given for a score of 15 in Table N. Now if we had used a deviation of 5.0 for a score of 15, the z would have been 2.24 and the tail probability would have been .013, appreciably too small.[1]

The need for a correction for continuity is thus clearly demonstrated by this example. The correction is made simply by deducting .5 from the deviation of the integral-score value from the mean. If one chooses particular score *points* on the normal-curve base line, of

[1] The student might check the probabilities for a score of 14 or greater, with and without correction for continuity, in comparison with the result from Table N.

course, no such correction is necessary. It is needed only when talking about integral-score *categories*. We shall discuss a similar need for correction in certain applications of chi square in the next chapter.

SOME OTHER APPLICATIONS OF BINOMIAL TESTS It would be unfortunate if we left the impression that the binomial model is useful only when $p = q = \frac{1}{2}$, although no doubt this is one of the more common cases. Let us return to the ESP problem mentioned in Chap. 7, where there is a probability of $\frac{1}{5}$ of making a correct chance response in guessing the ESP cards. Let us suppose a sample run of 50 guesses, for which the model is the binomial $(\frac{1}{5} + \frac{4}{5})^{50}$. The mean Np is 10, and the SE is $\sqrt{50 \times .2 \times .8} = 2.83$. We may use a normal distribution with a mean of 10 and an SE of 2.83. This time, let us ask a different kind of question: What scores higher than 10 will indicate significance at the .05 and .01 levels?

The z deviations in a one-tail test must be 1.64 and 2.33, at the .05 and .01 levels, respectively. The product of each of these times the SE gives the deviation x in each case, with $1.64 \times 2.83 = 4.6$ and $2.33 \times 2.83 = 6.6$. Added to 10, these deviations give score points of 14.6 and 16.6. Remembering the need for a correction for continuity, we conclude that an integral score of 15 is needed for significance at the .05 level, and a score of 17 for significance at the .01 level.[1]

DEPARTURES FROM RANDOM CONDITIONS Other common applications occur in connection with multiple-choice tests (with two to five possible responses) and similar situations of various kinds. A word of caution should be offered concerning such applications, however. Experience tends to show that in the absence of knowledge, human beings do *not* always guess at random. They sometimes exhibit patterns of responses or pattern habits. With such biases present, hypotheses based upon chance distributions must be advanced with caution and sometimes are precluded.

The presence of bias cannot easily be detected, but one kind of evidence of it would be a "significant" deviation in an "unreasonable" direction, as when in a guessing situation a significantly large number of *wrong* responses or judgments occur. Goodfellow has shown in connection with "experiments" on telepathy over the radio, for example, that when an audience made five successive guesses of "black" versus "white," there were a number of common sequence patterns.[2] Alternations occur less frequently than one would expect by chance, runs are avoided, and certain initial responses may be fa-

1 By overlooking the slight discrepancies between the limits 14.6 and 16.6 and the actual score limits of 14.5 and 16.5.

2 Goodfellow, L. D. The human element in probability. *Journal of General Psychology,* 1940, **24,** 201–205.

vored, sometimes in response to an incidental cue that the experimenter overlooks. The presence of such nonrandom effects is bothersome, but experimental controls that may help to reduce their occurrence can be executed. There is probably enough randomness in a wide range of behavior to permit profitable use of statistical tests that depend upon that condition.

EXERCISES

1 Suppose that we ask an observer to arrange a series of weights in rank order from heaviest to lightest, the differences between them being very small. What is the probability that he could place them in perfect rank order by random guessing? For each set of weights, regardless of the number of weights in the set, there is only one way of achieving the correct order. The total number of ways for a set is $k!$ (k factorial), where k is the number of weights in the set.

a. For the cases in which k is 3, 4, 5, 6, and 7, in turn, determine the total number of ways in which the weights could be ordered.
b. State the probability for the case of perfect ordering for each set.
c. Which of these probabilities lead to rejection of the random-guessing hypothesis at the .05, .01, and .001 levels?

2 In a discrimination-learning experiment, a rat can make two alternative responses, one of which is correct in each trial. The correct response is on the right in random sequence. During a particular set of 12 trials, the rat is correct in 9.

a. Set up a binomial model that should be used in testing the null hypothesis.
b. State the hypothesis.
c. Determine the probability of a score as high as 9 (in a one-tail test), using the appropriate binomial model (see Table N in the Appendix).
d. Do the same, using the best-approximating normal distribution.

3 An observer is told that he will hear one of three speech sounds in a series of stimulations. He is given sets of three sounds each for identification of the "right" sound, with a total of 30 such trials. How many correct responses must he give in the 30 trials if we are to regard his success as significant at the .05 and the .01 levels in a one-tail test?

a. State the binomial model that applies (without expanding it).
b. Make a test by using the most appropriate normal-curve model. What are the mean and SD for that model?

4 A certain four-choice test contains 40 items.

a. How large a score must a student make before you can feel that he probably knows something about the subject matter of the examination? Define "probably" statistically.
b. How large a score must he make before you can feel that he definitely knows something about the subject? Define "definitely" statistically.

5 In the kind of polling problem mentioned in this chapter, with a hypothetical mean of 50.0 percent, in a sample of 144 interviewees, determine what percentage points mark off the critical regions with alphas of .05 and .01:

a. In a one-tail test.
b. In a two-tail test.
c. Why is the correction for continuity not applied in solving these problems?

6 On the polling problem, again, with $p = .50$ and $N = 144$:
a. Determine the beta probabilities for the one-tail test in the upper tail, with alpha of .05, at H_a values of .45, .50, .55, .60, .65, and .70.
b. Derive the power function and plot it on graph paper.

7 In an examination composed of five-choice items, how many items would you need to include in order to have confidence at the .95 and the .99 levels (in a one-tail test):
a. That a score of 30 percent right indicates some knowledge of the subject matter of the examination?
b. Answer the same question with respect to 25 percent correct responses.

8 Also in connection with the polling problem, assuming a one-tail test (with $\sigma = .5$ and with $z = 1.645$ at the .05 level):
a. Determine the size of sample needed with $\alpha = .05$ and $\beta = .05$ and with the alternative hypothesis H_a at .55.
b. Do the same, with $H_a = .51$.
c. Repeat parts *a* and *b*, not specifying the probability of a type II error.
d. What beta probabilities are associated with the solutions in *c*?

ANSWERS

1 a. k: 6, 24, 120, 720, 5,040.
b. p: $1/6$, $1/24$, $1/120$, $1/720$, $1/5{,}040$.
c. The p for four weights indicates significance beyond the .05 level; that for five weights, beyond the .01 level; only that for seven weights indicates significance beyond the .001 level.

2 a. Binomial model: $(1/2 + 1/2)^{12}$.
b. The rat's choices of responses are in purely random sequence.
c. Binomial solution: $p = .073$ (in a one-tail test).
d. Normal-curve solution: $\mu = 6$; $s_f = 1.73$; $z = 1.45$; $p = .074$.

3 a. Binomial: $(1/3 + 2/3)^{30}$; $\mu = 10$; $s_f = 2.58$.
b. Mean = 10.0; SD = 2.58; approximate integral scores required: 15 and 17, respectively.

4 $\mu = 10$; $s_f = 2.74$; score points: 14.5 and 17.4; integral scores: 15 and 18, respectively.

5 a. Critical percentage points: 56.8 and 59.6, at the .05 and .01 levels, respectively.
b. Critical percentage points at the .05 level: 41.9 and 58.1; at the .01 level: 39.4 and 60.6.
c. The critical points pertain to the hypothetical normal-curve model only, not to obtained integral scores.

6 a. The beta probabilities: .998, .950, .672, .225, .025, and .001, for hypothesized values .45 to .70, respectively.
b. The plotted values should be the complements of the betas just listed, i.e., $1 - \beta$ in each case.

7 a. N: 44 and 87, at the .05 and .01 levels, respectively.
b. N: 174 and 347, at the .05 and .01 levels, respectively.

8 a. With H_a at .55 $N = 1{,}082.41$, or 1,083.
b. With H_a at .51, $N = 27{,}060$.
c. With H_a at .55, $N = 271$; with H_a at .51, $N = 6{,}765$.
d. Associated with the latter N's are beta probabilities of .50 and .50.

11 Chi square

Although the statistical tests covered thus far are quite varied and their applications are numerous, they do not provide for all our needs. One reason is that they are limited to the evaluation of one statistic or one difference at a time. In this chapter and the next two, we shall examine a considerably expanded repertoire of statistical tests. The first of these to be considered is the versatile statistic called *chi square*.

General features of chi square

Chi square is used with data in the form of frequencies, or data that can be readily transformed into frequencies. This includes proportions and probabilities. One important feature of chi square is its additive property, which makes possible the combination of several statistics or other values in the same test. Thus, a hypothesis involving more than one set of data can be tested for significance.

THE BASIC NATURE OF CHI SQUARE

The fundamental nature of chi square can be very simply, if not completely, explained on the basis of what is already known about z, the standard score or measure. When there is one degree of freedom, chi square is identical with z^2, or

$$\chi^2 = z^2 = \frac{(X - \mu)^2}{\sigma^2} \qquad \text{(Mathematical relation of } \chi^2 \text{ to } z^2\text{, with 1 df)} \qquad (11.1)$$

where X is any measurement in a normally distributed population, μ is its mean, and σ is its standard deviation. Now suppose that we have a sampling situation in which there are k mutually independent measures of X. There are also k mutually independent z values and k mutually independent χ^2 values. It is a most useful property of χ^2 that a sum of k mutually independent chi-square values is also a χ^2, with k degrees of freedom.[1] In terms of an equation,

$$\chi^2 = \Sigma z^2 = \sum \frac{(X - \mu)^2}{\sigma^2}$$

where it is understood that k values are summed.

CHI SQUARE AS A SAMPLING STATISTIC Like z, chi square can also be used as a sampling statistic. Just as z has a sampling distribution, so has chi square. The sampling distribution of z is normal, but since χ^2 is related more directly to z^2, its sampling distribution is definitely not normal. Something will be said about its sampling distribution below. Here we shall pursue further the relation of χ^2 to z.

Suppose we have taken a very limited opinion poll in a small sample of married male graduating seniors in a certain university. Of the 40 men who were questioned, 28 felt that it is a good idea for undergraduates to be married, and 12 disagreed. Could these frequencies have arisen from a population in which the opinion is evenly divided on the experimental question? The null hypothesis for this instance of hypothesis testing is a 50-50 division.

With the 50-50 hypothesis, the expected frequency in a sample of 40 is Np, which equals $40 \times .5 = 20$. We therefore assume a population distribution of frequencies with a mean of 20 and a variance of Npq, where $p = q = .5$. In using chi square, the mean is known as an *expected frequency* f_e. With it is to be compared an *obtained frequency* f_o. Does the obtained frequency of 28 differ significantly from the frequency of 20, to be expected on the basis of the null hypothesis?

THE z TEST AND THE CHI-SQUARE TEST If we make the ordinary $\bar{z}$ test, we have the usual operations to perform. Using the new symbols for the numerator,

$$z = \frac{f_o - f_e}{\sqrt{Npq}} = \frac{28 - 20}{\sqrt{40 \times .5 \times .5}} = 2.53$$

This z is a little short of significance at the .01 level.

From the relationship indicated in formula (11.1),

$$\chi^2 = z^2 = \frac{(f_o - f_e)^2}{Npq}$$

For this kind of problem, where $p = q = .5$, the expression Npq can be written as $f_e/2$, since $f_e = Np$ and $q = 1/2$. This change gives us a formula for chi square, which reads

1 For a rather detailed mathematical development of chi square, see Lewis, D. *Quantitative methods in psychology.* New York: McGraw-Hill, 1960.

$$\chi^2 = \frac{2(f_o - f_e)^2}{f_e} \qquad \text{(Chi square in testing a null hypothesis for two frequencies in alternative categories)} \qquad (11.2)$$

It should be noted that the ratio of the square of the difference (between f_o and f_e) to f_e is the basic mathematical definition of chi square. More accurately stated, χ^2 is the *sum* of such ratios. Implied in formula (11.2) is actually a sum of two ratios of identical value, because of the 2 in the numerator. Relating this to the illustrative data, it should be pointed out that there are actually two observed frequencies, one f_o of 28 men and another f_o of 12. The double use of the same difference here illustrates another peculiarity of chi square—the attention to negative as well as to positive deviations. Here they have identical values, except for algebraic sign, but we shall find that elsewhere they may differ in value. The use of deviations in both directions also calls our attention to the fact that basically we have made a two-tail test, just as we often do in making a z test.

Let us see what chi square is for the polling data. Applying formula (11.2),

$$\chi^2 = \frac{2(28-20)^2}{20} = \frac{128}{20} = 6.4$$

This value is interpreted by reference to the sampling distribution of chi square with 1 df.[1] Like the distribution of *t*, the frequency distribution of chi square differs in shape, depending upon the number of df. Table E in the Appendix gives the chi squares corresponding to extreme proportions under the χ^2 distribution curves. The first row in Table E gives the distribution with 1 df, which applies to our illustrative problem. Looking across the row, we find that the obtained chi square of 6.4 is a little below 6.635, which is the value of chi square above which are .01 of the chance-generated chi squares. The inference that we may not reject the null hypothesis at the .01 level is the same as for the z test. This is not surprising, of course, since with 1 df, $\chi^2 = z^2$. Taking the square root of the chi square of 6.4, we find a z of 2.53, which checks with the value obtained earlier.

If the chi-square and z tests give identical answers regarding significance, it might be asked why we need anything but z. The quickest answer is that the relation to z holds only for 1 df; there are numerous hypotheses to be tested where there is more than 1 df. The z test is limited to the case of 1 df; chi square is not.

The more versatile nature of chi square arises from its additive property, the secret of which lies in the squaring of the deviations. This virtue is analogous to the additivity of the squared coefficient of correlation r^2, where *r* itself lacks that property. We shall soon see that many

1 Although we have summed two chi squares, they are not *mutually independent;* one is free to vary, but the other is then determined in each case.

sets of observations provide a number of frequency values, for each of which a null hypothesis can be stated and a test of the departures of the obtained frequencies from those expected from the null hypothesis is needed. Chi square enables us to make such combined tests, or tests of combined data.

COMPUTATION OF z AND χ^2 FROM OBSERVED FREQUENCIES In a set of two observed frequencies such as we have in the illustrative polling problem, it is possible to compute z or χ^2 without finding the expected frequency f_e, where $Np = N/2$. First, let us express all values appearing in formula (11.2) in terms of the two observed frequencies f_1 and f_2, where f_1 is arbitrarily the larger of the two. The difference $f_o - f_e$ is half the difference between f_1 and f_2. That is, $f_o - f_e$, without algebraic sign, equals $(f_1 + f_2)/2$. The expected frequency f_e is equal to $N/2$, which also equals $(f_1 + f_2)/2$. Substituting these values in formula (11.2),

$$\chi^2 = \frac{2(f_o - f_e)^2}{f_e} = \frac{2(f_1 - f_2)^2/4}{(f_1 - f_2)/2}$$

$$= \frac{(f_1 - f_2)^2}{f_1 + f_2} \quad \text{(Chi square computed from two observed frequencies in alternative categories)} \qquad (11.3)$$

Since with 1 df, $z = \chi$, taking square roots of both sides of (11.3) gives

$$z = \frac{f_1 - f_2}{\sqrt{f_1 + f_2}} \quad \text{(Standard deviate } z \text{ computed from observed frequencies in two alternative, mutually exclusive categories)} \qquad (11.4)$$

And since $f_1 + f_2 = N$, we have the interesting conclusion that under these circumstances, $z = (f_1 - f_2)/\sqrt{N}$. The student should check on the applicability of these formulas by computing χ^2 and z for the polling data.

A CHI SQUARE FOR THREE SIMULTANEOUS COMPARISONS

Let us carry our exploration of the nature of chi square a bit further by bringing in more clearly the additive principle. Let us use another polling problem, in which two groups of individuals respond to a question by giving one of three responses. The data are fictitious but perhaps realistic. Thirty men and thirty women in selected samples of students were asked the question, "Should the average woman graduate work for a postgraduate degree?" Each student answered by saying "Yes," "No," or "Undecided." The frequencies of these three responses for men and women are listed in Table 11.1.

The major interest in data such as these would be in whether there is a genuine sex difference in reactions. There are three ways in which the two sexes could be compared, each in terms of one of the three response categories. We could find three chi squares, one for each pair of frequencies for the three responses, and we could sum the three chi squares to determine whether the *set* of three frequencies for the men

TABLE 11.1 Numbers of men and women students who responded "Yes," "No," and "Undecided" to the question, "Should the average woman graduate work for a postgraduate degree?"

Response	f_o			f_e		$f_o - f_e$		$(f_o - f_e)^2$		$(f_o - f_e)^2/f_e$		
	M	*W*	*Both*	*M*	*W*	*M*	*W*	*M*	*W*	*M*	*W*	*Both*
Yes	9	15	24	12	12	−3	+3	9	9	0.75	0.75	1.50
No	12	2	14	7	7	+5	−5	25	25	3.57	3.57	7.14
Undecided	9	13	22	11	11	−2	+2	4	4	0.36	0.36	0.72
Sums	30	30	60	30	30	0	0			4.68	4.68	9.36 = χ^2

is significantly different from the *set* of frequencies for the women. Let us make the test of the null hypothesis (no sex difference) for all the frequencies taken together. For the computation of a chi square from a table with any number of cells there is a more general, standard formula, which reads

$$\chi^2 = \sum \frac{(f_o - f_e)^2}{f_e} \qquad \text{(General computing formula for chi square)} \qquad (11.5)$$

The first step is to find the expected frequencies that correspond to the obtained frequencies. Since the total numbers of men and women are identical, the null hypothesis should lead us to expect the same number of "Yes" responses for the two sexes, the same number of "No" responses, and the same number of "Undecided" responses. We have to consider the number of each kind of response that was given by the two sexes combined. There were 24 "Yes" responses, which, evenly divided, gives f_e's of 12 for each of the sexes (see Table 11.1). The 14 "No" responses divide evenly to give 7 and 7. The 22 "Undecided" responses give 11 and 11. The f_e's in each column should sum to 30, and they do.

Next we find the deviations $f_o - f_e$, which are numerically 3, 5, and 2, with opposite signs for men and women, for the three response categories, respectively. They should sum to zero in each column, and they do. Squared, they give 9, 25, and 4, respectively, for both men and women. Dividing each of the squared discrepancies by its corresponding f_e and summing the pair of ratios in each row, we find a chi square for each row. These chi-square values appear in the last column of Table 11.1. The sum of the three is 9.36, which is the chi square for the whole table of six frequencies. This value indicates how far the six obtained frequencies depart from the frequencies we should have expected if there were no sex difference.

INTERPRETATION OF AN OBTAINED CHI SQUARE In order to draw any inference concerning whether or not a chi square of 9.36 indicates a significant departure from the null hypothesis, we have to relate that value to the appropriate sampling distribution of chi square. Reference to Table E in the Appendix will give us the basis for a decision, but we have to consider first the number of degrees of freedom in order to know which line in Table E is appropriate to this problem.

In deciding on the number of df for this problem, we can be guided somewhat, but not entirely, by the previous illustrative problem. There we had two frequencies and one degree of freedom. There only one of the two frequencies was free to vary, the other being determined, because $f_1 + f_2$ must add up to N. Applying this reasoning to the data in Table 11.1, we might expect to find three degrees of freedom because there are three pairs of frequencies, each pair adding up to a fixed total. There is another kind of restriction to be considered, however, which is provided by the sums of the columns. Once two *pairs* of frequencies are established, the third pair is determined as well. The conclusion is that the obtained chi square has 2 df. In Table E, we find that for 2 df, in the second row of chi squares, a value as large as 9.36 could occur by random sampling alone a little less than once in a hundred times ($p < .01$). We can reject the null hypothesis of no sex difference with a high degree of confidence.

Looking now at the three chi squares for the three rows, we see that it was the sex difference in "No" responses that was almost entirely the source of the overall significant chi square. In fact, taking only the "No" responses, with one degree of freedom, we find a difference significant beyond the .01 level, the chi square of 7.14 exceeding the required 6.635 (from Table E). If these were genuine data, a natural psychological inference (to be distinguished from a statistical inference) would be that men are decidedly more in favor of denying women the privilege of working for postgraduate degrees, but when it comes to favoring the privilege, or being undecided about it, we cannot say that they are less inclined than women.

THE SAMPLING DISTRIBUTIONS OF CHI SQUARE

Table E is based upon the sampling distributions of chi square, with different numbers of df from 1 to as high as 30. The shapes of sampling distributions of χ^2 may be seen in Fig. 11.1, where the curves for 1, 2, 4, 6, and 10 df are shown. A distribution of z is normal. Squaring z to obtain chi square (with 1 df) does two things. It makes all values positive, so that instead of having a distribution symmetrical about zero, we have a distribution all on the positive side of zero. Obviously, the second effect is to change the shape of the distribution by piling up the frequencies near 0 and 1 and spreading the larger values.

Remember that for higher degrees of freedom, each chi square comes from a summing of chi squares, each with 1 df. Where the summed chi squares are mutually independent, they come in chance combinations. As the number summed in each set increases, the chances of zero sums decrease, and the mode of the distribution moves away from zero. It cannot easily be proved here, but with 2 or

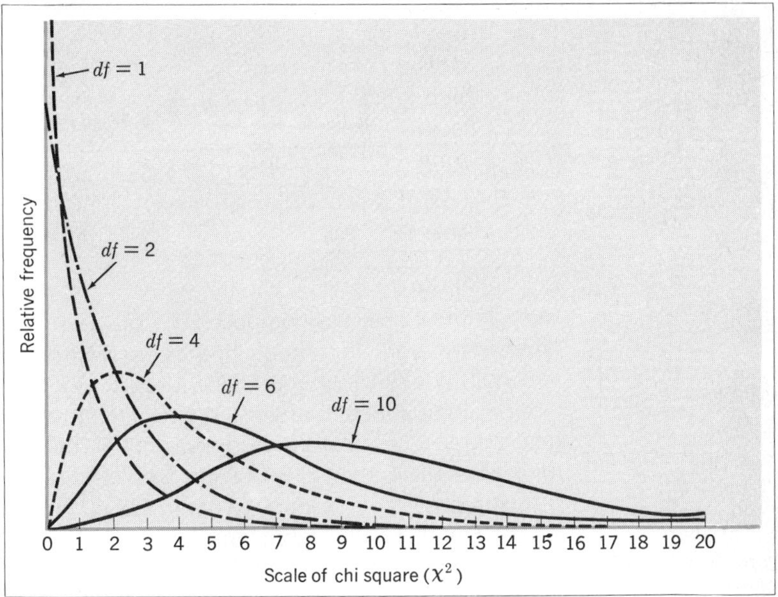

FIGURE 11.1 Sampling distributions of chi square for various degrees of freedom. (After Lewis, D. *Quantitative methods in psychology*. New York: McGraw-Hill, 1960.)

more df the principle is that the mode is at a chi-square value equal to the quantity (df − 2). The mode for 2 df is at zero; the mode for 10 df is at 8 (see Fig. 11.1). By 30 df the χ^2 distribution is virtually symmetrical, and it approximates the normal form.

So much for the general features of chi-square distributions; of more critical importance are the regions for rejections of null hypotheses. As indicated earlier, a chi-square test is a two-tail test. Although the tail area under the curve is only at the higher end of the range, remember that the squaring of z that is implied combines extreme negative cases of z with the extreme positive ones. If a one-tail test *should* be wanted for a particular case, it can be made by cutting the probabilities given in Table E in the Appendix in half. A one-tail test is hardly logical, however, unless one is dealing with a clear case of a simple outcome that can go in only one of two opposite directions. Most sets of data to which chi square is applied are too complex for application of a one-tail test.

Chi square in contingency tables

Consider the data in Table 11.2, which is an example of a contingency table because it sets forth two possibly related variables—intelligence level and marital status. Whether or not an individual in the data is married may be contingent upon intelligence. In the table we have

TABLE 11.2 Solution of chi square in a contingency table for data relating marital status to normal versus feebleminded status with respect to intelligence

Marital status	f_o Feebleminded	f_o Normal	f_o Both	f_e Feebleminded	f_e Normal
Married	84	111	195	97.5	97.5
Unmarried	122	95	217	108.5	108.5
Sum	206	206	412	206	206

two samples: one was composed of 206 young American males who, when they were in school, had been regarded as feebleminded in terms of IQ. Their IQs were in the range 60 to 69. The other group was composed of 206 men of similar age (in their twenties) whose IQs were near 100.[1] At the time the study was made, the proportions of married men were .408 and .539, for the feebleminded and normal groups, respectively. One question we could ask is, "Is this difference in marital status statistically significant?"

Another way of posing the same question is to ask whether the married and unmarried groups differ significantly with respect to intelligence. Another, more comprehensive question is, "Is there any correlation between being married and level of intelligence in this combined population?" Being married or unmarried and being normal or feebleminded are two genuine dichotomies (discrete groups), calling for the special correlation coefficient known as phi, or r_ϕ (see Chap. 14). The phi coefficient for these data is .13. Is this small coefficient significantly different from zero? Such a question normally suggests a t test such as we saw in Chap. 8. But the usual t test pertaining to an ordinary Pearson r does not apply to r_ϕ. We can apply a chi-square test, however, as will be demonstrated next.

CHI SQUARE AS A TEST OF INDEPENDENCE

The null hypothesis for a contingency table such as Table 11.2 is that there is no correlation: the two variables (marital status and intelligence) are independent in the population in question. The application of a chi-square test to the data in Table 11.2 is the same as for the data in Table 11.1. On the basis of the fact that there are equal numbers of feebleminded and normal subjects, we should expect to find both the married and the unmarried to be equally divided with respect to intelligence category, with f_e's of 97.5 for the married group and f_e's of 108.5 for the unmarried group. The differences $f_o - f_e$ are all 13.5, with two negative and two positive values. The difference squared is the same for all four cells (182.25). The chi square for the table as a whole is 7.10.

The number of degrees of freedom appropriate for this chi-

[1] Baller, W. R. A study of the present status of adults who were mentally deficient. *Genetic Psychology Monographs*, 1936, **18**, 165–244.

$f_o - f_e$		$(f_o - f_e)^2$		$(f_o - f_e)^2/f_e$		
Feeble-minded	*Normal*	*Feeble-minded*	*Normal*	*Feeble-minded*	*Normal*	*Both*
13.5	−13.5	182.25		1.87	1.87	3.74
−13.5	13.5	182.25		1.68	1.68	3.36
0.0	0.0			3.55	3.55	7.10

square distribution is 1. The reason for this is similar to that in the preceding polling problem. It can also be seen from the fact that with the four marginal totals established, after one cell frequency has been found, the other three are completely determined. With 1 df, a chi square of 7.10 is significant beyond the .01 level, at which level we find a χ^2 of 6.635 in Table E in the Appendix.

DEGREES OF FREEDOM IN A CONTINGENCY TABLE

The general rule about the number of degrees of freedom in a contingency table of any size, where the expected frequencies are determined from the marginal totals, is that

$$df = (r - 1)(k - 1) \quad \text{(Number of degrees of freedom in a contingency table of } r \text{ rows and } k \text{ columns)} \tag{11.6}$$

where r is the number of rows and k is the number of columns. For the four-cell table, with an r of 2 and a k of 2, df = 1, as we have seen.

COMPUTING EXPECTED CELL FREQUENCIES

Thus far, the computing of the expected frequencies f_e has been simple because of an even division of marginal frequencies in one of the variables. Very often the marginal frequencies are not evenly distributed, and a more general procedure is needed. A method for computing the expected cell frequencies in a contingency table of any number of rows and columns is illustrated by Table 11.3, which is a limited 3 × 3 table. Let the f's with double subscripts stand for the obtained cell frequencies. The sums of the rows are symbolized by f_a, f_b, and f_c, and the sums of columns by f_1, f_2, and f_3. The expected frequency for any cell in row r and column k can be found by the formula

$$f_{e(rk)} = \frac{f_r f_k}{N} \quad \text{(Expected frequency for a cell in row } r \text{ and column } k\text{)} \tag{11.7}$$

Thus, the expected frequency corresponding to f_{b3} would be derived from the product (f_b) (f_3) divided by N. The expected frequency for the married-normal subgroup in Table 11.2, in row A and column 2, is equal to

$$\frac{(195)(206)}{412} = 97.5$$

TABLE 11.3
Schema and symbols for computation of expected cell frequencies in a contingency table

Rows	Columns			Sums of rows
	1	2	3	
A	f_{a1}	f_{a2}	f_{a3}	f_a
B	f_{b1}	f_{b2}	f_{b3}	f_b
C	f_{c1}	f_{c2}	f_{c3}	f_c
Sums of columns	f_1	f_2	f_3	N

Let f_r stand for a sum of any row, for example, f_a, f_b, . . . , etc.
Let f_k stand for a sum of any column, for example, f_1, f_2, ...

CHI SQUARE WHEN FREQUENCIES ARE SMALL

When we apply chi square to a problem with 1 df and when any f_e frequency is less than 10, we should apply a modification known as *Yates's correction for continuity*. This correction consists in reducing by .5 each obtained frequency that is greater than expected and in increasing by the same amount each frequency that is less than expected. This has the effect of reducing the amount of each difference between obtained and expected frequency to the extent of .5. The result is a reduction in the size of chi square.

The correction is needed because a computed chi square, being based on frequencies (which are whole numbers), varies in discrete jumps, whereas the chi-square table, representing the distributions of chi square, gives values from a continuous scale. When frequencies are large, this correction is relatively unimportant, but when they are small, a change of .5 is of some consequence. The correction is particularly important when chi square turns out to be near a point of division between critical regions.

AN EXAMPLE OF YATES'S CORRECTION In a public-opinion poll conducted some years ago, attitudes toward radio newscasts were sampled.[1] Some 43 interviewees in one sample were asked the question, "Do you find it easier to listen to news than to read it?" The sample had been stratified into higher and lower socioeconomic status, 19 being in the former and 24 in the latter. The numbers responding "Yes" to the question in the two groups were 10 and 20, respectively. The problem to be investigated was whether there was a real difference between the two groups in their opinions on the question.

The data have been arranged in the usual manner in Table 11.4. Two of the expected frequencies are less than 10. Let us carry through the computations first *without* Yates's correction and then with it to see what difference it will make in the conclusion.

[1] From Cantril, H. The role of the radio commentator. *Public Opinion Quarterly*, 1939, **3**, 654–662.

TABLE 11.4 Computation of chi square for responses of two socioeconomic groups to preference for radio news to reading a newspaper

Response	*Obtained frequencies*			*Expected frequencies*		
	Lower	*Higher*	*Both*	*Lower*	*Higher*	*Both*
Yes	20	10	30	16.74	13.26	30
No	4	9	13	7.26	5.74	13
Both	24	19	43	24	19	43

Without the correction, the cell deviations would all equal 3.26. This value squared is 10.63. Applying formula (11.5) and solving, we find that chi square equals 4.76, which is significant between the .05 and .01 levels. *With* the correction, the cell deviation in all cells is 2.76 (rather than 3.26), which squared is 6.72. Here chi square becomes 3.43 and thus fails to reach the .05 level of significance. The correction will not always make a difference of this kind in the conclusion, but it should be used in a problem like this.

It should be noted that the correction of .5 is applied to *all* cells in the table even though only one or two frequencies are small. Note also that it is low *expected* frequencies that determine whether the correction should be applied, not low observed frequencies. It is also applied only to instances of 1 df, including 2 × 2 and 1 × 2 tables. In larger tables the need for the correction is not so great, and it would be complicated to apply. It is also possible to combine categories in such a way as to get rid of small expected frequencies. Examples of this will be seen later.

TESTING SIGNIFICANCE BY DIRECT COMPUTATION OF PROBABILITY There are lower limits to utilizable frequencies, below which even Yates's correction is inadequate. If any expected frequency is less than 2, we should not apply the computing formulas for chi square, even with the correction. If there is 1 df and there is a frequency less than 2, it is still possible to answer the question, "Given the marginal frequencies, what is the probability that distributions among the four cells could be as extreme as this one, or more extreme?" The probability, and hence the level of significance, can be determined without computing chi square.

For the special, but common, case of a fourfold table in which two equal groups of observations are being compared, Table M in the Appendix will serve to answer the question of statistical significance. It was designed for the following very common type of problem. Let us say that an experimental group of 30 individuals is administered a dose of dramamine sulfate and a control group of 30 individuals is administered a placebo before a rough flight in an airplane. Of the experimental group, 5 become airsick and 25 do not; of the control group, 18 become airsick and 12 do not.

In Table M, each row pertains to two groups, each of a certain size N_i. In the illustrative problem, $N_i = 30$. To use Table M, locate the row that applies, in this case the row for $N_i = 30$. Next, find the column headed with the number that corresponds to the smallest frequency in the fourfold table. In this problem that frequency is 5, the number in the experimental group who became airsick. Given these two values, 30 and 5, we ask the question, "How many cases are needed in the other group corresponding to the smallest cell frequency to achieve chi squares significant at the .05 and .01 levels?" Corresponding to the frequency of 5 is the frequency of 18 airsick cases in the control group. Table M tells us that it would take 13 cases of airsickness in this group to be significant at the .05 level and 16 cases of airsickness to be significant at the .01 level. The obtained frequency of 18 exceeds both those values and is therefore a basis for concluding that we have significance beyond the .01 level.

The values in Table M are based upon exact probabilities up to an N_i of 20 and upon solutions by formula with Yates's correction for N_i's greater than 20.[1]

OTHER WAYS OF COMPUTING CHI SQUARE IN A 2 × 2 TABLE

In a fourfold-table problem, since the difference $f_o - f_e$ is the same for all cells, the formula for chi square can be written

$$\chi^2 = (f_o - f_e)^2 \sum \left(\frac{1}{f_e}\right) \qquad \text{(Chi square in a } 2 \times 2 \text{ contingency table)} \qquad (11.8)$$

That is, chi square equals the common difference squared times the sum of the reciprocals of the four f_e's. As applied to the marital-status problem,

$$\begin{aligned}\chi^2 &= 13.5^2 \left(\frac{1}{97.5} + \frac{1}{97.5} + \frac{1}{108.5} + \frac{1}{108.5}\right)\\ &= 182.25\ (.01026 + .01026 + .00922 + .00922)\\ &= 7.10\end{aligned}$$

If the data are arranged in a 2 × 2 table as shown in Table 11.5, another convenient formula for the computation of chi square is

$$\chi^2 = \frac{N(ad - bc)^2}{(a+b)(a+c)(b+d)(c+d)} \qquad \text{(Alternative formula for chi square in a } 2 \times 2 \text{ table)} \qquad (11.9)$$

Applied to the opinion-poll data,

$$\chi^2 = \frac{43[(10)(4) - (20)(9)]^2}{(30)(19)(24)(13)} = 4.74$$

1 For dealing with contingency tables where the sums of columns N_i are not equal, exact probabilities can be computed by methods described by Walker, H. M., and Lev, J. *Statistical inference.* New York: Holt, 1953. Pp. 104–108. A useful table is also provided in Siegel, S. *Nonparametric statistics for the behaviorial sciences.* New York: McGraw-Hill, 1956.

TABLE 11.5 Symbolic arrangement of data in a 2 × 2 contingency table illustrated by the public-opinion data

Variable II \ *Variable I*	Lower	Higher	Both
Higher	*b*	*a*	*a* + *b*
Lower	*d*	*c*	*c* + *d*
Both	*b* + *d*	*a* + *c*	*N*

Response \ *Socioeconomic group*	Lower	Higher	Both
Yes	20	10	30
No	4	9	13
Both	24	19	43

The answer is within rounding error of that computed earlier by formula (11.2).

The last solution was done without Yates's correction. The same formula with Yates's correction incorporated reads

$$\chi^2 = \frac{N(|ad - bc| - N/2)^2}{(a+b)(a+c)(b+d)(c+d)} \qquad \text{[Same as (11.9), with Yates's correction]} \qquad (11.10)$$

Note that the difference $ad - bc$ is always taken as positive, as indicated by the vertical lines enclosing it.

OTHER GENERAL COMPUTING FORMULAS FOR CHI SQUARE

It is possible to compute chi square by using formulas that require less information than is involved in the use of formula (11.5). By expanding the expression $(f_o - f_e)^2$ and simplifying, we arrive at the equation

$$\chi^2 = \sum \left(\frac{f_o^2}{f_e}\right) - N \qquad \text{(Formula for chi square without finding } f_o - f_e\text{)} \qquad (11.11)$$

Making use of the fact that from formula (11.7), $f_e = f_r f_k / N$, substituting this expression in (11.11), and using f_{rk}^2 for f_o^2 we have

$$\chi^2 = N\left[\sum \left(\frac{f_{rk}^2}{f_r f_k}\right) - 1\right] \qquad \text{(Formula for chi square without finding } f_e\text{'s)} \qquad (11.12)$$

where f_{rk}, f_r, and f_k are defined as in (11.7).[1]

CHI SQUARE IN OTHER THAN 2 × 2 TABLES

In the very first example, we examined the case of chi square computed from a two-cell table. We shall now consider cases in which either *r* or *k*, or both, exceed 2. For such a case, let us consider the data in Table 11.6, which has three categories each way. The object of the study which developed this particular set of data was to determine whether the ability to distinguish between the taste qualities of three popular brands of cola beverages is related to amount of cola-drinking experience. Each of 79 individuals had been tested with pairs of

[1] From S. W. Brown. Simplified computation formulas for χ^2 and related statistics. *Educational and Psychological Measurement,* 1964, **24,** 219–221. For complete accuracy, the ratios in formula (11.12) should be carried to four or five decimal places.

TABLE 11.6
A chi-square test of independence between amount of cola-drinking experience and ability to discriminate between popular brands*

	f_{rk}				f_{rk}^2		
	0–2	3–6	7+	f_r	0–2	3–6	7+
Heavy	10	14	3	27	100	196	9
Medium	7	9	10	26	49	81	100
Light	8	12	6	26	64	144	36
f_k	25	35	19	79			

* From Thumin, F. J. Identification of cola beverages. *Journal of Applied Psychology*, 1962, **46,** 359. Used by permission of the author and publisher. The three experience categories are defined as follows: Heavy = seven or more colas consumed

sample cola brands; each achieved a score, which was the number of correct identifications he made. On the basis of the scores, the subjects were divided into three categories. The same subjects were also grouped in categories as "heavy," "medium," or "light" cola drinkers, according to the number of times per week they said they drank any colas. It is a reasonable psychological hypothesis that ability to discriminate should be related positively to amount of experience.

Let us test the alternative, null hypothesis—that the two variables are independent—by computing a chi square, applying formula (11.12). The work is laid out by steps in Table 11.6: first we find the f_r and f_k marginal sums, we square the cell frequencies to find f_{rk}^2, we find the $f_r f_k$ product for each of the nine cells of the table, and then we compute the ratios $f_{rk}^2/f_r f_k$ and their sum, which is 1.0694. It should be noted that four decimal places are retained in the ratios through this point. Deducting 1.0 from this figure leaves only .0694, which, multiplied by N, gives chi square. The final product is 5.48. With four degrees of freedom [according to formula (11.6)], Table E in the Appendix tells us that this χ^2 fails to reach significance at the .05 level. There are at least 5 chances in 100 that so large a χ^2 could have arisen by chance with a genuinely zero correlation between amount of cola-drinking experience and ability to discriminate between brands.

COMBINING COLUMNS OR ROWS It was pointed out earlier that in a 2×2 contingency table, when expected frequencies are small, we may apply the correction for continuity. This leaves open the question of what to do in larger contingency tables with small expected frequencies. In larger tables, f_e's as small as 5 may be tolerated. When a frequency is less than 5, the best remedy is to note the row or column (whichever has generally smaller frequencies) in which the small frequency occurs and to combine that array (row or column) with one of its neighbors. Choosing the neighbor that also has smaller frequencies would be a good policy, but other commonsense considerations should also be given some weight.

In the contingency table (Table 11.6), the smallest f_e, equal to 3,

$f_r f_k$			$f_{rk}^2/f_r f_k$			Row sums
0–2	3–6	7+	0–2	3–6	7+	
675	945	513	.1481	.2074	.0175	.3730
650	910	494	.0754	.0890	.2024	.3668
650	910	494	.0985	.1582	.0729	.3296
			.3220	.4546	.2928	1.0694

per week; medium = three to six; and light = fewer than three. The subjects were asked to identify by taste a sample which was one of three well-known brands. The score is the number of correct identifications.

appears in the last column, where the $f_r f_k$ product is 513. Dividing this by N (which is 79) gives an f_e of approximately 6. This value is not small enough to require combining arrays. If it were, the best combination to make would be the last two columns. After two arrays have been combined, the computed chi square is generally smaller, but there is compensation in the fact that the number of df is also smaller. It is possible that making such combinations will change one's decision as to the statistical significance of data.

Some special applications of chi square

CHI SQUARE WHEN PROPORTIONS ARE CORRELATED

Many of the applications of chi square involve the comparison of two proportions or percentages, as we have seen. In the examples given thus far, the two proportions were uncorrelated, for they were derived from different observations or individuals. We shall now consider some applications of chi square when proportions are correlated.

TEST FOR TWO CORRELATED PROPORTIONS In Chap. 9, we used a z test for a difference between two correlated proportions. Since with 1 df χ^2 is equal to z^2, we might expect a very direct estimate of chi square by squaring both sides of formula (9.10) (p. 162). This expectation is correct, and the formula is

$$\chi^2 = \frac{(b-c)^2}{b+c} \qquad \text{(Chi square for a difference between two correlated proportions)} \qquad (11.13)$$

where the symbols are as defined in Table 9.3 (p. 162).

It should be noted here, as in Table 9.3, that b and c indicate the numbers of cases that change categories between a first and second application of the experiment. Either the same individuals or matched individuals must be involved so that the numbers of changes may be counted. The illustrative problem in Chap. 9 involved 100 students who had attempted to answer two items. If there is correlation between the items, there is also correlation between the proportions. The number of changing individuals denoted by b (answering the first correctly but not the second) was 5. The number of changing individuals denoted by c (answering the second correctly but not the first) was 15. Applying formula (11.13),

$$\chi^2 = \frac{(5 - 15)^2}{5 + 15} = 5.00$$

which is significant between the .05 and .01 points.

In small samples, Yates's correction should be incorporated in formula (11.13). This involves deducting 1 from the difference, where the difference is regarded as positive, before squaring.

TEST FOR MORE THAN TWO CORRELATED PROPORTIONS
test for differences between more than two correlated proportions is described by McNemar.[1]

SIGNIFICANCE OF A COMBINATION OF TESTS

Sometimes we have made a z, t, or F test in several similar, independent samples. Perhaps the sampling statistic was not significant beyond the adopted probability level in any sample, and yet the deviations from the value indicated by the null hypothesis were all in the same direction. In other instances, perhaps some of the samples gave significant results and some did not. Were the significant ones merely high *chance* deviations? Some method is obviously needed to make a single test of all the data.

If we happened to know that certain sets of the data came by random sampling from the same population and if the means and variances within those sets proved to be homogeneous, we should be justified in pooling the sets and making new tests of significance, with enlarged df and more power. But this is probably not the most efficient way, even if we already have the necessary information regarding homogeneity. There are ways of considering in combination the results of several significance tests already applied to the samples individually, one of which will be described.

A CHI-SQUARE TEST OF COMBINED PROBABILITIES It has been demonstrated that there is a mathematical way of transforming a probability into a chi square.[2] In general, $\chi^2 = -2 \log_e p$, with 2 df. In this method, then, we need to know the value of the probability attached to each obtained sampling statistic. This can be found, of course, from tables of distributions of z, t, or F, whichever test we are applying.

Where several probabilities are involved, we can transform each into a chi square and then sum those chi squares and also sum their corresponding degrees of freedom. Because of the additive property of

[1] McNemar, Q. *Psychological statistics.* (4th ed.) New York: Wiley, 1969.

[2] Jones, L. V., and Fiske, D. W. Models for testing the significance of combined results. *Psychological Bulletin,* 1963, **50** 375–382.

chi square, the sum is also a chi square with combined df. The computing formula is

$$\chi^2 = -4.605\ \Sigma \log p_i \qquad (11.14a)$$

$$\chi^2 = -4.605 \log (p_1 p_2 p_3 \ldots p_k) \qquad (11.14b)$$

(Chi square for a combination of probabilities)

where p_i = probability that a deviation as large as the one obtained could occur by chance. The constant −4.605 represents the product of −2 times the constant 2.3026, which is needed because we are using common logarithms rather than Napierian logarithms. The sum has $2k$ degrees of freedom, where k is the number of tests made. It will be noted from the two forms of the equation that we may obtain the logarithm of each probability first and then sum them (11.14*a*), or we may find the product of all the probabilities and then find the one logarithm of the product. The latter solution is simpler when k is small.

Suppose that we have derived three estimates of correlation between the same pair of variables in three samples. In each sample we have tested the hypothesis of no correlation and have obtained a z or a t ratio. The probability for so large a z or t value would be found by reference to the normal or the Student distribution, respectively.[1] *The probability associated with a one-tail test is the one to use in formula* (11.14).[2]

In Table 11.7 we have, first, three coefficients of correlation from three independent samples. Each was based upon a sample in which $N = 50$. The N's need not be equal in order to apply this chi-square test. The SE of an r of zero when $N = 50$ is .143. Each r deviates from zero by the number of z units shown in the second column. One-tail tests give the probabilities in column 3. The logarithms of these probabilities are given in column 4.

As the student who remembers his algebra will recall, the four digits to the right of the decimal point are found in the table of logarithms (see Table J in the Appendix). The negative number at the left of each decimal point comes from the fact that each probability is a value less than 1.0. The rule is to make this number one more (in the

1 For probabilities from Student's distribution, see Walker and Lev, op. cit., Table IX. For all but small samples, the normal distribution will serve.

2 See Gordon, M. H., Loveland, E. H., and Cureton, E. E. An extended table of chi-square. *Psychometrika*, 1952, **17**, 311–316.

TABLE 11.7
Chi square for a combination of three probabilities

r_i	z_i	p_i	$\log p_i$
.294	2.056	.02	−2 + .3010
.222	1.552	.06	−2 + .7782
.168	1.175	.12	−1 + .0792

$\Sigma \log p_i = -5 + 1.1584 = -3.8416$

negative direction) than the number of zeros to the right of the decimal point in p_i. The summing of these logarithms is done for the two components separately, after which an algebraic sum of the two component sums is found. The sum of the logarithms is a numerical value of -3.8416. Multiplying this by -4.605 from formula (11.14*a*), we find a chi square of 17.69. Reference to the chi-square table with 6 df shows this to be significant beyond the .01 level. Thus, a correlation that failed to be significant beyond the .01 level in any of the three samples is found to be in that region when the tests are combined.

Several restrictions and qualifications with regard to the use of this composite test should be noted. The combined tests must be based upon independent samples. The probabilities to be used should be from one-tail tests. If in the end a two-tail test is wanted, we must double the probability attached to the obtained chi square.

If several parallel tests of samples have been made, the combination that is tested should not be a selected one, for example, those with highest p_i values only. All legitimate single-sample tests that are clearly parallel should be included. If the deviation from the null hypothesis happens to be in the opposite direction for any of the samples, for those samples use q_i ($q_i = 1 - p_i$, where p_i is the smaller tail area) instead of p_i, but include such a sample.

EXERCISES

In each of the following exercises, state what hypothesis is being tested, the number of df involved, your statistical inference, and your experimental conclusion.

1 In polling at random 45 interviewees, we find that 32 favor a certain routing of a freeway and that the rest do not. Is it likely that the population concerned is evenly divided on the proposition?

2 Compute a chi square for the contingency table in Data 11*A*.

DATA 11*A*
Numbers of two groups differing in ability who passed a certain test item

Outcome	*Low group*	*High group*	*Both groups*
Passed	**48**	**62**	**110**
Failed	**52**	**38**	**90**
Both	**100**	**100**	**200**

3 Compute a chi square for the contingency table in Data 11*B*.

4 In an experimental group of 15 who were inoculated, two developed colds within a specified time period, whereas in a control group of the same size, nine developed colds.

a. Determine chi square, with and without Yates's correction.

b. Make a test using Table M in the Appendix.

5 Of 13 identical-twin pairs, 10 pairs contained two criminals each, and the remaining pairs contained one criminal each. Of 17 fraternal-twin pairs, three pairs contained two criminals each, and the re-

DATA 11*B* Numbers of persons in two groups, depressed and not depressed in temperament, who responded in each of three categories to the question, "Would you rate yourself as an impulsive individual?"

Group	*Response*			
	Yes	*?*	*No*	*Total*
Depressed	72	45	133	250
Not depressed	106	35	109	250
Both	178	80	242	500

maining pairs contained one criminal each. Set up a contingency table and compute a chi square.

6 On the application of a certain test before therapy, 25 members of an experimental group were above the general median score, and 15 were below. After therapy, 16 were above the median on the same test, and 24 were below. Eleven were above the median both before and after. Set up a contingency table and compute chi square.

7 In three pairs of independent samples, differences between means $\bar{X}_1 - \bar{X}_2$ equaled 2.4, 1.7, and 5.2. The probabilities (one-tail tests) associated with these differences were .12, .35, and .015, respectively. What is the probability that a combination of differences as large as these could have occurred by chance?

ANSWERS

1 $\chi^2 = 8.02$; df = 1.

2 $\chi^2 = 3.96$; df = 1.

3 $\chi^2 = 10.12$; df = 2.

4 a. $\chi^2 = 7.03$ (without Yates's correction); df = 1.
$\chi^2 = 5.17$ (with correction).
b. From Table M, $p \leqq .05$.

5 $\chi^2 = 8.27$ (with correction); df = 1.

6 $\chi^2 = 4.26$ (without correction); df = 1.
$\chi^2 = 3.37$ (with correction).

7 $\chi^2 = 14.74$; df = 6.

12 Nonparametric, or distribution-free, statistics

Apart from the main stream, some statistical procedures have been developed especially to take care of the experimental situation in which samples are small and the form of the population distribution is not normal. Some of these statistics will now be described.

Before an investigator resorts to them, however, he should consider whether any of the more powerful tests can be used. Some of the distribution-free, or nonparametric, methods have lower power to detect a real difference as significant. When there is any choice, therefore, we should prefer a parametric test, except where a quick, rough test will do. Even where there seems to be no choice, we can sometimes create one, as will be seen in the following discussion.

TRANSFORMATION OF MEASUREMENT SCALES

Sometimes the nonnormal distribution in a population is caused by an inappropriate measuring scale or by restrictions that result in distorted scales. For example, distributions of simple reaction times are generally skewed positively. This comes about because there is some minimal time below which the reacting subject cannot go and because there is no restriction at the other side of the distribution. The effects of these restrictions are felt to some extent over a wide range, for the distribution is not simply truncated.

The question posed by such a situation is whether, by some transformation, we can convert the measurements into values on a new scale in which the distribution *is* normal. One justification for such a transformation would be an assumption that the underlying psychological variable or trait *is* normally distributed, if only we had an appropriate scale on which to measure it. Such reasoning is not necessary, however. Tests made of statistics on transformed scales lead to conclusions that hold for the natural phenomena under investigation. We saw this in connection with the transformation of r to Fisher's Z.

One way to transform the reaction-time measurements is to find the logarithm of each value. This would condense the larger measurements into smaller scale ranges relative to the smaller measurements and thus reduce, if not eliminate, skewing. With the measurements thus transformed to log form, we could proceed to apply parametric tests, even in small samples.

Other nonlinear transformation procedures exist. One is the conversion of proportions or percentages into corresponding angle values in degrees of arc. In sampling, these are normally distributed, whereas extreme proportions are not. For an excellent discussion of the subject of transformation, see Mueller.[1]

Tests of differences with correlated data

THE SIGN TEST One of the simplest tests of significance in the nonparametric category is the sign test. Let us say that we have two parallel sets of measurements that are paired off in some way. The data in Table 12.1 are 10 of the successive pairs of knee-jerk measurements presented originally in Table 9.2. The hypothesis to be tested is that they arose from random sampling from the same population. If this hypothesis is true, half the changes from T to R should be positive, and half should be negative. Another way of stating the null hypothesis is to say that the median *change* is zero.

1 Mueller, C. G. Numerical transformations in the analysis of experimental data. *Psychological Bulletin,* 1949, **46,** 198–223.

TABLE 12.1 Application of the sign test to 10 pairs of the knee-jerk data from Table 9.2

T*	R	Sign of $T - R$
19	14	+
19	19	(0)
26	30	−
15	7	+
18	13	+
30	20	+
18	17	+
30	29	+
26	18	+
28	21	+

*T = knee-jerk measurement under tension; R = measurement under relaxation.

There are 10 pairs of observations; therefore, 10 changes are involved. But note that one change is zero. Since we cannot include this as either positive or negative, it is discarded, leaving nine changes for the test. The hypothesis now calls for 4.5 positive differences, whereas we obtained eight. Is this a significant deviation?

The obvious test to make is based upon the binomial distribution for $p = .5$ and $N = 9$. On this basis, 8 or more plus signs could occur by chance 10 times in 512 trials (1 chance in 512 for exactly 9, plus 9 chances for exactly 8). For a one-tail test this deviation is significant with p equal to approximately .02. For a two-tail test we double the probability, as usual, which gives a departure significant at the .04 level. We would make a two-tail test if our alternative hypothesis were that these results did not come from the same population, i.e., with respect to central value. We would make a one-tail test if the alternative hypothesis at the start were to expect the T values to be higher than the R values.

Table N in the Appendix will be useful in applying this sign test, since cumulative tail proportions for the binomial distribution (where $p = .5$) are given for each value of N up to 25. For cases in which the number of pairs is greater than 25, the normal-curve approximation may be used, as described in Chap. 10.

The assumptions involved in making the sign test include mutual independence of the *differences*. The two parallel sets of values may or may not be correlated. Nothing is assumed concerning the shape of the distribution or the equality of variances. The differences need not even be measured accurately, but the *direction* of each difference should be experimentally established.

One weakness of the sign test is that it does not use all the available information. If the measurements are on a scale of equal units, on

TABLE 12.2
Application of the sign-rank test of differences, using the knee-jerk data

T*	R	$T - R$	*Rank of absolute difference*	*Ranks with minority sign*
19	14	+5	4.5	
19	19	0		
26	30	−4	3	−3
15	7	+8	7.5	
18	13	+5	4.5	
30	20	+10	9	
18	17	+1	1.5	
30	29	+1	1.5	
26	18	+8	7.5	
28	21	+7	6	
				$T = -3$

*T = knee-jerk score under tension; R = score under relaxation.

which differences may be compared for size as well as for direction, the sign test ignores the information provided by size. It is said that, except for very small samples, the sign test is only about 60 percent as powerful as a *t* test would be for the same data, where both apply. This difference in power could be compensated for by increasing the size of sample. If we had applied the sign test to the entire data in Table 9.2, we should have found that 18 out of 25 signs were positive. By the use of the binomial distribution, this would indicate a deviation significant near the .02 level (one-tail test), which agrees with the result from the smaller sample of 10 pairs. In Chap. 9, however, the *z* test for the same complete data was significant almost to the .001 point in a one-tail test. The difference in sensitivity of the two tests in this particular illustration is appreciable.

THE SIGN-RANK TEST OF DIFFERENCES

Let us use as an illustration of the sign-rank test of differences the same data to which the sign test was just applied in Table 12.1.[1] The 10 pairs of knee-jerk measurements under tensed and relaxed conditions are repeated for convenience in Table 12.2. Here the numerical differences, with algebraic signs, are also listed. Unlike the sign test, however, this test utilizes the additional information of *sizes* of differences. As in the sign test, however, we cannot use zero differences, since the differences must be classified according to algebraic sign.

Having the differences with their algebraic signs, we first forget the signs and rank the differences according to size only, giving the smallest difference a rank of 1. There are two differences of 1. We do not know which one to call rank 1 and which rank 2, and so we give each of them an average rank of 1.5. The next smallest difference is 4, which is given a rank of 3, and so on until all nonzero differences are ranked.

Next, we consider the algebraic signs of the differences. We single out all differences whose signs are in the minority. If there are fewer negative than positive signs, as here, we select all ranks corresponding to the differences having that sign. There is only one negative difference in Table 12.2. We put this rank with a negative sign in the last column. We sum this column to give a statistic T.

The hypothesis tested is that the differences are symmetrically distributed about a mean difference of zero. If this were true, T would coincide with the mean of such sums of randomly selected ranks $\bar{T}$, which is also half the sum of N successive ranks and which would be given by the formula

$$\bar{T} = \frac{N(N+1)}{4} \qquad \text{(Mean of sums of ranks)} \qquad (12.1)$$

[1] The sign-rank test and the composite-rank test, to be described later, have been attributed to Wilcoxon. See Wilcoxon, F. *Some rapid approximate statistical procedures.* Stamford, Conn.: American Cyanamid Co., 1949.

The deviation obtained is $T - \bar{T}$. Wilcoxon has supplied a table giving the T values significant at the .05, .02, and .01 levels (see Table O in the Appendix). Reference to Table O indicates that the obtained T of -3 (the algebraic sign does not matter in the use of the table) is significant at the .02 level (a two-tail test) when we have nine differences involved.

It will be seen that the outcome of this test agrees with that from the sign test for the same data. There will not always be this much agreement, and when there is not, the result of the sign-rank-difference test should be regarded as more dependable, since it rests upon more information.

For samples larger than 25, a standard deviation and a z ratio can be computed, and z can be interpreted in terms of the normal distribution. For a sample of size N,

$$s_t = \sqrt{\frac{N(N+1)(2N+1)}{24}} \tag{12.2}$$

and z is equal to $(T - \bar{T})/s_t$.

Tests of differences with uncorrelated data

THE MEDIAN TEST The median test involves finding a common median for the combination of the two samples being compared, as a first step. Next, the numbers of cases above and below the common median are counted in each sample, resulting in a fourfold contingency table, as the one in Table 12.3. The observations are not paired or correlated, and the N may differ in the two samples. Equal N's would make the test easier to apply, as will be seen. Finally, we find the chi square for the contingency table. With equal numbers of observations in each sample, we can conveniently use Table M in the Appendix for a test of significance without computing chi square.

The median of the 14 observations in Table 12.3 is 9.5. Values of

TABLE 12.3
Application of the median test to two samples under conditions A and B

Samples	
A	*B*
14	5
13	7
10	6
12	5
15	11
9	8
9	10
$Mdn = 9.5$	

Contingency table samples

	A	*B*	*Both*
10+	5	2	7
9–	2	5	7
	7	7	

TABLE 12.4
Application of the median test to more than two samples

Samples		
D	*E*	*F*
2	10	12
7	7	15
5	12	9
6	14	16
8	9	14
3	8	
	10	
N_i 6	7	5

Mdn = 9.0

Contingency table

	D	*E*	*F*	*All*
10+	0	4	4	8
9–	6	3	1	10
All	6	7	5	18

10 and above are easily segregated from those of 9 and below, as shown in the fourfold table. With such small frequencies, we should not compute chi square for this table. Reference to Table M indicates that chi square is not significant, *p* being greater than .05 (in a two-tail test).

The hypothesis tested is that the median is the same for both populations. Since the samples are likely to be small in making this test, exact probabilities should be obtained, or Table M should be used. If a one-tail test is wanted, then a more exact *p* should be estimated, and this *p* should be divided by 2.

A MEDIAN TEST WITH MORE THAN TWO SAMPLES Suppose that we have three samples, each from its own treatment or set of conditions. We want to test the homogeneity of their central values. For example, consider the three samples in Table 12.4.

The median of all 18 observations is 9.0. Since we have some 9s in the lists, we cannot make the point of dichotomy at exactly 9. In such a situation we make it as near the median as we can. Let it be the point 9.5. We then set up a contingency table, such as the one in Table 12.4. From these data, chi square is 7.82. With 2 df this chi square is significant near the .02 point. We reject the null hypothesis and say that the three medians are not homogeneous.

THE COMPOSITE-RANK METHOD

When the observations are not paired, another method is to produce a single rank order for all values in the two samples. If the two samples came from the same population, in a single ranking the sums of the ranks belonging to the two samples should be equal, except for sampling errors. The composite-rank test is concerned with the departure of the two sums from equality.

Consider the two samples of seven cases each, in Table 12.5, obtained under conditions *A* and *B*. We assign the numerically lowest ranks to the lowest values. There are two lowest scores of 5, each of which receives a rank of 1.5. The score of 6 then receives a rank of 3, and so on, until the highest score of 15 receives a rank of 14 (which equals N unless there are ties for top place).

The sums of the ranks for conditions A and B, which we shall call R_a and R_b, are 71.5 and 33.5, respectively. The check for these sums is that they should add up to $N(N+1)/2$, where there are N ranked values. In this case, $71.5 + 33.5 = N(N+1)/2$.

We select the smaller of the two sums, which happens to be R_b in this problem, as our sampling statistic. It is distributed about the mean of the sums, which is given by formula (12.1), and will be called $\bar{R}$. For values of N_i (number in each sample) not greater than 20 and for samples of equal size ($N_a = N_b = N_i$), Wilcoxon has provided tables of the values of significant R's, which are to be found in Table P in the Appendix. With seven pairs of observations ($N_i = 7$), an R of 33.5 is significant between the .02 and .01 levels, a bit closer to the .02 level (in a two-tail test).

For the application when N_i exceeds 20, the R's significant at the three levels may be computed by the formulas

$$R_{.05} = \bar{R} - 1.960\sqrt{\frac{N\bar{R}}{12}}$$

$$R_{.02} = \bar{R} - 2.326\sqrt{\frac{N\bar{R}}{12}} \qquad \text{(Values of statistic R significant at three levels)} \qquad (12.3)$$

$$R_{.01} = \bar{R} - 2.576\sqrt{\frac{N\bar{R}}{12}}$$

where $\bar{R}$ = mean of the sums of ranks and the radical expression is the standard deviation of the sampling distribution of R.

THE MANN-WHITNEY U TEST There is a generalization of the R test just described that takes care of samples of unequal size. For this more general case we have the Mann-Whitney U test. The hypothesis being tested is the same as that for the R test, and the operations through to the finding of the sums of the ranks are the same as well. When N_a and N_b are both as large as 8, a z test can be used, and z can be computed by the formula

$$z = \frac{2R_i - N_i(N+1)}{\sqrt{N_a N_b (N+1)/3}} \qquad \text{(z value for an obtained sum of ranks for a U test)} \qquad (12.4)$$

where R_i = one of the sums of ranks
N_a and N_b = replications in samples A and B, respectively
N = total number of cases = $N_a + N_b$
N_i = number of cases corresponding to R_i

The hypothesis being tested is that one set of measures, as a group, is equal to another. Statistically, the H_0 being tested is that the

TABLE 12.5 Application of the *R* test of a difference, based upon the sum of ranks

Measurements		*Ranks*	
A	*B*	*A*	*B*
14	5	13	1.5
13	7	12	4
10	6	8.5	3
12	5	11	1.5
15	11	14	10
9	8	6.5	5
9	10	6.5	8.5
		Σ 71.5	33.5
		R_a	R_b

obtained U_i minus the U to be expected for a particular combination of N_a and N_b is zero. The expected U is equal to $N_aN_b/2$. U_i is given by the formula

$$U_i = N_aN_b + \frac{N_i(N_i + 1)}{2} - R_i \qquad \text{(The Mann-Whitney } U \text{ statistic)} \qquad (12.5)$$

where $U_i =$ the Mann-Whitney U statistic computed from one of the sums of ranks R_i and where other symbols are as defined above. Deducting the expected U from the obtained U_i and multiplying through by 2, we obtain the numerator of formula (12.4). The denominator is also 2 times the standard error of U, the formula for which has a 12 in its denominator rather than 3 as shown in formula (12.4). The z's computed from R_a or R_b should be the same in absolute value, differing only in algebraic sign.

Let us apply formula (12.4) to a problem with data similar to those in Table 12.5. In this problem, N_a from sample A is 10, and N_b from sample B is 8. All 18 measurements were ranked together, as was done with other data, in Table 12.5. The sum for sample A (R_a) was 123, and that for sample B (R_b) was 48. The sum of these two values is 171, which gives us one check, since the total sum of the ranks, which is given by $N(N + 1)$, also equals 171. Using the smaller R_i and applying formula (12.4), we have

$$\begin{aligned} z &= \frac{2(48) - 8(19)}{\sqrt{(10)(8)(19)/3}} \\ &= \frac{96 - 152}{\sqrt{1520/3}} \\ &= -2.49 \end{aligned}$$

For a normal distribution for this z, the difference between the two sets of ranks—and thus that between the two sets of measurements—appears to be significant beyond the .05 level (in fact, not far from the .01 level) in a two-tail test. The algebraic sign of z here does not matter.

For very small samples, one or both of which have N's less than 8, the U statistics should be computed, using formula (12.5). The statistical significance of the departure of the obtained U from that expected, that is, a departure of zero, can be decided by reference to tables constructed for the Mann-Whitney test. The more useful parts of those tables are given as Table S in the Appendix. The table is designed for interpretation of the smaller U_i; the decision would apply to the larger U_i as well. The table is designed for one-tail tests, which means doubling probabilities for two-tail tests.

Consider a problem in which 15 cases were selected at random from data like those in Table 9.2, 5 of which were from the knee-jerk measurements under the tensed condition and 10 from the measurements under the relaxed condition. The sum of the ranks under the tensed condition was 56.0, and that under the relaxed condition was 64.0. Since the former is smaller, we compute U_i in terms of the set from which it came. Applying formula (12.5), where $N_a = 5$ and $N_b = 10$,

$$U_a = 5(10) + \frac{(5)(6)}{2} - 56$$

$$= 9$$

Now the mean of the U values for any particular combination of N_a and N_b is equal to $N_aN_b/2$. For our illustrative problem, the mean of the U's is 25.0. The question is whether the obtained U of 9 is sufficiently lower than the mean to justify rejection of the hypothesis H_o, which is that $U - \bar{U} = 0.0$. The critical U values that would deviate from $\bar{U}$ to justify rejection of H_o at the alpha levels of .05 and .01 (in one-tail tests) are given in Table S. The larger N_i is to be found among the headings of the columns, and the smaller N_i among the headings of the rows. In the part of the table for which alpha is .01, we find that for N_i's of 10 and 5, respectively, the critical U value is 6. That is, it would take a U of 6 *or smaller* to be significant at the .01 level in a one-tail test, or at the .02 level in a two-tail test. In the part of the table for which alpha is .05, we find a critical U of 11. The obtained U is smaller than that, and so we may conclude that there are fewer than 5 chances in 100 that so small a U could have arisen by random sampling from the same population, with a predicted difference in a certain direction between the two samples. For a two-tail test, the obtained U of 9 can lead to rejection of H_o at the .10 level.

A word of caution is needed. It may happen that the obtained U is not the smaller of the two that could be computed from the same data.

This point can be checked by use of the relation

$$U_a = N_a N_b - U_b \qquad \text{(Relation between } U_a \text{ and } U_b \text{ in the same pair of distributions)} \qquad (12.6)$$

In the data under discussion, $N_a N_b = (5)(10) = 50$. From this result, where $U_a = 9$, we know that U_b should be equal to 41. This conclusion could be checked by applying formula (12.5) to R_a, which is 64.0. It is good practice to compute both U's and to see whether $U_a + U_b$ equals $N_a N_b$. At least formula (12.6) should be used in order to determine whether the U obtained first is the smaller one. Table S contains only critical U's on the low side, i.e., below the mean.

In the case of even smaller samples, when both N_a and N_b are below 9, Exact probabilities for U values for particular combinations of N_i can be found in tables provided by Siegel.[1]

EXERCISES

In each of the following exercises, state the hypothesis being tested, your statistical inference, and the experimental conclusion.

1 Apply the sign test to the first 15 pairs of measurements in Table 9.2 (p. 153).

2 Apply the sign-rank difference test to the same data as in exercise 1.

3 In three samples the observations were:

A: 9, 7, 2, 10, 8, 5
B: 10, 15, 12, 11, 16, 6
C: 18, 15, 14, 20, 10, 13

a. Apply the median test to all three samples together.
b. Apply the median test to each pair of samples, using the same median value as in part *a*.

4 Apply the composite-rank test to the three pairs of samples given in Exercise 3.

5 Assume that the following two sets of scores were obtained from a memory test given to random samples of 8 college freshmen (*A*) and 9 sixth-grade children (*B*).

A: 31, 38, 20, 33, 26, 36, 25, 34
B: 24, 21, 30, 24, 32, 19, 18, 6, 23

a. What is the U value to be expected if hypothesis H_o is correct? In other words, what is $\bar{U}$?
b. Compute the R_i sums and the two U_i statistics, applying available checks.
c. Using Table S in the Appendix, reach a decision regarding hypothesis H_o, in terms of both one- and two-tail tests.

6 Using the data given in exercise 5:

a. Derive a z ratio for U_a and U_b, for another kind of test of the difference between the two samples.
b. Interpret your results. How does the decision reached using this test compare with that reached using critical U values?

[1] Siegel, op. cit., p. 115.

ANSWERS

1 For a one-tail test, from Table N in the Appendix, $p = .09$.
2 $T = 16.5$, with p between .02 and .05.
3 χ^2 (A versus B versus C) = 0.34; df = 2.
χ^2 (A versus C) = 3.38 (with Yates's correction); df = 1.
χ^2 (B versus C) = 0.00 (with Yates's correction); df = 1.
χ^2 (A versus C) = 5.49 (with Yates's correction); df = 1.
4 R_a (A versus B) = 25.5; $.02 < p < .05$; $\bar{R} = 39.0$.
R_a (B versus C) = 31.0; $p > .05$.
R_a (A versus C) = 21.5; $p = .01$.
5 a. $U = 36 = N_a N_b/2$.
b. $R_a = 97$; $R_b = 56$; $U_a = 11$; $U_b = 61$.
c. For N's of 9 and 8, a U of 11 barely reaches significance at the .01 level in a one-tail test or at the .02 level in a two-tail test.
6 a. $z = 2.41$ for either R_i.
b. z fails to reach significance at the .01 level in a two-tail test.

13 Analysis of variance

It frequently happens in research that we obtain more than two sets of measurements on the same experimental variable, each under its own set of conditions, and we want to know whether there are any significant differences between the sets. We could, of course, pair off two sets at a time, pairing each one with every other one, and test the significance of the difference between means, or other statistics, in each pair.

Perhaps the variation of condition has been a qualitative one; for example, we have test scores for children from each of five neighboring states, or we have simple-reaction-time measurements under four different verbal instructions. Every other variable thought to be significantly related in a causal way to the experimental variable has been held constant. Perhaps the variation is a quantitative one, for example, retention scores obtained after different proportions of time have been spent in memorizing by the anticipation method versus the reading method, or arithmetic scores of children who have devoted different proportions of class time to drill in number operations versus concrete applications of numbers.

One practical problem involved in testing for significance of differences is the amount of labor involved. Five samples involve 10 pairs; six samples involve 15 pairs; 10 samples involve 45 pairs; and so on. There is a possibility that none of the differences between pairs would prove to be significant. In meeting this situation, it would be desirable to have some overall test of the several samples to tell us whether the differences considered simultaneously are significant.

There are more important logical and statistical reasons for wanting a single composite test. If we happened to have as many as a hundred differences to be tested, and if we found one of them significant at the .01 level and approximately five of them significant at the .05 level, we should actually conclude that *none* of the differences is significant. We could even have a few more than these meeting the significance standards, all occurring as a result of chance. We should suspect even the large differences of being due to chance unless we have an excess number of them. A simultaneous test should be of such a nature that we can conclude whether the whole distribution of obtained sampling statistics could have happened by chance.

There is still another statistical reason for wanting to treat the data together. If we tested each pair separately, we would use as an estimate of the population variance only the data from the two samples involved. If we make the null hypothesis apply to *all* the samples—that they all arose by random sampling from the same population—we could use *all* the data from which to make a much more stable estimate of the population variance. We should have to assume, of course, that the variances from the different samples are homogeneous.

Although some attention was given to problems of composite tests of significance in Chaps. 11 and 12, the methods described there have limited application. The reason is that when we can make the appropriate assumptions, there are more powerful parametric tests available. These come under the general heading of *analysis of variance*.

Analysis in a one-way classification problem

Consider again the case in which we have several samples of the same general character and we want to determine whether there are any significant differences between the means. The basic principle of such a test is to determine whether the sample means vary further from the population mean than we should expect, in view of the variations of single cases from their means.

TWO ESTIMATES OF POPULATION VARIANCE In a single subsample, the amount of expected variation of single cases from the population mean is indicated by the statistic s^2, which is an estimate of the population variance, or parameter σ^2. The variation of randomly sampled means about the population mean is indicated by the SE of the mean, squared, which is denoted by $\sigma_{\bar{X}}^2$ and is computed by the ratio σ^2/n, where n is the size of each sample.[1] If we multiply this ratio by n, we obtain σ^2, the population variance.

In other words, we have a way of estimating the population variance from the variance among means. If there is no significant vari-

1 In connection with analysis of variance, we shall use n to stand for the number of cases in a subsample and N to stand for the number of cases in all subsamples in the problem combined. We shall deal first with the case in which all the subsample n's are equal.

ation among the means, if they arose by random sampling from the same population (or from populations with equal means), the population variance estimated from them should be essentially the same as that estimated from the single observations. The test for determining the significance of the differences between two variances is the *F* test, which was described in Chap. 9. *F* is a ratio of two variances. It is necessary to have two independent estimates of the population variance in order to form an *F* ratio. With appropriate df applied to the two variances being compared, we can interpret *F* as significant or not.

BETWEEN-SETS SUM OF SQUARES AND BETWEEN-SETS VARIANCE

Our attention will be directed next to the operations by which the two estimates of population variances are achieved, one from the means and one from the single observations. We have already seen that there is a basis for estimating the population variance from the means. The computational steps will now be described.

Suppose that we have *k* samples, or sets, of *n* cases each, where *n* is constant. For each of the *k* means we should have the deviation

$$d_i = \bar{X}_i - \bar{X}_t \quad \text{(Deviation of a set mean from the grand mean)} \tag{13.1}$$

where $\bar{X}_i$ = mean of set *i*, where sets vary from 1 to *k*, and $\bar{X}_t$ = the grand mean, mean of means, or mean of all observations in all sets combined.

If we square all the deviations d_i and sum the squares, we should be on the way to finding the variance of the means about the estimated population mean, where $\bar{X}_t$ is our best estimate of that population mean. This variance is actually the *variance error* of the mean, which is the square of the SE of the mean. This variance is not exactly what we want. We want an estimate of the variance of *individual observations* about the population mean, not the variance of the means.

We ordinarily compute an estimated variance from a sum of squares of deviations of single observations. The sum of squares that we want, derived from the means, is given by the expression $n\Sigma d_i^2$. This statement can be made more reasonable by saying that each d_i value is shared by all *n* cases in the set in which it appears. It is as if we gave all the cases in that set the same deviation value. In estimating the variance of individual observations from the mean, we need as many deviations as there are observations. Thus the expression $n\Sigma d_i^2$ is an estimate of the sum of squares of deviations of observations from the population mean. Since it is derived from the means, it is called the *between-sets sum of squares.*

Looking at the matter a little more mathematically, it is as if we regard every observed measurement that we obtain in the experiment as a summation of three independent components: $\bar{X}_t + d_i + e_r$. Since it is a constant, $\bar{X}_t$ makes no contribution to variance. The term d_i is the set deviation from the mean, i.e., $\bar{X}_i - \bar{X}_t$, and is the same for every d_i case within a set, but probably differs from set to set. The term e_r is a random error that differs from one observation to another. In the present context, e_r is a deviation of an observation from a set mean. The combination of d_i and e_r gives to each observation its unique value. Combined, they are the total deviation of an observation from $\bar{X}_t$.

As will be explained in Chap. 16, the variance of a sum of two independent variables is the sum of the variances of those two variables. Since both d_i and e_r are deviations from means, the means of their squares are estimates of variances; the former of *between-sets* variance and the latter of *within-sets* variance.

A *mean* of squares implies division of the sum of squares by the number of values squared. In estimating population variances, in order to allow for bias we divide instead by the number of degrees of freedom. There are k deviations d_i involved, so we have $k-1$ degrees of freedom. One degree of freedom is lost in using the computed grand mean X_t to find the deviations. The *between-sets mean square* MS_b is therefore computed by the formula

$$MS_b = \frac{SS_b}{k-1} = \frac{n\Sigma d_i^2}{k-1} \qquad \text{(Between-sets mean square)} \qquad (13.2)$$

where SS_b = sum of squares between sets.

WITHIN-SETS SUM OF SQUARES AND WITHIN-SETS VARIANCE

If we may assume that the variances within the different samples are equal except for random fluctuations, we may combine the sums of squares from all sets in order to obtain from this source an estimation of the population variance. As we combine sums of squares we also combine degrees of freedom by which to divide the sum of squares. In each set sample the number of df is $n-1$. In k samples combined we have $k(n-1)$ df. This can also be expressed as $(N-k)$ df, since $N = kn$. One df is lost for each set mean used in finding the within-sets deviations.

In terms of a formula, the *within-sets mean square* (MS_w) is computed from the *within-sets sum of squares* by the equation

$$MS_w = \frac{SS_w}{k(n-1)} = \frac{\Sigma x^2}{k(n-1)} = \frac{\Sigma x^2}{N-k} \qquad \text{(Within-sets mean square)} \qquad (13.3)$$

where SS_w = within-sets sum of squares and x = a deviation of an observation from its set mean.

THE SOLUTION OF A ONE-WAY ANALYSIS OF VARIANCE PROBLEM

In Table 13.1 we have four sets of values ($k = 4$, $n = 5$, $N = 20$), each value being the mean of a series of settings made by a different individual on the Galton bar (an instrument for matching lines for length).[1] Each subject was asked to adjust a constant horizontal line of 115 mm and was asked to adjust another line until the two seemed equal. Four sets were obtained under four different conditions (A_1, A_2, A_3, A_4) selected by the experimenter. A particular condition is often called a *treatment,* in the context of analysis of variance. Is it likely that the observations all came by random sampling from the same general "population" of adjustments, or were there systematic differences sufficient to say that the data are really not homogeneous among the sets?

The following steps are carried out in the solution of the type demonstrated in Table 13.1:

step 1 Compute the sums and means of the sets, the grand total ΣX, and the grand mean $\bar{X}_t$.

step 2 For every set, compute the deviations from the set means $\bar{X}_i$. These are equal to $(X - \bar{X}_i)$ and are denoted by x.

step 3 Square the deviations within sets to find each x^2. Sum these to obtain Σx^2, the sum of squares of deviations within sets.

step 4 For each set, compute d_i, which equals $\bar{X}_i - \bar{X}_t$.

step 5 Square each d_i and find $n\Sigma d_i^2$.

With all these calculations completed (see Table 13.1), we have the values for use in formulas (13.2) and (13.3). The Σx_s^2 is 69.20, and the $n\Sigma d_i^2$ is 82.80. Dividing these by the appropriate numbers of degrees of freedom, we obtain the mean squares.

For this purpose we set up Table 13.2. Listing first the sum of squares for between sets and degrees of freedom that go with it, we obtain the ratio 27.60 as the variance estimated from the d's. For the corresponding values for within sets, we find the value 4.325 as an estimate of variance from the x's. The ratio of the between-sets mean square to the within-sets mean square gives an F ratio $27.6/4.325 = 6.38$. The between-sets estimate of variance is more than six times the within-sets estimate.

1 The reason for selecting a different individual randomly from the same population for each observation is that we wish to generalize the conclusions to the population in general; we could not do this if one observer were used for all conditions. If the same set of observers had been used under each of the four treatments, this procedure would have complicated the experimental design, requiring some other kind of statistical treatment of the results, such as will be discussed later in this chapter. Each observer was assigned at random to one of the four experimental conditions, so that the within-sets variance would represent random error and serve as an estimate of experimental error variation. This setup is referred to as a randomized-groups design.

TABLE 13.1 Work sheet for the analysis of variance in four sets of measurements on the Galton bar for four experimental conditions

	Experimental conditions (treatments)				
	A_1	A_2	A_3	A_4	
The measurements (X)					
	114	119	112	117	
	115	120	116	117	
	111	119	116	114	
	110	116	115	112	
	112	116	112	117	
ΣX_i	562	590	571	577	2,300 ΣX
$\bar{X}_i$	112.4	118.0	114.2	115.4	115.0 $\bar{X}_t$
Deviations within sets (x)					
	+1.6	+1.0	−2.2	+1.6	
	+2.6	+2.0	+1.8	+1.6	
	−1.4	+1.0	+1.8	−1.4	
	−2.4	−2.0	+0.8	−3.4	
	−0.4	−2.0	−2.2	+1.6	
Squares of deviations within sets (x^2)					
	2.56	1.00	4.84	2.56	
	6.76	4.00	3.24	3.56	
	1.96	1.00	3.24	1.96	
	5.76	4.00	0.64	11.56	
	0.16	4.00	4.84	2.56	
	17.20	14.00	16.80	21.20	69.20 Σx^2
Deviations of set means from grand mean (d_i)					
d_i	−2.60	+3.00	−0.80	+0.40	
d_i^2	6.76	9.00	0.64	0.16	16.56 Σd_i
nd_i^2	33.80	45.00	3.20	0.80	82.80 $n\Sigma d_i^2$

THE EXPECTED MEAN SQUARE

In Chap. 7 we defined probability as a long-term average, that is, the relative frequency of occurrence of an event taken over time. Such long-term averages from experiment to experiment are also called expected values, or expectations. They are denoted by E(X), which is read "the expected value of X."

Expected values may be obtained for other statistics as well as probabilities. For instance, the expected value of a sample mean $\bar{X}$ will be the population mean μ, that is,

$$E(\bar{X}) = \frac{1}{k} \Sigma\mu_i = \mu \tag{13.4}$$

Similarly, the expected value for the sample variance will be the population variance.

$$E(s^2) = \frac{1}{k} \Sigma\sigma_i^2 = \sigma^2 \tag{13.5}$$

The expected mean squares are used to specify the parameters estimated for each source of variance.

In analysis of variance the expected values of the mean squares, which are called the expected mean squares, specify the parameters that are estimated without bias by the experimental mean squares. They also provide the basis for determining which mean squares will be used for tests of hypotheses in analysis of variance.

TEST OF SIGNIFICANCE FOR A ONE-WAY ANALYSIS For a one-way analysis of variance, we define F as

$$F = \frac{MS_b}{MS_w} \tag{13.6}$$

and this F will have $(k - 1)$ df for the numerator and $k(n - 1)$ df for the denominator.

The null hypothesis tested by formula (13.6) is that we have k independent random samples from the same normally distributed population so that

$$E(\bar{X}_i) = \mu \tag{13.7}$$

for all k treatments, and that

$$E(s_i^2) = \sigma^2 \tag{13.8}$$

for all k treatment variances.

If formula (13.8) is true, then MS_w is an unbiased estimate of the common population variance σ^2; that is, $E(MS_w) = \sigma^2$. If formulas (13.7) and (13.8) are true, then it can be shown[1] that MS_b is also an unbiased estimate of σ^2; that is, $E(MS_b) = \sigma^2$. However, if formula (13.7) is not true, then, as we shall see, $E(MS_b) > E(MS_w)$. The tabled values of $F = MS_b/MS_w$ are those that would be expected to occur, at various levels of significance, when both formulas (13.7) and (13.8) are true. Consequently, if the obtained value of F is greater than the tabled value for some defined level of significance, so that we may regard it as an improbable value when the null hypothesis is true, we may decide to reject the null hypothesis. In doing so, we may conclude that the treatment means are not all estimates of the same common population mean μ, under the additional assumption that

1 For a proof, see Edwards, A. L. *Experimental design in psychological research.* (4th ed.) New York: Holt, Rinehart and Winston, 1972.

the treatment variances are homogeneous; that is, the separate variance estimates for the k samples are all estimates of a common population variance.

The expected value of the treatment mean square is

$$E(MS_b) = \sigma^2 + \frac{n\Sigma d_i^2}{k-1} \tag{13.9}$$

and the F ratio is

$$F = \frac{E(MS_b)}{E(MS_w)} = \frac{\sigma^2 + \dfrac{n\Sigma d_i^2}{k-1}}{\sigma^2} \tag{13.10}$$

Only if all μ_i's equal the same value, μ, will $\Sigma d_i^2 = 0$. If they are not, then the expected value of the numerator of the ratio defined by formula (13.10) will be larger than the expected value of the denominator. The null hypothesis tested by the sample value of $F = MS_b/MS_w$, in other words, is that $\Sigma d_i^2 = 0$, and this will be true only if the expected values of the k treatment means are all equal to the same population value. Sample values of F significantly greater than 1.0 thus provide evidence that the treatment population means are not all equal to the same value μ.

INTERPRETATION OF THE *F* RATIO The significance of an F is determined by reference to Snedecor's table (Table F in the Appendix). In using this table, we have to consider the two different df values. For the numerator of F (almost always the larger variance) we look for df_1 at the top of a column. For the denominator of F we look for df_2 at the left of a row. In our illustrative problem, there is a df_1 of 3, which can be found at the head of a column, and a df_2 of 16, which can be found at the left of a row. At the cell where the appropriate column and row intersect we find that it takes an F of 3.24 to be significant at the .05 point and an F of 5.29 to be significant at the .01 level, when the df are 3 and 16. The obtained F is greater than that required for significance at the .01 point, which is sufficient reason for rejecting the null hypothesis.

ASSUMPTIONS ON WHICH AN *F* TEST RESTS

As usual, a statistical decision is sound to the extent that certain assumptions have been satisfied in the data that are used. In analysis of variance there are usually four stated requirements:

1. The sampling within sets should be random, which, as usual, means observations that are mutually independent and have equal opportunity to occur.
2. The variances from within the various sets must be approximately equal. The within-sets mean square is commonly the denominator of F ratios, and consequently much depends upon its accuracy. Much vari-

ation among set variances leads to suspicion of an inaccurate estimate of the population variance from within sets.

3 Observations within experimentally homogeneous sets should be from normally distributed populations. *F* is mathematically a ratio of two chi squares, each divided by its appropriate df. It will be recalled from Chap. 11 that chi square requires normally distributed populations.

4 The contributions to total variance must be additive. We have already seen that it was necessary to assume independence of the deviations between and within sets in order to say that the total variance is a simple sum of the two contributing variances.

Later, we shall consider what may happen when these assumptions are not satisfied. It was said before that the obtained *F*, significant at the .01 point, indicates that the means of sets are significantly different, i.e., that somewhere among them, at least, there are significant differences. This is not to say that it is necessarily the means that are significantly different. In cases where the population variances are not equal, a significant *F* might be due in part to this fact. Conclusions about the means would then be in some doubt. But if we can eliminate the possibility of unequal variances, a significant *F* must indicate significant differences somewhere among the means.

SOME COMPUTATIONAL CHECKS

It will be noted in Table 13.2 that we have recorded the total sum of squares and the number of df for the same. These values have been found by summing the components in both instances, i.e., component sums of squares and component df. The total sum of squares is a composite of two independent factors—that derived from deviations of means and that derived from deviations within sets. From the additive feature of sums of squares, it follows that

$$\Sigma x_t^2 = \Sigma\Sigma x_i^2 + n\Sigma d_i^2 \qquad \text{(subscript } t \text{ stands for "total")}$$

where x_t = deviation of X from the grand mean $(X - \bar{X}_t)$
x_i = deviation of X from a set mean $(X - \bar{X}_i)$
d_i = deviation $(\bar{X}_i - \bar{X}_t)$

The double summation sign before x_i^2 indicates that the within-sets deviations are squared and summed for each set and that these sums

TABLE 13.2 The total variance in the Galton-bar data subdivided into two components

Components	*Sum of squares*	*Degrees of freedom*	*Mean square*
Between sets	**82.80**	**3**	**27.60**
Within sets	**69.20**	**16**	**4.325**
Total	**152.00**	**19**	

are then summed over all sets. The relationship can also be represented by $SS_t = SS_b + SS_w$.

If SS_t is computed from the complete data, it can be used as a check, for it should exactly equal the sum of the two component sums of squares. The number of df to be associated with SS_t is $N - 1$, which should equal the sum of the two different df values for between-sets and within-sets sums of squares. In Table 13.2 we find these checks satisfied.

FORMATION OF AN *F* RATIO

In analysis of variance generally, the numerator of the variance ratio is the estimate of variance that arises from variations whose sources we are testing. The denominator is the estimate that arises from variations whose sources are usually unknown. The latter variance is sometimes referred to as the *error term*. We assume that random sampling is the only source of the variations involved in this term. Sometimes the denominator is called the *residual term*, since its source is all that is left over after other sources have been accounted for.

It will almost always happen that the numerator term is larger and that F is therefore greater than 1.0. We are thus dealing with the right-hand tail of the F distribution in our interpretation of F. We have a one-tail test. Should F on rare occasions turn out to be less than 1.0, the conclusion is merely that we do not reject the null hypothesis. There is no need to consult the F table for this kind of outcome, or even to compute an F ratio at all.

THE LIMITING CASE OF TWO SETS

When we have only two sets of observations, as when we compare two means for significance of their difference, we can still make an F test. The between-sets variance has associated with it only 1 df. For this particular situation, when n_1 and n_2 are equal, $\bar{X}_1 - \bar{X}_2 = 2d_i$ and $d_i = (\bar{X}_1 - \bar{X}_2)/2$. Therefore, d^2, which is needed to find the SS_b, equals $(\bar{X}_1 - \bar{X}_2)^2/4$. The Σd_i^2, which is needed in formula (13.2), is equal to $2d_i^2$, or $(\bar{X}_1 - \bar{X}_2)^2/2$. Thus, for the between-sets mean square we have

$$MS_b = \frac{n(\bar{X}_1 - \bar{X}_2)^2}{2} \quad \text{(Mean square for between sets when there are two sets)} \tag{13.11}$$

To illustrate, let us take the largest difference between means in Table 13.1. Those two means are 112.4 and 118.0, with a difference of 5.6. Applying formula (13.11), we find 78.4 for the mean square for between sets. The within-sets sum of squares is a sum of 17.2 and 14.0, from Table 13.1. With 8 df, the within-sets mean square is 3.9. The F ratio is $78.4/3.9 = 20.10$. With 1 and 8 df, from Table F in the Appendix we see that F is significant beyond the .01 point.

It has been proved that with 1 df for the between-sets variance, $F = t^2$. In this same problem, then, $t = \sqrt{20.10} = 4.48$. If we compute t by means of formula (9.4) (see p. 157), we arrive at the same value.

COMPUTATION OF MEAN SQUARES FROM ORIGINAL MEASUREMENTS

Just as we can compute standard deviations, and therefore variances, from original measurements without computing each deviation from the mean [see formula (5.4), p. 72], so we can calculate the necessary constants for an analysis of variance. Such an approach requires squaring the original measurements as well as ΣX.

For pencil-and-paper calculations, it is convenient to reduce the observed numbers to a more workable size by coding them. The three-place numbers in Table 13.1 can be conveniently coded by deducting the constant 110, the lowest measurement, leaving the remainders shown at the left in the top half of Table 13.3. The mean squares are not in the least affected by this kind of transformation. Intuitively, one can see that the coded numbers maintain exactly the same distances from one another and from the means as before coding. The coded values are called X' in Table 13.3.

For the general solution, without knowing deviations x or d, the sums of squares we need are found by the following procedures. The between-sets sum of squares is given by the formula

$$SS_b = n\Sigma d_i^2 = \sum \frac{(\Sigma X)_i^2}{n} - \frac{(\Sigma X)^2}{N} \quad \text{(Sum of squares for between sets)} \tag{13.12}$$

TABLE 13.3 Solution of an analysis of variance from original measurements (without determining deviations from means).

	Experimental conditions				
	A_1	A_2	A_3	A_4	
Measurements (reduced) (X')					
	4	9	2	7	
	5	10	6	7	
	1	9	6	4	
	0	6	5	2	
	2	6	2	7	
$(\Sigma X')_i$	12	40	21	27	100 $\Sigma X'$
					5.0 $\bar{X}'_t$
$(\Sigma X')_i^2$	144	1,600	441	729	2,914 $\Sigma(\Sigma X')^2$
Squared measurements (X'^2)					
	16	81	4	49	
	25	100	36	49	
	1	81	36	16	
	0	36	25	4	
	4	36	4	49	
$(\Sigma X'^2)_i$	46	334	105	167	652 $\Sigma(\Sigma X'^2)_i$

The within-sets sum of squares is given by

$$SS_w = \Sigma x_i^2 = \Sigma(\Sigma X^2)_i - \frac{\Sigma(\Sigma X)_i^2}{n} \quad \text{(Sum of squares for within sets)} \tag{13.13}$$

The total sum of squares is given by

$$SS_t = \Sigma x_t^2 = \Sigma(\Sigma X^2)_i - \frac{(\Sigma X)^2}{N} \tag{13.14}$$

The steps called for in applying these formulas are:

step 1 Sum the measurements X for each set to obtain $(\Sigma X)_i$ for each set (see Table 13.3). Sum these values to obtain ΣX.

step 2 Square the sums of the scores to obtain $(\Sigma X)_i^2$ for each set. Sum these values to find $\Sigma(\Sigma X)_i^2$

step 3 Square all measurements to find the X^2 values. Sum these values to obtain ΣX^2.

Applying the three formulas, by formula (13.12), we have

$$SS_b = n\Sigma d_i^2 = \frac{2{,}914}{5} - \frac{10{,}000}{20} = 582.8 - 500 = 82.8$$

By formula (13.13) we have

$$SS_w = 652 - \frac{2{,}914}{5} = 652 - 582.8 = 69.2$$

and by formula (13.14),

$$SS_t = \Sigma x_t^2 = 652 - \frac{10{,}000}{20} = 652 - 500 = 152$$

A check for accuracy of computations is to see that $SS_b + SS_w = SS_t$. The check is satisfied, for $82.8 + 69.2 = 152$. A comparison of these values with those in Table 13.2 will show that we have arrived at the same sums of squares. From here on, the computation of mean squares and F is the same as before.

WHEN SAMPLES ARE OF UNEQUAL SIZE

The procedures described thus far apply to the special, but not unusual, case in which all set samples are of equal size. Experiments can be planned that way, but sometimes available data do not fit that specification. With a little modification of the formulas, we can take care of problems in which n varies.

For the between-sets sum of squares,

$$SS_b = \Sigma n_i(\bar{X}_i - \bar{X}_t)^2$$

$$= \Sigma \frac{(\Sigma X)_i^2}{n_i} - \frac{(\Sigma X)^2}{N} \quad \text{(Between-sets sum of squares when samples vary in size)} \tag{13.15}$$

where n_i = number of cases in a specified set
$\overline{X}_i$ = mean of that set
$\overline{X}_t$ = mean of all observations

For all expressions involving subscript i, the summation is made over k sets.

For the within-sets sum of squares,

$$SS_w = \Sigma x_i^2 = \Sigma(\Sigma X_i^2) - \sum \frac{(\Sigma X)_i^2}{n_i}$$

(Within-sets sum of squares when samples vary in size) (13.16)

For the total sum of squares the formula is the same as when we have samples of equal size; hence formula (13.14) will apply for the general case.

The degrees of freedom are the same as in the case of equal n's for the total and between-sets sums of squares. The df for the within-sets sum of squares equal $\Sigma(n_i - 1)$.

More detailed testing of differences between set means

Thus far, we have seen that an F test gives us only an overall answer regarding the significance of a whole collection of differences between means. There are times, however, when we want more information than that; we may have a special interest in certain differences. Before an F test is made, if there should be a few specific mean differences in which we are particularly interested, we can make special t tests for them if there are not too many. In fact, we can test simultaneously as many as k-1 differences, for that many can be regarded as independent. There is also room for a little more latitude in testing mean differences, for we can combine two or more sets of data to test means of combined sets to be compared, as will be illustrated.

If, for lack of any prior interest in specific mean comparisons, an F test has been made and if F turns out to be significant at the chosen level of alpha, we can still make some t tests of a special kind in order to see where the basis for the significant F lies. We must be aware of the fact that in applying ordinary t tests to all pairs of means, some of them should be expected to reach significance at the prescribed level, as the F test has told us. The question is whether the number of such "significant" t statistics could have arisen by chance in a multiple-set situation. Special t values must be used to fit this situation, values to be called t'. Incidentally, it is possible to compute more than $k(k$-1)/2

TABLE 13.4
A set of orthogonal comparisons and coefficients for $k = 4$ groups

Comparison d_i	$\bar{X}_1$	$\bar{X}_2$	$\bar{X}_3$	$\bar{X}_4$	$\bar{X}_1$	$\bar{X}_2$	$\bar{X}_3$	$\bar{X}_4$	Σa^2_i
	Values of coefficients				*Notation for coefficients*				
c_1	1	0	−1	0	a_{11}	a_{12}	a_{13}	a_{14}	2
c_2	0	−1	0	1	a_{21}	a_{22}	a_{23}	a_{24}	2
c_3	−1	1	−1	1	a_{31}	a_{32}	a_{33}	a_{34}	4

t' values because we can combine sets of data in various ways, as we shall see, for such post hoc comparisons.[1]

PLANNED COMPARISONS OF GROUP MEANS

In a one-way classification analysis of variance for k groups, $k-1$ independent group-mean comparisons, each with one degree of freedom, can be made. Each comparison involves two or more of the treatment means. We shall illustrate the case where the number of observations per group is the same.

Table 13.2, p. 231, shows the results for the analysis of variance for the four conditions under which the measurements on the Galton bar were made. Table 13.4 shows the coefficients for one set of orthogonal comparisons that could be made on $k = 4$ groups. The first comparison determines whether there is a significant difference between the means for A_1 and A_3. The second comparison determines whether there is a significant difference between the means for A_2 and A_4. The third comparison determines whether the means for A_1 and A_3, weighted equally, differ significantly from the means for A_2 and A_4, similarly weighted. At the right of the table is the notation for the coefficients that are used to compute the values for determining the comparisons (or contrasts as they are sometimes called).

A comparison is a linear function of means and is represented by the equation

$$c_i = a_{i1}\bar{X}_1 + a_{i2}\bar{X}_2 + \cdots + a_{ik}\bar{X}_k \qquad \text{(Comparison of group means)} \qquad (13.17)$$

To obtain the comparisons, we multiply the group means by the coefficients, which results in the following comparisons for the example:

$$c_1 = 1(\bar{X}_1) + 0(\bar{X}_2) - 1(\bar{X}_3) + 0(\bar{X}_4) = \bar{X}_1 - \bar{X}_3$$
$$c_2 = \bar{X}_4 - \bar{X}_2$$
$$c_3 = \bar{X}_2 + \bar{X}_4 - \bar{X}_1 - \bar{X}_3$$

The computation of the c_i values is shown in Table 13.5.

1 The distinction between planned and post hoc comparisons is also referred to as a priori versus a posteriori comparisons.

TABLE 13.5 Computation of the orthogonal comparisons of the Galton-bar set means based on the coefficients in Table 13.4

Comparison	$\bar{X}_1$ 112.4	$\bar{X}_2$ 118.0	$\bar{X}_3$ 114.2	$\bar{X}_4$ 115.4	Value of c_i
c_1	1	0	−1	0	−1.8
c_2	0	−1	0	1	−2.6
c_3	−1	1	−1	1	6.8

REQUIREMENTS FOR ORTHOGONAL COMPARISONS

In order for a set of comparisons to be orthogonal (independent), the sum of the products of the corresponding coefficients for any pair of comparisons must equal zero. This requirement may be represented by the equation

$$a_{i1}a_{j1} + a_{i2}a_{j2} + \cdot \cdot \cdot + a_{ik}a_{jk} = 0 \quad \text{(Sum of products of coefficients equals zero for orthogonal comparisons } d_i \text{ and } d_j\text{)} \qquad (13.18)$$

Note that all comparisons shown in Table 13.5 are mutually orthogonal since the sums of the products of the coefficients of all combinations of pairs of comparisons are equal to zero. For comparisons c_1 and c_2, for example,

$$(1)(0) + (0)(-1) + (-1)(0) + (0)(1) = 0$$

A second requirement for a comparison is that the coefficients sum to zero; i.e.,

$$\Sigma a_i = 0 \quad \text{(Sum of coefficients for a comparison equals zero)} \qquad (13.19)$$

It will be noted that the coefficients for each comparison indicated in Table 13.5 sum to zero.

The set of comparisons shown in Table 13.5 is only one of several possible orthogonal sets for $k = 4$. Other possible comparisons for $k = 4$ are shown in Table 13.6 (p. 240). We may select sets of up to three comparisons that are orthogonal to one another to represent the independent hypotheses concerning mean differences for $k = 4$ that we wish to test.[1]

The orthogonal comparisons for any set of $k > 2$ means should be determined by the experimental interests of the investigator and specified at the time the experiment is being planned. It is not necessary to make all the $k - 1$ orthogonal comparisons, only those of inter-

[1] For example, we could select the set 3 versus 4, 2 versus 3 + 4, and 1 versus 2 + 3 + 4; or the set 1 + 2 versus 3 + 4, 1 + 3 versus 2 + 4, and 1 + 4 versus 2 + 3.

est. The significance of any set of planned orthogonal comparisons—equal to, or less than, $k - 1$, where k is the number of treatment groups—may be tested simultaneously if they have been planned in advance.

STANDARD ERRORS AND *t* RATIOS

For a comparison c_i, the corresponding standard error is obtained from

$$s_{c_i} = \sqrt{\text{MS}_w \left(\frac{a_{i1}^2}{n_1} + \frac{a_{i2}^2}{n_2} + \cdots + \frac{a_{ik}^2}{n_k} \right)} \qquad \text{(Standard error of a comparison)} \qquad (13.20)$$

For equal numbers of cases per group, the formula reduces to

$$s_{c_i} = \sqrt{\frac{\text{MS}_w}{n} \Sigma a_i^2} \qquad \text{(Standard error of a comparison where the number of observations per group is uniform)} \qquad (13.21)$$

And for the comparison between two means, e.g., $\bar{X}_1 - \bar{X}_3$ (where the corresponding coefficients are 1 and -1 and $\Sigma a_i^2 = 2$), the formula for the standard error of the difference between two means, where the n's are equal, becomes

$$s_{c_{\bar{x}}} = \sqrt{\frac{2\text{MS}_w}{n}} \qquad \text{(Standard error of the difference between two means with equal } n\text{'s)} \qquad (13.22)$$

The value for the MS_w for the example can be obtained by refering to the Galton-bar analysis in Table 13.2, and it is equal to 4.325. The value for the Σa_i^2 for each comparison is shown at the right of Table 13.4. These values are

$$\Sigma a_1^2 = 2$$
$$\Sigma a_2^2 = 2$$
$$\Sigma a_3^2 = 4$$

Substituting these values in formula (13.21) gives the following standard errors for the three comparisons:

$$s_{c1} = \sqrt{\frac{4.325\ (2)}{5}} = \sqrt{1.730} = 1.315$$
$$s_{c2} = \sqrt{\frac{4.325\ (2)}{5}} = \sqrt{1.730} = 1.315$$
$$s_{c3} = \sqrt{\frac{4.325\ (4)}{5}} = \sqrt{3.460} = 1.860$$

The test of significance is made via a t ratio, each t having the number of degrees of freedom associated with the within-sets mean square, where

$$t_i = \frac{C_i}{S_{C_i}} \quad (t \text{ ratio for the significance of comparison}) \quad (13.23)$$

The t ratios for the three comparisons are

$$t_1 = \frac{-1.8}{1.315} = -1.369$$

$$t_2 = \frac{-2.6}{1.315} = -1.977$$

$$t_3 = \frac{6.8}{1.860} = 3.650$$

Looking up the values in Table D in the Appendix for 16 df indicates that only the last one is significant, and it is significant beyond the .01 level.

POST HOC COMPARISONS BY THE *S* METHOD

If we do *not* have a set of $k - 1$ orthogonal comparisons that we have planned in advance, we may want to do some "looking around" in the data and make several comparisons of interest to us after completing the one-way analysis of variance. To justify doing this by the S method, the overall F test should be significant at an α level set in advance.[1]

Table 13.6 shows the possible comparisons for four group means. A comparison is still defined as in formula (13.17) with the coefficients summing to zero as indicated in formula (13.19). The values needed to compute the t ratio are defined in the same way as they were for the orthogonal comparison [see formulas (13.21) and (13.23)]. The values for each comparison have been entered in Table 13.6.

A larger t ratio is required to achieve a specified level of significance by the S method than is required for independent (orthogonal) comparisons. The t ratio is evaluated for significance by comparing it with

$$t' = \sqrt{(k-1)F} \quad (t \text{ value for post hoc comparisons by the } S \text{ method}) \quad (13.24)$$

where F is the tabled value of F for $k - 1$ df for the numerator and the number of df associated with the MS_w for the denominator. The computation of the values leading up to them and the t ratios for the data in Table 13.1 are shown in the right half of Table 13.6. Table F in the Appendix indicates that an F ratio of 3.24 is required for significance at the .05 level and that an F ratio of 5.29 is required for significance at the .01 level for 3 and 16 df. The corresponding values for t' are 3.118 and 3.984.

Comparing the values in the last column of Table 13.6 with t' in-

1 Scheffé, H. A method for judging all contrasts in the analysis of variance. *Biometrika*, 1953, **40,** 87–104.

TABLE 13.6 Possible comparisons for $k = 4$ group means

Comparison	$\bar{X}_1$ 112.4	$\bar{X}_2$ 118.0	$\bar{X}_3$ 114.2	$\bar{X}_4$ 115.4	Σa_i^2	c_1	s_{c_i}	t
1 vs. 2	1	−1	0	0	2	−5.6	1.315	−4.259
1 vs. 3	1	0	−1	0	2	−1.8	1.315	−1.369
1 vs. 4	1	0	0	−1	2	−3.0	1.315	−2.281
2 vs. 3	0	1	−1	0	2	3.8	1.315	2.890
2 vs. 4	0	1	0	−1	2	2.6	1.315	1.977
3 vs. 4	0	0	1	−1	2	−1.2	1.315	−.913
1 vs. 2 + 3	2	−1	−1	0	6	−7.4	2.278	−3.248
1 vs. 2 + 4	2	−1	0	−1	6	−8.6	2.278	−3.775
1 vs. 3 + 4	2	0	−1	−1	6	−4.8	2.278	−2.107
2 vs. 1 + 3	−1	2	−1	0	6	9.4	2.278	4.126
2 vs. 1 + 4	−1	2	0	−1	6	8.2	2.278	3.600
2 vs. 3 + 4	0	2	−1	−1	6	6.4	2.278	2.809
3 vs. 1 + 2	−1	−1	2	0	6	2.0	2.278	.878
3 vs. 1 + 4	−1	0	2	−1	6	.6	2.278	.263
3 vs. 2 + 4	0	−1	2	−1	6	−5.0	2.278	−2.195
4 vs. 1 + 2	−1	−1	0	2	6	.4	2.278	.176
4 vs. 1 + 3	−1	0	−1	2	6	4.2	2.278	1.844
4 vs. 2 + 3	0	−1	−1	2	6	−1.4	2.278	−.615
1 + 2 vs. 3 + 4	1	1	−1	−1	4	.8	1.860	.043
1 + 3 vs. 2 + 4	1	−1	1	−1	4	−6.8	1.860	−3.656
1 + 4 vs. 2 + 3	1	−1	−1	1	4	−4.4	1.860	−2.366
1 vs. 2 + 3 + 4	3	−1	−1	−1	12	−10.4	3.222	−3.288
2 vs. 1 + 3 + 4	−1	3	−1	−1	12	12.0	3.222	3.724
3 vs. 1 + 2 + 4	−1	−1	3	−1	12	−3.2	3.222	−.993
4 vs. 1 + 2 + 3	−1	−1	−1	3	12	1.6	3.222	.497

dicates that eight comparisons are significant at the .05 level and two are significant at the .01 level. This statement with regard to the significance of the comparisons in Table 13.6 is made with probability equal to, or greater than, $1 - \alpha$, or .95 for the first conclusion and .99 for the second conclusion.[1]

1 There are several other procedures that are used in making multiple comparisons of set means. These include Duncan's multiple-range test, the Newman-Keuls method for making all comparisons in a set of group means, and Dunnett's test for the difference between the mean of a control group and each treatment group. Summaries and comparisons of these methods are given by Edwards, op. cit., and by Winer, B. J. *Statistical principles in experimental design.* (2d ed.) New York: McGraw-Hill, 1971.

ESTIMATING THE STRENGTH OF ASSOCIATION

In addition to determining the statistical significance between group means it is frequently desirable to estimate the degree of relationship accounted for between the treatment variable used to divide the subjects into groups and the measurement variable used to evaluate the effect. It is important to do this because with a large-enough sample, even a trivial difference may prove statistically significant. All too often investigators are satisfied to establish the significance of their results without also investigating whether an appreciable portion of the variance in the measurement variable is related to the variable which serves as a basis for classifying the individual into the treatment groups.

The measure of association can be viewed either as an indication of the proportion of the variance in the measurement variable that is related to the treatment variable or as the reduction in uncertainty indicated by the size of the variance within groups relative to the overall variance in the dependent variable for the combined groups. One of the best indicators of uncertainty about the value of a variable is the size of the variance of its distribution.

We can estimate the degree of association between the two variables with fixed categories in the treatment variable by *omega* square:

$$\omega^2{}_{\text{est}} = \frac{(k-1)\,(F-1)}{(k-1)\,(F-1)+kn} \tag{13.25}$$

(Estimated proportion of variance in the measurement variable accounted for by its relationship with the treatment variable in a one-way analysis)

If we apply formula (13.25) to the Galton-bar data, we have

$$\omega^2{}_{\text{est}} = \frac{(4-1)\,(6.38-1)}{(4-1)\,(6.38-1)+4(5)} = \frac{16.14}{36.14} = .45$$

This result indicates not only that the overall relationship is significantly different from zero but also that approximately 45 percent of the total variance in the Galton-bar scores is related to the conditions under which they were obtained.

If a negative value is obtained for $\omega^2{}_{\text{est}}$, it may be set equal to zero. One should use the ω^2 estimate only after a significant F test ensures a nonnegative estimate.[1] Even where the outcome of an experiment is statistically significant, if the proportion of variance accounted for by the treatments is small, it may not be worthwhile to spend additional time and effort in pursuing the matter further.

1 The ω^2 coefficient is a general index of relationship between two variables and is related to the intraclass correlation, the coefficient of concordance, and the correlation ratio, all to be introduced later. The index which is more directly comparable to a correlation coefficient is ω, that is, the square root of ω^2.

For samples from two populations where the test of significance has been made by the t test, the estimate can be obtained by

$$\omega^2_{\text{est}} = \frac{t^2 - 1}{t^2 + n_1 + n_2 - 1} \tag{13.26}$$

In this case also, we use ω^2 only after a significant t test has ensured a nonnegative estimate.[1]

Analysis in a two-way classification problem

In the preceding kind of problem the sets of data were differentiated on the basis of only one experimental variation. There was only one principle of classification, one reason for segregating data into sets.

In a two-way classification, there are two distinct bases of classification. Two experimental conditions are allowed to vary from trial to trial. There may be several trials or replications under each treatment. In the psychological laboratory, different artificial airfield landing strips, each with a different pattern of markings, may be viewed through a diffusion screen to simulate vision through fog at different levels of opaqueness. In an educational problem, four methods of teaching a certain geometric concept may be applied by five different teachers, each one using every one of the four methods. There would therefore be 20 combinations of teacher and method. Let us suppose that equal numbers of randomly assigned pupils receive learning scores under each combination. When the treatments consist of all possible different combinations of one level from each experimental condition, and we have an equal number of observations for each experimental condition, the experiment is referred to as a complete factorial design with equal replications.

TABULATION OF DATA IN A TWO-WAY CLASSIFICATION PROBLEM

For an illustration of the procedure in this type of problem, we shall assume an experiment on the relation of scores on a certain psychomotor test to the size of a target at which the examinee must aim. In conducting the experiment, it is convenient to use three testing machines simultaneously in order to reduce the overall testing time. It is known that there are some, usually small, individual differences between machines in this test, to the extent that it would be risky to attach one target size to only one machine throughout the tests. Under that arrangement, machine differences might make it appear that there were differences attributable to target differences or might by chance negate those differences. The four target sizes (A_1, A_2, A_3,

[1] The user of ω^2 is cautioned that the interpretation of an obtained value applies only to the treatment categories used in the experiment. It may differ appreciably if categories are added to, or deleted from, the set employed.

TABLE 13.7 Scores of 60 students earned on three different machines of a psychomotor test, each with the target size varied in four steps

Target size		*Machines*			*Sums for target size*	*Means for target size*
		B_1	B_2	B_3		
		6	4	4		
		4	1	2		
		2	5	2		
A_1		6	2	1		
		2	3	1		
	Σ	20	15	10	45	
	$\bar{X}$	4	3	2		3
		8	6	3		
		3	6	1		
		7	2	1		
A_2		5	3	2		
		2	8	3		
	Σ	25	25	10	60	
	$\bar{X}$	5	5	2		4
		7	9	6		
		6	4	4		
		9	8	3		
A_3		8	4	8		
		5	5	4		
	Σ	35	30	25	90	
	$\bar{X}$	7	6	5		6
		9	7	6		
		6	8	5		
		8	4	7		
A_4		8	7	9		
		9	4	8		
	Σ	40	30	35	105	
	$\bar{X}$	8	6	7		7
Sums for machines		120	100	80	300	
Means for machines		6	5	4		5

A_4) were combined with the three machines (B_1, B_2, B_3) systematically. There were therefore 12 target-machine combinations with five observed scores obtained with each combination. The scores (which are fictitious and have been provided for the sake of a good illustration) are tabulated in Table 13.7. This arrangement is typical and convenient for the operations of analysis of variance. The sums and means, as given, are also needed in the variance solution.

It should be emphasized that in the two-way problem to be illustrated, the observations within sets come from entirely different individuals assigned at random to the target-machine combinations. If the same individuals were used for more than one treatment, there would be some correlation introduced between sets of observations, and the two-way model as demonstrated would not apply. There are other statistical models available to take care of data that involve correlated observations, such as in the matched-groups and repeated-measures designs treated later in this chapter.

DIFFERENT MODELS IN FACTORIAL DESIGNS

At this point, one should know that in a two-way factorial experimental design different kinds of variables for classification introduce different models.[1] The main distinction of concern to us here is between fixed categories and randomly chosen categories.

Fixed categories are arbitrarily chosen by the investigator. Sometimes they are chosen for him because there are no other possibilities—the two sexes, the two major political parties, right versus left arrangement of stimuli, and so on. If the variable is a quantitative one, such as amount of practice in learning, time of exposure, or hours of food deprivation, the investigator is likely to select a limited number of constant values, covering as much of the range as he thinks necessary. In the illustrative experiment, target size is one of those fixed-category variables, with four arbitrarily chosen target sizes.

Randomly chosen categories represent a sampling approach. Here the choice of machines is an example. The three selected were from a pool of possible machines that could have been used. Examples from other sources are selected individuals, picked at random, each of whom is to undergo the various kinds of treatment as determined by the second experimental variable. Each person might be given doses of a drug of different concentrations. There might be replications within treatments (several trials for each person), or there might be no replications, in which case there is only one experimental measurement for each cell of the table of data from the two-way factorial experiment.

From the kinds of combinations of variables that are possible, we have three different models. With both variables having fixed categories, we have a *fixed model* or *fixed-effects model*. Both sets of categories are arbitrarily chosen. When both variables have randomly

[1] A factorial experimental design has two or more conditions varied systematically, such as instructional set, method, or amount of practice.

selected categories, we have a *random model*. The latter offers the widest range of possible generalizations from the findings of the experiment, since the categories follow the statistical principle of random sampling from universes. With fixed categories, generalizations are restricted to the kinds of categories chosen. A third kind of model is called the *mixed model* because it has both fixed and random sets of categories.[1] Besides the experimental implications of random versus fixed categories that have just been touched upon, we shall see later that these differences in models are clearly related to the way in which *F* ratios are formed. The differences in models have no bearing upon how we extract information about mean squares from the data in a two-way analysis of variance, the operations of which we shall consider next.

THE SOURCES OF VARIANCE IN A TWO-WAY CLASSIFICATION PROBLEM

If we chose, we could proceed to perform an analysis of variance based upon the model of the one-way classification problem already demonstrated. That is, we could take the 12 sets as if they represented categories based upon a single principle and test the 12 means collectively to see whether they could have arisen by random sampling from the same population. There are, however, logical objections to such a solution in a problem of this kind.

Suppose we did carry through the solution proposed and found an *F* ratio that indicated significance beyond the .01 point. We should not know whether this was due primarily or solely to the differences between targets, to the differences between machines, or to both possible sources. Suppose, on the contrary, that the *F* ratio indicated no significant differences between sets. We should not be sure that one of the experimental variations, perhaps target size, was not actually producing real variations that were either covered over or counteracted by the effects of the other experimental variations. We should have what is called a *confounding* of effects. We need some method that will segregate the variations associated with each of the experimental variables so that any significant differences at all will have a chance to emerge in the *F* test and so that we shall know to which source to attribute any significant differences found.

INTERACTION VARIANCE The procedure about to be described makes possible this kind of segregation of the sources of variations. As a result, we can then determine whether differences between means owe their divergencies to target size, to machine differences, or to both. Nor is that all; when there are two possible sources of variations, there is also a possibility of what is called *interaction variance*.

The phenomenon is well named. Interaction variations are those attributable not to either of two influences acting alone but to joint ef-

1 For more detailed treatment of this subject, see McNemar, Q. *Psychological statistics.* (4th ed.) New York: Wiley, 1969; Lindquist, E. F. *Design and analysis of experiments in psychology and education.* Boston: Houghton Mifflin, 1953; Edwards, A. L., op. cit.

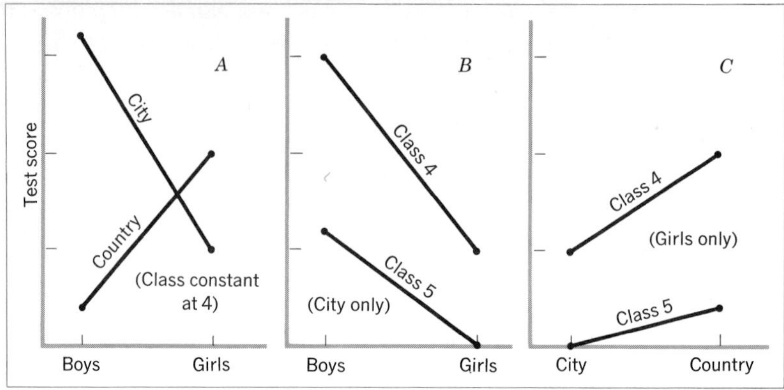

FIGURE 13.1 Illustration of the phenomenon of interaction of variables. Note that in some instances the relation of one variable to another depends upon the influence of a third variable (especially in Case *A*) and that in other instances the relation is very much the same (as in Case C). (Based on data from Young, H. B., Taguri, R., Tesi, G., and Montemagni, G. Influence of town and country upon children's intelligence. *British Journal of Educational Psychology,* 1962, **32,** 151–158, with permission of the authors.)

fects of the two acting together. If it turned out that the larger the target, the larger the scores tended to be, that would be one direct and isolable effect. It is one of the so-called main effects. If there were systematic machine differences, so that among the three there was a "most difficult" one (i.e., one that yielded lower mean scores) and an "easiest" one (i.e., one that yielded higher mean scores), that would be another main effect. There may be effects of target size and machine over and above these main effects. It is conceivable, but not very probable, that one machine, apart from its general difficulty, gains in difficulty by virtue of its having one size of target rather than others. It may be the coincidence of machine and target size that produces systematic variation in one direction from the general mean of scores. If this occurred, it would be an example of interaction variance.

Interaction variance might be more reasonably expected in a combination of teacher and instruction method, of kind of task and method of attack by the learner, and of kind of reward when combined with a certain condition of motivation. Interaction variance and main effects are illustrated in Fig. 13.1. The illustrative problem is one that could conceivably have been subjected to treatment as a three-way factorial design. Students in the fifth grade were given the same psychological test (Raven's Progressive Matrices) in city and country populations, composed of both boys and girls, at five socioeconomic levels. A 15 percent random sample included some Italian schools.[1]

[1] From data presented by Young, H. B., Taguiri, R., Tesi, G., and Montemagni, G. Influence of town and country upon children's intelligence. *British Journal of Educational Psychology, 1962,* **32,** 151–158.

If we hold one of the three variables constant at one of the categories, we can have a two-way factorial problem of the kind with which we are at present concerned. If we concentrate on children of socioeconomic class 4, we have a study of mean differences of city versus country children and of boys versus girls. Part *A* of Fig. 13.1 shows a plot of the means so as to demonstrate relationships between scores and sex in city versus country children. It will be noted that the differences are in opposite directions, which illustrates a drastic interaction effect that might well prove to be significant in an *F* test. Whether or not boys are superior to girls on the average in the test depends very much upon whether the children in question are from the city or the country. If city and country children were treated in a combined sample, there would be practically no difference in means. In a two-way analysis the main effect for sex would very likely be insignificant. There would be a small main effect from city versus country (from city versus country means when boys and girls are combined), but it, too, might prove to be insignificant. The interaction effect here is dominant.

By contrast, parts *B* and *C* of Fig. 13.1 show some main effects but little interaction effect. In part *B* (city children only), the trends of sex differences in means are similar in direction, if differing slightly in slope. If they were parallel, the interaction variance would be nil. The lines need not run in opposite directions, as they do in part *A*, to indicate interaction variance. Difference in slope would suffice, provided the *F* ratio proved to be significant. In part *B*, there is a fairly obvious main effect for sex, from the downward slope of the lines, and also for class, from the general levels of the lines in the field. In part *C*, with sex constant for the category of girls, interaction is less than in *B*; the main effect for classes is fairly evident, but, from the slopes of the lines, the main effect for city versus country is of doubtful significance.

Although we could treat the total data as three two-way analysis problems, treatment as a three-way analysis not only reveals the three main effects and the three interaction effects, two at a time—sex by urban-rural, socioeconomic level by sex, and socioeconomic level by urban-rural—but also provides the opportunity to test for a *triple interaction* effect—sex by urban-rural by socioeconomic level. Some of the deviations of means from the grand mean might be attributable to the combined effect of all three sources, over and above effects from two at a time, or of the variables taken singly. The principles demonstrated in the following two-way analysis apply, by extension, to more complex factorial designs to be presented in a subsequent section of this chapter.

COMPONENT SOURCES IN A TWO-WAY ANALYSIS As in the one-way analysis problem, we may write a basic equation for the two-way problem. It takes the form

$$X = \bar{X}_t + d_a + d_b + d_{ab} + e_r$$

where an obtained measurement X is conceived as being made up of several independent components. $\bar{X}_t$ is the grand mean and contributes nothing to variance. d_a is a contribution to the deviation of the single observation from $\bar{X}_t$ owing to its membership in one of the categories of variable A. Similarly, d_b is a deviation due to variable B. d_{ab} is the source of interaction variance. In each observation, after accounting for main effects, membership in a certain combination of A and B categories determines another increment, positive or negative. As earlier, e_r is the random-error contribution. The sums of squares derived from the four different sources add up to the total sum of squares for all the X's.

ESTIMATION OF THE VARIANCE FROM DIFFERENT SOURCES

Two solutions will be described, one using deviations of observed values and of means of sets, from the various appropriate means, and the other using original measurements and means. An attempt will be made to summarize the operations in terms of formulas, but some readers may find it easier to follow the examples as models rather than to apply the formulas. The system of symbols employed in the formulas is given in Table 13.8. This table provides only three columns and three rows, but it could be extended in the directions shown to take care of any number of columns and rows.

THE SOLUTION BASED UPON DEVIATIONS

In what follows, consistent with the symbols in Table 13.8, a subscript k stands for a particular column (we might have used c for column, but there would be a danger of confusing this with a particular row—row C), and r stands for a particular row. There are four rows (A_1, A_2, A_3, and A_4) and three columns (B_1, B_2, and B_3) in the psychomotor-test problem (see Table 13.7). The symbol X_{ij} stands for any one observation in row r and column k, and $\bar{X}_{rk}$ stands for a mean of the five observations in a cell described as being in row r and column k. In the following, n stands for the number of observations within each set; in the illustrative problem $n = 5$.[1] The number of the row is symbolized by r, and the number of the column by k. The subscript t refers to the total distribution, all sets combined. Thus $\bar{X}_t$ stands for the mean of all 60 scores.

The total sum of squares is given by the equation

$$SS_t = \Sigma(X_{ij} - \bar{X}_t)^2 \qquad (13.27)$$

[1] The methods of analysis described in this and subsequent sections of this chapter require that we have an equal number of observations for each set. For methods for dealing with unequal cell frequencies see more advanced texts such as Winer, op. cit.

Applied to the data of Table 13.7,[1]

$$SS_t = (6-5)^2 + (4-5)^2 + (4-5)^2 \quad \text{(from first row of Table 13.7)}$$
$$+ \cdot \cdot \cdot \cdot \cdot \cdot \cdot \cdot \cdot \cdot \cdot \cdot \cdot \cdot$$
$$+ (9-5)^2 + (4-5)^2 + (8-5)^2 \quad \text{(from last row of observations in Table 13.7)}$$
$$= 1^2 + (-1)^2 + (-1)^2$$
$$+ \cdot \cdot \cdot \cdot \cdot \cdot \cdot \cdot \cdot \cdot \cdot \cdot \cdot \cdot \cdot$$
$$+ 4^2 + (-1)^2 + 3^2$$
$$= 374 \quad \text{(Total sum of squares)}$$

The sum of squares between rows is given by the equation

$$SS_r = nk[\Sigma(\bar{X}_r - \bar{X}_t)^2] \tag{13.28}$$

Applied to the data of Table 13.7,

$$SS_r = 5 \times 3[(3-5)^2 + (4-5)^2 + (6-5)^2 + (7-5)^2]$$
$$= 15[(=2)^2 + (-1)^2 + 1^2 + 2^2]$$
$$= 15 \times 10$$
$$= 150 \quad \text{(Sum of squares between rows)}$$

The sum of squares between columns is given by the equation

$$SS_k = nr[\Sigma(\bar{X}_k - \bar{X}_t)^2] \tag{13.29}$$

Applied to the data of Table 13.7,

$$SS_k = 5 \times 4[(6-5)^2 + (5-5)^2 + (4-5)^2]$$
$$= 20[1^2 + (-1)^2]$$
$$= 20 \times 2$$
$$= 40 \quad \text{(Sum of squares between columns)}$$

The interaction variance can be estimated in several ways. Perhaps the most common way is to derive it from the sum of squares between all sets, eliminating the sums of squares between columns and between rows. We already know the last two sums of squares. We proceed next to compute the sum of squares between sets. The formula is similar to the numerator of formula (13.2) but with different notation to fit the new system:

$$SS_{rk} = n[\Sigma(X_{rk} - X_t)^2] \tag{13.30}$$

The symbol SS_{rk} refers to the sum of the squared differences between the set means and the total mean $\bar{X}_t$. The subscript *rk* implies that all rows and all columns are involved. Applied to the illustrative data,

1 The subscript in X_{ij} indicates each of the 20 rows of measurements in turn (*i*) and each of the 3 columns in turn (*j*).

TABLE 13.8 Symbolic scheme for the values in a tabulation preparatory to analysis of variance in a two-way classification problem

Rows			*Columns* B_1	B_2	B_3	*Sums of rows* ΣX_r	*Means of rows* $\overline{X}$
A_1	1						
	2						
	3						
	n						
		Σ	ΣX_{11}	ΣX_{12}	ΣX_{13}	ΣX_1	
		$\overline{X}$	$\overline{X}_{11}$	$\overline{X}_{12}$	$\overline{X}_{13}$		$\overline{X}_1$
A_2	1						
	2						
	3			X_{ij}			
	n						
		Σ	ΣX_{21}	ΣX_{22}	ΣX_{23}	ΣX_2	
		$\overline{X}$	$\overline{X}_{21}$	$\overline{X}_{22}$	$\overline{X}_{23}$		$\overline{X}_2$
A_3	1						
	2						
	3						
	n						
		Σ	ΣX_{31}	ΣX_{32}	ΣX_{33}	ΣX_3	
		$\overline{X}$	$\overline{X}_{31}$	$\overline{X}_{32}$	$\overline{X}_{33}$		$\overline{X}_3$
Sum of columns		(ΣX_k)	ΣX_1	ΣX_2	ΣX_3	ΣX_{ij}	
Means of columns		$(\overline{X}_k)$	$\overline{X}_1$	$\overline{X}_2$	$\overline{X}_3$		$\overline{X}_t$

Let X_{ij} = **any one of the cell entries**
$\overline{X}_{rk}$ = **any one of the set means**

$$
\begin{aligned}
SS_{rk} &= 5[(4-5)^2 + (3-5)^2 + (2-5)^2 && \text{(from first row of means)}\\
&\quad + \cdots\cdots\cdots\cdots\\
&\quad + (8-5)^2 + (6-5)^2 + (7-5)^2] && \text{(from last row of means)}\\
&= 5[(-1)^2 + (-2)^2 + (-3)^2\\
&\quad + \cdots\cdots\cdots\cdots\\
&\quad + 3^2 + 1^2 + 2^2]\\
&= 5 \times 42\\
&= 210 && \text{(Sum of squares between means of sets)}
\end{aligned}
$$

If we remove from the entire sum of squares for the 12 set means the sum of squares attributable to columns and to rows, we have left the interaction sum of squares. In terms of a formula,

$$SS_{r\times k} = SS_{rk} - SS_k - SS_r \qquad (13.31)$$

in which $SS_{r\times k}$ (the subscript reads "r times k," for reasons that will be explained) stands for the interaction sum of squares. For the illustrative problem,

$$\begin{aligned} SS_{r\times k} &= 210 - 40 - 150 \\ &= 20 \qquad \text{(Interaction sum of squares)} \end{aligned}$$

Another, more direct way of deriving interaction sums of squares utilizes the formula

$$SS_{r\times k} = n[\Sigma(\bar{X}_{rk} - \bar{X}_k - \bar{X}_r + \bar{X}_t)^2] \qquad (13.32)$$

in which $\bar{X}_k$ is the mean of the column in which each particular $\bar{X}_{rk}$ appears and $\bar{X}_r$ is the mean of its row. For the illustrative problem,

$$\begin{aligned} SS_{r\times k} &= 5[(4 - 3 - 6 + 5)^2 + (3 - 3 - 5 + 5)^2 \qquad \text{(from first row of means)} \\ &\quad + \cdots \\ &\quad + (6 - 7 - 5 + 5)^2 + (7 - 7 - 4 + 5)^2] \qquad \text{(from last row of means)} \\ &= 5[0^2 + 0^2 + \cdots + (-1)^2 + 1^2] \\ &= 5 \times 4 \\ &= 20 \qquad \text{(Interaction sum of squares; alternative solution)} \end{aligned}$$

The sum of squares within sets is computed by the formula

$$SS_w = \Sigma(X_{ij} - \bar{X}_{rk})^2 \qquad (13.33)$$

This formula, with new symbols, requires the same operations as formula (13.3), given in connection with the single-classification problem. Applied to the psychomotor problem,

$$\begin{aligned} SS_w &= (6 - 4)^2 + (4 - 4)^2 + (2 - 4)^2 + (6 - 4)^2 + (2 - 4)^2 \qquad \text{(from set } A1) \\ &\quad + \cdots \\ &\quad + (6 - 7)^2 + (5 - 7)^2 + (7 - 7)^2 + (9 - 7)^2 + (8 - 7)^2 \qquad \text{(from set } D3) \\ &= 164 \qquad \text{(Sum of squares within sets)} \end{aligned}$$

We can now check the solution of this by deducting all previously computed sums of squares from the total sum of squares, and we have

$$374 - 40 - 150 - 20 = 164$$

We could compute SS_w by this elimination process without going through the arduous arithmetic involved in using (13.33), but for checking purposes it is very desirable to derive all the component sums of squares separately and then check the results.

DEGREES OF FREEDOM Before taking the next important step of estimating the population variance from these different sources, we need, as usual, the degrees of freedom. Starting with the largest source, the total sum of squares, we have, as usual, $(N - 1)$ df, or 59. This figure is to be subdivided among the contributing components. The sum of squares among the means of sets should have allotted to it the number of sets minus 1, or $12 - 1 = 11$ df. These 11, in turn, are to be allotted to three sources. Rows have the number of row observations (row means) minus 1, or $4 - 1 = 3$.

By analogy, columns have $3 - 1 = 2$. This leaves 6 out of the 11 for interaction. This 6 degrees is 3×2, the product of the df for rows and columns, each source taken separately. This is consistent with the idea of interaction itself, whose contributions to variations may be regarded as the products of two sources. This is why we use the subscript $r \times k$ when referring to interaction. Having taken care of the special sources of variations, we are left with a remainder, $59 - 11 = 48$, which gives the df left for within-sets sums of squares. This number of df may also be determined directly from a summation of df within sets. Since there are 12 sets and each contains 4 df, we have $12 \times 4 = 48$ df for the residual variance.

In terms of symbolic descriptions, the degrees of freedom may be given as follows:

Source	*Degrees of freedom*
Between rows	$r - 1$
Between columns	$k - 1$
Interaction	$(r - 1)(k - 1)$
Within sets	$N - rk = rk(n - 1)$
Total	$N - 1$

THE F RATIOS We now face the question of how to constitute the three F ratios, F_r for main effect from rows (target size, in the illustrative problem), F_k for main effect from columns (machines), and $F_{r \times k}$ for interaction effect. In answering this three-part question, we must take into account the fact that a mixed model is involved. Let us consider a systematic solution to the problem for all three kinds of models which applies when $n > 1$. The appropriate error terms for the F ratios are based on the expected mean squares.

F RATIOS FOR DIFFERENT MODELS For the fixed model, the categories having been determined on a logical or experimental basis and not by sampling, the error term to use in all three F's is the residual or within-sets mean square. In other words, we extend to the two-way analysis the same kind of operation that was applied in the one-way analysis.

If the model is a random one, the categories in each variable having been obtained by random sampling from a universe of possible categories, the appropriate error term for testing the significance of *interaction effect* is the within-sets mean square. Whether or not the interaction F turns out to be significant, the appropriate error term, and the denominator for F for testing both main effects, is the interaction mean square. In so using the mean square from interaction, we wish to be sure that any significant main effect does not derive some of its possible significance from interaction effects themselves. We could not be sure about this if the within-sets mean square were used as the error term.

As we might expect, the mixed model presents some special questions. In the past, there has not been complete agreement on which error term to use in forming F ratios for the two main effects, but the present consensus seems to favor the following principles. There is only one error term that could be used in an F test for interaction, namely, the within-sets mean square. In testing the significance of the main effect that has the random categories, the mean square for within sets is considered to be the appropriate error term. But for testing the significance of the main effect with the fixed categories, the error term should be the mean square for interaction, regardless of whether or not the F for interaction has been found to be significant. The reason is that some of the interaction effects may have contributed to the variance of the main effect from the fixed-effects variable. Thus, if we used the within-sets mean square in connection with this variable's F test, and if the F were significant, we could not be sure that the significance was due only to the main effect. Since the interaction mean square is usually larger than the within-sets mean square, the use of the interaction mean square as the error term gives a smaller F, with less chance of being significant. But its use avoids the risk of having an ambiguous F to be interpreted.

F RATIOS FOR THE PSYCHOMOTOR-TEST PROBLEM In the illustrative problem, as Table 13.9 shows, the F for interaction proved to be so near 1.0 that there was no question regarding its lack of significance. The F_k, for machines, was 5.85, which was larger than the F required for significance at the .01 level, with 2 and 48 df. Had this F been derived from genuine data, we could conclude that the suspicion that the machines present psychological tasks of different degrees of difficulty, regardless of the size of target, was well founded. As should have been expected in a genuine experiment, the F ratio for target size

TABLE 13.9
Sources of variance in the psychomotor-test data analysis, with F ratios

Source	*Sum of squares*	*Degrees of freedom*	*Mean squares*
Target size (A)	**150**	**3**	**50.0**
Machine (B)	**40**	**2**	**20.0**
Interaction ($A \times B$)	**20**	**6**	**3.33**
Within sets	**164**	**48**	**3.42**
Total	**374**	**59**	

	Required F $p = .05$	Required F $p = .01$
F **for interaction** $= \frac{3.33}{3.42} = 0.97$	**2.30**	**3.20**
F **for machines** $= \frac{20.0}{3.42} = 5.85$	**3.19**	**5.08**
F **for targets** $= \frac{50.0}{3.33} = 15.00$	**4.76**	**9.78**

is well beyond the F required for significance at the .01 level. The insignificant F for interaction would indicate that a target's difficulty does not depend upon the machine upon which it appears.

REMOVAL OF SOURCES OF VARIATION

It may illuminate the concepts of different kinds of variance and the way in which they contribute to total variance in the sample if we separate them in another way.

Table 13.10*A* shows the 12 means of sets for the psychomotor-test data. Variations among them are due to the three possible sources—target differences, machine differences, and the interaction of the two. The possible effects of target size are most apparent in the means of the rows—3, 4, 6, and 7. The possible effects of machine differences are most apparent in the means of the columns—6, 5, and 4. The possible interaction variance is obscured. Possibly it contributes to the means of both rows and columns; we do not know. Let us strip away first the variations attributable to machines and then those attributable to targets and see what variations are left. We have eliminated the contributions of random errors e_r by averaging within cells.

The mean of all observations is 5. Any deviation of a column mean from 5 indicates a constant error for a particular machine. Machine 1 gave a mean of 6, indicating that it had a constant error of +1. Machine 2 apparently had no constant error, while machine 3 had a constant error of −1. If we deduct from each cell or set mean in column 1 the amount of constant error involved for machine 1, we should presumably remove from the means in column 1 the influence of machine 1 as a source of variation. We can do likewise for col-

TABLE 13.10 Analysis of the between-sets sums of squares in the psychomotor-test data into three components by successive removal of contributing sources of variation

Row	Column B_1	B_2	B_3	Σ	$\bar{X}$
A. Original matrix of means of sets					
A_1	4	3	2	9	3
A_2	5	5	2	12	4
A_3	7	6	5	18	6
A_4	8	6	7	21	7
Σ	24	20	16	60	
$\bar{X}$	6	5	4		5
B. With variations associated with machines removed					
A_1	3	3	3	9	3
A_2	4	5	3	12	4
A_3	6	6	6	18	6
A_4	7	6	8	21	7
Σ	20	20	20	60	
$\bar{X}$	5	5	5		5
C. With variations associated with target size also removed; only interaction variance remaining					
A_1	5	5	5	15	5
A_2	5	6	4	15	5
A_3	5	5	5	15	5
A_4	5	4	6	15	5
Σ	20	20	20	60	
$\bar{X}$	5	5	5		5

umn 3, deducting the constant error of −1, which is equivalent to adding +1 to each mean. We need do nothing for column 2. The results of these operations are shown in Table 13.10*B*. The means of the columns are now all 5, to agree with the composite mean $\bar{X}_t$. The means of the rows have been unaffected (they are still 3, 4, 6, and 7) because the changes in one column are compensated for by changes in the reverse direction in another column. The cell values in Table 13.10*B* still have in them the variance attributable to targets and to interaction variance.

Next we remove the target variance. The constant errors for rows are −2, −1, 1, and 2, respectively. Deducting these from the values in their respective rows of Table 13.10*B*, we have the results in subtable C. The means of the rows as well as of the columns are now all 5. But within four cells there are departures from 5. These possibly are the interaction deviations, depending upon whether or not they prove to be significant. Machine 2 seems to favor high scores when coupled with target A_2 and to favor low scores when coupled with target A_4. Machine 3 has a reverse tendency. But the F for interaction showed these deviations to be insignificant. This indicates that the deviations were such as might have happened as a result of random assignment of subjects to different conditions, rather than as a result of the dependence of a target's difficulty upon the machine on which it appeared.

THE SOLUTION FROM ORIGINAL MEASUREMENTS

We now give the formulas and their applications for the solution of sums of squares without computing means and deviations. With small integral numbers to start with, or numbers coded to such magnitude, these procedures are often more convenient than those utilizing deviations. The first solution, with deviations, is more meaningful to the beginner. In the following exposition, each formula will be stated and then immediately applied to the psychomotor-test data.

Total sum of squares:

$$SS_t = \sum X_{ij}^2 - \frac{(\Sigma X_{ij})^2}{N} \tag{13.34}$$

$$= 6^2 + 4^2 + 4^2 \qquad \text{(from first row of Table 13.7)}$$
$$+ \cdots\cdots$$
$$+ 9^2 + 4^2 + 8^2 \qquad \text{(from last row of Table 13.7)}$$
$$- \frac{(300)^2}{60}$$
$$= 1{,}874 - 1{,}500 = 374 \qquad \text{(Total sum of squares)}$$

Sum of squares between sets:

$$SS_{rk} = \frac{\Sigma(\Sigma X_{rk})^2}{n} - \frac{(\Sigma X_{ij})^2}{N} \tag{13.35}$$

$$= \tfrac{1}{5}[(20^2 + 15^2 + 10^2 \qquad \text{(from first } \Sigma \text{ row of Table 13.7)}$$
$$+ \cdots\cdots\cdots$$
$$+ 40^2 + 30^2 + 35^2)] \qquad \text{(from last } \Sigma \text{ row of Table 13.7)}$$
$$- \frac{(300)^2}{60}$$
$$= 1{,}710 - 1{,}500$$
$$= 210 \quad \text{(Sum of squares between means of sets)}$$

Sum of squares between rows:

$$SS_r = \frac{\Sigma(\Sigma X_r)^2}{nk} - \frac{(\Sigma X_{ij})^2}{N} \tag{13.36}$$
$$= [\tfrac{1}{15}(45^2 + 60^2 + 90^2 + 105^2)] - 1{,}500$$
$$= 1{,}650 - 1{,}500$$
$$= 150 \qquad \text{(Sum of squares between rows)}$$

Sum of squares between columns:

$$SS_k = \frac{\Sigma(\Sigma X_k)^2}{nr} - \frac{(\Sigma X_{ij})^2}{N} \tag{13.37}$$
$$= [1/20(120^2 + 100^2 + 80^2)] - 1{,}500$$
$$= 1{,}540 - 1{,}500$$
$$= 40 \qquad \text{(Sum of squares between columns)}$$

Sum of squares for interaction:

$$SS_{r\times k} = SS_{rk} - SS_r - SS_k \tag{13.38}$$
$$= 210 - 150 - 40$$
$$= 20 \qquad \text{(Sum of squares for interaction)}$$

Sum of squares within sets:

$$SS_w = SS_t - SS_{rk} \tag{13.39}$$
$$= 374 - 210$$
$$= 164 \qquad \text{(Sum of squares within sets)}$$

It will be noted that the correction factor $(\Sigma X_{ij})^2/N$ is the same in all the equations in which it appears and need be computed only once.

The sums of squares found by this method are seen to be identical with those found by the preceding method. The estimation of the population variance from each source and the application of the F test are the same as before (see Table 13.9).

ESTIMATING STRENGTH OF ASSOCIATION FROM A TWO-WAY ANALYSIS OF VARIANCE

Estimations of the degree of association for the main effects, and their interaction, in a two-way fixed-effects analysis may be obtained from an index similar to the one used in connection with a one-way analysis of variance. We can estimate these values for ω^2 by

$${}_k\omega^2{}_{\text{est}} = \frac{SS_r - (r-1)MS_w}{MS_w + SS_t} \tag{13.40}$$

(Estimate of proportion of variance in the measurement variable accounted for by the row effect)

$${}_r\omega^2{}_{\text{est}} = \frac{SS_k - (k-1)MS_w}{MS_w + SS_t} \tag{13.41}$$

(Estimate of proportion of variance in the measurement variable accounted for by the column effect)

$${}_{r\times k}\omega^2{}_{\text{est}} = \frac{SS_{r\times k} - (r-1)(k-1)(MS)_w}{(MS)_w + SS_t} \tag{13.42}$$

(Estimate of proportion of variance in the measurement variable accounted for by the interaction)

For the data in Table 13.9, the estimated value of ω for target-size effects is

$$_r\omega^2{}_{\text{est}} = \frac{150 - (4 - 1)(3.42)}{3.42 + 374.0} = .37$$

The estimate indicates that an appreciable proportion (.37)[1] of the psychomotor score is associated with target size. Of course, this statement is made on the basis of an inference from the sample data, and we cannot be sure that it holds in the population. The significance level assures us that the probability that such statements are not true in the population is rather small, the estimates of the strength of association being "best guesses" of the degree of association in the population. A significant F test is an indication that there probably are significant differences. A careful examination of the data and the conditions of the experiment is needed to make the meaning of the experiment clear. That is why an estimate of the amount in addition to the presence of the significant effects is helpful and informative in the interpretation of the results of an experiment.

A three-way classification factorial design

Factorial analysis-of-variance designs are not limited to the investigation of two-way classification analyses, but may have three or more bases for classifying the observations. The procedures for a three-way classification problem will be discussed and it should then be possible for the reader to generalize the procedures to any level of factorial design which may be of interest.

If, to the variables, four target sizes and three machines that were used in connection with the illustration for the two-way classification problem, we add a third variable, level of background illumination, with two categories, bright and dim, we have a $4 \times 3 \times 2$ factorial design. This arrangement, as shown in Figure 13.2, may be visualized as a $4 \times 3 \times 2$ box. The entries shown are sums, as will be explained in the next section.

BASIC EQUATION FOR A THREE-WAY ANALYSIS

The basic equation for the three-way, fixed-effects model design has the form:

$$X = \bar{X}_t + d_a + d_b + d_c + d_{ab} + d_{ac} + d_{bc} + d_{abc} + e_r$$

As in the two-way analysis (see page 248), X is thought to be

1 Since the machines and interaction variances involve a random variable, their strength of association is estimated by the intraclass correlation (see p. 270), which is the same in general form and meaning as ω^2, except that since it applies to the random effects, slightly different estimations of the variances apply. The F ratio for the interaction is less than 1, and so the estimate of the strength of association for it would be set equal to zero in any case.

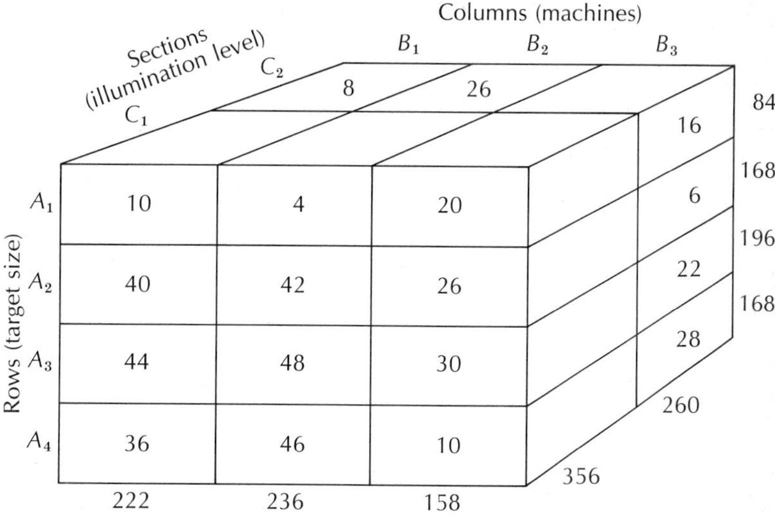

FIGURE 13.2 Representation of a 4 × 3 × 2 three-way analysis of variance. The values for the treatment combinations (only those that are visible are shown) are the sums of five observations each. Sums for the 4 rows, 3 columns, and 2 sections are given along the edges. The sums include cell values not shown (see Table 13.11), as well as those shown.

made up of a number of independent components. The terms added for the three-way analysis are d_c, a deviation due to variable C, d_{ac} and d_{bc}, the two-way interactions of variable C with the other two variables, and d_{abc}, the three-way interaction of the three variables.

In the illustration there are $4 \times 3 \times 2 = 24$ treatment combinations. In a factorial design with $n = 5$ observations per set, we have a total of 120 scores. The total sum of squares with 119 df can be partitioned into the following sources and degrees of freedom:

	Source	*df*
Main effects	A	**3**
	B	**2**
	C	**1**
Two-way interactions	$A \times B$	**6**
	$A \times C$	**3**
	$B \times C$	**2**
Three-way interaction	$A \times B \times C$	**6**
Within	**Error**	**96**

The number of degrees of freedom for the three-way interaction is the product of the A, B, and C df's. The data for the treatment sets may be tabulated as in Table 13.11.

TABLE 13.11 Scores in a $4 \times 3 \times 2$ factorial design experiment with 4 target sizes (A), 3 machines (B), and 2 levels of illumination (C) (each cell entry represents the sum of five observations)

	Level of illumination								
	C_1				C_2				
	Machine				Machine				Σ of
Target size	B_1	B_2	B_3	Σ_1	B_1	B_2	B_3	Σ_2	$\Sigma_1 + \Sigma_2$
A_1	10	4	20	34	8	26	16	50	84
A_2	40	42	26	108	22	32	6	60	168
A_3	44	48	30	122	28	24	22	74	196
A_4	36	46	10	92	34	14	28	76	168
Σ	130	140	86	356	92	96	72	260	616

THE CALCULATION OF SUMS OF SQUARES FROM ORIGINAL MEASUREMENTS

The calculation of sums of squares is done in the same way as in the previous examples, except that now each observation has three subscripts, i.e., X_{ijk}. The formula for the total sum of squares is given by:

$$SS_t = \Sigma X_{ijk}^2 - \frac{(\Sigma X_{ijk})^2}{N} \tag{13.43}$$

Let us assume that the total sum of squares has already been calculated and has been found to have a value of 1,981.467.

The sum of squares between sets (using the subscript s to indicate the categories in the third variable, which we will call "sections") is

Sum of squares between sets:

$$SS_{rks} = \frac{\Sigma(\Sigma X_{rks})^2}{n} - \frac{(\Sigma X_{ijk})^2}{N} \tag{13.44}$$

$= \frac{1}{5}(10^2 + 4^2 + 20^2 + 8^2 + 26^2 + 16^2$ (from last row, Table 13.11)

$+ \cdot \cdot \cdot \cdot \cdot \cdot \cdot$

$+ 36^2 + 46^2 + 10^2 + 34^2 + 14^2 + 28^2)$ (from first row, Table 13.11)

$- \frac{(616)^2}{120}$

$= 3{,}945.600 - 3{,}162.133$

$= 783.467$ (Sum of squares between sets)

It is this sum of squares, 783.467, with 23 df, that we will analyze into the columns, rows, sections, and the four interaction sums of squares.

From the data in Table 13.11, we may set up a table for the rows by sections (A by C) variables summed over the columns, as shown in Table 13.12. Since the columns (B) have three categories, each sum in the table is based on $3 \times 5 = 15$ observations. The two sums, 356 and

TABLE 13.12
The row-by-section ($A \times C$) table with the cell entries summed over the three categories of B (each cell entry is the sum of 15 observations)

	C_1	C_2	Σ
A_1	34	50	84
A_2	108	60	168
A_3	122	74	196
A_4	92	76	168
Σ	356	260	616

260, are the sums of C_1 and C_2, respectively. Each of these two sums is based on $4 \times 15 = 60$ observations.

Sums of squares between sections (C):

$$SS_s = \frac{(\Sigma X_s)^2}{nrk} - \frac{(\Sigma X_{ijk})^2}{N} \tag{13.45}$$

$$= \frac{1}{60}(356^2 + 260^2) - \frac{(616)^2}{120}$$

$$= 3{,}238.933 - 3{,}162.133$$

$$= 76.800 \qquad \text{(Sum of squares between sections)}$$

Sum of squares between rows (A):

$$SS_r = \frac{(\Sigma X_r)^2}{nks} - \frac{(\Sigma X_{ijk})^2}{N} \tag{13.46}$$

$$= \frac{1}{30}(84^2 + 168^2 + 196^2 + 168^2) - \frac{(616)^2}{120}$$

$$= 3{,}397.333 - 3{,}162.133$$

$$= 235.200 \qquad \text{(Sum of squares between rows)}$$

The rows by sections ($A \times C$) interaction sum of squares may also be obtained from Table 13.12, where the cell entries are summed over the three categories of B. First, we calculate the sum of squares between the eight sums entered in the cells of the table, remembering that each sum is based on fifteen observations. The $A \times C$ interaction is obtained by subtracting the sum of squares for A and the sum of squares for C, which we previously calculated, from the sets sum of squares. The sets sum of squares is calculated by:

$$SS_{rs} = \frac{1}{15}(34^2 + 50^2 + \cdot\cdot\cdot + 92^2 + 76^2) - \frac{(616)^2}{120}$$

$$= 3{,}568.000 = 3{,}162.133 = 405.867$$

Sum of squares for rows × sections ($A \times C$) interaction:

$$SS_{r\times s} = SS_{rs} - SS_r - SS_s \tag{13.47}$$

$$= 405.867 - 235.200 - 76.800$$

$$= 93.867 \qquad \text{(Sum of squares for rows-by-sections interaction)}$$

TABLE 13.13
The row-by-column ($A \times B$) table with the entries summed over the two categories of C (each cell entry is the sum of ten observations)

	B_1	B_2	B_3	Σ
A_1	18	30	36	84
A_2	62	74	32	168
A_3	72	72	52	196
A_4	70	60	38	168
Σ	222	236	158	616

As a general formula, when we have a two-way table with rows corresponding to the categories of one variable and columns to the categories of another variable, the interaction sum of squares is obtained by:

$$SS_{r\times k} = SS_{rk} - SS_r - SS_k \tag{13.48}$$

We now go back to Table 13.11, and set up another table from which we calculate the row-by-column ($A \times B$) interaction. In Table 13.13, the entries are summed over the levels of C. Since C has two levels, each sum in the table is based on $2 \times 5 = 10$ observations. The sums for B_1, B_2, and B_3 are the column sums, each based on $4 \times 10 = 40$ observations.

Sums of squares between columns (B):

$$SS_k = \frac{\Sigma(\Sigma X_k)^2}{nrs} - \frac{(\Sigma X_{ijk})^2}{N} \tag{13.49}$$

$$= \frac{1}{40}(222^2 + 236^2 + 158^2) - \frac{(616)^2}{120}$$

$$= 3{,}248.600 - 3{,}162.133$$

$$= 86.467 \qquad \text{(Sum of squares between columns)}$$

To obtain the row-by-column ($A \times B$) interaction sum of squares, we first calculate the sum of squares for sets from the entries in Table 13.14.

$$SS_{rk} = \frac{1}{10}(18^2 + 30^2 + 36^2 + \cdots + 70^2 + 60^2 + 38^2) - \frac{(616)^2}{120}$$

$$= 3{,}588.000 - 3{,}162.133 = 425.867$$

TABLE 13.14
The column-by-section ($B \times C$) table with the entries summed over the four categories of A (each cell entry is the sum of 20 observations)

	B_1	B_2	B_3	Σ
C_1	130	140	86	356
C_2	92	96	72	260
Σ	222	236	158	616

Following the general formula for a row-by-column interaction, and using the sums of squares for rows (*A*) and columns (*B*) previously calculated, we have the following formula.

Sum of squares for row-by-column ($A \times B$) interaction:

$$SS_{r\times k} = SS_{rk} - SS_r - SS_k \qquad (13.50)$$
$$= 425.867 = 86.467 - 235.200$$
$$= 104.200 \qquad \text{(Sum of squares for row-by-column interaction)}$$

To find the column-by-section ($B \times C$) interaction sum of squares, we set up Table 13.14, $B \times C$ summed over the categories of *A*. Since *A* has four categories, each sum in the table is based on $4 \times 5 = 20$ observations. First we calculate the sum of squares among the eight sums in the table. The sets sum of squares is:

$$SS_{ks} = \frac{1}{20}\left[(130^2 + 140^2 + \cdots + 96^2 + 72^2)\right] - \frac{(616)^2}{120}$$
$$= 3{,}338.000 - 3{,}162.130 = 175.867$$

The column-by-section ($B \times C$) sum of squares is obtained by the general formula for a two-way interaction, using the sum of squares for columns (*B*) and sections (*C*) previously calculated:

Sum of squares for column-by-section ($B \times C$) interaction:

$$SS_{k\times s} = SS_{ks} - SS_k - SS_s \qquad (13.51)$$
$$= 175.867 - 86.467 - 76.800$$
$$= 12.600 \qquad \text{(Sum of squares for column-by-section interaction)}$$

We can calculate the three-way column-by-row-by-section ($A \times B \times C$) interaction directly from the data, or obtain it as a remainder by subtraction. The method of direct calculation will be illustrated later, but now we obtain the $A \times B \times C$ interaction by subtraction since we know that the total between-sets sum of squares (SS_{rks}) is equal to the summation of the sums of squares for *A*, *B*, *C*, $A \times B$, $A \times C$, $B \times C$, and $A \times B \times C$. Since we have calculated the first six of these seven sums of squares, we can obtain the last one by subtraction.

The total between-sets sum of squares is 783.467 and the sum of the first six sums of squares listed above is $235.200 + 86.467 + 76.800 + 104.200 + 93.867 + 12.600 = 609.134$. The three-way sum of squares, obtained by subtraction, for $A \times B \times C = 783.467 - 609.134 = 174.333$.

F-RATIOS FOR THE THREE-WAY ANALYSIS

The sums of squares we have just calculated are listed in Table 13.15. The mean squares were obtained by dividing the sums of squares by their respective df's. Applying the principles for determining the appropriate error term we gave in the illustration on page 253, for the

TABLE 13.15 Sources of variance in the 4 × 3 × 2 factorial experiment with *F*-ratios

Source of variation	*Sum of squares*	*Degrees of freedom*	*Mean square*	*F*
Target size (A)	235.200	3	78.400	4.51*
Machines (B)	86.467	2	43.335	3.47*
Levels of illumination (C)	76.800	1	76.800	2.45
$A \times B$	104.200	6	17.367	1.39
$A \times C$	93.867	3	31.289	2.51
$B \times C$	12.600	2	6.300	
$A \times B \times C$	174.333	6	29.056	2.33*
Within sets	1198.000	96	12.478	
Total	1981.467	119		

$p \leqslant .05$

mixed model, where A and C are fixed and B is a random variable, we have the following error terms:

Source of variation	*Type of variable*	*Error mean square for test of significance*
A	Fixed	$A \times B$
B	Random	Within
C	Fixed	$B \times C$
$A \times B$		Within
$A \times C$		$A \times B \times C$
$B \times C$		Within
$A \times B \times C$		Within

The appropriateness of these error terms is determined most directly from a consideration of the expected mean squares, but one guide, in line with the principles for constituting F ratios given on page 253 for a mixed model, is that those sources that include one or more random variables are tested against the within mean square, and those sources containing fixed variables only are tested against the next-higher-order interaction containing the fixed and random variables. Applying these principles, we obtain the F ratios shown in the last column of Table 13.15.

The F ratios in Table 13.15 are interpreted with reference to the numbers of degrees of freedom involved. The F values that are significant at $\alpha = .05$ level have been starred. The interpretation of the significant values of F follows the same considerations that we applied previously. We conclude that the null hypotheses of no differences between means due to target size and machine may be rejected at

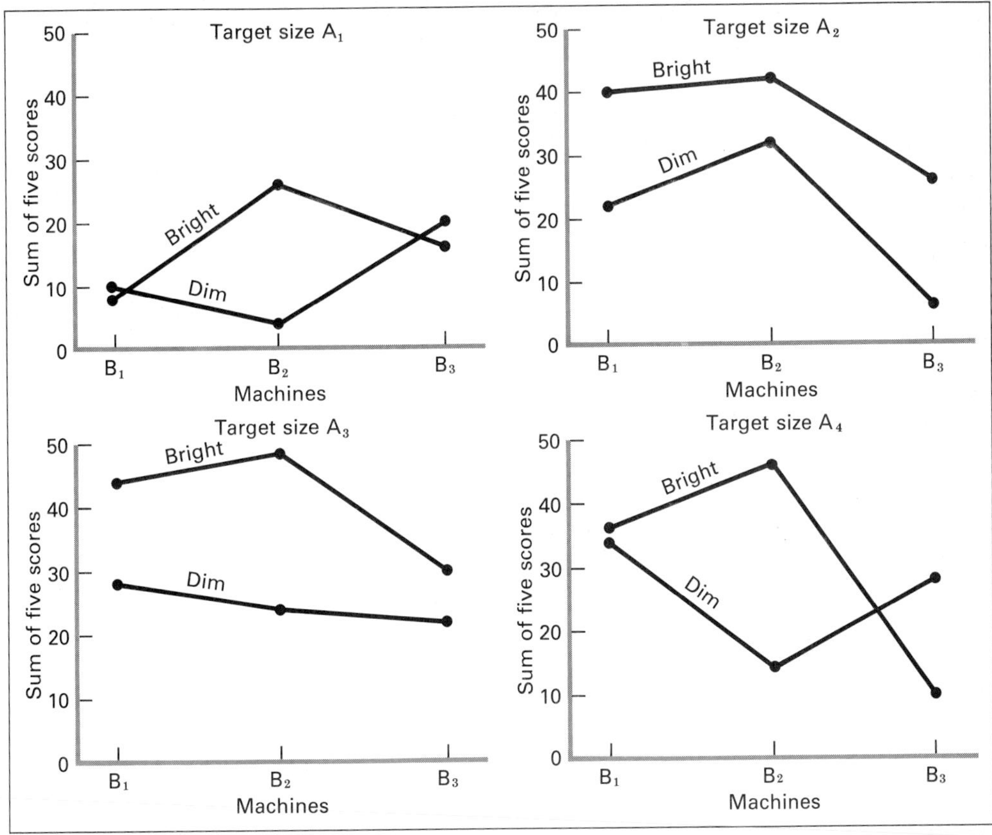

FIGURE 13.3 Interaction of variable B (Machines) with variable C (bright and dim levels of Illumination) for each of the four variable A target sizes. The A × B × C interaction is significant (See Table 13.15), so the relationship of B and C depends on the influence of the third variable A, Target size.

the $p < .05$ level, but the means for the two levels of illumination do not differ significantly. None of the two-way interactions are significant at the .05 level.

The three-way ($A \times B \times C$) interaction is significant, indicating that the two-variable interactions are not the same for the different categories of the third variable. Figure 13.3 illustrates the $B \times C$ interactions for each of the four categories of A.

CALCULATION OF THREE-VARIABLE INTERACTION SUM OF SQUARES

As mentioned previously, the three-way interaction sum of squares can be calculated directly from the data. The method is general and can be used also in obtaining higher-order interaction sums of squares.

To obtain the $A \times B \times C$ interaction sum of squares, we calculate the $A \times B$ sum of squares separately for each of the two levels of C, as in Table 13.11. We then sum them and subtract the two-way $A \times B$ interaction sum of squares previously computed.

$$SS_{r\times k\times s} = \Sigma(SS_{r\times k})_s - SS_{r\times k} = \Sigma C(A \times B) - A \times B \tag{13.52}$$

To compute the sum of squares $A \times B$ for each category of C, we follow the same general procedure as we followed previously for obtaining the sum of squares from a two-way table. First we obtain the sets sum of squares for a table, then we compute the sums of squares for the rows and for the columns and subtract them from the sets sum of squares, i.e.,

$$SS_{r\times k} = SS_{rk} - SS_r - SS_k. \tag{13.53}$$

For the 12 sets of C_1:
Sum of squares between sets for C_1:

$$SS_{rk} = \frac{1}{5}(10^2 + 4^2 + 20^2 + \cdots + 36^2 + 46^2 + 10^2) - \frac{(356)^2}{60}$$
$$= 2{,}641.600 - 2{,}112.267 = 529.333$$

Sum of squares between rows of C_1:

$$SS_r = \frac{1}{15}(34^2 + 108^2 + 122^2 + 92^2) - \frac{(356)^2}{60}$$
$$= 2{,}411.200 - 2{,}112.267 = 298.933$$

Sum of squares between columns of C_1:

$$SS_k = \frac{1}{20}(130^2 + 140^2 + 86^2) - \frac{(356)^2}{60}$$
$$= 2{,}194.800 - 2{,}112.267 = 82.533$$

Sum of squares for rows × columns ($A \times B$) for C_1 (by formula 13.53).

$$SS_{r\times k} = 529.333 - 298.933 - 82.533 = 147.867$$

We then perform the same calculations for C_2:
Sum of squares between sets for C_2:

$$SS_{rk} = \frac{1}{5}(8^2 + 26^2 + 16^2 + \cdots + 34^2 + 14^2 + 28^2) - \frac{(260)^2}{60}$$
$$= 1{,}304.00 - 1{,}126.667 = 177.333$$

Sum of squares for rows of C_2:

$$SS_r = \frac{1}{15}(50^2 + 60^2 + 74^2 + 76^2) - \frac{(260)^2}{60}$$
$$= 1{,}156.800 - 1{,}126.667 = 30.133$$

Sum of squares for columns of C_2:

$$SS_k = \frac{1}{20}(92^2 + 96^2 + 72^2) - \frac{(260)^2}{60}$$
$$= 1{,}143.200 - 1{,}126.667 = 16.533$$

Sum of squares for rows by columns ($A \times B$) for C_2 (by formula 13.53):

$$SS_{r \times k} = 177.333 - 30.133 - 16.533 = 130.667$$

Summing the interactions for $A \times B$ for each of the two levels of C, we have:

$$\Sigma C(A \times B) = 147.867 + 130.667 = 278.534$$

We have already calculated the $A \times B$ interaction sum of squares averaged over the categories of C for these data. As indicated in Table 13.15, it is equal to 104.20. The sum of squares for the three-way interaction, $A \times B \times C$, may be obtained by subtracting the $A \times B$ interaction sum of squares from the sum of the interactions of $A \times B$ for the separate categories of C, as was indicated in formula (13.52):

$$A \times B \times C = \Sigma C(A \times B) - A \times B \tag{13.54}$$

Substituting in (13.54), we obtain

$$A \times B \times C = 278.53 - 104.20 = 174.33$$

which checks, within rounding errors, with the value obtained by subtraction for the $A \times B \times C$ sum of squares.

The $A \times B \times C$ interaction sum of squares could also have been obtained by calculating the $A \times C$ interactions for each level of B, summing and then subtracting the $A \times C$ interaction obtained from the two-way table from which the $A \times C$ interaction sum of squares was previously obtained. Alternatively, the $A \times B \times C$ interaction sum of squares could have been obtained by summing the $B \times C$ interactions over the levels of A, and then subtracting the previously determined $B \times C$ interaction sum of squares. We used the $A \times B$ combination since variable C had the fewest categories and thus required the fewest tables and the least amount of calculation.

GENERAL METHODS FOR OBTAINING INTERACTION SUMS OF SQUARES FOR THREE OR MORE VARIABLES

In general, a three-variable interaction sum of squares can be obtained by:

$$\begin{aligned} A \times B \times C &= \Sigma A(B \times C) - B \times C \\ &= \Sigma B(A \times C) - A \times C \\ &= \Sigma C(A \times B) - A \times B \end{aligned}$$

and a four-variable interaction sum of squares can be obtained by:

$$\begin{aligned} A \times B \times C \times D &= \Sigma A(B \times C \times D) - B \times C \times D \\ &= \Sigma B(A \times C \times D) - A \times C \times D \\ &= \Sigma C(A \times B \times D) - A \times B \times D \\ &= \Sigma D(A \times B \times C) - A \times B \times C \end{aligned} \tag{13.55}$$

A similar set of equations may be written for interactions involving five variables, and so forth.

Some special analysis-of-variance methods

A TWO-WAY CLASSIFICATION ANALYSIS WITHOUT REPLICATIONS

Occasionally the kind of research problem arises in which there are two experimental variations but only one observation for each combination of conditions. This kind of problem will be illustrated by the use of ratings. The data in Table 13.16 will be utilized.

In these data, three raters have given their ratings of each of seven individuals on a single trait. The procedure of analysis is much like that previously illustrated when there are replications. The main difference is that the interaction and error effects are not segregated from one another here, since there is no basis for doing so. The error term is derived from this combined source.

The total sum of squares is computed as in formula (13.34). Applied to the data of Table 13.16,

$$SS_t = 720.00 - 618.86 = 101.14$$

The sum of squares between rows is given by the formula

$$SS_r = \frac{\Sigma(\Sigma X_r)^2}{k} - \frac{(\Sigma X_{ij})^2}{kr} \qquad (13.56)$$

Applying the data of Table 13.16, we have

$$SS_r = \frac{2{,}010}{3} - \frac{12{,}996}{21} = 670.00 - 618.86 = 51.14$$

The sum of squares between columns is given by

$$SS_k = \frac{\Sigma(\Sigma X_k)^2}{r} - \frac{(\Sigma X_{ij})^2}{rk} \qquad (13.57)$$

TABLE 13.16 Application of analysis of variance in a two-way classification without replication

Ratee	Rater I	Rater II	Rater III	ΣX_r	$(\Sigma X_r)^2$
1	5	6	5	16	256
2	9	8	7	24	576
3	3	4	3	10	100
4	7	5	5	17	289
5	9	2	9	20	400
6	3	4	3	10	100
7	7	3	7	17	289
ΣX_k	43	32	39	114 ΣX_{ij}	2,010 $\Sigma(\Sigma X_r)^2$
$(\Sigma X_k)^2$	1,849	1,024	1,521	12,996 $(\Sigma X_{ij})^2$	
				$\Sigma(\Sigma X_k)^2 = 4{,}394$	$\Sigma X_{ij}^2 = 720$

TABLE 13.17
Estimated variances and F ratios from the data of Table 13.16

Source	*Sum of squares*	*df*	*MS*	*F*	*p*
Ratees (rows)	**51.14**	**6**	**8.52**	**2.48**	**>.05**
Raters (columns)	**8.85**	**2**	**4.425**	**1.29**	**>.05**
Remainder $(r \times k)$	**41.15**	**12**	**3.43**		
Total	**101.14**	**20**			

Applied to the data of Table 13.16, this gives

$$SS_k = \frac{4{,}394}{7} - 618.86 = 8.85$$

The sum of squares for the remainder is obtained by deducting the last two sums of squares from the total sum of squares. We therefore have for the remainder sum of squares,

$$SS_e = 101.14 - 51.14 - 8.85 = 41.15$$

We are now ready to estimate variances and compute F ratios. The work is summarized in Table 13.17. Both F ratios prove to be insignificant. We therefore do not reject the hypothesis that there are no differences between raters and between ratees. There may be such real differences, but our F tests fail to indicate them. We should not be very surprised to find no significant differences between raters, except as some of them show marked errors of leniency in rating and some do not. We *should* be surprised, however, not to find significant differences between ratees, for individual differences in most traits are the almost universal finding. With a larger sample, the statistical test might have been sensitive enough to yield a significant F for ratees.

The smallness of sample, however, is not the whole story behind the insignificant F's. Note that it was stated at the beginning of this section that the error term includes contributions from interaction. If the interaction effects are of sufficient importance, they inflate the variance computed from the residual sum of squares and thus reduce the size of both F ratios. We know that there are often *halo errors,* which can be defined statistically as interaction effects—between rater and ratee. We should not be able to segregate this interaction effect without having independent replications, which would be difficult to obtain, or without having ratings made by the same raters of the same ratees on other traits.[1]

Another reason for the small variance among ratees is the lack of agreement among the raters. Some of this can be attributed to halo

1 For further treatment of ratings by analysis of variance, see Guilford, J. P. *Psychometric methods.* (2d ed.) New York: McGraw-Hill, 1954. Pp. 281–288.

errors. In the extreme case, if there were zero correlations among the raters' ratings, the means of the ratees would tend toward equality, or no variance at all. The higher the intercorrelation of raters, the greater will be the variance estimated from among ratees. To this problem of interrater correlation we turn next.

INTRACLASS CORRELATION

Using the data of Table 13.11, we can apply the information already extracted about variances, from which to compute correlations between raters. The intercorrelation thus obtained is known as an *intraclass correlation*. It is given by the formula

$$r_I = \frac{MS_r - MS_e}{MS_r + (k-1)\,MS_e} \qquad \text{(Intraclass correlation among } k \text{ series)} \qquad (13.58)$$

where MS_r = mean square or variance between rows, where each row stands for a person
MS_e = mean square for residuals, or error
k = number of columns

For the data of Table 13.16,

$$r_I = \frac{8.52 - 3.43}{8.52 + 2(3.43)} = .33$$

This result indicates that the degree of agreement among the three raters is .33. If we take the intercorrelations of raters to be an indication of reliability of ratings, we can say that the typical reliability of a single rater's ratings is of the order of .33. The actual correlations between single pairs of raters might vary considerably from this figure because of sampling errors in such a small sample.

We expect the value of r_I to range from zero to 1.0, although in the unlikely event that MS_e exceeds MS_r, formula (13.58) could yield a negative value. We can use

$$F = \frac{MS_r}{MS_e} \qquad \text{(}F\text{ ratio for testing the hypothesis that } \rho_I = 0\text{)} \qquad (13.59)$$

with $(r-1)$ and $(r-1)(k-1)$ df to test the hypothesis $\rho_I = 0$. We have already seen in Table 13.17 that this F does not reach the .05 point of significance, so we are not justified in rejecting the null hypothesis.

If we want to estimate the reliability of the sum or the mean of these three raters' ratings in this population, a modified formula is available:

$$r_{II} = \frac{MS_r - MS_e}{MS_r} \qquad \text{(Intraclass correlation of a sum or average)} \qquad (13.60)$$

Applied to the same data,

$$r_{II} = \frac{8.52 - 3.43}{8.52} = .60$$

From this we infer that if we averaged the three ratings for each ratee and could correlate the set of averages with a similar set of averages from comparable raters, the result would be about .60. Averaging reduces the relative importance of errors of measurement, leaving the relationships enhanced. This principle of reliability will be discussed more fully in connection with the Spearman-Brown estimate of reliability in Chap. 17.

COEFFICIENT OF CONCORDANCE

If the measurements are in the form of rankings instead of ratings, the amount of agreement among k sets of them can be expressed by a coefficient analogous to intraclass r, known as the *coefficient of concordance*. This coefficient was developed by Kendall and is represented by the symbol W.[1] W is an index of agreement of k sets of rankings on the same r individuals or objects. By way of illustration, the ratings by three raters of the seven ratees in Table 13.16 were converted to the rankings shown in Table 13.18.

It will be noted that there are a number of tied ratings in Table 13.16 and that ties have been assigned the average of the ranks they would otherwise occupy. W corrected for tied ranks would be slightly higher, but there must be a very large number of ties in order for the correction to produce an appreciable effect on the value of W.[2]

The formula for the coefficient is

$$W = \frac{SS_r}{SS_t} \qquad \text{(Coefficient of concordance)} \qquad (13.61)$$

The sum of squares for rows in the numerator may be obtained from a formula similar to (13.55) applied to ranks, the main difference being that the second, or "correction," term can be computed directly from the number of rows (r) and columns (k). The formula is

$$SS_r = \frac{\Sigma(\Sigma X_r)^2}{k} - \frac{kr(r+1)^2}{4} \qquad \text{(Sum of squares for rows computed from ranks)} \qquad (13.62)$$

The total sum of squares in the denominator of formula (13.61) is obtained directly from the number of rows and columns by

$$SS_t = \frac{k(r^3 - r)}{12} \qquad \text{(Sum of squares for total computed from ranks)} \qquad (13.63)$$

1 Kendall, M. G. *Rank correlation methods*. London: Griffin, 1948.

2 For further information see Siegel, S. *Nonparametric statistics for the behavioral sciences*. New York: McGraw-Hill, 1956, also Mosteller, F., and Rourke, R. E. K. *Sturdy statistics: nonparametrics and order statistics*. Reading, Mass.: Addison-Wesley, 1973.

TABLE 13.18 Ranks of the three sets of ratings in Table 13.16

Ratee	Rater A (Rating)	A Rank	B (Rating)	B Rank	C (Rating)	C Rank	ΣX_r	$(\Sigma X_r)^2$
1	(5)	5	(6)	2	(5)	4.5	11.5	132.25
2	(9)	1.5	(8)	1	(7)	2.5	5.0	25.00
3	(3)	6.5	(4)	4.5	(3)	6.5	17.5	306.25
4	(7)	3.5	(5)	3	(5)	4.5	11.0	121.00
5	(9)	1.5	(2)	7	(9)	1	9.5	90.25
6	(3)	6.5	(4)	4.5	(3)	6.5	17.5	306.25
7	(7)	3.5	(3)	6	(7)	2.5	12.0	144.00
Σ							84.0	1125.00

For the ranks in Table 13.18,

$$W = \frac{375 - 336}{84} = .4643 \qquad \text{(or rounded, .46)}$$

A test of the significance of W, as k becomes large (at least 7), can be made with the following formula, which gives a statistic that is distributed approximately as χ^2 with $r - 1$ degrees of freedom:[1]

$$\chi^2 = k(r-1)W \qquad \text{(Significance test for } W\text{)} \tag{13.64}$$

The average of the rank-difference correlations[2] between pairs of raters is not equal to W, but can be calculated from it by the formula

$$\bar{r}_\rho = \frac{kW - 1}{k - 1} \qquad \text{(Average rank correlation computed from } W\text{)} \tag{13.65}$$

For the example,

$$\bar{r}_\rho = \frac{3(.4643) - 1}{2} = \frac{.3929}{2} = .1965 \qquad \text{(or rounded, .20)}$$

If we wish to have an estimate of the reliability of a sum or mean of the k sets of rankings, it can be obtained from the value for $\bar{r}_\rho$ by

$$r_{KK} = \frac{k\bar{r}_\rho}{1 + (k-1)\bar{r}_\rho} \qquad \text{(Expected correlation of a set of } k \text{ rankings with a comparable set)} \tag{13.66}$$

1 Since $k = 3$ in this illustration we have not applied the χ^2 test to it. For a table giving the significance at the .05 and .01 points, for values of k from 3 to 7 and for selected values of r from 3 to 20, see Edwards, A. L. *Statistical methods.* (3d ed.) New York: Holt, Rinehart and Winston, 1973. Table XI.

2 See the next chapter for an explanation of the rank-difference correlation r_ρ.

TABLE 13.19 Ranks of the 20 measurements in Table 13.1

Set I		*Set II*		*Set III*		*Set IV*	
Measurement	*Rank*	*Measurement*	*Rank*	*Measurement*	*Rank*	*Measurement*	*Rank*
114	13.5	119	2.5	112	16.5	117	5
115	11.5	120	1	116	8.5	117	5
111	19	119	2.5	116	8.5	114	13.5
110	20	116	8.5	115	11.5	112	16.5
112	16.5	116	8.5	112	16.5	117	5
ΣX_i	80.5		23.0		61.5		45.0
$(\Sigma X_i)^2$	6480.25		529.00		3782.25		2025.00
						$\Sigma(\Sigma X_i)^2$	12,816.50

For the data in the example,

$$r_{kk} = \frac{3(.1965)}{1 + 2(.1965)} = \frac{.5895}{1.3930} = .42$$

From this we can infer that if we were to average the three ranks for each ratee and correlate these averages with the averages in a comparable set, we should expect the resulting value to be approximately .42.

ONE-WAY CLASSIFICATION ANALYSIS OF VARIANCE OF RANKS

For the case in which the measurements are in terms of ranks or they can be ordered over all k groups, a one-way analysis of ranked data has been developed by Kruskal and Wallis.[1] The measurements of the four groups on the Galton bar in Table 13.1 have been rank-ordered as shown in Table 13.19, where the largest value has been given a rank of 1.

A test of the null hypothesis—that the k independent samples are from the same population—is made using the statistic H, which is distributed approximately as chi square with $k - 1$ degrees of freedom as n_s becomes large (at least five cases per group).[2]

$$H = \left(\frac{12}{N(N+1)}\right)\left(\Sigma \frac{(\Sigma X_i)^2}{n_i}\right) - 3(N+1) \tag{13.67}$$

(H statistic for one-way analysis of variance of ranks)

1 Kruskal, W. H., and Wallis, W. A. Use of ranks in one-criterion variance analysis. *Journal of the American Statistical Association,* 1952, **47,** 583–621.

2 For three groups and a total N of up to 15 cases, probability values of H can be obtained from Siegel, op. cit., Table O.

where n_i is the number of cases per set and N is the total number of cases in all sets.

For the example in Table 13.14,

$$H = \left(\frac{12}{20(21)}\right)\left(\frac{12{,}816.50}{5}\right) - 3(21) = \frac{12(2563.30)}{420} - 63 = 10.273$$

Reference to Table E in the Appendix for 3 df indicates that the obtained value is significant at the .02 level, and so the null hypothesis may be rejected at this level. It is not required that the n_i's be equal.

If there is a large number of tied ranks, H may be corrected by dividing it by $1 - \Sigma T/(N^3 - N)$, where T equals $(t^3 - t)$ and t is the number of observations tied at a given rank. The correction has the effect of increasing the value of H, and the effect is very slight unless the number of ties is very large.

The H test may be useful when the assumption of homogeneity of variances is obviously not tenable and it is desired to test the hypothesis that the sample means are from populations with equal means.

RANDOMIZED MATCHED-GROUPS DESIGN

The randomized matched-groups design is another instance of a two-way classification analysis without replications, such as was considered in an earlier section. The main function of this design is to reduce heterogeneity on the measurement variable, yielding a smaller error variance and a more sensitive test of significance. This objective is accomplished by matching the subjects on a control variable related to the variable being analyzed. Examples of control variables that might be used for matching subjects are age, IQ, and scores on a pretest, where these variables are correlated with the measurement variable. The scores on the related variable are used to set up levels, or "blocks," the number of subjects at each level being equal to the number of groups. The subjects at each level are then assigned at random, one per experimental group.[1]

If the control variable is unrelated to the measurement variable, the matched-groups design is less efficient than a one-way analysis of variance of the data, since the heterogeneity of the dependent variable is not reduced, but fewer degrees of freedom are available for the denominator of the F ratio than in the one-way analysis of variance, due to the assignment of some of the subjects to the levels of the control variable.

Table 13.20 presents the data from a randomized matched-groups-design experiment in which three methods of teaching a unit of American history were tried out. The three methods were lecture, film,

[1] The design being discussed here is referred to in some textbooks as a *randomized-blocks* design.

TABLE 13.20 Data from a matched-groups-design experiment using three teaching methods where the 30 students have been grouped at 10 levels on the basis of their scores on a previous history-achievement test and assigned at random to one of the methods groups

	Teaching method				
Level	A_1	A_2	A_3	ΣX_r	$(\Sigma X_r)^2$
1	25	24	26	75	5,625
2	24	23	25	72	5,184
3	24	23	26	73	5,329
4	22	23	24	69	4,761
5	22	21	22	65	4,225
6	22	22	23	67	4,489
7	22	20	23	65	4,225
8	20	19	22	61	3,721
9	20	17	19	56	3,136
10	19	18	20	57	3,249
ΣX_k	220	210	230	660	43,944
ΣX_k^2	4,874	4,462	5,340	ΣX_{ij}	$\Sigma(\Sigma X_r)^2$
$(\Sigma X_k)^2$	48,400	44,100	52,900	435,600 $(\Sigma X_{ij})^2$	
		$\Sigma(\Sigma X_k)^2 = 145,400$		$\Sigma X^2_{ij} = 14,676$	

and discussion. The 30 students involved were assigned to 10 levels on the basis of their scores on a previous history-achievement test, and the three students at each level were randomly assigned to the three methods groups. Random assignment can be accomplished by use of a table of random numbers, by means of numbered playing cards, or by some other method that ensures chance assignment. The computations are carried out as in the earlier section for a two-way classification analysis without replications. First, the total sum of squares is computed, as in formula (13.34) (p. 256). Applied to the data in Table 13.20,

$$SS_t = 14,676 - 14,520 = 156$$

The sum of the squares between columns is given by

$$SS_k = \frac{\Sigma(\Sigma X_k)^2}{r} - \frac{(\Sigma X_{ij})^2}{rk} \tag{13.68}$$

Applied to the data in Table 13.20, this gives

$$SS_k = \frac{145,400}{10} - 14,520 = 20$$

TABLE 13.21
Estimated variances and F ratios for the data in Table 13.20

Source		*Sum of squares*	*df*	*Mean square*	*F*	*p*
Methods	(columns)	**20**	2	**10.00**	**22.73**	<**.05**
Levels	(rows)	**128**	9	**14.22**		
Remainder	($r \times k$)	**8**	**18**	**.44**		
Total		**156**	**29**			

The sum of squares between rows is given by the formula

$$SS_r = \frac{\Sigma(\Sigma X_r)^2}{k} - \frac{(\Sigma X_{ij})^2}{kr} \tag{13.69}$$

For the data of Table 13.20, we have

$$SS_r = \frac{43{,}944}{3} - 14{,}520 = 128$$

The sum of the squares for the remainder is obtained by deducting the last two sums of squares from the total sum of squares. Therefore, for the remainder sum of squares we have

$$SS_{r \times k}{}^2 = 156 - 128 - 20 = 8$$

In Table 13.21 are shown the sums of squares we have just calculated and the degrees of freedom associated with them. Dividing each sum of squares by its degrees of freedom, we obtain the mean squares and the F ratio shown in the table.

In our example we have an F value equal to 22.73 with 2 and 18 df. For $\alpha = .05$, $F = 22.73$ exceeds the tabled value of F, and the null hypothesis is therefore rejected. We conclude that the means for the three teaching-method groups do differ significantly.

The randomized matched-groups design accomplishes many of the same objectives as an analysis-of-covariance design, e.g., the goal of reduction of heterogeneity on the analyzed variable so that there will be less residual or error variance.[1] The smaller residual variance involved yields a more sensitive test of the differences between the group means. The analysis in a matched-groups design is a more direct approach than the analysis of covariance, since the heterogeneity of the dependent variable is reduced by direct rather than statistical control. Thus the design is not so dependent on the underlying assump-

[1] For treatments of analysis of covariance, see Edwards, op. cit., 1972; Lindquist, op. cit.; Ferguson, G. A. *Statistical analysis in psychology and education.* (4th ed.) New York: McGraw-Hill, 1976; or Winer, B. J. op. cit.

tions of both correlation and analysis of variance, as is the analysis-of-covariance design.

REPEATED-MEASURES DESIGN In some studies the individual subject is used to form the level or block, and is given each of the treatments. If the order in which each subject receives the treatments is randomized, we have a randomized matched-groups, repeated-measures design. If, however, all the subjects take the treatments in the same order, we have merely a matched-groups, repeated-measures design. Where several treatments are administered to the same subjects the possibility of "carryover" effects cannot, of course, be ignored. For two treatment groups, a t test may be run for the significance of the difference between correlated means (see p. 158).

TEST FOR TREND ON AN ORDERED VARIABLE If the experimental groups are ordered along a variable such as intensity level, amount of reward, number of reinforcements, dosage of drugs, number of trials, or complexity of stimulus, we may want to determine the shape of the trend of the group means. If the groups are evenly spaced along the ordered variable and are equal in size, we can use orthogonal coefficients to test whether the group means (or sums) have a linear relationship, or one of some other form, to the ordered variable.

The procedure is similar to the orthogonal comparisons made to test hypotheses concerning mean differences (see pp. 237 to 239), except that the appropriate coefficients are obtained from a table of orthogonal polynomial coefficients. This can best be explained by an illustration.

Suppose, for example, that we have an experiment in which the ordered variable is size of type, with the six sizes designated as 1, 2, 3, 4, 5, and 6. There are 60 subjects, 10 randomly assigned to each group, and the measurement variable is accuracy of apprehension. The results of an analysis of variance of these accuracy scores are given in Table 13.22 and the group sums are shown in Table 13.23. These group sums are also plotted against the size-of-type variable in Fig. 13.4. Inspection of the figure indicates that a straight line would be a reasonably good fit of the trend between size of type and accuracy of apprehension. The possibility also exists that an S-shaped

TABLE 13.22
Summary of analysis of variance

Source	*Sum of squares*	*df*	*Mean square*	*F*
Size of type	1108.33	5	221.67	11.97
Within	1000.00	54	18.52	
Total	2108.33	59		

TABLE 13.23
Sums and orthogonal coefficients for the linear, quadratic, and cubic components for $k =$ 6 size-of-type groups

Comparison	*Group sums*						Σa_i^2	C_i
	1	*2*	*3*	*4*	*5*	*6*		
	100	*110*	*120*	*180*	*190*	*210*		
Linear	−5	−3	−1	1	3	5	**70**	**850**
Quadratic	5	−1	−4	−4	−1	5	**84**	**50**
Cubic	−5	7	4	−4	−7	5	**180**	**−250**

curve (or cubic trend), corresponding to a third-degree equation (the number of times the sign of the polynomial coefficients changes determines the degree), might also fit the curve. We obtain the polynomial coefficients for the linear, quadratic, and cubic components from Table T in the Appendix. It will be noted that the coefficients in each row sum to zero and that for any specified value of k, the sum of the products of the coefficients for any two degrees of curvature is also zero. The comparisons based on them therefore meet the requirements for orthogonality previously specified [see formulas (13.18 and 13.19), p. 237].

The value for a comparison is obtained by multiplying the group sums by the coefficients for the comparisons as indicated by

$$C_i = a_{i1}\ \Sigma X_1 + a_{i2}\ \Sigma X_2 + \cdot\cdot\cdot + a_{ik}\ \Sigma X_k \tag{13.70}$$

(Comparison for group sums)

For the linear comparison,

$$C_{\text{lin}} = (-5)(100) + (-3)(110) + (-1)(120) + (1)(180) + (3)(190) + (5)(210) = 850$$

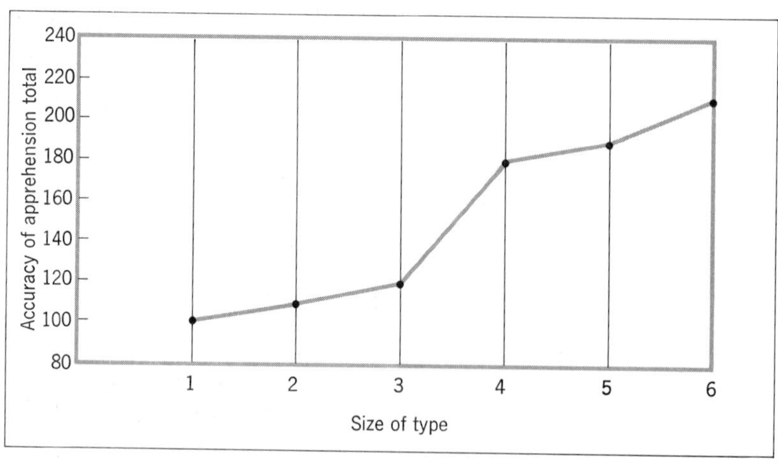

FIGURE 13.4
The totals for 10 subjects at 6 different sizes of type for accuracy-of-apprehension scores

The mean square for a comparison C_i with 1 df is given by

$$MS_C = \frac{C_i^2}{n\Sigma a_i^2} \quad \text{(Mean square for a comparison of group sums with 1 df)} \tag{13.71}$$

Since $\Sigma a_{\text{lin}}^2 = 70$ and $n = 10$ observations for each sum,

$$MS_{C_{\text{lin}}} = \frac{(850)^2}{(10)(70)} = 1032.14$$

The F test for a trend component is

$$F = \frac{MS_C}{MS_w} \quad (F \text{ for significance of trend component}) \tag{13.72}$$

The F defined by formula (13.72) has 1 df for the numerator and the degrees of freedom associated with MS_w for the denominator. For the linear trend,

$$F = \frac{1032.14}{18.52} = 55.73$$

With 1 and 54 df the obtained F ratio is significant beyond the .01 level, and the linear trend is therefore statistically significant.

The sum of squares for deviations from linearity may be obtained from

Deviations from linearity sum of squares
= between sum of squares − linear sum of squares (13.73)

Since for the example this is

$$1108.33 - 1032.14 = 76.19$$

there is still the possibility that the quadratic or cubic trend component may be statistically significant. Where the mean square for deviations from linearity is larger than the error mean square, we can test it for significance, with $k - 2$ df for the numerator, by

$$F = \frac{\text{mean square for deviations from linearity}}{\text{within mean square}} \tag{13.74}$$

(F test for significance of deviations from linearity)

For the example,

$$F = \frac{76.19/4}{18.52} = \frac{19.05}{18.52} = 1.03$$

This F ratio with 4 and 54 df is not significant at the .05 level; therefore the deviations from linearity are not significant.

Since we are dealing with orthogonal comparisons, it is possible that although the overall mean square for deviations from linearity is not significant, the mean square for one or more higher-order trends may be significant. If we are interested in testing for the quadratic trend, we can refer to Table 13.23, to obtain the value for C_{quad}, which is 50. The value for $MS_{C_{quad}}$ computed by formula (13.71) is 2.98, and the F ratio computed by formula (13.72) is only .16. This F is, of course, not significant, and the increase in predictability due to the quadratic component is not significantly different from zero. Similarly for the cubic trend, the value for C_{cubic} from Table 13.23 is -250. $MS_{C_{cubic}}$ is 34.72, and the F statistic is 1.87, which is not significant for 1 and 54 df at the .05 level. The cubic component does not add significantly to the predictability.

None of the remaining variations associated with higher-order components is significant, and it may be concluded that the form of the best-fitting curve is a first-degree (linear) equation.

ESTIMATION OF LINEAR CORRELATION An estimate of the linear correlation between the treatment and measurement variables can be obtained from

$$r = \sqrt{\frac{\text{linear sums of squares}}{\text{total sums of squares}}} \qquad \text{(Linear correlation between the ordered and measurement variables)} \qquad (13.75)$$

For the example,

$$r = \sqrt{\frac{1032.14}{2108.33}} = .70$$

This value estimates the degree of linear correlation between the size-of-type and accuracy-of-apprehension variables.

TESTS FOR TRENDS WITH REPEATED MEASURES If the ordered variable in the repeated-measures design is the number of trials in a learning experiment, we may test our hypothesis concerning the form of the learning curve by the orthogonal comparison method (see pp. 277 to 280), taking into account the fact that our observations are correlated from trial to trial. Table 13.24 gives the results of an analysis of variance of learning data. Each of the 10 sub-

TABLE 13.24
Analysis of variance of learning data

Source	*Sum of squares*	*Degrees of freedom*	*Mean square*	*F*
Trials (t)	**103.94**	**6**	**17.32**	**18.23**
Subjects (s)	**90.00**	**9**	**10.00**	
Subjects × trials ($s \times t$)	**51.30**	**54**	**.95**	
Total	**245.24**			

jects was given seven trials on a learning task. The total, trial, and subject sums of squares are computed in the usual manner for obtaining the total, column, and row sums of squares. The subject × trial (remainder) sum of squares is obtained by subtracting the row and column sums of squares from the total sum of squares. For the present example,

$$\text{Subject} \times \text{trial} = 245.24 - 103.94 - 90.00 = 51.30$$

This small subject × trial sum of squares indicates that there is relatively little interaction between these two variables and that the form of the learning curve is similar for the various subjects. Using the subject × trial mean square as our estimate of experimental error for the denominator, the *F* ratio of 18.23 for 6 and 54 df is highly significant, indicating significant differences between the trial totals (or means).

A plot of the trial totals is given in Figure 13.5. Inspection of the figure indicates that a straight line would represent the trend, but there is also the possibility that an S-shaped (cubic) curve might provide a better fit over the range of trials included in the study. The computations of the mean squares for the linear, quadratic, and cubic trends are shown in Table 13.25. The trial totals shown in the table are the sums over the observations for the 10 subjects. The coefficients for the seven trials corresponding to the linear, quadratic, and cubic trends, are obtained from Table T in the Appendix for the set $k = 7$. The entries in Table 13.25 in the column labeled C_i represent the numerical values of the comparisons. For example, the *linear* comparison is obtained by

$$(-3 \times 2) + (-2 \times 5) + (-1 \times 7) + (0 \times 18) + (1 \times 28) + (2 \times 31) + (3 \times 33) = 166$$

The entries in the column headed Σa_i^2 represent the sum of the squared coefficients in the corresponding row, and they may be either

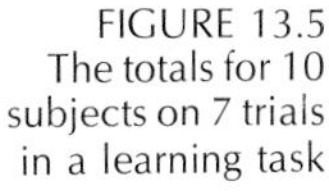
FIGURE 13.5
The totals for 10 subjects on 7 trials in a learning task

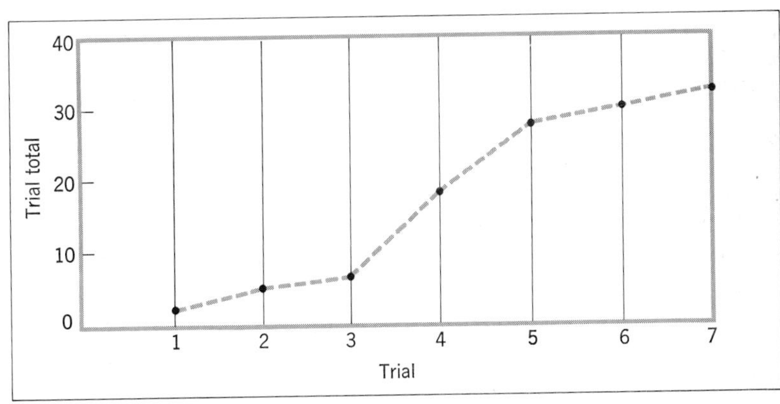

TABLE 13.25
Sums and orthogonal coefficients for the linear, quadratic, and cubic components for $k = 7$ trials on a learning task

Comparison	*Trial sums*							Σa_i^2	C_i
	1	*2*	*3*	*4*	*5*	*6*	*7*		
	2	*5*	*7*	*18*	*28*	*31*	*33*		
Linear	−3	−2	−1	0	1	2	3	28	166
Quadratic	5	0	−3	−4	−3	0	5	84	−2
Cubic	−1	1	1	0	−1	−1	1	6	−16

computed or obtained from Table T in the Appendix. The mean square for the linear components of the trend of trial totals is given by

$$MS_{lin} = \frac{C_{lin}^2}{n\Sigma a^2} \qquad \text{(Mean square for the linear-trend component)} \tag{13.76}$$

For the linear component we have

$$MS_{lin} = \frac{(166)^2}{10(28)} = 98.41$$

A test of significance of the linear trend is given by

$$F = \frac{MS_{lin}}{MS_{s \times t}} \qquad \text{(}F\text{ test for the linear-trend component)} \tag{13.77}$$

For the example data this gives

$$F = \frac{98.41}{.95} = 103.59$$

The linear trend, with 1 and 54 df, is significant beyond the .01 level.

The numerical value for the mean square for the quadratic comparison is zero to two decimal places. The mean square corresponding to the cubic comparison is obtained from

$$MS_{cubic} = \frac{C_{cubic}^2}{n\Sigma a^2} \qquad \text{(Mean square for the cubic-trend component)} \tag{13.78}$$

For the cubic component this is

$$MS_{cubic} = \frac{(-16)^2}{10(6)} = 4.27$$

A test of significance of the cubic trend is given by

$$F = \frac{MS_{cubic}}{MS_{s \times t}} \qquad \text{(}F\text{ test for cubic-trend component)} \tag{13.79}$$

For the example this gives

$$F = \frac{4.27}{.95} = 4.49$$

This value for F with 1 and 54 df is significant at the .05 level and indicates that within the range of trials included in the experiment, the cubic component does add significantly, at the .05 level, to the predictability obtainable from the linear trend.

The remaining variation due to higher-order-trend components, with $k - 4$ df, is

$$\text{Sum of squares}_{\text{higher order}} = \text{sum of squares}_{\text{trials}} - (\text{sum of squares}_{\text{lin}} + \text{sum of squares}_{\text{quad}} + \text{sum of squares}_{\text{cubic}}) \tag{13.80}$$

For the illustrative data this gives

$$\text{Sum of squares}_{\text{higher order}} = 103.94 - (98.41 + .00 + 4.27) = 1.26$$

The corresponding mean square is

$$\text{MS}_{\text{higher order}} = \frac{1.26}{3} = .42$$

The F ratio for the higher-order-trend components is

$$F = \frac{\text{MS}_{\text{higher order}}}{\text{MS}_{s \times t}} = \frac{.42}{.95} = .44 \tag{13.81}$$

Since this ratio is less than 1, it can be assumed that the higher-order components are negligible.

It may be concluded that a linear equation would fit the learning-task data over the seven trials well but that a cubic equation would represent the trend better and would add significantly to the accuracy of prediction of total (or mean) learning performance.

General comments on analysis of variance

ASSUMPTIONS TO BE SATISFIED IN APPLYING ANALYSIS OF VARIANCE

Like most statistics, those involved in analysis of variance have been derived on the basis of mathematical reasoning. Such reasoning starts with assumptions. If those assumptions are satisfied within certain limits of tolerance, the results in terms of F ratios may be interpreted as described in this chapter. If those assumptions are not sufficiently approximated, there is considerable risk that the conclusions may be faulty.

The four major assumptions to be satisfied in applying analysis of variance were presented earlier in this chapter. They include independent observations within sets (random assignment), equal variances within sets, normal distributions of population values within sets, and additivity of component contributions to variances.

Some extensive studies by Norton on sampling problems in analysis of variance have thrown considerable light upon what happens to *F* when distributions of populations are not normal and when variances are not equal.[1] With artificial populations of 10,000 cases, Norton varied the shape of distribution in various ways, making it leptokurtic, rectangular, markedly skewed, and even J shaped. Other populations were normally distributed, but variances were 25, 100, and 225 in different cases—in other words, differing markedly—the standard deviations being 5, 10 and 15, respectively.

One general finding was that *F* is rather insensitive to variations in shape of population distribution. This is consistent with the known principle that distributions of means (sampling distributions) approach normality even though populations are not normally distributed.[2] Another general finding was that *F* is somewhat sensitive to variations in variances of populations, but that only marked differences in variance are serious. Tests of homogeneity of variances in sets of data have been offered, but none is regarded as very satisfactory. A suggestion was made earlier in this chapter that a rough test would consist of applying an *F* test to any pair of variances suspected of being significantly different. Even if some pair shows a significant difference (which is not likely, with small samples), one may proceed with analysis of variance, but should then discount significance levels somewhat. If *F* proves to be significant at the .05 level, this result may actually indicate significance at levels .04 to .07; one significant at the .01 level may actually be significant from the .005 to the .02 levels. If anything, the significance is likely to be lower (probability higher) than that indicated by the *F* tables.

In other words, it can be concluded that the *F* test is relatively insensitive to heterogeneity of variances when we have the *same number of subjects* per set. As Box[3] has pointed out, because the *F* test is insensitive to nonnormality and because with equal *n*'s it is also insensitive to inequality of variances, it can be safely used under most conditions to indicate that the means in a set differ significantly when a significant *F* ratio is obtained.

1 Cited by Lindquist, op. cit.

2 This principle is referred to as the "central limit theorem."

3 Box, G. E. P. Non-normality and tests of variances. *Biometrika,* 1953, **40,** 318–355; also Some theories on quadratic forms applied to the study of analysis of variance problems: I. Effect of inequality of variance in the one-way classification. *Annals of Mathematical Statistics,* 1954, **25,** 209–302.

GENERAL UTILITY OF ANALYSIS OF VARIANCE

As should be clear from the numerous applications of analysis of variance procedures in this chapter, under this heading comes a very versatile collection of statistical aids in research. Not all uses were treated in this chapter. Two additional applications will be seen in Chaps. 15 and 16. Beyond all this, there are problems of four-way classification, and so on. Involved in them are interactions of quadruple or of even higher order. It is one of the strongest virtues of analysis-of-variance methods that they enable us to detect significant contributions from interactions of experimental variables.

Not the least of the merits of analysis of variance is the rather strict set of requirements it imposes in the designing of experiments. Since the experimenter knows that after his observations have been made and the data are in hand he needs to make certain tests of hypotheses, he is motivated to take special care to design his experiment in such a way that the appropriate statistical tests may be carried out. Although this chapter has presented quite a variety of experimental designs, it has by no means exhausted the topic. The reader is referred to books more concerned with the subject of experimental design.[1]

For many years, while analysis-of-variance methods were being developed, there were some undesirable limitations. The investigator could test his null hypotheses, answer the question of whether or not anything over and above chance samplings was at work in his experiment, and reach a "yes" or "no" decision. But if F tests showed that the answer was in the affirmative, he had no further evidence concerning the degree of relationship between his variables. From the number of instances given in this chapter in which some index of correlation could also be found, the reader can see that this situation has been corrected to a large extent. But some investigators, rather than selecting a very limited number of separated values on the quantitative variables that they are relating, prefer to study a problem with whole ranges on their experimental variables by using a correlational approach. Then an index of strength of relationship is always obtained, and it is very easy to test the hypothesis of zero correlation. This approach limits the possibility of finding interaction effects, but there are other ways of achieving that goal, e.g., by fractionating data in various ways. At any rate, we see that many bridges are being constructed between the correlational and analysis-of-variance approaches.

1 See Cochran, W. G., and Cox, G. M. *Experimental designs.* New York: Wiley, 1950; Maxwell, A. E. *Experimental designs in psychological and medical sciences.* New York: Wiley, 1958; Winer, op. cit.; Campbell, D. T., and Stanley, J. C. *Experimental and quasi-experimental designs for research.* Chicago: Rand McNally, 1963.

EXERCISES

The values in Data 13*A* are measurements of the lower threshold for hearing tones, given in cycles per second. Assume that they came from different groups of male and female observers, under each of four conditions, with 0, 4, 8, and 12 min of nearly complete relaxation just before each set of observations was made.

1 Using the four sets of observations made by the male observers only, apply an F test to determine whether there were systematic changes in threshold level associated with length of rest period. Estimate variances by using deviations from means. Interpret your results statistically and psychologically.

2 Make a similar F test for the data derived from the female observers, using the formulas for the original measurements. Estimate the proportion of the variance of the threshold measurements associated with length of rest period.

3 Treat the entire table of Data 13*A* as a two-way classification problem. Make F tests to determine the significance of the three separate sources of variance. Defend your use of error terms. Interpret your results.

4 Estimate the proportion of variance in the lower-threshold measurements in Data 13*A* associated with each of the three separate sources of variance.

5 Remove each source of variance in Data 13*A* step by step, as was demonstrated in Table 13.10.

6 Compute an F ratio for the analysis of Data 13*B*. State your conclusions.

7 Compute an intraclass correlation between raters and also between averages of ratings, using Data 13*B*.

8 The summary table given below is for a one-way analysis of variance based on n (10) subjects assigned at random to each of k (4) method groups.

Components	*Sum of squares*	*Degrees of freedom*	*Mean square*	*F*
Between groups	**62.55**	**3**	**20.85**	**8.69**
Within groups	**86.40**	**36**	**2.40**	
Total	**148.95**	**39**		

Assume that the following set of orthogonal comparisons represents your prior hypotheses and test the comparisons for significance.

Comparison	X_1	$\bar{X}_2$	X_3	X_4
	14.6	*17.2*	*12.4*	*13.6*
C_1	1	1	−1	−1
C_2	−1	1	−1	1
C_3	−1	1	1	−1

9 Of the possible comparisons for the $k = 4$ groups, for which the summary table is presented in the preceding problem, how many are significant at the .05 level and at the .01 level by the *S* method?

10 Rank-order the ratings for each of the raters in Data 13*B* and compute the coefficient of concordance for the three raters; find the average intercorrelation of the ranks; and determine the estimated reliability of the summed or averaged ranks.

11 Using the four sets of observations for the male and female observers in Data 13*A* (but disregarding the sex variable), rank-order the 32 measurements and determine whether there were systematic differences for the four conditions, using the Kruskal-Wallis *H* test.

12 The 30 subjects in Data 13*C* were placed at 10 levels of three subjects each on the basis of their IQ scores. The three subjects at each level were assigned at random to one of three memorization-method groups. The error scores have been rearranged at each level in accordance with methods groups. Treat the data using analysis of variance.

13 For the four conditions, for the males only in Data 13*A*, determine whether the linear component of the trend is significant. Determine whether the deviations from linearity are significant.

14 Assuming that *A*, *B*, and *C* in Data 13*D* are fixed variables in a $2 \times 2 \times 2$ factorial experimental design with $n = 8$, complete the analysis of variance.

15 In a three-way classification analysis of variance with $n > 1$: (*a*) what is the appropriate error term for testing the sources of variance if *A*, *B*, and *C* are fixed variables, (*b*) when testing the row sum of squares when rows and sections are fixed and columns are random, and (*c*) when testing the $r \times s$ interaction?

16 In a $3 \times 4 \times 2$ factorial experiment, with 10 observations for each treatment combination, set up the summary analysis-of-variance table showing the sources of variation and the df associated with each source. How would you calculate the triple interaction sum of squares?

ANSWERS

1 $n\Sigma d^2 = 62.76$; $SS_w = 125.0$; $F = 2.01$ (df = 3, 12).

2 $n\Sigma d^2 = 34.75$; $SS_w = 31.00$; $F = 4.48$ (df = 3, 12); $\omega^2_{est} = .40$.

3 $SS_t = 325.5$; $SS_{rk} = 169.5$; $SS_r = 72.0$; $SS_k = 90.5$; $SS_{r\times k} = 7.0$; $SS_w = 156.0$; F (between rows) = 11.08 (df = 1, 24); F (between columns) = 4.64 (df = 3, 24); F (interaction) < 1.0 (df = 3, 24).

4 ${}_r\omega^2_{est} = .20$; ${}_k\omega^2_{est} = .21$; ${}_{r\times k}\omega^2_{est}$ set at 0.

5 Means of columns and rows constitute the necessary checks.

6 $SS_r = 61.14$; $SS_k = 3.71$; $SS_t = 79.14$; $SS_e = 14.29$; F (for rows) = 8.56 (df = 6, 12); F (for columns) = 1.56 (df = 2, 12).

7 $r_I = .72$; $r_{II} = .88$.

8 $c_1 = 2.9$; $c_2 = 1.9$; $c_3 = .7$; $t_1 = 5.918$; $t_2 = 3.878$; $t_3 = 1.429$; df = 36.

9 Fourteen comparisons are significant at the .05 level for $t' = 2.929$, and 13 are significant at the .01 level for $t' = 3.625$.

10 $W = .75$; $\bar{r}_\rho = .625$; $r_{kk} = .83$.

11 $H = 7.50$, df = 3.

12 $F = 22.73$, df = 2, 18.

13 $F_{\text{lin}} = 5.88$ (df = 1, 12); F for deviations from linearity < 1.0.

14 $F_r < 1.0$; $F_k = 7.39$; $F_s < 1.0$; $F_{r\times k} < 1.0$; $F_{r\times s} < 1.0$; $F_{k\times s} = 2.41$; $F_{r\times k\times s} < 1.0$; df = 1, 56.

15 Within; $r \times k$; $r \times k \times s$.

16

Source	*df*
A	2
B	3
C	1
$A \times B$	6
$A \times C$	2
$B \times C$	3
$A \times B \times C$	6
Within	**216**
Total	**239**

DATA 13*A*
Data in a two-way classification

Observers	*Condition* B_1	B_2	B_3	B_4
A_1 **(Male)**	19	21	24	24
	12	16	18	26
	17	17	22	21
	20	18	18	17
A_2 **(Female)**	15	16	15	18
	15	19	19	19
	14	17	16	18
	12	14	18	17

DATA 13*B*
Ratings of seven individuals by three raters in a particular trait

Ratees	Raters		
	I	*II*	*III*
1	3	4	5
2	5	5	5
3	3	3	5
4	1	4	1
5	7	9	7
6	3	5	3
7	6	5	7

DATA 13C
Error scores for 30 subjects at 10 IQ levels

IQ level	*Method*		
	A_1	A_2	A_3
1	11	10	12
2	10	9	11
3	10	9	12
4	8	9	10
5	8	7	8
6	8	8	9
7	8	6	9
8	6	5	8
9	6	3	5
10	5	4	6

DATA 13*D*
Scores in a 2 × 2 × 2 fixed-effects factorial experiment

	C_1		C_2	
	B_1	B_2	B_1	B_2
A_1	7	9	4	4
	5	8	7	6
	8	3	9	2
	8	7	6	4
	7	7	9	2
	6	3	6	4
	5	2	7	4
	2	5	4	7
A_2	6	4	5	1
	9	6	7	6
	5	3	6	4
	6	6	5	6
	4	5	7	4
	6	7	8	8
	5	9	7	5
	9	5	8	5

THREE

RELATIONS AND PREDICTIONS

14 Special correlation methods and problems

Pearson's product-moment coefficient is the standard index of the amount of correlation between two variables, and we prefer it whenever its use is possible and convenient. But there are data to which this kind of correlation method cannot be applied, and there are instances in which it can be applied but in which, for practical purposes, other procedures are more expedient. The Pearson coefficient is most defensibly computed when the two variables X and Y are measured on continuous metric scales and the regressions are linear (see Chap. 15). Many data are in terms of frequencies of cases having certain attributes—they are on nominal scales (see Chap. 2). Less often, two continuously measured variables bear to each other a relationship that is curved rather than linear. This chapter will describe some procedures that take care of these irregular situations and other situations where shortcut methods are used to advantage in estimating a Pearson r.

Even when we can apply the product-moment correlation method, however, there are many circumstances which may give rise to a somewhat atypical estimate of correlation or one that does not apply to the population in which we are interested. Samples may be heterogeneous, they may be restricted in variability, or they may be forced into a smaller number of categories than we need for good estimates of correlation, estimates free from errors of grouping. These and other common irregularities in the sampling situation or in the data call for special corrective steps and for special interpretive action. It is impossible to anticipate all the peculiarities of data that the reader may encounter, but the more common exceptions to ideal correlation conditions will be touched upon.

Spearman's rank-difference correlation method

Especially when samples are small, a convenient procedure is Spearman's rank-difference method. It can be applied as a quick substitute when the number of pairs, or N, is less than 30. It should be applied when data are already in terms of rank orders rather than interval measurements.

THE COMPUTATION OF A SPEARMAN RANK-DIFFERENCE CORRELATION

If we have data in terms of measurements or scores, it is first necessary to translate them into rank orders. The procedure is demonstrated by means of the data in Table 14.1. There we have 15 pairs of scores for 15 individuals who responded to sets of cartoons and limericks by judging their humor values, each on a five-point scale. The score in each case is the sum of the points each individual assigned to the set. We could correlate these scores in the usual manner, described in Chap. 6, but the rank-difference method will be found to be shorter. The following steps are necessary for the application of formula (14.1):

step 1 Rank the individuals in the first variable (cartoon score) and call these ranks R_1. Give the highest score a rank of 1, the next highest a rank of 2, and so on. If there are ties in the scores, average the ranks they should occupy. For example, two persons have scores of 41. They would be 12 and 13, and so we call both 12.5. Next are two scores of

TABLE 14.1
A rank-difference correlation between humor scores in reactions to cartoons and to limericks

Cartoon score	*Limerick score*	R_1	R_2	D	D^2
47	75	11	8	3	9.00
71	79	4	6	2	4.00
52	85	9	5	4	16.00
48	50	10	14	4	16.00
35	49	14.5	15	0.5	0.25
35	59	14.5	12	2.5	6.25
41	75	12.5	8	4.5	20.25
82	91	1	3	2	4.00
72	102	3	1	2	4.00
56	87	7	4	3	9.00
59	70	6	10	4	16.00
73	92	2	2	0	0.00
60	54	5	13	8	64.00
55	75	8	8	0	0.00
41	68	12.5	11	1.5	2.25
					171.00
					ΣD^2

35, which should occupy ranks of 14 and 15, and so the ranks are 14.5 and 14.5. If the person with the lowest score is not tied with anyone, that person's rank should be equal to N (in this case 15). This relationship should serve as a check as to accuracy of ranking, although it will not reveal reversals of ranking somewhere along the line.

step 2 Rank the second list of measurements and call the ranks R_2. The three ties at a score of 75 should occupy ranks 7, 8, and 9. Averaging, we call them all ranks of 8, leaving 7 and 9 out of the list.

step 3 For every pair of ranks (for each individual), determine the difference (D) in the two ranks, with no attention paid to algebraic signs, for they are all going to be squared anyway.

step 4 Square each D to find D^2.

step 5 Sum the squares to find ΣD^2. The sum in the illustrative problem is 171.00.

step 6 Compute the coefficient r_ρ (r sub rho) by means of the formula

$$r_\rho = 1 - \frac{6\Sigma D^2}{N(N^2 - 1)} \qquad \text{(Spearman's rank-difference coefficient of correlation)} \qquad (14.1)$$

where ΣD^2 = sum of the squared differences between ranks and N = number of squared measurements in the data. In the illustrative problem,

$$r_\rho = 1 - \frac{6 \times 171}{15 \times 224}$$
$$= .69$$

INTERPRETATION OF A RANK-DIFFERENCE COEFFICIENT The rho coefficient is closely equivalent to the Pearson r that would be computed from the original measurements. It is actually the Pearson r that would be obtained from correlating the rank values for the same basic data. The r_ρ values are systematically a bit lower than the corresponding Pearson-r values, but the maximum difference, which occurs when both coefficients are near .50, is less than .02.

SIGNIFICANCE OF SPEARMAN'S RHO COEFFICIENT There is no generally accepted formula for estimating the standard error of an r_ρ coefficient. Therefore we cannot determine confidence limits. When N is as large as 30 we can test the hypothesis of a zero correlation by using the following standard-error formula:

$$s_{r_\rho} = \frac{1}{\sqrt{N - 1}} \qquad \text{(Standard error of } r_\rho \text{ when the population correlation is zero)} \qquad (14.2)$$

Under these conditions the sampling distribution may be assumed to be normal and we may estimate a z ratio by the formula

$$z = r_\rho \sqrt{N - 1} \qquad (14.3)$$

When N is less than 31 the use of Table K in the Appendix is recommended; this table gives the coefficients that are significant at the .10, .05, and .01 levels in two-tail tests. The obtained rho coefficient of .69 is significant beyond the .02 level in a two-tail test.

A BRIEF EVALUATION OF THE RANK-DIFFERENCE CORRELATION

Although there is no good estimate of the standard error of r_ρ, there is good reason to believe that its sampling stability is almost as good as for the Pearson r of the same size. It can therefore be used as an estimate of r. In view of the fact that r_ρ is usually computed only in small samples, in which low correlations cannot be accurately determined, its chief use is in testing the null hypothesis.

Kendall has developed a ranking method of correlation called τ (tau), which rests on no particular assumptions.[1] It has numerous applications especially the testing of hypotheses, but bears no close relation to the traditional family of product-moment correlations.

The correlation ratio

The correlation ratio is a very general index of correlation particularly adapted to data in which there is a curved regression. Among test scores, linear relationships are apparently the almost universal type of regression. Normality, or near normality, in both distributions correlated is almost sufficient in itself to promote linearity. Outside the sphere of psychological and educational tests, however, or when nontest variables are correlated with test scores, we sometimes encounter curved trends in the scatter diagram. The means of the columns do not progressively increase as we go up the X scale. They may increase slowly at first and then rapidly later, they may increase to a maximum in the center and then decrease, or other systematic divergencies from linearity may be apparent.

NONLINEAR REGRESSIONS

A common instance of nonlinear relationship is found when we correlate performance scores with chronological age. Typically, performance, as measured, increases most rapidly from ages five to ten and thereafter shows a slackening in upward trend through the teens. If we follow the progression still further, we find typically a maximal performance somewhere in the twenties, with slow decline to the forties and an increasing rate of decline thereafter. If we included all ages from five to seventy-five in our correlation study and if we computed the usual Pearson r between age and scores, the r would probably prove to be near zero. On such a correlation diagram, the scattering of

[1] Kendall, M. G. *Rank correlation methods*. London: Griffin, 1948. Also Siegel, op. cit., pp. 213–223.

points would be considerably dispersed from any straight line that we might try to draw through the data, slanting upward or slanting downward. Nevertheless, inspection would show a relationship between age and performance that takes into account the waxing and waning of ability within the span of ages studied.

We might break the chart in two and treat by themselves the years during which there is improvement and the years during which there is decline. We should be able to compute a positive correlation for the earlier span and a negative correlation for the later span by assuming straight-line trends. But these would be of doubtful significance and certainly would not do justice to the full strength of relationships, even within the two segments of life-span. The reason is that the trends still deviate from straight lines. Curvature has been overlooked, and to that extent the index of correlation is perhaps markedly underestimated.

TWO REGRESSION LINES AND TWO CORRELATION RATIOS The scatter diagram in Fig. 14.1 represents a sample of relationship between performance score in a form-board test and chronological age between five and fourteen years inclusive. Here the score is time required for completion; hence a high number indicates poor performance, and the trend is downward. But the relationship obviously drops most rapidly during the first three years and settles down to slight changes from year to year during the later years. Two regression lines are drawn in

FIGURE 14.1 Scatter diagram for a correlation-ratio problem

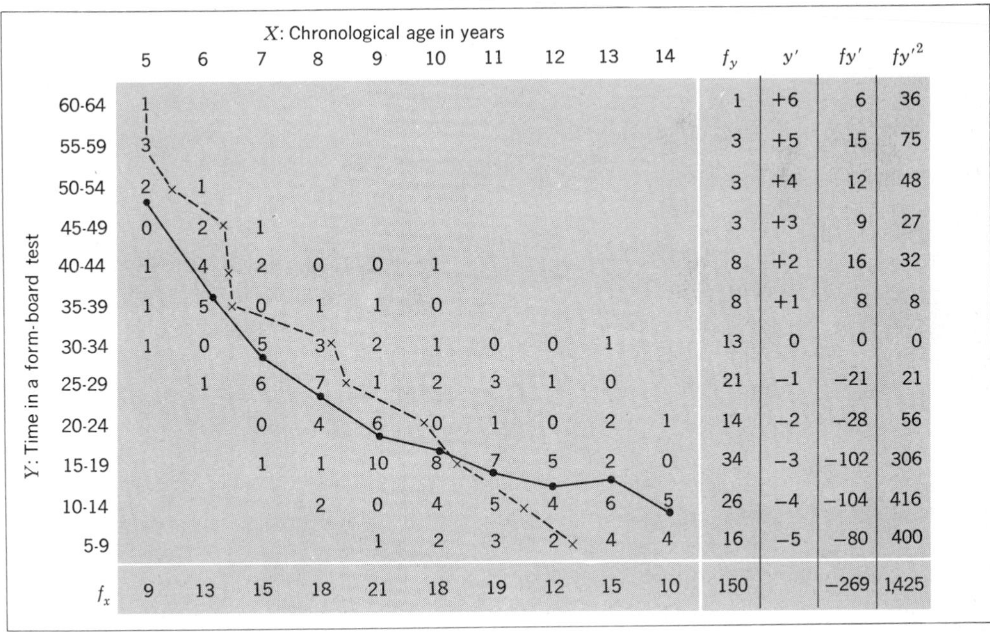

Y: Time in a form-board test \ X: Chronological age in years	5	6	7	8	9	10	11	12	13	14	f_y	y'	fy'	fy'^2
60-64	1										1	+6	6	36
55-59	3										3	+5	15	75
50-54	2	1									3	+4	12	48
45-49	0	2	1								3	+3	9	27
40-44	1	4	2	0	0	1					8	+2	16	32
35-39	1	5	0	1	1	0					8	+1	8	8
30-34	1	0	5	3	2	1	0	0	1		13	0	0	0
25-29		1	6	7	1	2	3	1	0		21	−1	−21	21
20-24			0	4	6	0	1	0	2	1	14	−2	−28	56
15-19			1	1	10	8	7	5	2	0	34	−3	−102	306
10-14				2	0	4	5	4	6	5	26	−4	−104	416
5-9					1	2	3	2	4	4	16	−5	−80	400
f_x	9	13	15	18	21	18	19	12	15	10	150		−269	1,425

the diagram to show the trends more clearly. The regression of test score on age is shown by the solid line that is drawn connecting the points, which are plotted at the means of the columns. The regression of age upon test score is shown by the dashed line and the means of the rows.

Just as we find two regression lines for a less than perfect correlation in Chap. 15, where linear regressions are treated, so here we find two regression curves, differing in shape as well as in slope. Accordingly, we have two correlation ratios, or eta coefficients, one for each of the regressions, and they will not necessarily be the same in value. This result differs from that in the case of linear correlation, where $r_{yx} = r_{xy}$.

The two correlation ratios are estimated by the formulas

$$r_{\eta_{yx}} = \frac{s_{y'}}{s_y} \quad \text{(Correlation ratio for the regression of } Y \text{ on } X\text{)} \tag{14.4a}$$

$$r_{\eta_{xy}} = \frac{s_{x'}}{s_x} \quad \text{(Same, for regression of } X \text{ on } Y\text{)} \tag{14.4b}$$

where $s_{y'}$ = standard deviation of the Y' values predicted from X
$s_{x'}$ = standard deviation of the X' values predicted from Y
s_y and s_x = standard deviations of the total distributions

The manner in which $s_{y'}$ and $s_{x'}$ are determined will be explained next.

THE COMPUTATION OF A CORRELATION RATIO

In a prediction problem of this sort, the best prediction of Y for any column is the mean of the Y's in that column. This prediction will have the smallest sum of squared deviations from the observed Y's in that column. So Y' for each column is the mean of that column. We therefore first compute the means of the columns. These are listed in column 3 of Table 14.2. Now if there were no correlation, no relationship between Y and X, these Y' values would lie along the level of the mean of *all* the Y values, which in this problem is 23.0. No predictions could then be made on the basis of knowledge of X values. For every column with its X value (midpoint), the most probable corresponding Y would be 23.0, and our margin of error would be indicated by s_y. It would be as large as it would be if we had no knowledge of X for each individual (see Chap. 15 for a more complete discussion of this point).

The more the means of the columns deviate from the mean of all the Y's relative to the overall deviations of Y's from their mean, the more accurate our predictions are. We are therefore interested in how far the Y' values do deviate from 23.0 here. Those deviations $(Y' - \overline{Y})$ are given in column 4 of Table 14.2. As usual, we square the deviations and find their mean as an indicator of how great their average is. The squared deviations $(Y' - \overline{Y})^2$ are given in column 5 of Table 14.2. But before finding a mean of the squared deviations, we weight

TABLE 14.2 The computation of a correlation ratio for the regression of time score on chronological age

(1) X' CA	(2) n_c	(3) Y' Time	(4) $Y' - \bar{Y}$	(5) $(Y' - \bar{Y})^2$	(6) $n_c(Y' - \bar{Y})^2$	
14	10	11.0	−12.0	144.00	1,440.00	
13	15	14.0	− 9.0	81.00	1,215.00	
12	12	14.5	− 8.5	72.25	867.00	
11	19	16.0	− 7.0	49.00	913.00	
10	18	18.1	− 4.9	24.01	432.18	
9	21	20.8	− 2.2	4.84	101.64	
8	18	25.1	+ 2.1	4.41	79.38	
7	15	31.3	+ 8.3	68.89	1,033.35	
6	13	40.5	+17.5	306.25	3,981.25	
5	9	49.8	+26.8	718.24	6,464.15	
Sum	150				16,544.96	$\Sigma n_c(Y' - \bar{Y})^2$
					111.040	$s^2_{y'}$
					10.54	$s_{y'}$

each one for a column by the number of cases in that column. The weighted, squared deviation for each column will be found in the last column of Table 14.2. Then the weighted, squared deviations are summed and we divide by $N - 1$, or 149, to find $s_{y'}{}^2$, which is 111.040. The positive square root of this value is 10.54, which is $s_{y'}$.

Remember that the $(Y' - \bar{Y})$ values are *not* the deviations of the observed points from the predicted Y values, because the larger such values are, the *lower* the correlation. We are interested here in the size of deviations of predicted Y values from the mean of all the Y values, and the *larger* these are, the *higher* the correlation. When the correlation is perfect, $s_{y'}$ is as large as s_y, for then $s_{y'}/s_y$ equals 1.0. When $s_{y'} = 0$, the s ratio is zero and r_η is zero. In this problem, $s_y = 12.58$. The correlation is therefore

$$r_{\eta_{yx}} = \frac{10.54}{12.58} = .838$$

THE SIGNIFICANCE OF AN ETA COEFFICIENT As with the Pearson r, the most common question concerning significance is whether or not eta is so small that it could have arisen by chance in random sampling from a population in which the correlation is zero. The standard error to go with this hypothesis is given by

$$s_{r_\eta} = \frac{1}{\sqrt{N - 1}} \quad \text{(Standard error of eta for testing the hypothesis that } r_\eta = 0\text{)} \qquad (14.5)$$

THE RELATION OF THE CORRELATION RATIO TO ANALYSIS OF VARIANCE

Those who have read Chap. 13 will find much that is familiar in the preceding paragraphs. If we regard the successive columns of data, which are really the result of a one-way classification on a quantitative variable (namely, chronological age), as sets, we have all the information we need to proceed with an analysis-of-variance solution (see Table 14.3). From Table 14.2, the sum of 16,544.96 will be recognized as the sum of squares between sets, since it is based upon the squared deviations of set means from the "grand" mean. The quantity 23,568.93 is the total sum of squares. This sum is found most conveniently here from what we already know. It is given by the product $(N-1)S_y^2$, which in this problem is $149 \times (12.58)^2 = 23{,}580.20$. The sum of squares within sets can be obtained from $SS_t - SS_b$, or $23{,}580.20 - 16{,}544.96 = 7{,}035.24$. All we need next is the numbers of degrees of freedom. For the between-sets variance (mean square) there are 9 (the number of sets minus 1). For the within-sets variance there are 140 (N minus the number of sets). The two estimates of the population variance—and also the F ratio, which is 36.6—are given in Table 14.3. Reference to Table F in the Appendix shows that this F is well above the F required for significance at the .01 level, which is about 2.5.

The relationship discussed here is of more academic than practical interest, for we already know that the eta coefficient is so high that there is little doubt that a relationship exists between chronological age and test score. Furthermore, the eta coefficient tells us a fact concerning the *degree* of relationship, which an F ratio does not convey. When the eta is near the lower margin of significance and a more rigorous test of significance is required, and when a decision is to be made as to whether or not there is *any* genuine relationship at all, then the F test has its advantages.

THE STANDARD ERROR OF ESTIMATE IN A NONLINEAR REGRESSION

The standard error of estimate here can be computed as from a Pearson r [see formulas (15.15a) and (15.15b), p. 352, substituting η for r], but it can also be obtained from the within-sets sum of squares we just computed. Dividing the within-sets sum of squares by $N-2$ and taking the square root of the results gives us an estimate of the

TABLE 14.3 An analysis of variance based upon statistics derived in the solution of a correlation ratio

Component	*Degrees of freedom*	*Sums of squares*	*Mean squares*
Between sets	**9**	**16,544.96**	**1,838.33**
Within sets	**140**	**7,035.24**	**50.25**
Total	**149**	**23,580.20**	

$$F = \frac{1{,}838.33}{50.25} = 36.6$$

standard error of estimate,

$$s_{yx} = \sqrt{\frac{SS_w}{N-2}} \qquad \text{(Standard error of estimate computed from } SS_w\text{)} \qquad (14.6)$$

For the illustrative problem,

$$s_{yx} = \sqrt{\frac{7{,}035.24}{150-2}} = \sqrt{47.5354} = 6.75$$

The standard error of estimate tells us how much dispersion there is of the obtained values (Y values in this case) around the predicted values (Y' values). The quantity 6.75 tells us that two-thirds of the time scores in the form-board test may be expected to be within 6.75 units of the predicted values when the predicted values are the means of the columns of the scatter diagram. Such an estimate is useful, however, only when the variances within columns are fairly uniform; in other words, when the data approach homoscedasticity (equal column variances).

A TEST OF LINEARITY OF REGRESSION

Often the curvature in regression is so slight that we do not know whether it is merely a chance deviation from linearity. We therefore want some statistical test to show whether or not the curvature is probably genuine. Several tests of nonlinearity have been proposed. The test currently most widely accepted is an F test based upon an analysis-of-variance approach. The computation of F in this instance is simple, requiring only knowledge of eta and the Pearson r for the same scatter plot and knowledge of the numbers of degrees of freedom. The formula is

$$F = \frac{(r_\eta^2 - r^2)(N-k)}{(1-r_\eta^2)(k-2)} \qquad (F \text{ test of linearity}) \qquad (14.7)$$

where k = number of columns (or rows). For the problem discussed above, the correlation ratio was found to be .838, and the Pearson r was found to be .763. By formula (14.7) we have

$$F = \frac{(.702244 - .582169)(150-10)}{(1-.702244)(10-2)}$$

$$= 7.06$$

In interpreting this F, the numbers of degrees of freedom are $(k-2)$ and $(N-k)$. Reference to Table F in the Appendix shows that the obtained F is significant well beyond the .01 point. Thus, the difference between r_η and r_{yx} is so great as to leave little doubt of nonlinearity.

The hypothesis tested here is that the regression of Y on X is linear. In more exact terms the hypothesis requires that the means of the columns all lie exactly on a straight line whose slope is determined by the Pearson r. Now if the actual form of regression were linear, sampling errors would cause the means of columns to deviate only slightly from the best-fitting straight line. The sampling distribution is of these deviations of the actual means of the columns, the Y values, from the regression line. These deviations are ordinarily sufficient to make the eta coefficient larger than the Pearson r computed from the scatter diagram. The question is whether the deviations are large enough to suggest that something over and above these chance deviations is involved. That is what the F test is supposed to tell us here. The F test should be applied to this particular use only when N exceeds k considerably.

AN EVALUATION OF THE CORRELATION RATIO

The chief advantage and use of the eta coefficient have been indicated and illustrated—it is employed to determine the closeness of relationship between two variables when the regression is clearly nonlinear. Although very few nonlinear regressions have been found in the correlation of measures of ability with one another, there are probably many more such relationships in psychology and education than has been realized.

CORRELATION COEFFICIENTS AS INDICES OF GOODNESS OF FIT Broadening the concept of correlation leads us to consider curves of learning and retention and many others. The eta coefficient assumes no particular type of functional relationship between Y and X. The type of relationship is defined by the actual, unsmoothed trend of the means of the columns (or rows). In this fact are both strength and weakness. Allowing the curvature of the regression to be as complex as the ups and downs in obtained class means make it, we find in eta the maximum size of correlation index for any set of data.

We might assume some kind of mathematical function for the data represented in Fig. 14.1—a hyperbola, parabola, logarithmic function, or some other. The goodness of fit, as indicated by some other correlation index of nonlinear correlation, would probably not be so high for any of these functions as the eta coefficient. Because the eta coefficient does allow the regression curve to follow the means of the columns, a certain amount of error or purely sampling variance undoubtedly gets into the deviations of column means from the general mean of the Y's; hence the eta is a somewhat inflated figure. When the actual regression is linear, the difference between eta and r computed for the same data tells us about how much inflation has occurred. When the regression is nonlinear, we have less ready evidence as to how much inflation there is. We should therefore discount any eta a little, particularly if the means of sets do not follow a smooth trend rather well. The smaller the sample, the more irregular the trend of the set means is likely to be, and therefore the greater the proportion of inflation in eta.

EXAMPLES OF NONLINEAR REGRESSIONS In addition to the functional relationships involved in learning and other phenomena, it is likely that when more is known about human traits that are not abilities—temperament, interests, attitudes, and the like—and their interrelations, we shall find many more examples of nonlinear regression. In the validation of test scores against vocational or other criteria of adjustment, more such examples have come to light. It has been known for some time that high "intelligence" may be just as bad prognostically as low "intelligence" in connection with proficiency in routine and repetitive job assignments. This result will probably be found more general than has been supposed. The reason it has not been more widely recognized is that somewhat shortened ranges of ability have been related to proficiency criteria. If the total range, from lowest to the very highest, is studied in relation to proficiency indices on various kinds of jobs (except those requiring the highest abilities), we may find the optimal ability to be somewhat short of the top in most cases. This means nonlinear regressions.

A number of instances have been known in which scores on temperament tests bore a relation to rated proficiency in such a way that the optimal position on the trait score was barely above average. The application of the Pearson r method sometimes shows a near-zero correlation in such instances, whereas an eta coefficient might be as high as .30 or even .50. The straight line, in other words, was a very poor fit to the regression of the data. This should stress the importance of plotting scatter diagrams more frequently than is ordinarily done; otherwise, important nonlinear regressions may be overlooked. It is possible that many a zero Pearson r reported in the literature conceals a significant nonlinear relationship.

THE ALGEBRAIC SIGN OF ETA Some writers regard it as a weakness of eta that its algebraic sign is always positive. The algebraic sign of r is meaningful in that it shows whether the general trend is upward or downward. In defense of eta it may be said that it tells us what we are most interested in knowing—the goodness of fit or closeness of relationship between two variables. If the overall trend is either upward or downward, we can readily perceive it by inspection of the scatter plot, and we can attach whatever sign is appropriate if we wish to do so. Some curved regressions—for example, U-shaped or inverted U-shaped—may yield a significant eta without any general trend away from the horizontal. In this case no sign is meaningful for eta.

DEPENDENCE OF ETA UPON THE NUMBER OF CATEGORIES A more serious weakness of eta is that its size depends upon the number of columns (or rows). The minimum number of classes that would show any curvature at all is three, but three might give a much-smoothed and distorted view of the real relationship. With too small a number of classes, therefore, we run the chance of obtaining an estimate of correlation that is too small. On the other hand, as we increase the number of classes, we make the means of the classes less stable, and as they fluctuate more, chance errors become more important in inflating eta. The limiting case would be classes so small that there is only one observation per class (assuming no duplicate measures on X), in which case the variance in the columns would be just as great as the overall variance in Y, and eta would equal 1.00.

Methods for correcting eta for number of classes have been proposed, but none can be recommended. The best rule is to keep the classes large enough so that means of classes are fairly stable and fall rather smoothly along a line in the scatter plot and yet to have enough classes to bring out clearly enough the shape of the regression. The size of sample has some bearing on this. The larger the sample, the larger the number of classes that can be tolerated. Very small samples would be unsuitable for the computation of eta at all. With large samples (100 and above) it is suggested that the number of classes range between six and twelve.[1]

THE USE OF MATHEMATICAL FUNCTIONS Better than the correlation-ratio approach, in research studies, is an effort to establish the form of a regression as some mathematical function and then to test the goodness of fit of data to that function by methods which we cannot go into here. There are other texts that treat this topic in some detail.[2]

The biserial coefficient of correlation

The biserial r is especially designed for the situation in which both of the variables correlated are continuously measurable but one of the two is for some reason reduced to two categories. This reduction to two categories may be a consequence of the only way in which the data can be obtained, as, for example, when one variable is whether or not a student passes or fails a certain standard. We can well assume a continuum along which individuals differ with respect to achievement required to pass this standard. Those whose achievement is above a certain crucial point pass, and those whose achievement is below that point fail.

Let us assume the standard is graduation from pilot training. Although not all graduates are equal in achievement and neither are all

1 For small samples, a statistic known as *epsilon* (a correlation ratio without bias) is recommended. See Peters, C. C., and Van Voorhis, W. R. *Statistical procedures and their mathematical bases.* New York: McGraw-Hill, 1940. Pp. 319*ff.*

2 Lewis, D. *Quantitative methods in psychology.* New York: McGraw-Hill, 1960.

eliminees, we know only whether each person belongs to one category or the other. It is as if the grouping were so coarse in this variable as to be confined to two class intervals rather than a dozen or so. If we are prepared to justify the assumption of normality of distribution in this dichotomized variable, we have a formula by which a coefficient of correlation can be computed.

COMPUTATION OF A BISERIAL *r* The principle upon which the formula for a biserial *r* is based is that with zero correlation, there would be no difference between means for the continuous variable, and the larger the difference between means, the larger the correlation. The general formula for biserial *r* is

$$r_b = \frac{\bar{X}_p - \bar{X}_q}{S_t} \times \frac{pq}{y} \qquad \text{(Biserial coefficient of correlation)} \qquad (14.8)$$

where $\bar{X}_p$ = mean of X values for the higher group in the dichotomized variable, the one having more of the ability on which the sample is divided into two subgroups

$\bar{X}_q$ = mean of X values for the lower group

p = proportion of cases in the higher group

q = proportion of cases in the lower group

y = ordinate of the unit normal-distribution curve at the point of division between segments containing p and q proportions of the cases (see Fig. 14.2)

S_t = standard deviation of the total sample in the continuously measured variable X

Table 14.4 presents typical data for computing a biserial correlation. The passing group and the failing group were distributed as shown. The proportions passing and failing are .65 and .35, respectively.[1] The y ordinate (from Table C in the Appendix) is .3704. The distribution of the total group is assumed to be normal as indi-

1 It is good practice to compute p and q to three significant digits rather than two as here.

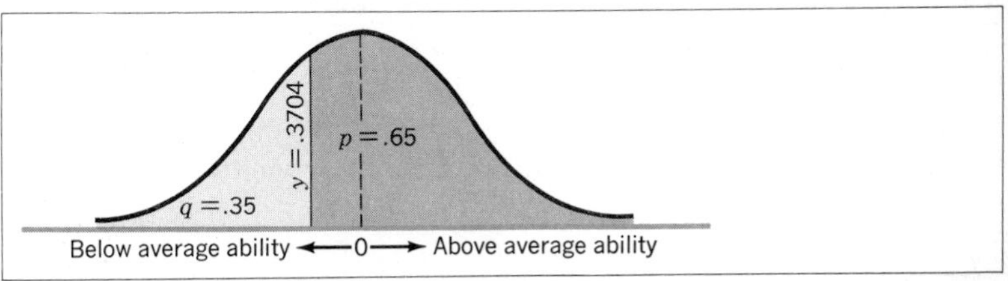

FIGURE 14.2 A normal distribution of the cases along the scale of ability to pass the course of training. The area to the right of the ordinate shown represents the 65 percent who graduated, and the area to the left represents the 35 percent who failed to graduate.

TABLE 14.4 Distribution of scores for two groups of students—those passing and those failing—also a combined distribution

	Scores											
	40–49	50–59	60–69	70–79	80–89	90–99	100–109	110–119	120–129	130–139	n	n/N
Passing students		1	3	10	27	30	26	21	7	5	130	.65 = p
Failing students	2	6	4	11	21	16	7	3	—	—	70	.35 = q
Total	2	7	7	21	48	46	33	24	7	5	200	1.00

cated in Fig. 14.2. The computation of the biserial r proceeds as follows:

$$r_b = \frac{98.27 - 83.64}{17.68} \times \frac{(.65)(.35)}{.3704} = .508$$

Table G in the Appendix is designed in part to supply several of the constants needed in the computation of the biserial r, by either formula (14.8) or formula (14.10), and the computation of its standard error. For given values of p, Table G supplies corresponding values of pq/y, p/y, and $\sqrt{pq}/y$.

THE STANDARD ERROR OF r_b In order to test the hypothesis that a biserial r comes from a population in which ρ_b is zero, if neither p nor q is less than .05, we may use the formula

$$s_{r_b} = \frac{\sqrt{pq}}{y\sqrt{N}} \qquad \text{(Standard error for testing the hypothesis of a zero correlation)} \qquad (14.9)$$

where the symbols are already defined. In the illustrative problem,

$$s_{r_b} = \frac{.4770}{.3704\sqrt{200}} = .091$$

Since the obtained r_b is greater than 1.96 times its standard error, we conclude that at the .05 level the obtained correlation would not very probably have arisen by chance from a population in which the correlation is zero.

ALTERNATIVE FORMULA FOR THE BISERIAL r

In many situations a more convenient formula for the biserial r is

$$r_b = \frac{\bar{X}_p - \bar{X}_t}{S_t} \times \frac{p}{y} \qquad \text{(Alternative formula for computing a biserial } r\text{)} \qquad (14.10)$$

where $\bar{X}_t$ is the mean of the total sample. The advantage of this formula over (14.8) is that it gives one less distribution with which to deal.

AN EVALUATION OF THE BISERIAL r

Since the biserial coefficient of correlation is a product-moment r and is designed to be a good estimate of the Pearson r, the same requirement must be satisfied as for the latter—linear regression—in addition to the unique requirement that the distribution of the values on the dichotomous variable, when continuously measured, be normal. This requirement of normality applies to the form of population distribution. Even if the sample distribution is not quite normal, the population distribution may still be.

The use of the quantities p, q, and y in formulas (14.8) and (14.10) directly implies the normal distribution of the dichotomized variable. Departures from normality, if marked, may lead to very erroneous estimates of correlation. With bimodal distributions, for example, it is possible that the computed r will prove to exceed 1.0. Bimodal and other nonnormal distributions are most likely to occur in heterogeneous samples—for example, in variables in which there is a significant sex difference and both sexes are included in a sample.

Some attention must also be given to the distribution on the continuous variable. Extreme skewness may indicate lack of linearity. The distribution need not be normal, but it should be unimodal and rather symmetrical. Oddly shaped distributions have been known to result in biserial r's greater than 1.0.

WHEN TO DICHOTOMIZE DISTRIBUTIONS There are instances in which the Y variable has been continuously measured but in which there are irregularities that preclude computing a good estimate of the Pearson r. In such cases the biserial r may be brought into service. One example of this would be a truncated distribution;[1] another would be when there are very few categories for the Y variable and it is doubtful whether they are equidistant on a metric scale. Still another would be the case of a markedly skewed sample distribution of Y values due to a defective measuring instrument.

Before computing r_b, of course, we need to dichotomize each Y distribution. In adopting a division point, it is well to come as near the median as possible. The reason for this will be made clear in the next paragraph. In all these special instances, however, we are not relieved of the responsibility of defending the assumption of the normal population distribution of Y. It may seem contradictory to suggest that when the obtained Y distribution is skewed, we resort to the biserial r, but note that it is the *sample* distribution that is skewed and the *population* distribution that must be assumed to be normal.

THE BISERIAL r IS LESS RELIABLE THAN THE PEARSON r Whenever there is a real choice between computing a Pearson r or a biserial r, however,

1 A truncated distribution is one that has been cut off at either end, with no cases with scale values beyond a certain limit.

one should favor the former, unless the sample is very large and computation time is an important consideration. The standard error for a biserial r is considerably larger than that for a Pearson r derived from the same sample. For $r_b = .00$, the standard error of r_b is at least 25 percent larger than that for r for the same size of sample. As p approaches 1.0 or 0.0, the ratio becomes larger until, when $p = .94$, it is as large as 2. This is why in the preceding paragraph it was recommended that dichotomies have the division point as near the median as possible. It also suggests that for the same dependability, we need larger samples for r_b than for r and that we should hesitate to compute r_b for very one-sided divisions of cases unless the sample is extremely large. This is reasonable from another point of view. Remember that prominent in the formula for r_b is the difference between means. This difference is not very stable unless each mean comes from a sample of sufficient size. Even if the sample totaled 1,000 cases, if only 1 percent of the cases were in one of the two categories, its mean would be based upon only 10 cases. Such a condition is not favorable for a reliable estimate of a mean of differences.

Point-biserial correlation

When one of the two variables in a correlation problem is a *genuine* dichotomy, the appropriate type of coefficient to use is the point-biserial r. Examples of genuine dichotomies are male versus female, being a farmer versus not being a farmer, owning a home versus not owning one, living versus dying, living in Boston versus not living in Boston, and so on. Bimodal or other peculiar distributions, although not representing entirely discrete categories, are sufficiently discontinuous to call for the point-biserial rather than the biserial r. Examples of this type are color blindness versus normal color vision, being alcoholic versus being nonalcoholic, and being a criminal versus being a noncriminal.

There are other variables, not fundamentally dichotomous and even normally distributed, which we should treat in practice as if they were genuine dichotomies. An outstanding example of this is a test item the response to which is scored as either right or wrong. No doubt those who answer the item correctly are not all equally capable in the trait or traits measured by the item. A total test score measuring the same trait would provide continuous gradations in trait levels. In testing practice, however, the kind of item described is limited to separating individuals into two groups, and only gross predictions can be made from responses to it. Such a variable is a good example with which to explain the computation of a point-biserial r.

If we gave a "score" of +1 to each person with a correct answer and a "score" of zero to each person with a wrong answer, in the item variable we would have only two class intervals, and we would treat them as if they were genuine categories. A product-moment *r* could be computed with Pearson's basic formula. The result would be a point-biserial *r*. Computer programs for giving Pearson *r*'s from score data automatically yield point-biserial *r*'s between continuous and dichotomized variables.

A special formula, which does not resemble the basic Pearson formula, reads

$$r_{pbi} = \frac{\bar{X}_p - \bar{X}_q}{S_t} \sqrt{pq} \quad \text{(The point-biserial coefficient of correlation)} \tag{14.11}$$

where the symbols are defined as in the formula for the ordinary biserial *r* [formula (14.8)]. The only differences between this formula and the one for the ordinary biserial *r* are that the numerator contains $\sqrt{pq}$ rather than *pq* and the constant *y* is missing from the denominator. For the same set of data, then, the ordinary biserial *r* would be $\sqrt{pq}/y$ times as large as r_{pbi}. In this ratio lies a feature of r_{pbi} to which we shall return soon.

Let us apply formula (14.11) to some data on the relation of body weight to sex membership. In a sample of 51 sixteen-year-old high school students, of whom 24 were male and 27 were female, the mean weights in kilograms were 67.8 and 56.6, respectively. The proportion of males is 27/51 = .471 and *q* is .529. The standard deviation of the combined distribution is 13.2. Solving the problem with formula (14.11),

$$r_{pbi} = \frac{67.8 - 56.5}{13.2} \sqrt{(.471)(.529)} = .42$$

The correlation between sex membership and body weight for 16-year-old high-school students is thus estimated to be .42.

SIGNIFICANCE OF A POINT BISERIAL *r*

The hypothesis of zero for a point-biserial *r* can be tested in two ways. Since r_{pbi} depends directly upon the difference between means $\bar{X}_p$ and $\bar{X}_q$, a significant departure from a mean difference of zero also indicates a significant correlation. A *t* test for the difference between means can therefore be used to test the significance of the departure of the coefficient of correlation from zero.

A direct test of the correlation coefficient can also be made, but only for the hypothesis of a correlation of zero. The test is the same as for a Pearson *r* [see formula (8.13), p. 142], and the interpretation can be made with reference to Student's distribution.[1] For the illustrative

[1] Perry, N. C., and Michael, W. B. The reliability of a point-biserial coefficient of correlation. *Psychometrika,* 1954, **16,** 313–325.

problem, for which $r_{pbi} = .42$ and $N = 51$, $t = 3.24$, which indicates a correlation significant beyond the .01 level. Table Q in the Appendix may also be used to determine whether an obtained r_{pbi} is significant.

When the population value of r_{pbi} is not zero, the mean of the t distribution is not zero, hence the determination of confidence limits for any obtained r_{pbi} is not a simple matter.[1]

ALTERNATIVE METHODS OF COMPUTATION FOR r_{pbi}

As for the ordinary biserial r, there is an alternative formula for computing r_{pbi}, which may be more convenient in many situations. It reads

$$r_{pbi} = \frac{\bar{X}_p - \bar{X}_t}{S_t} \sqrt{\frac{p}{q}} \qquad \text{(Alternative formula for the point-biserial } r) \qquad (14.12)$$

Some formulas that make unnecessary the computation of p and q are:

$$r_{pbi} = \frac{(\bar{X}_p - \bar{X}_q)\sqrt{N_p N_q}}{NS_t} \qquad (14.13)$$

$$r_{pbi} = \frac{\bar{X}_p - \bar{X}_t}{S_t} \sqrt{\frac{N_p}{N_q}} \qquad (14.14)$$

(Other alternative formulas for the point-biserial r)

where N_p and N_q are the frequencies in the two categories.

AN EVALUATION OF THE POINT-BISERIAL r

Since the r_{pbi} coefficient is not restricted to normal distributions in the dichotomous variable, it is much more generally applicable than r_b. When there is doubt about computing r_b, the point-biserial r will serve. For this reason, it should probably be used more than it is. Although it is a product-moment r in value, r_{pbi} is rarely comparable numerically with a Pearson r, or even with an ordinary biserial r, when computed from the same data. Under special circumstances, to be described below, it may be used as a basis for making an estimate of the Pearson r. In regard to the continuous distribution, when r_{pbi} is computed, the same requirements apply as in computing the Pearson r or the biserial r—they include a rather symmetrical, unimodal continuous distribution.

MATHEMATICAL RELATION OF r_{pbi} TO r_b If r_{pbi} were computed from data that actually justified the use of r_b, the coefficient computed would be markedly smaller than r_b obtained from the same data. Even if the one variable is actually continuous but not normally distributed (in which case we might better utilize r_{pbi}), the latter would give an underestimate of the amount of correlation. As was pointed out before, r_b is

1 For methods of determining confidence limits in this situation, see Perry, N.C., and Michael, W. B. A tabulation of the fiducial limits for the point-biserial correlation coefficient. *Educational and Psychological Measurement,* 1954, **14,** 715–721.

$\sqrt{pq}/y$ times as large as r_{pbi} when they are computed from the same data. This ratio varies from about 1.25 when $p = .50$ to about 3.73 when p (or q) equals .99 (see Table G in the Appendix).

In terms of formulas,

$$r_b = r_{pbi}\frac{\sqrt{pq}}{y} \qquad (14.15a)$$

$$r_{pbi} = r_b\frac{y}{\sqrt{pq}} \qquad (14.15b)$$

(Conversion of one biserial r into the other when normality of distribution exists)

When the dichotomous variable is normally distributed without reasonable doubt, it is recommended that r_b be computed and interpreted. If there is little doubt that the distribution is a genuine dichotomy, r_{pbi} should be computed and interpreted. When in doubt, the point-biserial r is probably the safer choice.

Tetrachoric correlation

A tetrachoric r is computed from data in which both X and Y have been reduced artificially to two categories. Under the appropriate conditions it gives a coefficient that is numerically equivalent to a Pearson r and may be regarded as an approximation to it.

ASSUMPTIONS UNDERLYING THE TETRACHORIC *r*

The tetrachoric r requires that both X and Y represent continuous, normally distributed, and linearly related variables. A problem in which the tetrachoric r may be computed is illustrated in Table 14.5, if we are willing to make the necessary assumptions. These data represent the numbers of students responding "Yes" and "No" to two questions in a personality inventory. Question I was, "Do you enjoy getting acquainted with most people?" and question II was, "Do you prefer to work with others rather than alone?" Out of 930 replies to both ques-

TABLE 14.5 Fourfold table from which a tetrachoric coefficient of correlation is computed

		Question I			
		No	*Yes*	*Total*	*Proportion*
Question II	*Yes*	**167** (*b*)	**374** (*a*)	**541**	**.582** (*p*)
	No	**203** (*d*)	**186** (*c*)	**389**	**.418** (*q*)
	Total	**370**	**560**	**930**	**1.000**
	Proportion	**.398** (*q'*)	**.602** (*p'*)	**1.000**	

tions, we have the numbers who responded similarly to the two questions (cells *a* and *d* in Table 14.5) and the numbers who responded differently (cells *b* and *c*). It is obvious that in the case of a perfect positive correlation, all the cases would fall in cells *a* and *d*. In a perfect negative correlation, they would fall in cells *b* and *c*. In a zero correlation, the frequencies would be proportionately distributed in the four cells.

The assumptions of continuity and normality of distribution, in this particular example, can be defended as follows: It is unlikely that all who respond "Yes" to either question do so with an equal degree of affirmation. It is similarly unlikely that those who respond "No" do so with an equal degree of negation. It is most likely that the answers to either question represent a continuum of behavior extending from strong affirmation at the one extreme to strong negation at the other. Continuity, and not a real dichotomy, is thus the probable state of affairs. If a continuum is granted, the general law of unimodal distribution approaching normality in psychological traits may be cited in defense of the other requirement.

Figure 14.3 illustrates the situation in which two continuous, normally distributed variables are dichotomized. With a substantial degree of correlation between *Y* and *X*, in a scatter plot the cases would be distributed as in the ellipse. Drawing the dividing lines at the score levels of *z* and *z′* partitions the cases into four groups. The normal-curve values that appear in formula (14.16) are represented. We take advantage of those values in computing a coefficient of correlation. The diagram should indicate the importance of fulfilling the assumption of normal distributions in both *X* and *Y*.

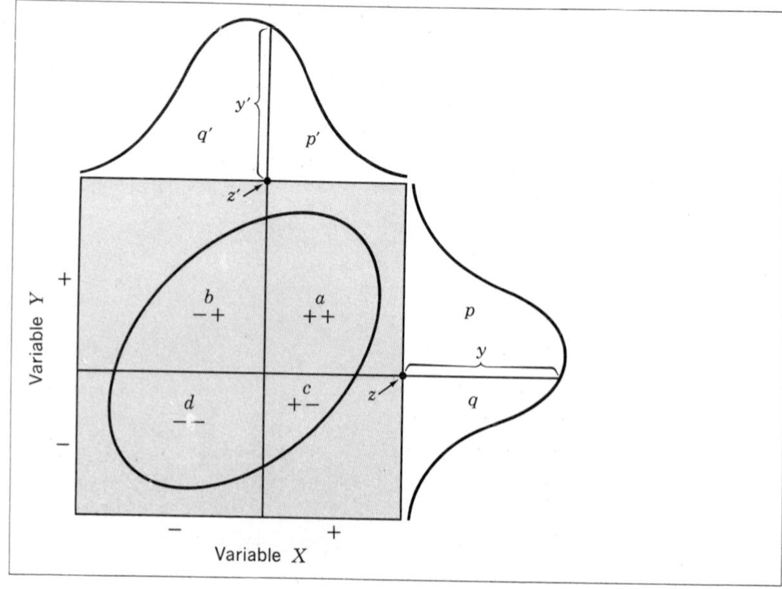

FIGURE 14.3 Theoretical normal bivariate distribution of cases divided into four categories by dichotomizing distributions on *X* and *Y*, illustrating the constants appearing in the equation for the tetrachoric *r*

THE EQUATION FOR THE TETRACHORIC *r*

The complete equation for the tetrachoric r is a long and complicated one, involving a series of terms many of which contain powers of r. With only the first few terms included, it reads

$$r_t + r_t^2 \frac{zz'}{2} + r_t^3 \frac{(z^2-1)(z'^2-1)}{6} + \cdots = \frac{ad-bc}{yy'N^2} \tag{14.16}$$

The symbols will be explained with reference to Table 14.5. The letters a, b, c, and d refer to the frequencies in the four cells of the fourfold table. r_t is given the subscript to indicate that it is a tetrachoric r. Numerically, under the appropriate conditions, it approximates a Pearson r.

In Table 14.5, it will be noted that the two-category distribution of all responses to question I is given in terms of proportions p' and q'. The distribution of all responses to question II is similarly given in terms of p and q. These proportions are required for finding the values for the y's and z's in formula (14.16). The symbols z and z' stand for the standard-score measurements on the base line of the unit normal-distribution curve at the points of division of cases in the two distributions. The symbols y and y' are ordinates corresponding to z and z' in the unit normal distribution.

METHODS OF ESTIMATING THE TETRACHORIC *r*

The solution for r_t by means of formula (14.16) is a formidable task and can be only an approximation at best. Consequently, a number of shortcuts have been devised for estimating it. One of these will now be described.

THE COSINE-PI FORMULA One approximation formula for r_t is known as the *cosine-pi formula*. In mathematical form, in terms of radians,

$$r_{\text{cos-pi}} = \cos\left(\pi \frac{\sqrt{bc}}{\sqrt{ad}+\sqrt{bc}}\right)$$

For most purposes, it is more convenient to deal in terms of degrees rather than radians. Substituting 180° for π, the equation reads

$$r_{\text{cos-pi}} = \cos\left(\frac{180°\sqrt{bc}}{\sqrt{ad}+\sqrt{bc}}\right) \quad \text{(Cosine-pi approximation to a tetrachoric } r\text{)} \tag{14.17}$$

By dividing the numerator and denominator by $\sqrt{bc}$, we obtain a formula that is more convenient for computation. It reads

$$r_{\text{cos-pi}} = \cos\left(\frac{180°}{1+\sqrt{ad/bc}}\right) \quad \text{(Cosine-pi formula in simpler form)} \tag{14.18}$$

where a, b, c, and d are the frequencies as defined in Table 14.5.

It is well to remember that a and d represent the like-signed cases (Yes-Yes and No-No) and that b and c represent the unlike-signed cases (Yes-No and No-Yes). When numbers are substituted, the expression within the parentheses reduces to a single number, which is an angle in terms of degrees of arc. The cosine of this angle is the estimate of r_t. The angle will vary between zero, when either b or c (or both) is zero, to 180 deg, when either a or d (or both) is zero. In the first case, when the angle is zero, the correlation is +1.00. When the product bc equals ad, the angle is 90 deg, the cosine of which is zero, and r_t is estimated to be .0.[1]

Applying the cosine-pi formula to the data in Table 14.5, we have

$$r_{\text{cos-pi}} = \cos\left(\frac{180°}{1 + \sqrt{(374)(203)/(167)(186)}}\right)$$

$$= \cos 70.24°$$

$$= .338$$

In this method, if the angle should prove to be between 90 and 180 deg, the correlation is negative. This can be anticipated by noting that the product bc is greater than ad. Angles over 90 deg are not listed in Table I in the Appendix. For an angle between 90 and 180 deg, deduct the angle from 180 deg, find the cosine of this difference, and give it a negative sign.

Table L in the Appendix provides a quick solution for $r_{\text{cos-pi}}$ to two decimal places. Only the ratio ad/bc (or its reciprocal bc/ad) need be known; compute whichever gives a value greater than 1.0. For the illustrative problem above, ad/bc equals 2.444. This lies between the given ratios 2.421 and 2.490, which indicates a correlation of .34.

LIMITATIONS TO THE USE OF THE COSINE-PI FORMULA It should be pointed out that formula (14.18) gives a very close approximation to r_t only when both variables X and Y are dichotomized at their medians.[2] As p and p' depart from .5, as p and p' differ from each other increasingly, and as r_t becomes very large, $r_{\text{cos-pi}}$ departs more and more from r_t and is systematically larger than r_t. For example, if $p = .5$ and $p' = .84$, when r_t is .79, $r_{\text{cos-pi}}$ is approximately .90. If both p and p' are within the limits of .4 to .6, however, when r_t is .50 the maximum discrepancy is approximately .02, and when r_t is .90 the maximum discrepancy is approximately .04, both in the direction of overestimation. In many situations we can control to a large extent the point of dichotomy and can see to it that p and p' are close to .5. When they are not, it would be best to use one of the graphic methods mentioned next.

GRAPHIC ESTIMATES OF TETRACHORIC r When a large number of tetrachoric r's must be computed, the Thurstone computing diagrams

[1] See Table I in the Appendix for cosines of angles.

[2] Bouvier, E. A., Perry, N. C., Michael, W. B., and Hertzka, A. F. A study of the error in the cosine-pi approximation to the tetrachoric coefficient of correlation. *Educational and Psychological Measurement,* 1954, **14,** 690–699.

can provide a considerable savings in labor.[1] These are highly recommended since they yield two-place accuracy with little effort after the fourfold table is reduced to the status of proportions throughout. From the computing diagrams, r_t for the two questions of Table 14.5 is estimated as +.34, which checks with previous estimates. Another graphic procedure has been published by Hayes.[2]

THE STANDARD ERROR OF A TETRACHORIC *r*

The tetrachoric *r* is less reliable than the Pearson *r*, being at least 50 percent more variable. It is most reliable when (1) *N* is large, as is true of all statistics, (2) r_t is large, as is true of other *r*'s, but also when (3) the divisions in the two categories are near the medians. The complete formula for estimating s_{r_t} is too complicated to be useful. But for testing the null hypothesis the simpler formula is[3]

$$s_{r_t} = \frac{\sqrt{pp'qq'}}{yy'\sqrt{N}} \quad \text{(Standard error for testing the hypothesis of zero correlation)} \tag{14.19}$$

For the 930 cases in the problem in Table 14.5,

$$s_{r_t} = \frac{\sqrt{(.582)(.602)(.418)(.398)}}{(.3905)(.3858)\sqrt{930}}$$
$$= .053$$

Since the obtained r_t, .34, is more than 2.6 times this standard error, we can reject at the .01 level the hypothesis that the two qualities represented by the two questions are uncorrelated in the population.

To attain the same degree of reliability in a tetrachoric *r* as in a Pearson *r*, one needs more than twice the number of cases in a sample. For estimating the *degree* of correlation by means of r_t, it is recommended that *N* be at least 200, and preferably 300. In smaller samples than these, even less than $N = 100$, a tetrachoric *r* can be used to test the null hypothesis.

SOME APPLICATIONS OF r_t TO BE AVOIDED

Many of the limitations of the tetrachoric *r* have already been pointed out. There are others that should not go unnoticed. It is well to avoid estimating r_t when the split in either *X* or *Y* is very one-sided—for example, a 95-5, or even a 90-10, division of the cases. The standard error is much larger in such situations.

Especially to be avoided is an attempt to estimate r_t when there is

[1] Chesire, L., Saffir, M., and Thurstone, L. L. *Computing diagrams for the tetrachoric correlation coefficient*. Chicago: University of Chicago Press, 1938.

[2] Hayes, S. P. Diagrams for computing tetrachoric correlation coefficients form percentage differences. *Psychometrika*, 1946, **11,** 163–172.

[3] For aids in estimating this s_{r_t} see Guilford, J. P., and Lyon, T. C. On determining the reliability and significance of a tetrachoric coefficient of correlation. *Psychometrika*, 1942, **7,** 243–249. Also, Hayes, S. P. Tables of the standard error of tetrachoric correlation coefficient. *Psychometrika*, 1943, **8,** 193–203.

TABLE 14.6 Illustrations of some unusual fourfold contingency tables in which computation of a tetrachoric r is questionable

A

200	0	200
90	110	200
290	110	400

B

80	110	190
150	0	150
230	110	340

C

85	15	100
95	105	200
180	120	300

a zero in only one cell. Table 14.6, *A* and *B*, illustrates two such examples. If r_t were computed for problem *A*, it would equal -1.0 (the zero is in cell *a*); if computed for problem *B*, r_t would equal $+1.0$. These results are in spite of the fact that nearly one-fourth of the cases belie the perfect correlations apparent by computation (90 cases out of 400 in *A* are out of line with the finding, and 80 cases in *B*).

These examples are perhaps somewhat rare, but zero frequencies are certainly not unheard of. Even scatters like that in *C* would probably give a false estimate of correlation. There is no zero, but there is an exceptionally small frequency (15) among much larger ones. In all three fourfold tables the distributions are such as to suggest nonlinear regressions if these broad categories were broken down into finer groupings. Such distributions as those in Table 14.6 are not proof of nonlinear regression, but they strongly suggest it. In general, a distribution in such a table should appear to be rather symmetrical along one diagonal axis or the other, depending upon whether the correlation is negative or positive. This holds true if the proportion p is somewhat near the proportion p', but even if they differ very much, asymmetry cannot be taken necessarily to mean curved regression.

The phi coefficient r_ϕ

When the two distributions correlated are genuinely dichotomous—when the two classes are separated by a real gap between them, and previously discussed correlational methods do not apply—we may resort to the *phi coefficient.* This coefficient was designed for so-called point distributions, which implies that the two classes have two point values or merely represent some qualitative attribute. Such a case would be illustrated by eye color (when limited to blue versus brown eyes), sex membership, living versus dead, and the like. The method can be applied, however, to data that are measurable on continuous variables if we make certain allowances for the continuity, with appropriate corrections and interpretations.

TABLE 14.7
Illustration of the correlation of attributes

		Intellectual status		
		Feeble-minded	*Normal*	*Both*
Marital status	*Married*	.204 (β)	.269 (α)	.473 (p)
	Unmarried	.296 (ϑ)	.231 (γ)	.527 (q)
	Both	.500 (q')	.500 (p')	1.000

The phi coefficient is a close relative of chi square, which is applicable to a wide variety of situations.

THE COMPUTATION OF PHI

To illustrate the use of phi, we shall use again some data that were previously employed with chi square (see Table 11.2). They are given here as we need them, in proportion form, in Table 14.7. The basic formula for computing the phi coefficient is

$$r_\phi = \frac{\alpha\vartheta - \beta\gamma}{\sqrt{pqp'q'}} \quad \text{(A computed phi coefficient)} \tag{14.20}$$

where the symbols correspond to the labeled cells in Table 14.7. The solution of r_ϕ for that table is

$$r_\phi = \frac{(.269)(.296) - (.204)(.231)}{\sqrt{(.473)(.527)(.5)(.5)}} = .1302, \text{ or } .13$$

THE RELATION OF PHI TO CHI SQUARE

Phi is related to chi square from a 2 × 2 table by the very simple equation

$$\chi^2 = Nr_\phi^2 \quad \text{(Chi square as a function of phi)} \tag{14.21}$$

and phi is derived from chi square by the equation

$$r_\phi = \sqrt{\frac{\chi^2}{N}} \quad \text{(Phi as a function of chi square)} \tag{14.22}$$

Applying formula (14.21) to the data of Table 11.2,

$$\chi^2 = (.412)(.016952) = 6.98,$$

(412 subjects)

which checks well with the solution for chi square in Chap. 11.

Since phi can be directly derived from chi square when the latter is applied to a 2 × 2 table, any of the formulas for chi square given for such tables in Chap. 11 apply toward its computation. Adapting formula (11.9) toward the computation of phi, we have

$$r_\phi = \frac{ad - bc}{\sqrt{(a+b)(a+c)(b+d)(c+d)}} \quad \text{(Phi computed from frequencies)} \qquad (14.23)$$

THE SPECIAL CASE OF PHI WHEN ONE DISTRIBUTION IS EVENLY DIVIDED When one of the distributions, let us say the one for which we use p' and q' as marginal proportions, is evenly divided ($p' = q' = .50$), the solution of r_ϕ is considerably simplified. The formula reads

$$r_\phi = \frac{\alpha - \beta}{\sqrt{pq}} \quad \text{(Phi from evenly divided proportions)} \qquad (14.24)$$

Applied to the data on marital status,

$$r_\phi = \frac{.269 - .204}{\sqrt{(.473)(.527)}} = .13$$

STATISTICAL SIGNIFICANCE OF PHI A test of the null hypothesis, of zero correlation, can be made through phi's relationship to chi square. If χ^2 is significant in a fourfold table the corresponding r_ϕ is also significant. The procedure, then, is to derive the χ^2 corresponding to the obtained r_ϕ by means of formula (14.21), then examine Table E in the Appendix to find whether the standard level of significance is met for 1 degree of freedom. In the data on marital status we find that a chi square of 6.96 is significant beyond the .01 point. Therefore the obtained phi of .13 is also significant.

AN EVALUATION OF THE PHI COEFFICIENT Phi is actually a product-moment coefficient of correlation that one would find by applying the Pearson-r formula to two parallel distributions of scores restricted to values 1 and 0. But such a limitation in scores often involves biased coefficients.

LIMITATIONS IN THE SIZE OF PHI Although phi can vary from -1.0 to $+1.0$, only under certain conditions can r_ϕ be as extreme as either of these normal limits. The reason for the common reduction in the size of phi is that a 2×2 table places some restrictions upon it. Only when p and p' (the two means are equal) can the r_ϕ coefficient equal 1.0. The greater the difference between means, the more reduction there is in the size of phi. For this reason, other coefficients treated in the next section are sometimes used instead of phi.

Correlations independent of the direction of measurement scales

Because of the biasing effects of differences between means in computing phi coefficients, several indices of relationship have been developed that are not affected by the direction of measurement. Means are in effect equated.

The first of these statistics was the *G* index of agreement,[1] to be symbolized consistently with other coefficients of correlation as r_g. It was designed particularly to correlate two persons, each of whom has answered *n* questionnaire items each by "Yes" or "No." Each item is regarded as providing a scale of measurement, with scores of 1 or 0 or of X or $-X$. For the sake of comparing two people as to similarity in their responding to the questionnaire, calling a "Yes" response 1 or 0 is really arbitrary, so we assume that each person is scored twice, reversing the scale for the item. The effect is that the means for all individuals are .5. Differences in means are thus eliminated. It may be recalled that under this condition a phi coefficient is unbiased.

THE *G* INDEX OF AGREEMENT, r_g

The r_g index of agreement is very easily computed. In a 2 × 2 contingency table, it is based upon the proportion of agreeing cases (α and ϑ) as compared with the proportion of disagreeing cases (β and γ), as those symbols are defined in Table 14.7 (p. 317). One formula is

$$r_g = 2(\alpha + \vartheta) - 1 \qquad \text{(The } G \text{ index of agreement)} \qquad (14.25)$$

Another formula, which uses all four proportions and shows more clearly what is being done, is

$$r_g = (\alpha + \vartheta) - (\beta + \gamma) \quad \text{(Alternative formula for the } G \text{ index)} \quad (14.26)$$

Applying either formula to the data in Table 14.7 gives an r_g of .13, which coincides with the value of phi in this particular case because the means p and p' are nearly equal. When the means are not equal, r_g and r_ϕ are likely to differ and sometimes they are very far apart.

A convenient computing formula for r_g that does not require finding the proportions is

$$r_g = \frac{(a + d) - (c + d)}{N} \qquad \text{(A } G \text{ index computed from cell frequencies)} \qquad (14.27)$$

where the symbols, *a*, *b*, *c*, and *d* are defined as in Table 14.5.

THE SIGNIFICANCE OF A *G* INDEX There is no known standard error for r_g but Lienert[2] has suggested a way of testing the null hypothesis, that the

1 Holley, J. W., and Guilford, J. P. A note on the *G* index of agreement. *Educational and Psychological Measurement,* 1964, **24,** 749–754.

2 Lienert, G. A. Note on tests concerning the *G* index of agreement. *Educational and Psychological Measurement,* 1972, **32,** 281–288.

population correlation is 0. He assumes that we have two categories of cases—agreements and disagreements—and that the observed cases are randomly and independently assigned to the two categories.[1] Such is the situation appropriate for a sign test, as described in Chap. 12. When the population correlation is zero, the sampling distribution of $(a + d)$ is assumed to be binomial with a mean of Np. With $p = .5$, the mean is $N/2$. Assuming a binomial distribution, the significance of the deviation of an obtained $(a + d)$ from the mean can be determined by the use of Table N in the Appendix.

If N is as large as 30, it may be assumed that the sampling distribution of $(a + d)$ approaches the normal form very closely. The standard error of this distribution is given by $\sqrt{Npq}$, which in this particular case is $\sqrt{N/4}$, or $\sqrt{N}/2$. The z test is made by using the formula

$$z = \frac{(a + d) - .5N}{.5\sqrt{N}} \qquad \text{(A } z \text{ deviate for testing the null hypothesis)} \qquad (14.28)$$

With certain simplifications this equation becomes

$$z = r_g\sqrt{N} \qquad \text{(}z\text{ ratio for testing departure of a } G \text{ index from zero)} \qquad (14.29)$$

For the data in Table 14.8, which come from Table 11.2, in which $r_g = .131$ and $N = 412$, z is calculated to be 2.66, a very significant deviation.

When N is between 20 and 30, the normal-curve approximation may still be used if there is correction for continuity. The correction is simply made by deducting .5 from the absolute value of the numerator in formula (14.28).

If samples are sufficiently large, the significance of a difference between two r_g values may be tested by using the difference between the two $(a + d)$ frequencies in the contingency tables, as one would do for testing the significance between any two frequencies in two samples (see Chap. 9). Lienert has provided other tests, including one for the homogeneity of a number of G indices.[2]

GENERALIZATION OF THE G INDEX

The G index has proven to be of considerable value for indicating similarities between individuals and also the conformity of persons to classes of individuals, the latter especially in factor-analytic studies. Consequently, a number of variations of this type of correlation have been developed, some of which apply to other than dichotomous scores.

TRANSFORMATION TO DICHOTOMOUS SCORES Suppose we have data like those in the following lists, which include evaluations, perhaps in the

[1] A restriction applies when the G index indicates correlation between trait variables. Each person's responses are not mutually independent.

[2] Lienert, op. cit.

form of ratings, of two individuals in ten different qualities:

X:	1	5	5	3	4	6	2	4	3	3
Y:	2	6	2	1	5	5	5	6	3	4

In order to use the method described next, it is necessary that the qualities and the scales be bipolar. That is, with traits like introversion-extraversion, honest-dishonest, friendly-hostile, and the like, each may be assumed to have a neutral point near the middle of a continuum. One could, of course, compute an ordinary Pearson r between two such parallel sets of values, but this would be inappropriate because for correlating two individuals the scores should be *ipsative*. The ipsative scores for a person are distributed about the mean for that particular individual, not about the mean of the population, as is true of *normative* scores. When the G index is applied to correlating individuals, in effect, the scores are ipsatized; means of persons become equal.

The computation of r_g utilizes dichotomous scores of 1 and 0. We may take advantage of the principle of the G index by dichotomizing the scores as given above. Since there is a neutral point for each scale, we can give a value of 1 to all scores above that point and 0 to all those below that point. In the lists of scores given above, scores of 4-6 receive a new value of 1 and scores of 1–3 a new value of 0. The transformed scores are therefore as follows:

X_d:	0	1	1	0	1	1	0	1	0	0
Y_d:	0	1	0	0	1	1	1	1	0	1

In these data the number of like-signed pairs of scores $(a + d)$ in the contingency table is 7, from which $r_g = .40$.

COHEN'S COEFFICIENT r_c A disadvantage of the simple method just described is that it does not make use of all the available information in the data. A method that utilizes all the given score values was suggested by Cohen.[1] It also assumes a neutral point in each scale. For a six-point scale like the ones used for the preceding illustration, the midvalue would be 3.5. Here we see the advantage of using an even number of categories in the scales; no deviations from the midpoint will be exactly zero. Treating the midpoint as the mean of both X and Y distributions, the correlation is computed by a formula very similar to that for the Pearson r. It is

$$r_c = \frac{\Sigma x_n y_n}{\sqrt{\Sigma x_n{}^2 \Sigma y_n{}^2}} \qquad \text{(Cohen's general index of agreement)} \qquad (14.30)$$

where $x_n = X - X_n$ (X_n being the neutral point for scale X)
$y_n = Y - Y_n$ (Y_n being the neutral point for scale Y)

1 Cohen, J. A profile similarity coefficient invariant over variable reflection. *Psychological Bulletin*, 1969, **71**, 281–284.

For the 10 pairs of scores given above, $\Sigma x_n y_n = 10.0$, $\Sigma x_n^2 = 20.5$, and $\Sigma y_n^2 = 30.5$, from which $r_c = .40$.

Other generalizations of the G index take care of cases in which not all measurement scales have the same number of categories or when certain scores are given extra weights in determining the correlations of persons because of known higher validity for those variables in discriminating individuals and hence also in discriminating types of individuals.[1]

Partial correlation

THE MEANING OF PARTIAL CORRELATION

A partial correlation between two variables is one that nullifies the effects of a third variable (or a number of other variables) upon both the variables being correlated. The correlation between height and weight of boys in a group where age is permitted to vary would be higher than the correlation between height and weight in a group at constant age. The reason is obvious. Because certain boys are older, they are both heavier and taller. Age is a factor that enhances the strength of correspondence between height and weight. With age held constant, the correlation would still be positive and significant because at any age, taller boys tend to be heavier.

If we wanted to know the correlation between height and weight with the influences of age ruled out, we could, of course, keep samples separated and compute r at each age level. But the partial-correlation technique enables us to accomplish the same result without so fractionating data into homogeneous age groups. When only one variable is held constant, we speak of a *first-order partial correlation.* The general formula is

$$r_{12.3} = \frac{r_{12} - r_{13}r_{23}}{\sqrt{(1 - r_{13}^2)(1 - r_{23}^2)}} \qquad (14.31)$$

(First-order partial coefficient of correlation)

In a group of boys aged twelve to nineteen, the correlation between height and weight (r_{12}) was found to be .78. Between height and age, $r_{13} = .52$. Between weight and age, $r_{23} = .54$. The partial correlation is therefore

$$r_{12.3} = \frac{.78 - (.52)(.54)}{\sqrt{(1 - .52^2)(1 - .54^2)}}$$
$$= .69$$

1 See, for example, Holley, J. W. On the generalization of the G index of agreement, G_o, for use with ordinal scores. *Scandinavian Journal of Psychology,* 1976, **17,** 149–152. Vegelius, J. *On various G index generalizations and their applicability within the clinical domain.* Doctoral dissertation, University of Uppsala, 1976.

With the influences of age upon both height and weight ruled out or nullified, then, the correlation between the two is .69.

As another example with three variables, the correlation between strength and height (r_{41}) in this same group was .58. The correlation between strength and weight (r_{42}) was .72. Although there is a significantly high correlation between strength and height, we wonder whether this is not due to the factor of weight going with height rather than to height itself. Accordingly, we hold weight constant and ask what the correlation would be then. Will boys of the same weight show any dependence of strength upon height? The correlation is given by

$$r_{41.2} = \frac{.58 - (.72)(.78)}{\sqrt{(1 - .72^2)(1 - .78^2)}} = .042$$

By partialing out weight, it is found that the correlation between height and strength nearly vanishes. We conclude, therefore, that height *as such* has no bearing upon strength and that only by virtue of its association with weight does it show any correlation at all.

SECOND-ORDER PARTIALS

When we hold two variables constant at the same time, we call the coefficient a *second-order partial r*. The general formula is

$$r_{12.34} = \frac{r_{12.3} - r_{14.3}r_{24.3}}{\sqrt{(1 - r_{14.3}^2)(1 - r_{24.3}^2)}} \tag{14.32}$$

(Second-order partial coefficient of correlation)

In using this formula, the subscripts should be modified to suit the choice of variables. Here we are assuming that we want to know the correlation that would occur between X_1 and X_2 with the effects of X_3 and X_4 eliminated from both. It is clear that this formula requires the previous solution of three first-order partial-correlation coefficients.

As an example of this partial, we may cite the case of correlation between strength and age, with height and weight held constant. We ask whether in a group of boys of the same height and weight, the older boys would be stronger. The raw correlation between age and strength was .29. The second-order partial also turned out to be .29. This means that it seemingly makes no difference whether or not we allow height and weight to vary; the relation between age and strength is the same within the range examined.

SOME SUGGESTIONS CONCERNING PARTIAL CORRELATION

Needless to say, unless the assumptions necessary for computing the Pearson r's involved are fulfilled, there is little excuse for using them as the basis for computing partial correlations. Rectilinearity of relationships is assumed. In a real sense, the use of partial correlation is a statistical substitute for experimental controls. Perhaps the most common instances in which it is useful are those involving the partialing out of such variables as chronological age and intelligence. It should be pointed out, however, that where the relationship is nonlinear, only the linear component of the regression is partialed out. Since partial correlation deals with three or more variables simultaneously, it is classified as a multivariate method, and it is related to the methods of multiple correlation and factor analysis, to be considered later. (See Chaps. 16 and 18.)

SIGNIFICANCE OF AN OBTAINED PARTIAL *r* The standard error of a partial coefficient of correlation takes into account the number of degrees of freedom. For testing the hypothesis that the correlation is zero, the formula is

$$(\text{When } \rho_{12.34.\ldots m} = 0) \qquad s_{r_{12.34\ldots m}} = \frac{1}{\sqrt{N - m}} \tag{14.33}$$

(SE for testing the hypothesis that a partial ρ is zero, where m is the total number of variables).

Other special problems

It is clear that the size of a coefficient of correlation depends to some extent upon how we compute it. More importantly, coefficients computed between the same two variables by the same procedure will vary not only from sample to sample but also from population to population. In reporting coefficients of correlation, a writer should state all the pertinent conditions that bear upon the size of the obtained coefficients, and any reader should be aware of the relevance of those conditions. A few of the more common circumstances that affect the sizes of r's will now be reviewed briefly.

THE VARIABILITY OF THE CORRELATED VARIABLES The size of r is dependent upon the variability of measured values in the correlated sample. The greater the variability, the higher the correlation, other things being equal. If the variability of either X or Y were zero, the correlation would be zero—the limiting case. Often we know the correlation between some predictive index, such as aptitude-test score, and achievement or some vocational criterion of success as derived from one group of individuals, but we shall often be applying the same index to groups with different ranges of ability, larger or smaller. What will be the effectiveness of predictions in the new groups?

In the selection of personnel by means of tests, research on selective instruments is constantly beset with this very practical problem.

New tests are put into use in the selection of personnel, and they correlate substantially with tests that are already being used in selection. The result is that those who go into training represent only a higher segment of the population from which selection is to be made by the new test. The validity of a test can be estimated only for this higher segment of restricted range. And yet it is the validity in the total tested population that is important to know, for it is that validity which indicates the full selective value of the test. The coefficient of validity is almost invariably smaller than it would be in an unrestricted group.

CORRECTION FOR RESTRICTION OF RANGE

One kind of restriction is produced by selection on the basis of the variable X_1. There is knowledge of standard deviations on X_1 for both restricted and unrestricted groups, and the correlation r_{12} is known in the restricted group.[1] The correlation r_u for the unrestricted group is estimated by

$$r_u = \frac{r_c(S_u/S_c)}{\sqrt{1 - r_c^2 + r_c^2(S_u^2/S_c^2)}} \qquad (14.34)$$

(Correlation estimated for an unrestricted range from one known in a restricted range)

where r_c = correlation within the restricted or curtailed group
S_c = standard deviation on X_1, on which restriction occurs
S_u = standard deviation on the same variable in the unrestricted group

To take an example, suppose that the selection test (X_1) correlates .30 with the criterion of achievement within the group selected for training on the basis of the same test. Suppose further, that the standard deviation in the unrestricted group (S_u) is 20 and that the standard deviation in the restricted group (S_c) is 10. The solution is

$$r_u = \frac{30(20/10)}{\sqrt{1 - .09 + .09(20^2/10^2)}}$$
$$= .53$$

The reader may be somewhat surprised at the rather radical change in correlation thus illustrated. To show that such a change is not unreasonable, some data may be cited from results obtained in the AAF Aviation Psychology Program during World War II.[2] An experimental group of more than 1,000 pilot students had been permitted to enter training without any selection whatever on the basis of the qualifying examination (which normally eliminated about 50

1 Other cases of restriction of range are treated in the fourth edition of this volume.

2 Thorndike, R. L. Research problems and techniques. *AAF Aviation Psychology Research Program Reports,* No. 3. Washington: GPO, 1947.

TABLE 14.8 Validity coefficients for selective tests and a composite score for the selection of pilot students with and without restriction of range

Variable	*Correlation in the total group (N = 1,036)*	*Correlation in the selected highest 13 percent (N = 136)*
Pilot stanine	.64	.18
Mechanical principles	.44	.03
General information	.46	.20
Complex coordination	.40	−.03
Instrument comprehension	.45	.27
Arithmetic reasoning	.27	.18
Finger dexterity	.18	.00

percent of the applicants) or the classification tests (which ordinarily eliminated many more). It could be determined, then, how the pilot stanine (a composite aptitude score) correlated with the graduation-elimination criterion in an unrestricted range after primary-school training. For an illustration of the effects of restriction, a high segment of the cases was selected to form a sample with restricted range. The results are given in Table 14.8, for the instance in which rather high, but not unknown, selection of the top 13 percent occurred. It can be seen that there were substantial correlations in the unrestricted sample but that the correlations within the restricted group often shrank close to zero. There was shrinkage not only in the correlation with the stanine but also in correlations of the criterion with tests that contributed to the stanine. On the whole, those tests which correlated highest with the stanine lost most, as one should expect.

EVALUATION OF THE CORRECTIONS FOR RESTRICTION It should be repeated that the problem of restriction in range is important and that if one wishes to avoid drawing wrong conclusions, when a substantial amount of selection has been made, one should apply correction procedures. Had the second (restricted) set of coefficients in Table 14.8 been taken seriously, it might have been concluded that, without knowledge to the contrary, the stanine and the tests were of near-zero validity.

It should be remembered that the formulas rest upon the assumption of normal distribution of the population on the variables used, and the Pearson product-moment r is presupposed. The use of the biserial r

or tetrachoric r as an estimate of the Pearson r in a restricted range raises considerable question when selection is severe. Experience tends to show, however, that when the biserial r is used as the validity coefficient, the formulas tend to underestimate the unrestricted correlation. The standard errors for these corrected coefficients are unknown, but it is probable that they are larger than those for Pearson r's of comparable size.

CORRELATIONS IN HETEROGENEOUS SAMPLES

Studies of the validity of tests have frequently been faulty from a number of standpoints. The use of school marks as criteria of success in training has a number of faults, one of which will concern us here.

One procedure that works against accurate estimates of validity is indiscriminately pooling marks from different subjects and from different instructors and treating them as if they were all the same kind of coin. A cursory inspection of grade distributions in any school will show that marks are by no means of constant value when obtained from different sources. Means and variances differ from set to set of data. The data are *heterogeneous.* Much of this is caused by variations in ideas about marking from instructor to instructor. This variation among sets of marks when they are collectively correlated with other measures is very likely to affect the apparent amount of correlation.

As an example, in six sections of freshman English, *within* sections the correlation between quiz averages for the semester and a final comprehensive examination ranged from .63 to .92, with an overall correlation within sections, *when intersection differences had been eliminated,* of .83. Yet when the six sections were combined, *with intersectional differences left in,* the correlation was reduced to .71. It was interesting to find that *between* sections the correlation was −.17, which means that there was a very slight tendency for sections with average lower achievement to be given a higher average quiz mark! This fact accounts for the reduction in correlation from .83 to .71 when sections were combined.[1]

Diagram II in Fig. 14.4 shows the situation just described, in somewhat exaggerated form. Diagram II is best understood by contrasting it with diagram I. In the latter we have a homogeneous combination of four subsamples drawn from the same population. The amount of correlation between X and Y within each subsample is indicated by the shape of a smaller ellipse. All the ellipses are of about the same shape, indicating about the same degree of correlation of X and Y *within* groups. The dots indicate the means of Y and X within each subsample. If we combine the four samples, we obtain a distribution described approximately by the large faint ellipse. Note in dia-

1 Further discussion of "within" versus "between" correlations when groups are combined will be found in Lindquist, E. F. *Statistical analysis in education research.* Boston: Houghton Mifflin, 1940.

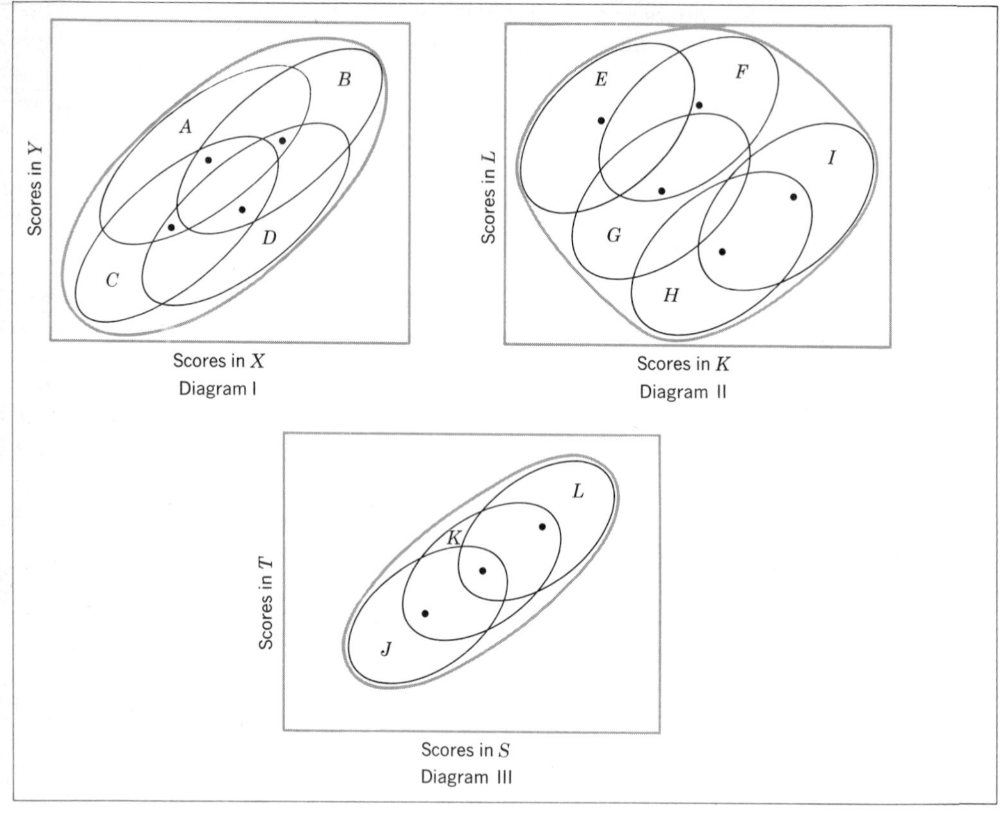

FIGURE 14.4 Bivariate distributions of cases in homogeneous and heterogeneous subsamples

gram I that the proportions of the large ellipse are about the same as those of each small ellipse, indicating the same level of correlation within the composite distribution as within each subsample. Note, also, that the distribution of the four means forms roughly an ellipse of similar proportions. If the correlation between means of *Y* and means of *X* differs from that within subsamples, the correlation of *X* and *Y* in the composite sample will differ from that within subsamples.

In diagram II of Fig. 14.4 we have a very different situation. While with each subsample the correlation between *K* and *L* is approximately the same, the subsamples did not arise from the same population so far as means are concerned. An ellipse drawn to enclose just the dots would slant in the direction that would ensure a negative correlation between means of *K* and means of *L*. The effect of this can be seen in the faint line enclosing all subsamples. Its form suggests approximately zero correlation. Such situations can really exist. In general practice, if it is doubtful that subsamples arise by random sampling from the same population, it would be best to compute correlations within subsamples separately and to average them, thereby obtaining a single value, or to apply equivalent procedures which we shall not take the space to describe here.[1] The hypothesis of homogeneity of samples can be examined by means of *t* tests or *F* tests as described in Chaps. 9 and 13.

THE CORRELATION OF AVERAGES

It was stated in an earlier chapter in connection with tests of significance of differences between statistics (Chap. 9) that the correlation between averages of samples is equal to the correlation between individual pairs of measurements. *This statement assumes random samples from a homogeneous population.* Diagram I in Fig. 14.4 illustrates this kind of situation and shows how an *r* obtained within one sample can be used as an estimate of a correlation between means. Diagram II shows how a correlation coefficient obtained within a single sample might be very misleading about the amount of correlation between means. This example shows an instance in which the correlation between means is decidedly lower than that within samples, one perhaps even reversed in sign.

The correlation between means could also be higher than that within samples, as diagram III shows. An example of this would be the correlation between IQ and salary. Correlating *individuals,* we should find some positive correlation, but because of great variations in salary at any single IQ value, the correlation might not be very high. If we divided people into sets according to vocation and correlated *average* IQ with *average* salary, the coefficient would probably be very high. This is because people of different IQ levels gravitate to certain occupations, and occupations as such have established characteristic salary scales. Other factors that make for individual differences in salary *within* occupations are thus minimized in importance. The sampling is biased the moment we divide groups along occupational lines.

AVERAGING COEFFICIENTS OF CORRELATION

As stated earlier, one solution to the problem of correlations in some heterogeneous samples is to estimate the correlation between *X* and *Y* within each subsample and then average the coefficients in order to obtain a single estimate of the population correlation. This would presumably describe the relation between *X* and *Y* throughout the composite sample, free from whatever sampling biases there may have been in segregating the subsamples. Before averaging coefficients, however, we must make the assumption that the several *r*'s did arise by random sampling from the same population—same with respect to the degree of correlation. It should go without saying, also, that we have correlated the same variables in all samples.

1 See Lindquist, op. cit.

There are several procedures sometimes used in averaging *r*'s. Coefficients of correlation are not values on a scale of equal metric units; they are index numbers. Differences between large *r*'s are actually much greater than those between small *r*'s. If the few sample *r*'s to be averaged, however, are of about the same value and if they are not too large, a simple arithmetic mean will suffice. If the *r*'s differ considerably in size and if they are large, it is best to use transformations to Fisher's *Z* coefficients. This procedure is illustrated in Table 14.9. It involves transforming each *r* into a corresponding *Z* (Table H in the Appendix may be used for this purpose), finding the arithmetic mean of the *Z*'s, and finally transforming the mean *Z* back to the corresponding mean *r*.

The results of Table 14.9 show differences to be expected in the use of an arithmetic mean of *r*'s and of corresponding *Z*'s. Samples *A* and *B* have the same range of *r*'s, those in *B* being merely .30 greater than those in *A*. In sample *A*, agreement is perfect in the results from the two methods. In sample *B*, the mean *r* by the *Z* method is .01 higher (.77, as compared with .76). In samples *C* and *D*, there is much more spread in the *r*'s averaged. For the *r*'s of moderate size, in sample *C*, the *Z* method gives a result only .01 greater than the simple mean of *r*'s. In the high coefficients, however, the difference is about .05.

There is serious question whether *r*'s differing as much as these would satisfy the belief that they came from the same population by random sampling and hence would be candidates for averaging. When a few *r*'s do satisfy this belief, the chances are that any discrepancy between a simple mean of *r*'s and an average obtained by the *Z* method would be small as compared with the standard error of *r*. If the *r*'s did come from the same population, a mean of several would be a much more reliable estimate of population correlation. With the

TABLE 14.9 Demonstration of averaging coefficients of correlation when *r*'s differ in range and in size

	Sample A		*Sample B*		*Sample C*		*Sample D*	
	Mean of r	*Z method*	*Mean of r*	*Z method*	*Mean of r*	*Z method*	*Mean of r*	*Z method*
	.45	.48	.75	.97	.35	.37	.65	.78
	.50	.55	.80	1.10	.55	.62	.85	1.26
	.42	.45	.72	.91	.68	.83	.98	2.30
	.38	.40	.68	.83	.50	.55	.80	1.10
	.55	.62	.85	1.26	.58	.66	.88	1.38
Σ	2.30	2.50	3.80	4.75	2.66	3.03	4.16	6.82
$\bar{X}_z$		.50		1.014		.606		1.364
$\bar{X}_r$	.46	.46	.76	.77	.532	.543	.832	.877

requirements satisfied, we could sum degrees of freedom from the different subsamples to represent the degrees of freedom of the mean r and interpret its significance accordingly.

WEIGHTING COEFFICIENTS IN AVERAGING One more requirement should be mentioned, particularly if the last operation, combining degrees of freedom, is to be carried out—that is, weighting the obtained r's in averaging them. The weight for each sample is its number of degrees of freedom $(N - 2)$. In using the Z method, the weights are applied to the Z's. The weight to be applied to a Z is its number of degrees of freedom, $(N - 3)$.

THE CORRELATION OF PARTS WITH WHOLES

We may sometimes want to correlate a part measurement, such as a part of a test battery or a test item, with the whole of which it is a part. The variance of the total is in part made up of the variance of the component, and that fact alone introduces some degree of positive correlation. The greater the relative contribution to the total variance made by the component, the more important is this "spurious" factor. Thus, a part-whole correlation is in part a correlation of a variable with itself. It is possible in a particular instance for the part to be totally *uncorrelated* with the remaining parts and yet correlated with the total. If it is negatively correlated with the remaining parts, it will be less negatively correlated with the total.

If each part contributes about the same amount of variance to the total or if the part is one of a great many, so that its proportion of contribution is relatively small, we can compare correlations between parts and total with some confidence that they are compared on a very similar basis. But if these conditions do not obtain, we should do better to correlate each part with a composite of all other parts. When such a composite is unknown or is hard to obtain, we can still estimate the correlation by means of the formula

$$r_{pq} = \frac{r_{tp}S_t - S_p}{\sqrt{S^2 + S_p^2 - 2r_{tp}S_tS_p}} \tag{14.35}$$

(Correlation of part with a remainder, when correlation of part with total is known)

where p = part score
t = total score
$q = t - p$; in other words, the total with the part excluded

In the correlation of test items each with the total score of the test of which it is a part, it is important to know about how much a part would correlate with the total when there is really no relationship at all. We can estimate this, but only under the conditions that each part has the same variance and there is zero intercorrelation among all parts. Under these special conditions the average amount of correlation of a part with the total is given by the equation

$$r_{pt} = \frac{1}{\sqrt{n}}$$ (Average correlation of a number of parts, of equal variance and zero intercorrelation, with their total) (14.36)

in which $n =$ number of parts.

INDEX CORRELATION

Correlations of index numbers, such as IQs from different tests, can give rise to what are sometimes called *spurious index correlations*. When there is a range of chronological ages in the data, this common factor in the denominator of the MA/CA ratio biases the correlation upward. There would be some positive correlation between IQs even if the MAs correlated zero. Table 14.10 shows by means of a fictitious example how this phenomenon works. For eight children who differ in chronological age from five to nine inclusive, MA values on two different tests are given. These are obviously selected children, since their mental ages are all either seven or eight. Note, however, how the IQs spread, from 160 through 78. The spread in IQs is almost entirely due to the spread in chronological ages, which should not be permitted to contribute to measures of intelligence. Where the mental ages correlate zero, the IQs correlate .92; this is due entirely to the CA variable.

CORRECTION IN r FOR ERRORS OF GROUPING

If, in computing a Pearson r by means of grouping data in class intervals, a small number of classes either way has been used, the estimate of correlation is lowered to some degree. In the limiting case, of two classes each way, the computed r is about two-thirds of the r that would have been obtained had there been no grouping. When the number of intervals is 10 both ways, r is about 3 percent underes-

TABLE 14.10 Demonstration of how index numbers may acquire a high degree of correlation because of a common denominator: an extreme case

Child	*CA*	*MA* I	*MA* II	*IQ* I	*IQ* II
A	5.0	7	8	140	160
B	5.5	8	8	145	145
C	6.0	7	7	117	117
D	6.5	8	7	123	108
E	7.5	8	8	106	106
F	8.0	7	8	88	100
G	8.5	8	7	94	82
H	9.0	7	7	78	78

Correlation between mental ages I and II = .00

Correlation between IQ's I and II = .92

TABLE 14.11 Correction factors for errors of grouping in the computation of Pearson's r when distributions are normal and midpoints of intervals stand for cases in the intervals

No. of intervals	*2*	*3*	*4*	*5*	*6*	*7*	*8*	*9*	*10*	*11*	*12*	*13*	*14*	*15*
Correction factor	**.816**	**.859**	**.916**	**.943**	**.960**	**.970**	**.977**	**.982**	**.985**	**.988**	**.990**	**.991**	**.992**	**.994**
Squared correction factor	**.667**	**.737**	**.839**	**.891**	**.923**	**.941**	**.955**	**.964**	**.970**	**.976**	**.980**	**.983**	**.985**	**.987**

timated. For any number of classes in X or in Y, we can correct for the error of grouping by dividing r by a constant corresponding to that number of classes. The correction is called for because errors of grouping yield overestimates of the standard deviations.

Table 14.11 supplies the list of constants given by Peters and Van Voorhis to be used in making corrections in r.[1] Correction is made for the number of categories or intervals in Y as well as in X. The correction factors are used in the following manner. Suppose that we have an obtained r of .61 in a problem with eight intervals in X and nine in Y. The correction factors for these numbers of intervals are .977 and .982, respectively. The correction is made by dividing the obtained r by the product of the two correction factors. In terms of a formula,

$$r_c = \frac{r}{c_x c_y} \quad \text{(Coefficient of correlation corrected for coarse grouping)} \qquad (14.37)$$

in which c_x and c_y are the correction factors for variables X and Y, respectively, based upon the number of class intervals in each. Applied to the correlation of .61 with eight and nine categories in X and Y,

$$r_c = \frac{.61}{(.977)(.982)} = .626 \text{ (or .63)}$$

When there are the same number of intervals in both X and Y, the correction factor is the same for both, and the factor squared would be called for in the denominator of formula (14.37). The factors squared are given for this purpose in Table 14.11.

It should be remembered that the correction factors given in Table 14.11 are designed especially for the situation in which the midpoint of an interval is the index number for cases in that interval, the intervals are equal in size, and the distributions are normal. For other, less common situations, see the reference below.[2]

1 Peters and Van Voorhis, op. cit., p. 398.
2 Peters and Van Voorhis, op. cit.

EXERCISES

1 By the rank-difference method:
a. Compute the correlations between the first 20 pairs of scores for variables I and II and for variables V and VI in Data 6*A* (p. 92).
b. Interpret your results and comment on the question of statistical significance of the two coefficients.

2 For Data 14*A* compute and interpret:
a. $r_{\eta_{yx}}$ and $r_{\eta_{xy}}$.
b. $s_{\eta_{yx}}$.
c. The *F* ratio for significance of eta.
d. The *F* ratio for a test of linearity (assume $r_{xy} = .75$).

DATA 14*A* Two-way tabulation of relationship between chronological age and score on a history knowledge test.

Y (*Score on a history test*)	*X* (*Chronological age*) 8	9	10	11	12	f_y
86–100			1	10	4	15
71–85		1	6	13	10	30
56–70	1	4	12	7	5	29
41–55	3	6	4			13
26–40	6	5				11
11–25	2					2
f_x	12	16	23	30	19	100

3 For Data 15*A* (p. 366):
a. Compute a correlation ratio for the prediction of *Y* from *X*.
b. Find the standard error for the obtained eta.
c. Compute a standard error of estimate.
d. Apply the *F* test of linearity, with r_{xy} taken as .67

4 In a biology course practicum examination, 80 students were given passing grades and 70 students were failed. The passing students had a mean of 88.11 in the final examination and the failing students a mean of 70.98. The standard deviation for the two groups combined was 42.36.
a. Compute r_b.
b. Determine whether or not r_b is probably different from a population correlation of zero.

5 Assuming that "pass-fail" represents a genuine dichotomy, using the data from the preceding exercise,
a. Compute r_{pbi}.
b. Determine whether this r_{pbi} is probably different from zero.
c. Estimate r_b from r_{pbi}.

6 For Data 14*B*:
a. Estimate a tetrachoric coefficient of correlation.
b. Determine whether the correlation is probably significantly different from zero.
c. If Thurstone's diagrams or any other computing aid is available, find another estimate of r_t for the same data.

DATA 14*B* Relationship between passing or failing in college and being above or below the median in high-school graduating classes

Status in high-school class	*Passing in all college courses*	*Failing in one or more courses*	*Total*
Above the median	340	40	380
Below the median	70	140	210
Total	410	180	590

7 Preparatory to computing an estimate of r_t, reduce each of the following tabulations to a fourfold table:
a. The frequencies in Data 11*B* (p. 211)
b. The frequencies in Table 6.4 (p. 86)
c. Data 15*A* (p. 366)

8 For Data 11*A* (p. 210):
a. Compute r_ϕ, using the different formulas given in this chapter.
b. From this obtained r_ϕ estimate a chi square. Compare with the chi square obtained in connection with Exercise 2 in Chap. 11.
c. Estimate a Pearson *r* also by computing a cosine-pi *r*. Compare the result with r_ϕ.

9 For Data 14C:
a. Compute r_ϕ and test for its significance.
b. Compute r_g and test for its significance.
c. Compare the results for *a* and *b*.

DATA 14C Two-way tabulation of 100 respondants to two questions on a census questionnaire

Response to Question 1	*Response to Question 2* No	Yes	
Yes	20	60	80
No	10	10	20
	30	70	100

10 a. Compute an r_g for dichotomized scores from Data 14*D*.
b. Compute r_c for the same data.

DATA 14*D* Ratings of two persons with respect to ten traits on 6-point scales

Person I:	3	0	5	2	3	0	4	4	1	2
Person II:	1	3	4	0	5	2	5	5	3	0

11 a. Compute the following partial *r*'s for Data 16*A* (p. 402): $r_{34.2}$, $r_{41.2}$, and $r_{51.2}$. Interpret the results.
b. Which of these coefficients has the most psychological or practical meaning? Which the least? Explain.

12 A stenographic-aptitude test (variable 7) was given to 100 applicants. The distribution of scores obtained had a mean $\bar{X}_1 = 55.0$ and a standard deviation S_1 of 16.0. Thirty-five of the applicants with the highest scores were hired. For these scores $\bar{X}_1 = 65.0$ and $S_1 = 10.0$. Their scores had a correlation $r_{12} = .40$ with a performance criterion (variable 2). Estimate the validity of the test for the entire group of applicants.

ANSWERS

1 a. $r_{\rho_{12}} = .18$ (for parts I and II); $r_{\rho_{56}} = .65$.
b. From Table K in the Appendix, $r_{\rho_{12}}$ is insignificant; $r_{\rho_{56}}$ is significant beyond the .01 level.

2 a. $r_{\eta_{yx}} = .79$; $r_{\eta_{xy}} = .77$; b. $s_{yx} = 11.63$; c. $F = 39.98$, 29.06; d. $F = 5.19$, 2.75.

3 a. $r_\eta = .660$; b. $s_{r_\eta} = .053$; c. $S_{yx} = 5.06$; d. $F = .12$.

4 a. $r_b = .25$; b. $s_{r_b} = .102$.

5 a. $r_{pbi} = .20$; b. $t = 2.72$; c. estimated $r_b = .25$.

6 a. $r_{\text{cos-pi}} = .82$; b. $s_{r_t} = .041$ (for $\rho_t = 0$).

7 a.

	Yes ?	+No
D	117	133
ND	141	109

b.

	55–79	80–99
.140–.189	9	8
.100–.139	15	20

c.

	0–14	15–23
21–38	16	45
6–20	39	13

8 a. $r_\phi = .14$; b. $\chi^2 = 3.92$; c. $r_{\text{cos-pi}} = .22$.

9 a. $r_\phi = .29$; $\chi^2 = 8.249$; b. $r_g = .30$, $z = 3.0$.

10 $r_h = .40$, $r_c = .53$

11 a. $r_{34.2} = .395$, $r_{41.2} = .466$, $r_{51.2} = .241$.
b. $r_{41.2}$ is most practical and most meaningful psychologically; $r_{34.2}$ is least meaningful or useful.

12 $r_{u_{12}} = .50$.

15 Prediction and accuracy of prediction

One of the most important fruits of scientific investigation and one of the most exacting tests of any hypothesis is the provision it makes for predictions. So important is this topic that it deserves to have considerable space devoted to it, particularly in a book of this sort, since statistical reasoning applies to all predictions in the behavioral sciences. Statistical ideas not only guide us in framing statements of a predictive nature but also enable us to say something definite concerning how trustworthy our predictions are—about how much error one should expect in the phenomenon predicted.

It is the purpose of this chapter and the next to illustrate the kinds of prediction that statistically oriented investigators make and how they do not blind themselves to their failures, but bring them clearly into the light.

GENERAL TYPES OF PREDICTION

Although in this volume we have generally emphasized continuous measurement, we have to recognize that the highest types of measurement (i.e., interval and ratio measurements, as defined in Chap. 2) often cannot be made and that data are sometimes merely classified in categories. With classification or nominal measurement, it is a matter of assigning attributes to cases rather than making quantitative evaluations on linear scales, for example, identifying individuals as to sex, race, or political-party affiliation. Although such data are not allocated to linear-scale positions, we can still make predictions from them as well as predict them from linear measurements. We thus have four general cases of prediction:

1. Attributes from other attributes—as when we predict incidence of divorce from race, political-party affiliation, or religious creed
2. Attributes from measurements—as when we predict divorce from scores on tests of ability or of other behavior traits
3. Measurements from attributes—as when we predict probable test scores from sex, socioeconomic status, or marital status
4. Measurements from other measurements—as when we predict academic achievement from aptitude scores

GENERAL WAYS OF EVALUATING ACCURACY OF PREDICTION

Predictions are obviously sound if they prove to be correct. The degree of correctness is indicated by *how often* or *how nearly* we hit the mark. In the case of predicting attributes, our success can be numerically indicated in terms of the percentages of "hits" or "misses." But a more accepted way is to ask how much better our predictions are, in the sense of how much we have reduced the errors, than they would have been had we not used the information at our disposal—in other words, if we had not tried to predict one thing from the knowledge of another.

In predicting measurements, whether from attributes or from other measurements, we ask a similar question. But whereas in predicting attributes for cases, we work in terms of the *number* of hits and misses, in predicting measurements, we work in terms of *by how much* on the average we have missed the mark. We compare this average deviation between fact and prediction with the average of the errors we should make without using our knowledge as a basis of prediction. We shall begin with the case of predicting measurements from attributes.

Predicting measurements from attributes

THE PRINCIPLE OF LEAST SQUARES

What would be the most accurate prediction of the weight of a sixteen-year-old youth? By "most accurate" we mean a weight that, if chosen to predict the weight of each sixteen-year-old selected at random from a certain population, would be closer to the facts in the long run than any other estimate would be. In other words, we want a predicted weight that would give us the smallest average deviation from the actual weights. We should find the difference between the actual weight of each person and our prediction in order to learn how good our prediction is for that particular person.

Statisticians have good reason to deal here in terms of the *squares* of the deviations rather than in terms of the deviations themselves. They demand a predicted measurement from which the sum of the squared deviations is a minimum. The prediction that will satisfy this requirement is the mean of the distribution, as was pointed out in Chap. 4.

PREDICTIONS APPLY TO SELECTED POPULATIONS

In answer to the question with which we started this discussion, the best prediction of the weight of a sixteen-year-old, if we lack better knowledge, is the mean weight of the population of which he or she is a member. If we want this to cover *all* sixteen-year-olds, we should see to it that the distribution from which we derive the mean is made up of a large sample in which both sexes, all races, and all socioeconomic and geographic groups are proportionally represented. We might, however, confine the question to sixteen-year-olds from the United States. We might further confine it to high school youths in one city or, even further, to one particular high school. Whatever our restriction in population, the predicted weight would apply only (except by chance) to that kind of population. Whenever we extend predictions to samples beyond the known population, we do so at the risk of enlarging errors of prediction.

ERRORS OF PREDICTION MEASURED BY THE STANDARD DEVIATION

In a certain high school in a certain city, a random sample of 51 sixteen-year-olds had weights distributed as shown in Fig. 15.1. For the sake of illustration, these sixteen-year-olds will be our *population*. What we say concerning predictions within this group will hold by analogy to larger, more inclusive populations. The mean of the 51 students' weights is 61.9 kg, and the standard deviation is 13.2. If the 51 students were now listed in alphabetical order and if, without seeing them, we used merely our knowledge of the mean, we should most nearly predict the actual weights if we wrote after each student's name "61.9 kg." The odds are about 2 to 1, as the interpretation of S goes, that our errors would be no greater than 13.2 kg either way from the predicted weight. The S of 13.2 kg may therefore be taken to

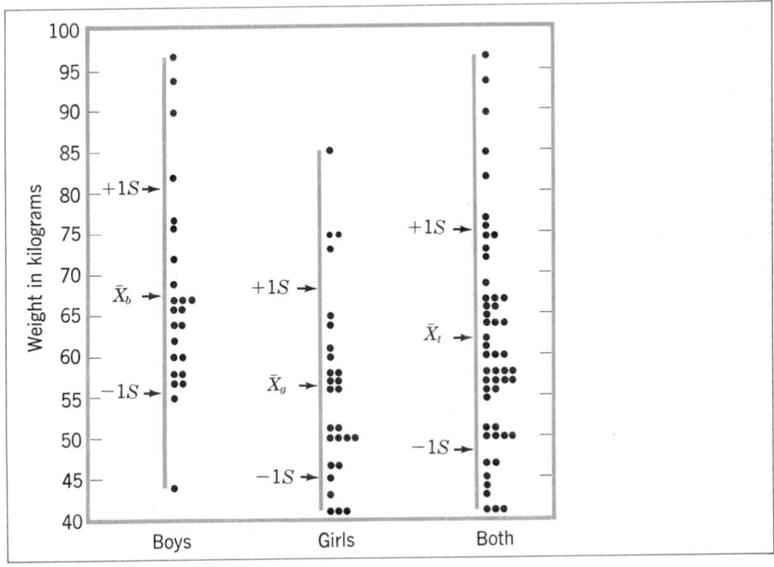

FIGURE 15.1 Distributions of sixteen-year-old high school boys and girls for weight in kilograms. Each dot represents an individual.

measure our margin of error in predicting single cases within the sample, when prediction is based only upon knowledge of the mean.

According to the *principle of least squares,* any other prediction we might make for all the individuals would yield a larger margin of error. We should not be very proud of our accuracy of prediction in this instance, and for practical purposes of making decisions for individuals where their weights are important factors, we should be seriously in error in many cases. But we could do less well in predicting the individuals' weights if we did not even possess the knowledge of their mean. Even if we knew the mean of sixteen-year-olds in general and used that as the predicted value, we should do worse, unless the mean of this small population coincided (by chance) with that of all sixteen-year-olds. In other words, by knowing two attributes of our population—a specified age and enrollment in a certain high school—and the mean weight associated with that attribute, we reduce the error of prediction to some extent.

PREDICTING WEIGHT FROM KNOWLEDGE OF SEX

Of the 51 cases in our population of sixteen-year-olds, 24 were boys and 27 were girls. Will it help to predict more accurately if we know each individual's sex? It should, since there is a sex difference in weights. Although many girls are heavier than many boys, the averages are distinctly different—67.8 for the boys and 56.6 for the girls. Using the attribute of sex to assist in predicting individual cases and following the principle of least squares, we should predict each boy's weight to be 67.8 kg and each girl's to be 56.6 kg.

How much will predictions now be improved? The margin of error of predictions for the boys is given by the S of their distribution, which is 12.6 kg, and the margin of error for the girls is given by an S of 11.3. From this information, we see that both boys' and girls' weights have been predicted more accurately than before (when the margin of error was 13.2) and that the girls' predicted weights are more free from error than the boys'.

To assess the gain in prediction from knowledge of sex membership, we may ask what the percentage of reduction in error of prediction is. For the boys, the change of .6 in the S is 4.5 percent, and for the girls, the change in S is 1.9, or 14.4 percent. Thus in predicting weight, we gain more accuracy from knowledge that the student is a girl.

THE STANDARD ERROR OF ESTIMATE

There is a way of summarizing the margin of error for all cases combined. It requires the computation of a *standard error of estimate,* which is a kind of summary of all the squared discrepancies of actual measurements from the predicted measurements.[1] In terms of a

[1] A standard error of estimate is more meaningful and more representative of all errors of prediction when the variances within different distributions are homogeneous, a condition that is known as *homoscedasticity.*

formula, the standard error of estimate is

$$S_{yx} = \sqrt{\frac{\Sigma(Y - Y')^2}{N}} \qquad \text{(Standard error of estimate)} \qquad (15.1)$$

where Y = measured value of a case we are trying to predict
Y' = predicted value for the case
N = total number of cases predicted

The subscript in S_{yx} tells us that we are predicting variable Y from knowledge of X. In the illustrative problem, Y is the variable of weight, and X is the variable of sex. The sum of the discrepancies squared (see Table 15.1) is 7,288.11, and so

$$S_{yx}{}^2 = \frac{7{,}288.11}{51} = 142.90$$

$$S_{yx} = 11.9$$

The standard error of estimate, in predicting weight on the basis of knowledge of sex, is 11.9. Using only the knowledge that this is a particular group of sixteen-year-olds with a mean of 61.9, the error of estimate was given by a standard deviation of 13.2. The margin of error using the information supplied by sex difference is 90.2 percent as large as that without using this information. The reduction in size of error of prediction is 9.8 percent, which is rather small but represents some gain.

In computing the standard error of estimate in this kind of problem, it is probably more natural to do so by finding the S's of the two part distributions separately and then combining them. They cannot be combined directly by simple addition or by averaging. It is the squared deviations in the two groups that must be combined. The sum of the squared deviations in each distribution can be found by the formula

$$\Sigma x_a{}^2 = N_a S_a{}^2 \qquad \text{(Sum of squares of discrepancies within one distribution)} \qquad (15.2)$$

where $\Sigma x_a{}^2$ = sum of the squared discrepancies between prediction and fact (or between measurements and the mean) in distribution A (one of the attribute distributions)
N_a = number of cases in distribution A
S_a = standard deviation of distribution A

When these sums of squared deviations are obtained from all component distributions (distributions A, B, C, etc.), they may be combined by simple addition to give $\Sigma(Y - Y')^2$. In other words,

$$\Sigma(Y - Y')^2 = \Sigma N_k S_k{}^2 \qquad \text{(Sum of squares of discrepancies in all distributions)} \qquad (15.3)$$

TABLE 15.1
Summary of the combinations of sums of squares from different subsamples

Distribution	N_k	S	S^2	N_kS^2
Boys	**24**	**12.65**	**160.02**	**3,840.48**
Girls	**27**	**11.30**	**127.69**	**3,447.63**
				7,288.11
				$\Sigma(Y - Y')^2$

where N_k = number of cases in any component distribution (distributions *A*, *B*, *C*, etc., in turn) and S_k = standard deviation of the same distribution.[1]

The work of computing $\Sigma(Y - Y')^2$ for the problem on weights of sixteen-year-olds may be summarized as in Table 15.1. From here on the computation of S_{yx} is exactly the same as previously demonstrated (p. 341).

OTHER PREDICTIVE INDICES MAY BE INTRODUCED

Other attributes may be brought in as aids in prediction. For instance, if different glandular constitution has a definite bearing on body weight (for example, thyroid functioning), we could subdivide each sex group into two or three categories as to glandular condition. The mean of each new subgroup would then become the prediction for members of that group. The deviations of actual weights from these means would be smaller, and the new standard error of estimate would be reduced in size.

If we were successful in singling out all the significant factors correlated with weight and could predict from all of them at the same time, theoretically we could reduce errors of prediction to approximately zero. We can probably never know what all the significant factors are from which weight can be determined, and if we did, it might be impossible to assign all the attributes to each individual. We are here speaking of the hypothetical limiting case. Any improvement in prediction approaches that limit. From a practical standpoint, it is always a question of whether the trouble of uncovering and using new descriptive attributes is justified by the gains in predictive accuracy that result.

ESTIMATION OF ERRORS OF PREDICTION IN THE POPULATION

Strictly speaking, the standard error of estimate computed for the weight-prediction problem applies to the sample only. It is a biased estimate of the margin of error that would occur in making predictions beyond this particular sample but in the same population. To estimate the standard error of estimate for the population, we need, as usual, to consider degrees of freedom, unless the sample is large. The formula

[1] It will be recognized that $\Sigma(Y - Y')^2$ is essentially a sum of squares from which the *within-sets* variance would be estimated in analysis of variance (see Chap. 13).

would be the same as (15.1) with the substitution of $N - k$ for N, where k is the number of categories from which predictions are made:

$$S_{yx} = \sqrt{\frac{\Sigma(Y - Y')^2}{N - k}} \qquad \text{(Standard error of estimate corrected for bias)} \qquad (15.4)$$

Applying this formula instead of formula (15.1), the corrected standard error of estimate is 12.2 rather than 11.9.

Predicting measurements from other measurements

When both known and predicted variables are measured on continuous scales and there is some correlation between them so that predictions are possible, we have a much more complicated problem. A complete treatment of it involves correlation methods, regression analysis, and other procedures.

THE CORRELATION DIAGRAM Our illustration of this kind of problem consists of two achievement examinations in a course on educational measurements. In Table 15.2, we have the two distributions grouped in class intervals and the measurements in each class interval broken down to form a distribution of its own in the other test. The class intervals for test X are listed along the top of Table 15.2, and the class intervals for test Y are listed along the left margin.

TABLE 15.2 Predicting scores in one test from known scores in another test

Test Y	*Test X*										
	60–64	*65–69*	*70–74*	*75–79*	*80–84*	*85–89*	*90–94*	*95–99*	f_y	$\bar{X}_{\text{row}}$	S_{row}
135–139								1	1	97.0	–*
130–134				1	1	0	1		3	83.7	6.61
125–129				1	0	2	1		4	85.8	5.45
120–124			1	4	4	6	2		17	83.2	5.67
115–119			7	5	7	2	1		22	78.6	5.72
110–114	1	4	2	9	4	2			22	75.9	6.56
105–109	1	1	2	5	1				10	74.0	5.56
100–104	1	3	0	1	1				6	70.3	6.87
95– 99		2							2	67.0	0.00
f_x	3	10	12	26	18	12	5	1	87 = N		
$\bar{X}_c$	107.0	105.5	114.9	114.5	116.4	120.3	124.0	137.0			
S_c	4.08	5.52	4.31	6.83	6.43	4.71	5.10	–*			

* The standard deviation of this array is indeterminate.

PREDICTION OF Y FROM X

As usual, we have here a double prediction problem: the prediction of a score in Y from a known score in X, and vice versa. Let us consider the prediction of Y from X first. For the individuals in any class interval in test X, the best prediction is the mean of the Y distribution in that column, in other words, the mean of the column ($\bar{X}_c$). The mean of each column in Table 15.2 is listed in the next to the last row. For the first column, $\bar{X}_c$ is 107.0. Any person receiving a score from 60 to 64 inclusive in test X will most probably earn a score near 107.0 in test Y. The other means of the columns are similarly interpreted. It will be noticed that there is a general upward trend in the $\bar{X}_c$'s as we go up the scale in test X, though there are two small inversions. In view of the small numbers of cases upon which these means are based, some inversions are not surprising.

The margin of error in predicting Y from X in each column is indicated by the standard deviation of that column. The S_c's are listed in the last row of Table 15.2. They remain fairly constant, but the range is from 4.08 to 6.83. The significance of the variations in S_c could be examined by making F tests (see Chap. 9).

The entire picture of predictions and their margins of errors within columns is shown graphically in Fig. 15.2. The dots indicate the positions of the column means, and the vertical lines running through them extend from $-1S_c$ to $+1S_c$. In each column, we expect two-thirds of the observed scores to lie within the limits of these lines.

STANDARD ERROR OF ESTIMATE

In order to obtain a single indicator of the goodness of the prediction of Y scores from X scores, we may compute a standard error of estimate as we did before when predicting measurements from attributes. The work is best organized as in Table 15.3. For every column, we list first N_c, the number of cases in that column. Second, we list S_c^2, the

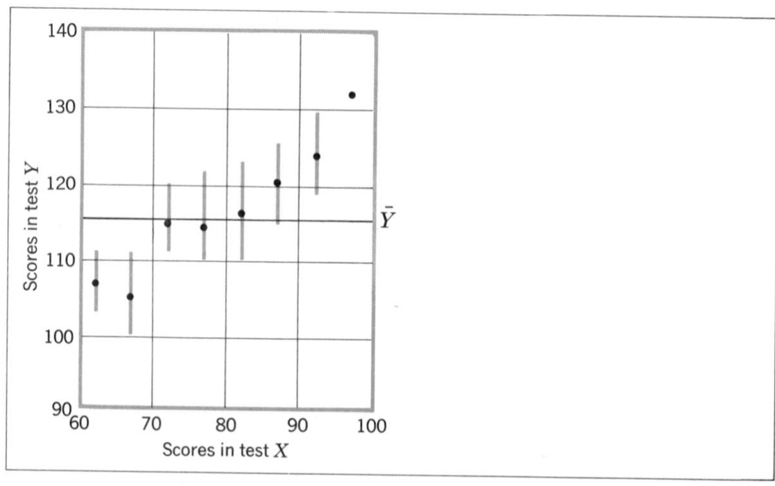

FIGURE 15.2 The most probable score in test Y corresponding to each midpoint score in test X; also the range between plus and minus one standard deviation within each column

squared S of the distribution in that column. Next we find the product of these two values for that column. The sum of these products for all columns yields $\Sigma(Y - Y')^2$, which we need for computing S_{yx}. This sum is 2,930.97. From here on the work follows formula (15.1).

$$S_{yx}^2 = \frac{2{,}930.97}{87} = 33.6893$$

$$S_{yx} = 5.80$$

The S of the entire distribution of Y scores is 7.85, and so there is a reduction in variability of 2.05, or 26.1 percent, a marked improvement in prediction, as such tests go. We may say that the forecasting efficiency for predicting Y scores from X scores as we did is approximately 26 percent.

PREDICTING *X* **FROM** *Y*

The predictions of X and Y are listed in Table 15.2 under X_{row} in the next to the last column. The most probable X score for any interval of Y scores is the mean of the row. The margin of error of the predictions is given in each case by S_{row}, and these appear in the last column of Table 15.2. To complete the picture of these predictions and their S's, Fig. 15.3 is presented. The standard error of estimate of the X scores S_{xy} (note the order of x and y in the subscript) is equal to 5.93. Since the total S of the X scores is 7.60, the reduction in error of prediction is 1.67, which is 22.0 percent. The forecasting efficiency in predicting X from Y in this problem is a bit lower than the forecasting efficiency (26.1 percent) in predicting Y and X.

The procedure for making predictions by using means of columns and rows is not used very much in practice. It was presented here because of the principles it illustrates, principles that underlie the regression methods to be described next. Readers will find that the

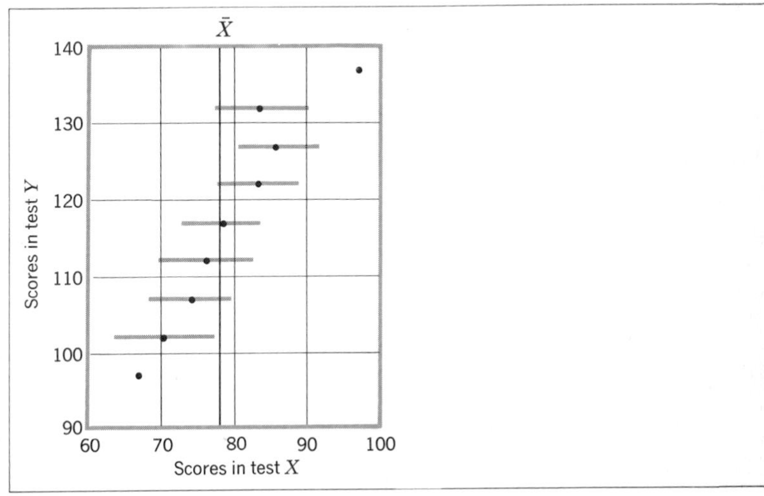

FIGURE 15.3 The most probable score in test X for each midpoint score in test Y; also the range between plus and minus one standard deviation within each row

TABLE 15.3
Computations of the standard error of estimate of Y scores from X scores

N_c	S_c^2	$N_c S_c^2$
3	16.67	50.01
10	30.45	304.50
12	18.58	222.96
26	46.63	1,212.38
18	41.36	744.48
12	22.22	266.64
5	26.00	130.00
		2,930.97
		$\Sigma(Y - Y')^2$

main principle for making predictions of measurements still holds—the principle of least squares. They will also find that the principles for testing accuracy of prediction—the standard error of estimate and the percentage of reduction of errors—also still apply. New ways of estimating them will be shown, and their relation to the coefficient of correlation will be explained. In addition, new ways of interpreting the usefulness of predictions will be demonstrated.

Regression equations

THE MEANING OF A REGRESSION EQUATION The main use of a regression equation is to predict the most likely measurement in one variable from the known measurement in another. If the correlation between Y and X were perfect (with a coefficient of +1.00 or −1.00), we could make predictions of Y from X or of X from Y with complete accuracy; the errors of prediction would be zero. If the correlation were zero, predictions would be futile. Between these two limits, predictions are possible with varying degrees of accuracy. The higher the correlation, the greater the accuracy of prediction and the smaller the errors of prediction.

When we use the means of columns of a scatter diagram as the most probable corresponding Y values, we are actually predicting Y's only from the midpoints of intervals on X, or, stated in another way, we are predicting the same Y value for a certain range of values on X. If we desire comprehensiveness, we have to be able to make predictions for *all* values of X. This the regression line and the regression equation enable us to do.

We found (see Figs. 15.2 and 15.3) that the means of the columns (and of the rows) tended to lie along a straight line, with some minor deviations from strict linearity. We shall now assume that the best predictions of Y from X lie exactly along a line that best fits the means

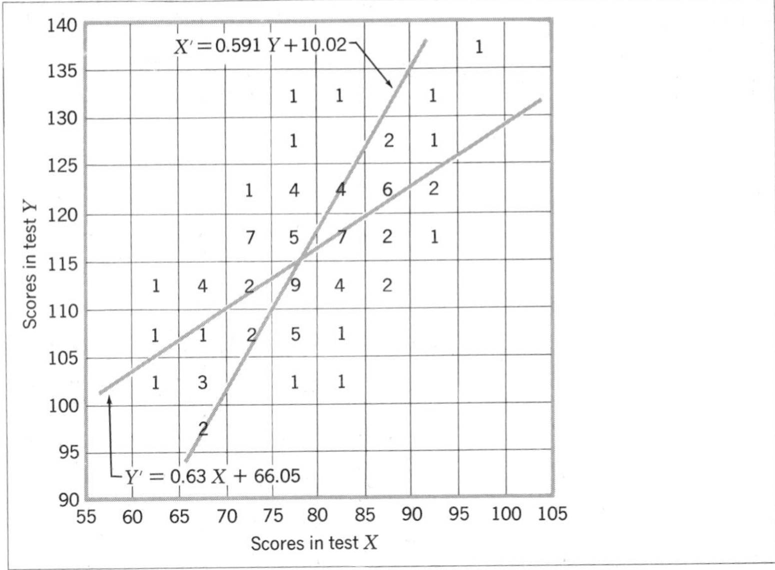

FIGURE 15.4 Scatter diagram for two examinations, with two regression lines represented and their equations

of the columns when those means are weighted according to the number of cases represented in each one. This is known as the *line of* best fit or the regression line. When predicting X from Y, we have another such line for the regression of X on Y. The two regression lines for the achievement-test data are pictured in Fig. 15.4. Only when a correlation is perfect will the two regression lines coincide throughout their lengths. The higher the correlation, the closer they lie to one direction, for either plus or minus. All such regression lines intersect at the point representing the means of X and Y. In this case they cross at $\bar{X} = 78.15$ and $\bar{Y} = 115.28$.

THE REGRESSION EQUATIONS AND REGRESSION COEFFICIENTS

From elementary algebra it should be remembered that the equation for a straight line, in general form is

$$Y = a + bX \qquad \text{(Equation for a straight line)} \tag{15.5}$$

Such an equation describes a straight line when the constants a and b are known. They are the *regression coefficients,* which must be obtained from the data we have. Ignoring for the moment the coefficient a, or assuming it to be zero, we have $Y = bX$, or Y equals b times X. We see from this that b is a ratio; it tells us how many units Y increases for every unit of increase in X. If $b = 2$, then for every unit increase in X, Y increases 2 units. If $b = 0.5$, Y increases half as rapidly as X. The b coefficient gives us the slope of the regression line.

The formula for computing b_{yx} from raw scores is

$$b_{yx} = \frac{N\Sigma XY - (\Sigma X)(\Sigma Y)}{N\Sigma X^2 - (\Sigma X)^2} \quad \text{(The } b \text{ coefficient for the regression of } Y \text{ on } X\text{)} \tag{15.6}$$

In terms of deviations from the means of X and Y the formula is

$$b_{yx} = \frac{\Sigma xy}{\Sigma x^2} \quad \text{(The } b \text{ coefficient of linear regression from deviations)} \tag{15.7}$$

If r is already known, the commonly useful formula is

$$b_{yx} = r_{yx}\left(\frac{S_y}{S_x}\right) \quad \text{(The } b \text{ coefficient computed from } r \text{ and } S\text{)} \tag{15.8}$$

In all these equations the subscripts in b_{yx} in that order imply that we are predicting Y from X.

In predicting X from Y we have a different b coefficient, which is given by the parallel formula

$$b_{xy} = r_{xy}\left(\frac{S_x}{S_y}\right) \quad \text{(Coefficient } b \text{ for linear regression of } X \text{ on } Y\text{)} \tag{15.9}$$

The coefficient of correlation is, of course, the same numerically in both cases; $r_{yx} = r_{xy}$. But the b's are different because in each case b is r times the ratio of the SD of the predicted variable to the SD of the variable predicted from. The predicted variable is commonly called the *dependent variable,* and the one predicted from the *independent* or *predictor variable*. Using a regression equation, the value on the dependent variable does depend upon the selected value on the independent variable.

The regression coefficient a is a constant that we must add in order to ensure that the mean of the predicted values will equal the mean of the observed values. As b_{yx} determines the slope of the regression line, a determines its general level. The two a coefficients are given by the similar formulas

$$a_{yx} = \bar{Y} - (\bar{X})b_{yx} \quad \text{(The } a \text{ coefficients in linear} \tag{15.10a}$$
$$a_{xy} = \bar{X} - (\bar{Y})b_{xy} \quad \text{regression equations)} \tag{15.10b}$$

The derivation of an entire linear regression equation is possible by using a composite formula, obtaining coefficients a and b in one operation, as follows:

$$Y' = r\left(\frac{S_y}{S_x}\right)(X - \bar{X}) + \bar{Y} \quad \text{(Equations with the two linear-regression} \tag{15.11a}$$
$$X' = r\left(\frac{S_x}{S_y}\right)(Y - \bar{Y}) + \bar{X} \quad \text{coefficients in terms of } r \text{ and } S\text{)} \tag{15.11b}$$

We use Y' and X' here rather than X and Y to show that they are predicted rather than obtained values. Predicted and obtained values should coincide only when the correlation is perfect.

Applying these formulas to the data of Table 15.2, we have

$$
\begin{aligned}
Y' &= .61\,\frac{7.85}{7.60}\,(X - 78.15) + 115.28 \\
&= (.61)(1.03)(X - 78.15) + 115.28 \\
&= .630X - 49.23 + 115.28 \\
&= .630X + 66.05
\end{aligned}
$$

$$
\begin{aligned}
X' &= .61\,\frac{7.60}{7.85}\,(Y - 115.28) + 78.15 \\
&= .591Y + 10.02
\end{aligned}
$$

Interpreting these equations, we may say that Y' increases .63 units for every unit increase in X and that X' increases .591 units for every unit increase in Y. One way of checking the accuracy of the computation of the regression equations is to substitute $\bar{X}$ in the first one, to see whether the predicted Y' is the mean of the Y's and to substitute $\bar{Y}$ in the second equation to see whether the predicted X' is the mean of the X's.

Another check, for the b coefficients only, is to use the equation

$$b_{yx}b_{xy} = r^2 \qquad \text{(Relation of the regression coefficients to } r^2\text{)} \qquad (15.12)$$

In the illustrative problem, $(.630)(.591) = .3723 = .61^2$.

THE CONCEPT OF REGRESSION

It may help in understanding regression equations as given in formulas (15.11 *a* and *b*) to note their historical origin. The idea of regression came first, and the correlation method followed. It began with Sir Francis Galton, who was making some studies of heredity. When he made a study of the relation between heights of children and heights of their parents, he started by plotting a scatter diagram, perhaps the first. In order to put parents and children on a common measuring scale, he transformed all heights to standard (z) scores. As the reader probably already knows, this meant stating each person's height as a ratio between the deviation from the group's mean to the standard deviation of the group's dispersion. The unit for the offspring's scale and also for the parents' scale was then 1S. Figure 15.5 shows the type of figure Galton drew.

Galton next computed the means of offspring's heights (in z scores) corresponding to certain fixed parents' heights (in z scores). As we saw in the example earlier in this chapter when the same operations were performed (but with raw scores), he found that the means of columns fell along a straight-line trend. It struck him as remarkable that the means of offspring's heights did not increase as rapidly as those of parents' heights. The mean heights of offspring deviated less from their general mean than the heights of the parents from which they came deviated from their mean. This "falling back" of heights of offspring toward the general mean has been called the *law of filial regression*, or more generally known as regression toward the mean. It is an illus-

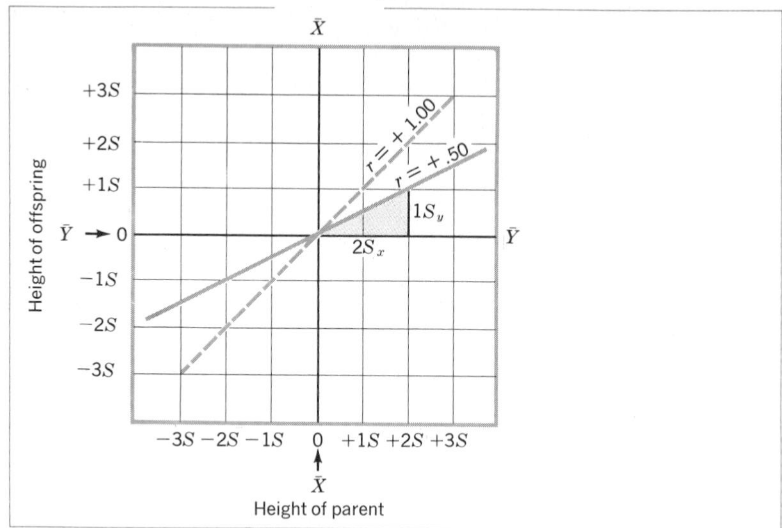

FIGURE 15.5 The relation of the Pearson coefficient of correlation to the slope of the regression line when scores in both *X* and *Y* are in standard-score units

tration of imperfect correlation. Had the correlation between children and parents in height been perfect, the regression would have been as shown by the dotted line in Fig. 15.5. The correlation was actually about +.50, and the obtained regression line was as shown.

ORIGIN OF THE COEFFICIENT OF CORRELATION

Galton wanted a single value which would express the amount of this regression phenomenon in any particular relationship problem. Karl Pearson solved the problem with the formula to which his name is attached. The steps were somewhat as follows. Galton's own idea was to use the slope of the regression line as the index of relationship because the steeper the slope, the closer the agreement between two variables. The slope of the regression line in Fig. 15.5, as in any coordinate plot, is the ratio of the increase in *Y* corresponding to a certain increase in *X*. From the plot we see that as *X* changes 2*S* (from the mean to +2*S*, as shown), *Y* changes only 1 *S*. The slope is ½, or .5. This was Galton's coefficient of regression, which received the symbol *r* for that reason. That symbol has remained. The Pearson *r* is the slope of the regression line when both *Y* and *X* are measured in standard-deviation units. In this case, it can be shown that

$$r_{yx} = \frac{\Sigma z_y z_x}{N} \qquad \text{(Pearson } r \text{ from standard measures)} \qquad (15.13)$$

In other words, *r* is an average of all the cross products of standard measures.

DERIVATION OF THE REGRESSION EQUATIONS

Since *r* is the slope of the regression line when standard measures are used, the equation for this situation is

$$z_{y'} = r z_x \qquad \text{(Regression equation with standard measures)} \qquad (15.14)$$

Here we use $z_{y'}$, the prime indicating a predicted value as distinguished from an obtained value. From this beginning, the interested reader can easily derive the regression equations of formulas (15.11 *a* and *b*), by substituting for the two standard measures given above the expressions $(Y' - \bar{Y})/S_y$ and $(X' - \bar{X})/S_x$.

PREDICTIONS FROM REGRESSION EQUATIONS

As an illustration of how a regression equation is applied in prediction, let us assume some values of X and find the corresponding Y' values. Because in the preceding methods of prediction we predicted Y's corresponding to midpoints of the intervals of X, let us do the same here for the sake of comparison, remembering that we might have chosen any values of X that we pleased, within the range of obtained values of X. Table 15.4 gives the X values and their corresponding Y' values. When X is 62, Y' is 105.1, and when $X = 97$, $Y' = 127.2$, etc.

It is interesting to compare these particular predictions with the means of the columns, which are given in the third row of Table 15.4. The discrepancies will be found very small as a rule. Granting that the column means are generally not very reliable because of small samples, we may feel more assurance in the Y' predictions because they are determined from the trend of the entire data rather than by small samples in separate columns. The predictions of X' from Y are given in the second section of Table 15.4 and are compared with the means of the rows as a matter of interest.

As a practical means of prediction, a graphic method will often be the most suitable procedure. If the regression lines are drawn as in Fig. 15.4 on cross-section paper, for any value of X on the base line, one can follow vertically up to the regression line and note the corresponding Y value at this point.

TABLE 15.4 Predictions of Y from X and X from Y by means of regression equations*

$Y' = 0.630X + 66.05$

If $X =$	62	67	72	77	82	87	92	97
$Y' =$	105.1	108.3	111.4	114.6	117.7	120.9	124.0	127.2
$\bar{X}_c =$	107.0	105.5	114.9	114.5	116.4	120.3	124.0	132.0

$X' = 0.591Y + 10.02$

If $Y =$	97	102	107	112	117	122	127	132	137
$X' =$	67.3	70.3	73.3	76.2	79.2	82.1	85.1	88.0	91.0
$\bar{X}_{row} =$	67.0	70.3	74.0	75.9	78.6	83.2	85.8	83.7	97.0

* The means of the columns and rows are obtained from Table 15.2.

THE STANDARD ERRORS OF ESTIMATE When we have predicted on the basis of regression equations, we can estimate the margin of error of prediction, as given by S_{yx} (or by S_{xy}), from the coefficient of correlation. The formulas are

$$S_{yx} = S_y \sqrt{1 - r_{yx}^2} \quad \text{(Standard error of estimate} \quad (15.15a)$$
$$S_{xy} = S_x \sqrt{1 - r_{xy}^2} \quad \text{computed from } r) \quad (15.15b)$$

It will be seen that the two equations are the same except for the use of S_y when we are predicting Y and of S_x when we are predicting X (for $r_{yx} = r_{xy}$). The two standard deviations are multiplied by the common factor $\sqrt{1 - r^2}$. This factor is always less than 1.00 and *gives us an estimate of the reduction in errors of prediction from knowledge of correlated measurements as compared with errors of prediction without that knowledge.* When r is zero, this factor equals 1.00, and then $S_{yx} = S_y$ and $S_{xy} = S_x$. In other words, when $r = 0$, there is no basis for prediction. When $r = 1.0$ (or -1.0), the factor reduces to zero, and so does the standard error of estimate. This coincides with the expectation that the margin of error of prediction is zero when the correlation is perfect.

INTERPRETATION OF AN OBTAINED STANDARD ERROR OF ESTIMATE The interpretation of the standard error of estimate when r is neither zero nor 1.00 is somewhat as follows. Like any standard deviation, S_{yx} can be referred to the normal curve of distribution, if the column and row distributions appear to approach the normal form. For the examination problem,

$$S_{yx} = 7.85 \sqrt{1 - .3721} = 6.22$$
$$S_{xy} = 7.60 \sqrt{1 - .3721} = 6.02$$

Unless there are obvious inequalities of variances within columns or rows, no matter in what part of the measuring scale we are predicting (within the range of obtained scores, naturally), we assume that the margin of error is the same.[1] When we predict Y from X, the average dispersion of observed measurements about Y' is given by an S of 6.22. We expect two-thirds of the observed cases to lie within the limits of plus and minus 6.22 from Y'. This situation is illustrated graphically in Fig. 15.6. There we have the regression line, along which the predicted Y's lie, and the dashed lines represent the limits of one S_{yx} to either side of it. Had we plotted a point for every individual, we should have expected about two-thirds of the dots to fall between the two dashed lines. To make a particular prediction, when $X = 90$, $Y' = 122.8$. The odds are 2 to 1 that any individual whose X score is 90 will not fall below 116.6 or go above 129.0, these scores

[1] In other words, homoscedasticity is assumed.

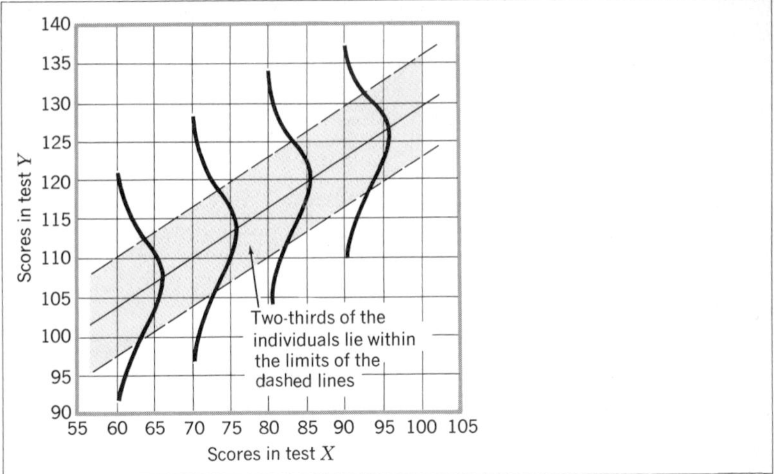

FIGURE 15.6 The line of regression of Y on X, showing the range of observed values expected in Y in separate categories of score values on X. Parallel dashed lines above and below the regression line at the vertical distance of one standard error of the estimate each way mark off the region within which we expect two-thirds of the observed values to be.

being one S_{yx} below and above the predicted Y, respectively. We could state other odds for a divergence of two S_{yx} either way or any other distance. It all depends upon our purposes.

We could prepare a similar diagram showing the limits of the middle two-thirds of the individuals about the line of regression of X on Y, and we could interpret the errors of prediction in a similar manner. It will be noted that the margin of error as given by S_{xy} is 6.02, or 0.2 smaller in predicting in the other direction, i.e., X from Y, but this is merely because S_x is smaller than S_y. The *percentage of error is the same in the two cases.* The ratio of S_{yx} to S_y is exactly the same as the ratio of S_{xy} to S_x, and that ratio is given by the factor $100\sqrt{1-r^2}$. This factor we shall meet again with a special name attached to it [see formula (15.21)].

THE REGRESSION LINE AS A MEAN

One way of looking at the regression line is to regard it as a moving average, a moving arithmetic mean. Like the arithmetic mean of any sample, the regression line satisfies the principle of least squares. The regression coefficients are so determined by the data that the sum of the squares of the deviations of observed points from the line is a minimum. Other lines might describe the trend of relationship nearly as well, but only the one line satisfies the principle of least squares. It is reasonable that if the line is a mean, the deviations from it should be measured by a standard deviation. That standard deviation is the standard error of estimate.[1]

1 For an excellent discussion of regression effects in research problems see Thorndike, R. L. Regression fallacies in the matched group experiment. *Psychometrika,* 1942, **7,** 85–102.

CORRECTION OF A STANDARD ERROR OF ESTIMATE FOR BIAS

In smaller samples (N less than 50) it would be well to make a correction in S_{yx} (or S_{xy}) before accepting it as an estimate of errors of prediction in a population. The change can be made by the formula

$$_cS_{yx} = S_{yx}\sqrt{\frac{N}{N-2}} \qquad \text{(Correction for bias in } S_{yx}) \qquad (15.16)$$

where N is the number of cases in the sample. The correction can be applied in the original computation of S_{yx}, as follows:

$$_cS_{yx} = S_y\sqrt{(1-r^2)\left(\frac{N}{N-2}\right)} \qquad (15.17)$$

THE ACCURACY OF A *b* REGRESSION COEFFICIENT

Like all descriptive statistics, the b coefficient in the regression equation has its sampling error. Its standard error is estimated by

$$s_{b_{yx}} = \frac{S_{yx}}{S_x\sqrt{N}} \qquad \text{(Standard error of a } b \text{ regression coefficient)} \qquad (15.18)$$

or by

$$s_{b_{yx}} = \frac{S_y}{S_x}\sqrt{\frac{1-r^2}{N}} \qquad (15.19)$$

The statistic $S_{b_{xy}}$ would be the same except for interchanging the x and y subscripts. For the examination problem,

$$s_{b_{yx}} = \frac{6.22}{(7.60)(9.3274)} = .088$$

The obtained b_{yx} of .63 is more than seven times the standard error of 088. There is very little chance that the true coefficient is zero. One could also test the significance of that coefficient's deviation from any other hypothesized value.

The correlation coefficient and accuracy of prediction

The chief index of goodness of prediction of measurements thus far in this discussion has been the standard error of estimate. It has been shown how the latter is closely related to the coefficient of correlation. As r increases, the standard error of estimate decreases. There are other ways in which r and some of its derivatives can be used to indicate accuracy of prediction. Three of the common derivatives are the *coefficient of alienation,* the *index of forecasting efficiency,* and the *coefficient of determination.* Each has its unique story to tell about the closeness of correlation between two things and about the utility of predictions.

THE COEFFICIENT OF ALIENATION

Whereas r indicates the strength of relationship, the *coefficient of alienation* k indicates the degree of *lack* of relationship. By formula,

$$k = \sqrt{1 - r^2} \quad \text{(Coefficient of alienation computed from } r\text{)} \tag{15.20}$$

Squaring both sides of this equation, we have

$$k^2 = 1 - r^2$$

And transposing, we have

$$k^2 + r^2 = 1.00$$

TABLE 15.5 Indicators of the importance of coefficients of correlation

r_{xy}	k_{xy} *Coefficient of alienation*	$100(1 - k_{xy})$ *Percentage reduction in errors of prediction of Y from X*	$100\ r_{xy}^2$ *Percentage of variance accounted for*
.00	1.000	0.0	0.00
.05	.999	.1	0.00
.10	.995	.5	1.00
.15	.989	1.1	2.25
.20	.980	2.0	4.00
.25	.968	3.2	6.25
.30	.954	4.6	9.00
.35	.937	6.3	12.25
.40	.917	8.3	16.00
.45	.893	10.7	20.25
.50	.866	13.4	25.00
.55	.835	16.5	30.25
.60	.800	20.0	36.00
.65	.760	24.0	42.25
.70	.714	28.6	49.00
.75	.661	33.9	56.25
.80	.600	40.0	64.00
.85	.527	47.3	72.25
.90	.436	56.4	81.00
.95	.312	68.8	90.25
.98	.199	80.1	96.00
.99	.141	85.9	98.00
.995	.100	90.0	99.00
.999	.045	95.5	99.80

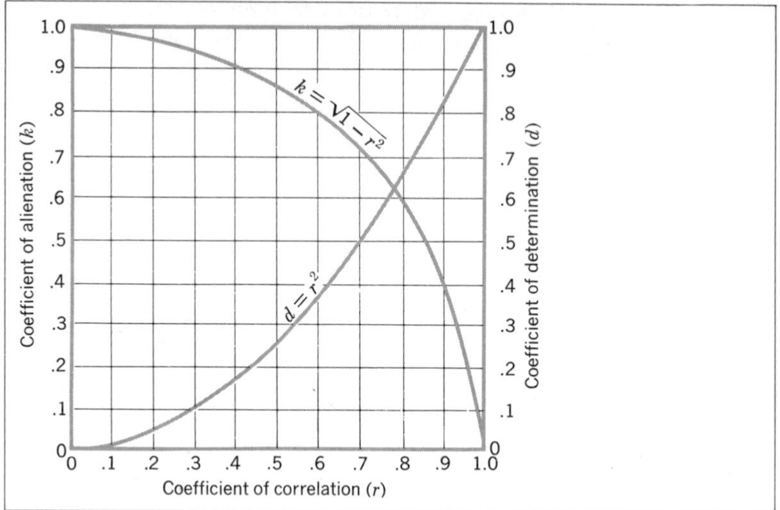

FIGURE 15.7 k (coefficient of alienation) and d (coefficient of determination) as functions of r (coefficient of correlation)

Thus, although we might have expected k plus r to equal 1.00, it is rather the sum of their squares that equals 1.00. If r is .50, k is *not* also .50 but .886. When r is .50, then, the degree of relationship is less than the degree of *lack* of relationship. It is when $r = .7071$ that relationship and lack of relationship are equal, for k also then equals .7071. Then $r^2 + k^2 = .50 + .50 = 1.00$. Other values of k for different sizes of r can be found in Table 15.5. Figure 15.7 shows pictorially the functional relationship between k and r. Students of mathematics will recognize the relationship $r^2 + k^2 = 1.00$ as the equation for a circle with a radius of 1.00. The diagram shows only positive values of r and k.[1]

Sometimes we wish to stress the independence of two variables rather than their dependence. In such instances, we present k as well as r. Besides being related to r, k is also related to other indices of goodness of prediction, to be mentioned next.

THE INDEX OF FORECASTING EFFICIENCY

In the formula for the SE of estimate $S_{yx} = S_y \sqrt{1 - r_{yx}^2}$, we can see that the factor with the radical $\sqrt{1 - r_{yx}^2}$ is really the coefficient of alienation. We could rewrite the formula as $S_{yx} = S_y k_{yx}$. If we were to multiply k by 100, we should have the percentage S_{yx} is of S_y. When $r = .61$, as in our recent illustration, $k = .7924$. The SE of estimate in this problem is 79.24 percent of the observed dispersion of observations. Our margin of error in predicting Y *with* knowledge of X scores is about 79 percent as great as the margin of error we should make *without* knowledge of X scores. For then we predict every Y to be the

[1] The relationship of k to r is the same as that of the sine of an angle to the cosine of that angle. Values of k corresponding to known values of r can be found by using Table I in the Appendix.

mean of the Y's, and the SE of the prediction then equals S_y. The *reduction* of margin of error is 100 minus 79.24, or 20.76 percent. The *index of forecasting efficiency* is defined as the percentage reduction in errors of prediction by reason of correlation between two variables. The general, simplified formula is

$$E = 100(1 - \sqrt{1 - r^2}) \qquad \text{(Index of forecasting efficiency)} \qquad (15.21)$$

or

$$E = 100(1 - k)$$

The calculation of E is facilitated by Table 15.5, where many of the E values are given for corresponding r's. Inspection will show that r must be as high as about .45 before E is 10 percent. When a test has a validity coefficient of .45, the size of errors of prediction, on the whole, is only 10 percent less than the size we should have without knowledge of test scores but with knowledge of the mean criterion measure. Taken at its face value, this does not seem much of a gain. There are situations, however, in which an even smaller gain might be of practical importance.

Better tests, with validity coefficients of .60, have an E of 20 percent, and still better tests, when $r = .75$, have an E of about 34 percent. Although these efficiencies may also seem small, we must treat them in a relative, not an absolute, sense. It is probable that the efficiency of predictions based upon the average unsystematic interview is less than 5 percent. With this as a base, the efficiency of tests looks much better.

Figure 15.8 shows graphically the functional relationship between E and r. The range of r's from .3 to .8 is marked off as representing the

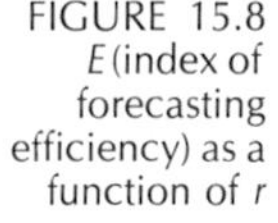

FIGURE 15.8 E (index of forecasting efficiency) as a function of r

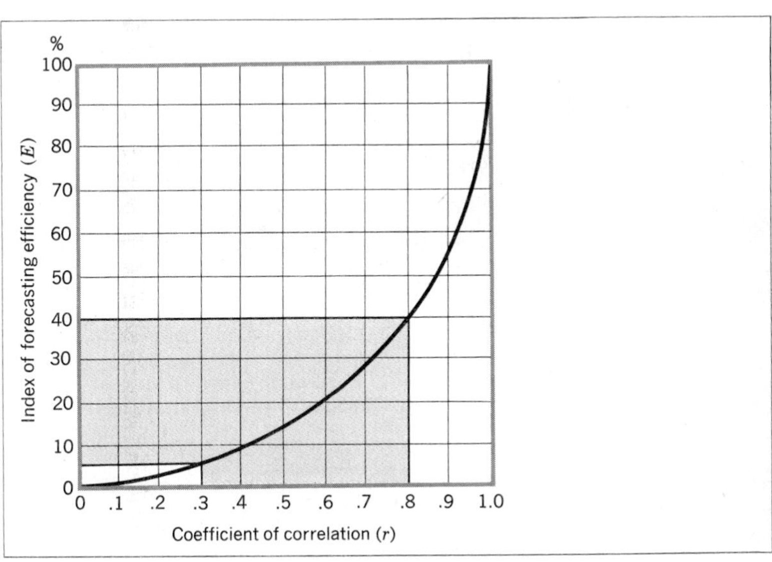

level of validity coefficients usually found for useful predictive instruments in psychological and educational practice. Tests rarely show correlations greater than .8 with practical criteria, and those correlating less than .3 are commonly regarded as being of limited value when used alone. In a test battery to which they add unique contributions they have been found useful.

Actually, the value of a test in the common use of assigning individuals to categories, as, for example, in deciding which individuals should be accepted and which rejected, is a very complicated matter. In this situation there is need to take into account the proportion of the individuals selected (the selection ratio) and the proportion (before selection) who would be successful (the success ratio).[1] There may also be a need to take into account the utility connected with a given dividing line, where utility depends upon payoff value and the costs of applying the testing procedures.[2] In this connection, statistics such as the index of forecasting efficiency are of no value. The coefficient of correlation, on the other hand, is of considerable value, in connection with the other variables just mentioned.

THE COEFFICIENT OF DETERMINATION

Another mode of interpretation of r is in terms of r^2, which is called the *coefficient of determination*. This statistic is also sometimes symbolized as d. When multiplied by 100, the coefficient r^2 gives us the percentage of the *variance* in Y that is associated with, determined by, or accounted for by variance in X. When $r = .50$, the percentage of the variance in Y that is accounted for by variance in X is 25, or one-fourth. To account for half the variance of any set of measurements, the r with another variable would have to be .7071.

The proportion of the variance in Y *not* determined by, or associated with, variance in X is given by k^2, which is called the *coefficient of nondetermination*. These statements about determination of Y by X are reversible and apply equally well to determination of X by Y, when the regression is linear. We should speak of *determination* of one thing by another, however, only when a causal relationship can be logically defended; otherwise, the expression *associated with* or *accounted for* (by way of prediction) is better. In Table 15.5, several of

1 The roles of selection ratio and success ratio in this connection were pointed out by Taylor, H. C., and Russell, J. T. The relationship of validity coefficients to the practical effectiveness of tests in selection. *Journal of Applied Psychology*, 1939. **23,** 565–578. Their treatment of the subject was described in the third edition of this volume.

2 For an excellent treatment of this subject from the point of view of decision theory, see Cronbach, L. J., and Gleser, G. C. *Psychological tests and personnel decisions*. Urbana, Ill.: University of Illinois Press, 1965. They make much of the principle that the payoff value of a test in selection is directly proportional to its validity coefficient, however, they make statements such as ". . . a test of validity .20 in one situation may be more beneficial than a test of validity .60 in another situation" (p. 53).

the $100r^2$ values are given for corresponding r's. Figure 15.7 presents graphically the functional relationships of d and k to r.

PREDICTED AND NONPREDICTED VARIANCES The coefficient of determination, as well as its relations to r, k, and other statistics, can best be clarified by introducing another new idea. The total amount of variance in the predicted variable Y is denoted by S_y^2. We can think of this variance as composed of two independent factors, the predicted and the nonpredicted portions. The predictions of Y, which we have called Y', have their dispersion and their variance, which are denoted by $S_{y'}$ and $S_{y'}^2$, respectively. The standard deviation $S_{y'}$ would be computed from the deviations of the predicted Y values from the mean of the Y values, $\bar{Y}$. The amount of nonpredicted variance is indicated by the square of the standard error of estimate, S_{yx}^2. This statistic is computed from the deviations of the obtained Y values from the regression line (or from the predicted Y' values). The two component variances of S_y^2 are therefore

$$S_y^2 = S_{y'}^2 + S_{yx}^2 \qquad \text{(Component variances in the predicted variable)} \qquad (15.22)$$

If we divide this equation through by S_y^2, we have everything in terms of proportions:

$$\frac{S_y^2}{S_y^2} = \frac{S_{y'}^2}{S_y^2} + \frac{S_{yx}^2}{S_y^2} = 1.0 \qquad \text{(Total variance as the sum of two proportions)} \qquad (15.23)$$

The first term on the right, $S_{y'}^2/S_y^2$, is the proportion of the variance in Y that is predicted, and the second term is the proportion of the variance that is not predicted. We have already defined r^2 as the proportion of predicted variance and k^2 as the proportion of nonpredicted variance. This means that r^2 equals $S_{y'}^2/S_y^2$ and k^2 equals S_{yx}^2/S_y^2, and that $r = S_{y'}/S_y$ and $k = S_{yx}/S_y$. We therefore have some new concepts of r and k. We can say that r is the ratio of the dispersion of predicted values to the dispersion of obtained values and that k is the ratio of the dispersion of errors to the dispersion of obtained values.

Predictions of attributes from measurements

We sometimes wish to decide, on the basis of known measurements, whether an individual should be expected to be in one category rather than another, having one attribute rather than another. He will be happily married after 10 years, or he will be divorced; he will vote for a Republican candidate or a Democratic candidate; he will develop a psychosis, or he will not; he will commit a crime, or he will not; and if he is on parole, he will succeed in keeping away from crime, or he will fail. Sometimes it is a matter of placing individuals in different categories in order to help to bring about improved adjustment or success. We shall consider here the use of a single continuously measured variable as the predictor, or independent variable, and class membership as the criterion, or dependent variable. We shall note two ways in which the predictions may be made, how the goodness of prediction may be evaluated, and some of the special problems encountered in this kind of prediction.

A SOLUTION BY LINEAR REGRESSION One approach to the prediction of membership in two categories is to take a cue from the use of regression equations, as we have seen in the prediction of continuous variables. In this case, we arbitrarily assign values of 0 and 1 to the two categories and treat this dependent variable as if it were continuous. With the two categories representing a genuine dichotomy, the point-biserial coefficient of correlation applies.

As an example of the prediction of two alternative attributes from measurements, let us use some data on sex differences in strength of handgrip measurements for 417 high school students, 171 boys and 246 girls. Prediction of sex membership is not a serious practical problem, but it will do for illustrative purposes. For deriving a regression equation, we need certain statistical information, which follows: The mean of .410 for sex membership comes from arbitrarily assigning a value of 1 to being a boy and 0 to being a girl, from which the mean is the proportion of boys ($\Sigma X/N = 171/417 = .410$). The standard deviation of the distribution for sex membership is given by the expression $\sqrt{pq}$, where $p = .410$ and $q = 1 - p$.

For the regression equation we need to find two constants, b_{yx} and a_{yx}, using formulas (15.6) and (15.10a):

$$b_{yx} = .763 \left(\frac{.492}{10.74}\right)$$
$$= .035$$
$$a_{yx} = .410 - (.035)(27.51)$$
$$= -.552$$

	Y *Sex membership*		*X* *Handgrip scores*
Mean	**.410**		**27.51**
Standard deviation	**.492**		**10.74**
Correlation coefficient		**.763**	

The regression equation then reads $Y' = .035X - .552$. We could use this equation, substituting selected values of handgrip score, to obtain a predicted Y. If Y' exceeds .5, the prediction is a boy; if it is smaller than .5, the prediction is a girl.

A simple operation makes it unnecessary to apply the equation more than once. That is, if we determine what handgrip-score point yields a prediction of exactly .5, we know that all scores above that critical value predict boy and that all scores below it predict girl. Substituting .5 for Y' in the obtained regression equation and solving for X, to give X_c,

$$.5 = .035X_c - .552$$

Transposing and changing signs,

$$.035X_c = 1.052$$

and

$$X_c = 30.06$$

From this result we may say that scores of 31 and above predict boy and that those of 30 and below predict girl.

GOODNESS OF PREDICTION IN CATEGORIES Whereas the accuracy of prediction of measurements on a continuous scale is often evaluated in terms of a standard error of estimate, or something related to it, in prediction of cases within categories we are more concerned about misplacements of cases. When an error does occur, it is to the extent of a whole point. One procedure, when there are two categories, is to set up a fourfold contingency table in which the cases with scores above and below the critical score represent one dichotomous variable and the true dichotomy represents the other variable. In the handgrip problem, the predicted categories are scores of 31 and above versus scores of 30 and below. Remembering that the 0-1 distinction applies to boys versus girls, with the boy category positive, the girls who were predicted to be boys on the basis of handgrip scores are known as "false positives," and the boys who were likewise predicted to be girls are known as "false negatives." These two cells of the contingency table represent the "misses" in prediction. For the purpose of setting up such a contingency table, we need score distributions for boys and girls separately, either ungrouped or finely grouped, so as to be able to apply the cutoff score accurately. We lack such distribution information here, and so we cannot use this approach to assessing goodness of prediction in this particular example.[1]

1 Other methods of predicting attributes from continuous measurements are described by Guilford, J. P., and Michael, W. B. *The prediction of categories from measurements.* Beverly Hills, Calif.: Sheridan Supply, 1949.

Prediction of attributes from other attributes

Sometimes both the dependent and independent variables are in the form of category membership; neither is in the form of continuous measurements. For example, this would be true in predicting marital adjustment from church membership, success on parole from marital status, or living versus committing suicide from membership in one racial group versus another. If both variables are dichotomies, we have a simple fourfold table of frequencies. There are other contingency tables with more than two categories in one or both variables. We need to consider how predictions should be made and how they should be evaluated.

PREDICTING FOR MAXIMUM PROBABILITY

As an example of prediction of attributes from other attributes, let us consider the data in Table 15.6. There we have the numbers of persons in a "depressed" group who responded by saying "Yes," "?," and "No" to the question, "Would you rate yourself as an impulsive individual?" and also the numbers of persons in a group described as "not depressed." The individuals in these two categories are the highest and lowest quarters of a sample of 1,000 students who were ranked in terms of a provisional score on a personality inventory. Although one can make an assumption of continuity of values on the response scale, all we have is information classifying the individuals into three categories; hence we may treat the responses as attributes. The separation of individuals in extreme quarters on the depression scale produces a definite break in continuity.

Table 15.6 provides us with two prediction problems. We can attempt to predict the verbal response to the question, knowing whether the person is in the depressed or the not-depressed group, or we can attempt to predict the group to which the person belongs, knowing in what response category that person is. Let us take the prediction of verbal responses first.

Considering first the depressed group by itself, we find that the largest number of its members respond with "No." Taking the members of the depressed group as they come along, we should predict for each of them the response "No." If all 250 came up for inspection, we should be correct 133 times out of 250, or 53.2 percent

TABLE 15.6 Distribution of responses to the question, "Would you rate yourself as an impulsive individual?" as given by two extreme groups of students

Group	Response			
	Yes	*?*	*No*	*Total*
Depressed	72	45	133	250
Not depressed	106	35	109	250
Both	178	80	242	500

of the time. For other samples from the same depressed population, we should expect a similar ratio of correct predictions. This illustration sets the pattern for all predictions of attributes from other attributes. The prediction observes the *mode,* or most frequent attribute, in the segment of the population chosen at the moment.

For the not-depressed group, the mode is also at the response "No." Hence that is our prediction also for them, and our percentage of accuracy is only 43.6 percent. However, it is higher than if we had predicted either the "Yes" or the "?" response. But since the prediction of response is in the same category for both the depressed and the not-depressed group, we have made no discrimination; we have gained nothing in the way of information.

The second prediction problem here is to reverse matters and predict group membership from knowledge of response category. We again apply the *principle of maximum probability.* All persons responding "Yes" we should predict to be members of the not-depressed group, since 106 actually are, as compared with 72 who are not. For those responding "?" the prediction is membership in the depressed group. The prediction is the same for those responding "No." Altogether, there are 284 correct predictions, or 56.8 percent. Without knowledge of which response each person would make to the question, but with knowledge that half the total sample are depressed and half are not, the expected number of chance successes is 250. Our predictions *with* knowledge of responses yield an excess of 34, or an indicator of forecasting efficiency of 13.6 percent. We can say that predictions with knowledge of responses to the question are 13.6 percent better than those made without this knowledge would be, since 34 is 13.6 percent of 250.

PREDICTION NOT EQUALLY GOOD IN THE TWO DIRECTIONS

It is now apparent that we can predict successfully group membership from knowledge of responses in this particular study, whereas we cannot successfully predict responses from knowledge of group membership. It is not always true, as here, that successful prediction is possible in one direction only, but it is quite common for prediction to be better in one direction than the other. This is a more serious matter in dealing with attributes than in dealing with measurements, for in the latter case, with linear regressions, the predictability is about equally good in the two directions, as indicated by the coefficient of correlation or by derivatives from it.

CORRELATION INDICES AND THE PREDICTION OF ATTRIBUTES

The correlation indices provide an alternative to the counting of false positives and false negatives, as suggested in dealing with predictions of attributes from measurements, and to the use of the percentage gain in making predictions. In the case of 2 × 2 contingency tables, the phi coefficient would be one natural coefficient to employ. This coefficient was described in Chap. 14, with its relation to chi square, which can be used to indicate departure of the data from a chance distribution of frequencies, given the marginal sums that we have.

If the contingency table is larger than 2 × 2, a chi-square test would give us an answer regarding significance for the whole table, but it would not tell us where that significance lies, unless we break down the table by rows or columns, or both, and compute a chi square for each combination of cells. Since possibilities for predictions, and goodness of prediction, vary for different directions and different sections of a contingency table, such statistical tests by sections are sometimes desirable. These points will now be illustrated for the depression data.

Because of the relation of phi to chi square in a 2 × 2 table, phi coefficients can be computed for such sections of a larger table. For the data in Table 15.6, we can test the relationship for each response separately as against other responses. With attention given to the "Yes" response, we can combine the other two columns to form a "not-yes" category. The same may be done for each of the other two responses. The phi's for the three responses prove to be .142, .055, and .095, respectively, corresponding to chi squares of 10.08, 1.49, and 4.61, respectively.

From these results we may conclude that the correlation corresponding to the "Yes" response is significant at the .01 level, that the correlation corresponding to the "?" response is insignificant, and that the correlation corresponding to the "No" response is significant at the .05 level. We could also conclude that the combination of "?" and "No" responses should be significantly predictive because it shares the phi of .142 with the "Yes" response. This information would be a basis for combining the "?" response with the "No" response in making predictions and in keying the question for assessment of depression.

However, Curtis[1] has pointed out that in making predictions in contingency tables, we need to consider the marginal frequencies and the relations between those for the criterion Y and the predictor X. He recommends that the phi coefficient be used only when neither distribution (of X or Y) is more unbalanced than a .6 to .4 split. For general use he recommends a coefficient *lambda,* which was developed by Goodman and Kruskal.[2] Lambda is the ratio of the decrease in the number of errors of prediction to the number of errors that would be made without knowledge of membership in category X. The formula reads

[1] Curtis, E. W. Predictive value compared with predictive validity. *American Psychologist,* 1971, **26,** 908–914.

[2] Goodman, I. A., and Kruskal, W. H. Measures of association for cross-classifications. III. Approximate sampling theory. *Journal of the American Statistical Association,* 1963, **58,** 310–364.

$$\lambda = \frac{\Sigma f_c - f_r}{N - f_r} \quad \text{(Coefficient of forecasting efficiency)} \tag{15.24}$$

where f_c = the sum of the largest cell frequencies from the columns and f_r = the largest row sum. Let us apply this formula to the data in Table 15.6, with interest in the goodness of prediction of being depressed versus not depressed from affirmative versus negative answer to the given question. In order to reduce the data to a 2×2 contingency table, let us combine the "?" responses with the "No" responses. The result is that the depressed subjects gave 72 "Yes" and 178 "No" responses, and the not-depressed subjects gave 106 and 144, respectively. These are the row values. The column values are 72 versus 106, and 178 versus 144.

$$\lambda = \frac{(106 + 178) - 250}{250}$$

$$= .136$$

Here it just happens that the two row sums are the same, so that value is used for f_r.

It may be noted that λ is essentially the same as the indicator of forecasting efficiency mentioned earlier. Lambda is simply a proportion, whereas the other statistic is a percentage. It may also be mentioned that when the means of both X and Y are .50, and only then, does phi equal lambda. In addition to the general bias in phi that was mentioned earlier, this is another reason for not using it to indicate goodness of category predictions.

EXERCISES

1 a. What is the most probable score for the passing and for the failing students represented in Table 14.4 (see p. 306)?
   b. What is the accuracy of prediction for each category?
   c. How much improvement is there from knowledge of category?

2 For Data 15*A*:
   a. Find the best prediction of score in the Opposites test corresponding to each midpoint score in the Mixed Sentences test.
   b. Estimate the margin of error for each prediction.
   c. Estimate the margin of error for the predictions taken as a whole.

3 For the information in Data 15*A*, the following statistics have been computed: $\bar{X} = 14.19$; $\bar{Y} = 21.65$; $S_x = 5.71$; $S_y = 6.73$; $r_{xy} = .651$.
   a. Find the two regression equations for Data 15*A*.
   b. Make a check for the accuracy of your b coefficients.

4 Using the appropriate regression equation:
   a. Make a prediction of score in the Opposites test corresponding to each midpoint score in the Mixed Sentences test.
   b. Compare these predictions with those obtained in Exercise 2 by drawing a diagram.

5 a. Compute the two standard errors of estimate for Data 15*A*.

DATA 15*A* A scatter diagram for two mental tests

Y (Opposites test in Army Alpha)	*X (Mixed Sentences test in Army Alpha)*								
	0–2	3–5	6–8	9–11	12–14	15–17	18–20	21–23	f_y
36–38								1	1
33–35							1	2	3
30–32				1	1	3	7	2	14
27–29						4	5	2	11
24–26			1	3	3	2	4	4	17
21–23			1		6	1	5	2	15
18–20		1	2	1	9	5	4		22
15–17	2	1	2	2	2	2	1		12
12–14	1	2	0	2	2	1			8
9–11	3	1	2	1	2				9
6– 8				1					1
f_x	6	5	8	11	25	18	27	13	113

b. What are the *amounts* of predicted and nonpredicted variance in Y?

c. What are the proportions of these two kinds of variance?

6 Using statistics derived from Data 15*A*:

a. Draw a diagram like Fig. 15.4, showing the two regression lines.

b. Draw a diagram like 15.6, showing the two limits set by the standard error of estimates S_{yx}.

7 From the same source of data, derive the statistics *k*, *E*, and r^2. Interpret your findings.

8 Using formulas (15.18) and (15.10), derive a regression equation for the first 10 pairs of scores for Parts V and VI in Data 6*A* (p. 92).

9 For the data represented in Fig. 15.1 (p. 339) adapt the problem of predicting sex from body weight. The predictor variable *X* is weight and the predicted variable *Y* is sex membership. The problem is to find a critical point on *X*, called X_c, that will make the best discrimination of cases in the two sex categories. Using the regression-equation approach, you need the following information: $r_{pbi} = .424$, $\bar{X} = 61.9$, $\bar{Y} = .4706$, $S_x = 13.2$, and $S_y = .4991$.

10 Using a critical point of 63.7 kg, estimate from Fig. 15.1 the frequencies of boys and girls above and below that point. Form a 2 × 2 contingency table.

a. From the table, report the numbers of false positives and false negatives.

b. Report the percentage of correct predictions and the phi, r_g, and λ coefficients.

11 In Data 15*B*, make predictions of whether students will report "Yes,"

DATA 15*B*
Relationship between walking in one's sleep and talking in one's sleep as reported by 1,794 students

		Walking in sleep			
		No	*?*	*Yes*	*Total*
Talking in sleep	**Yes**	400	9	88	497
	?	194	21	3	218
	No	1,069	3	7	1,079
	Total	1,663	33	98	1,794

"?," or "No" to the question about talking in their sleep when they make each of the same responses to the question about walking in their sleep.

a. What is the percentage of correct predictions for each of the predictor categories, and what is the overall percentage?
b. What is the percentage of correct predictions that could be made without knowledge of responses to the question about walking?
c. What is the forecasting efficiency?
d. Can you predict responses to the question about walking in one's sleep from knowledge of responses to the question about talking in one's sleep? Explain.

12 a. Compute a chi square for the 3 × 3 contingency table in Data 15*B* and state the number of degrees of freedom involved.
b. Combine the "Yes" and "?" categories for the variable of walking in one's sleep, compute a chi square, state the number of df, and interpret your results.

ANSWERS

1 a. Means: 98.3; 83.6. SD's: 16.27; 16.19.
b. Gains: for passing group, 8.0 percent; for failing group, 8.4 percent; for both combined ($S_y = 17.68$ and $S_{yx} = 16.22$) 8.3 reduction in amount of error.

2 a. Means of columns: 12.5, 14.2, 17.1, 18.2, 19.5, 23.2, 25.7, 28.2.
b. Standard deviations of columns: 2.7, 3.1, 5.0, 7.1, 4.7, 5.7, 4.9, 4.6.
c. $S_{yx} = 5\ 07$

3 a. $Y' = .767X + 10.76$; $X' = .552Y + 2.23$.
b. $b_{yx}b_{xy} = .424 = r^2$.

4 a. Y′: 11.5, 13.8, 16.1, 18.4, 20.7, 23.0, 25.3, 27.6.

5 a. $S_{yx} = 5.11$; $S_{xy} = 4.33$;
b. $S_{y'} = 19.19$; $S_{yx}{}^2 = 26.10$.
c. $r_{xy}{}^2 = .4238$; $k^2 = .576$.

7 With $r = .651$, $k = .759$, $E = 24.1$, $r^2 = .424$.

8 $\bar{X}_5 = 22.9$; $\bar{X}_6 = 27.7$; $X_5' = .651X_6 + 4.87$; $X_6' = 9.45X_5 + 6.06$; $b_{65}b_{56} = .6152 = r^2$.

9 Regression equation: $S = .0159X - .5198$ (where S is sex membership); $X_c = 63.7$ kg.

10 a. False positives, 6; false negatives, 9.

b. Percentage of correct predictions, 70.6; $r_\phi = .41$; $r_g = .41$; $\lambda = .375$.

11 a. Percentages: 89.8 (100 × 88/98); 63.6 (100 × 21/33); and 64.3 (100 × 1,069/1,663), for predictions from "Yes," "?," and "No," respectively.

b. Percentages without knowledge: 60.1 (100 × 1,079/1,794); with knowledge, 65.7 (100 × 1,178/1,794); gain with knowledge, 9.3 (100 × 5.6/60.1).

c. $\lambda = .138$.

12 a. For the 3 × 3 table, $\chi^2 = 287.95$, with 4 df.

b. For the 3 × 2 table, $\chi^2 = 178.75$, with 2 df.

16 Multiple prediction

Multiple correlation

INDEPENDENT AND DEPENDENT VARIABLES

Thus far we have been dealing with correlations between two variables at a time and the prediction of some variable Y from another variable X, or vice versa. Actual relationships between measured variables in psychology and education are by no means so simple as that. One variable is found associated with, or dependent upon, more than one other variable at the same time. When we can think of some variables as being causes of another one, or even when we merely want to predict that one variable from our knowledge of several others correlated with it, we call the one variable the *dependent* variable and the ones upon which it depends the *independent* variables. The independent variables are so called because we can manipulate them at will or because they vary by the nature of things, and consequently we expect the dependent variable to vary accordingly.

Whether or not people like a certain color depends upon several factors: its hue (e.g., yellow, red, or purple), its brilliance (e.g., light, medium, or dark), and its chroma (saturation or density). The affective value of the color also depends upon its area, its use, and its background. We are here naming independent variables upon which the affective value of a color depends. Insofar as each one is a determinant of agreeableness of color, it will exhibit some correlation individually with affective value. The size of any one of these correlations will depend upon the relative strength of that factor and also upon how well the other factors have been neutralized, as they should be in an ordinary bivariate experimental situation.

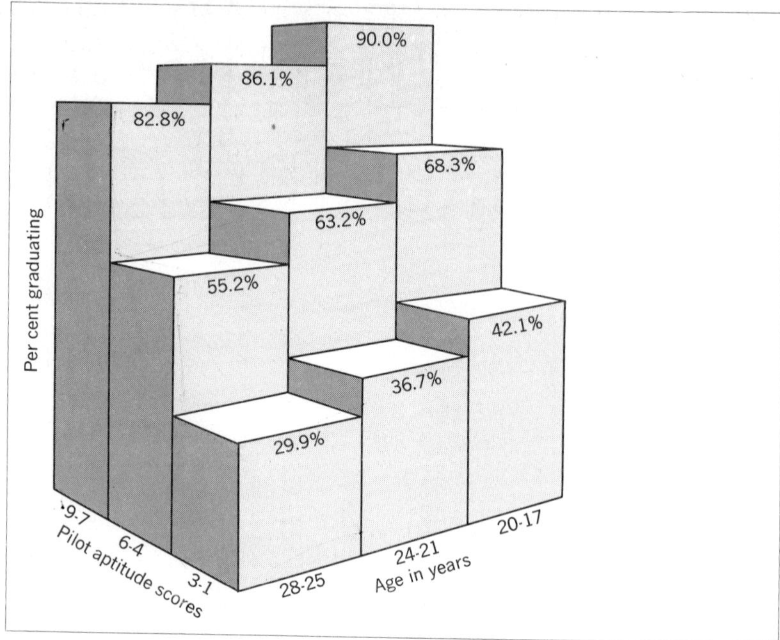

FIGURE 16.1 A multiple regression, with percentage graduating from pilot training as a function of both chronological age and aptitude score (the latter variables measured in very broad categories). (Adapted from an unpublished report of Headquarters, AAF Flying Training Command, Fort Worth, Tex.)

A GRAPHIC PICTURE OF MULTIPLE DEPENDENCE

The idea of dependence of one variable upon two others is illustrated by Fig. 16.1. That illustration shows how the dependent variable, success in pilot training, is related both to aptitude scores and to chronological age. It requires a three-dimensional figure to show these relationships. The vertical dimension represents the dependent variable. Here it is measured in terms of percentage of graduates—not an ordinary way of measuring, but one, nevertheless, that shows the principles involved. The two independent variables are represented as sides of the base, at right angles to each other. The scale of chronological age is shown reversed for convenience, since the correlation between age and the training criterion was negative. Both independent variables are shown here in very coarse categories for the sake of a simpler diagram.

By noting rows of blocks (left to right), we can see how graduation rate changes with age for a relatively constant level of aptitude. By noting the columns of blocks (front to back), we can see how graduation rate changes with aptitude score for a relatively constant age level. The term *constant* covers an unusual range in this illustration, but with finer grouping on age and aptitude we should expect similar trends. It is obvious that the regressions of the criterion on aptitude are much steeper than those of the criterion on age. The difference would be even more apparent if we had the criterion in terms of a properly graded measurement scale. The correlation between aptitude scores and the criterion was much higher (approximately .55) than that between age and the criterion (approximately −.10). A very rough appreciation of the joint predictive value of aptitude score and age can be seen by noting the change of height from the lowest block (29.9 percent) to the highest (90.0 percent). This change may be compared with changes across columns alone or across rows alone. From this comparison we should expect better prediction from both independent variables than from either alone.

THE COEFFICIENT OF MULTIPLE CORRELATION The coefficient of multiple correlation indicates the strength of relationship between one variable and two or more others combined with optimal weights. The multiple correlation is related to the intercorrelations between independent variables as well as to their correlations with the dependent variable. The interdependency of the determinants suggested for affective value of colors is probably not so apparent as that of determinants related to achievement in college algebra. Here we can think of such predictive variables as intelligence-test scores and high-school marks, which, being related, duplicate one another to some extent in predicting achievement in college algebra. Hours of study and interest also are interrelated and thus are not completely independent determinants of success in algebra.

A MULTIPLE-CORRELATION PROBLEM Table 16.1 presents some data that call for a multiple-correlation solution. Four of the variables (X_2, X_3, X_4, and X_5) are measures of supposed determinants of academic success among college freshmen. X_1 is the

TABLE 16.1 Intercorrelations between variables, including one index of scholarship and four predictive indices ($N = 174$)*

Variable	X_1	X_2	X_3	X_4	X_5
X_1		.465	.583	.546	.365
X_2	.465		.562	.401	.197
X_3	.583	.562		.396	.215
X_4	.546	.401	.396		.345
X_5	.365	.197	.215	.345	
$\bar{X}$	73.8	19.7	49.5	61.1	29.7
S	9.1	5.2	17.0	19.4	3.7

X_1 = an average grade for the first semester in university
X_2 = arithmetic test in the Ohio State Psychological Examination, Form 10 (a college aptitude test)
X_3 = analogies test in the same examination
X_4 = an average grade in high-school work
X_5 = student interest inquiry (measuring breadth of interest)

* These data were abstracted from the *Ohio State College Bulletin, 58,* by L. D. Hartson, and have been used by permission.

dependent variable—average freshman marks. It is customary to designate the dependent variable by X_1.

An examination of Table 16.1 shows that the analogies-test scores and high school averages have the highest correlations with X_1, whereas the interest scores X_5 have the lowest. The highest *intercorrelations* are between X_2, X_3, and X_4. All represent abilities of one kind or another, and their correlations with X_5 (interests) are generally lower. This suggests that the interest scores will contribute something to the prediction of college marks that will not have been already contributed by the other variables; therefore, it may be worthwhile to include X_5 in the battery of predictive indices or predictors.

THE SOLUTION OF A THREE-VARIABLE PROBLEM

We first take the simplest case of multiple correlation, that between the dependent variable and two independent variables. In the general problem given by the data in Table 16.1, we may ask what the correlation is between freshman marks on the one hand and the two variables—analogies-test scores and high-school averages—on the other. The simplest general formula for this case is

$$R_{1.23}^2 = \frac{r_{12}^2 + r_{13}^2 - 2r_{12}r_{13}r_{23}}{1 - r_{23}^2} \qquad \text{(Square of coefficient of multiple correlation with three variables)} \qquad (16.1)$$

where $R_{1.23}$ = coefficient of multiple correlation between X_1 and a combination of X_2 and X_3. *Notice that this formula gives R^2, the square root of which is R.*

The immediate example calls for finding $R_{1.34}$ rather than $R_{1.23}$. To use formula (16.1), we need merely to substitute the subscripts 3 and 4 for 2 and 3. The solution is

$$R_{1.34}^2 = \frac{(.583)^2 + (.546)^2 - 2(.583)(.546)(.396)}{1 - (.396)^2}$$

$$= \frac{.339889 + .298116 - .252108}{1 - .156816}$$

$$= .45766$$

$$R_{1.34} = .677$$

THE MULTIPLE-REGRESSION EQUATION

A multiple-prediction problem calls for a regression equation that involves all three variables, in other words, a multiple-regression equation. From such an equation, we can predict an X_1 value for every individual. The correlation between these predicted values (X_1') and the obtained ones (X_1) will be .677. This is another interpretation of a coefficient of multiple correlation.

For the three-variable problem, the regression equation has the general form $X_1' = a + b_{12.3}X_2 + b_{13.2}X_3$. As in previously presented regression equations, the coefficient a is a constant and must be calculated from the data. Its function is to ensure that the mean of the

X_1' values coincides with the mean of the X_1 values. The b coefficients serve the same purpose here as in the simple, two-variable equation. The coefficient $b_{12.3}$ is the multiplying constant, or weight, for the X_2 values, and $b_{13.2}$ is the weight for the X_3 values. The value of $b_{12.3}$ tells how many units X_1' increases for every unit increase in X_2, when the effects of X_3 have been nullified or held constant. The value of $b_{13.2}$ tells how many units X_1 increases for every unit increase in X_3, with the effects of X_2 held constant.

The particular b weights, as computed by the formulas given below, are the *optimal* weights. They ensure the maximum correlation between predicted and obtained X values. The solution, with the obtained b weights, satisfies the principle of least squares in that the sum of the squares of discrepancies between the X_1 values and the X_1' values will be a minimum.

SOLUTION OF THE b COEFFICIENTS We do not find the b coefficients directly from the correlations but indirectly through the beta coefficients. Beta coefficients are called *standard partial regression coefficients*—"standard" because they would apply if standard measures, such as $(X - \bar{X})/S$, were used in all variables, and "partial" because, as in the case of the coefficient of partial correlation (see Chap. 14), the effects of other variables are held constant. The $b_{12.3}$ and $b_{13.2}$ are known as *partial regression coefficients* because they, too, are weights that presuppose that other independent variables are held constant. They are given by the formulas[1]

(Partial regression coefficients)

$$b_{12.3} = \left(\frac{S_1}{S_2}\right) \beta_{12.3} \tag{16.2a}$$

$$b_{13.2} = \left(\frac{S_1}{S_3}\right) \beta_{13.2} \tag{16.2b}$$

The betas are found by the formulas

(Standard partial regression coefficients)

$$\beta_{12.3} = \frac{r_{12} - r_{13}r_{23}}{1 - r_{23}^2} \tag{16.3a}$$

$$\beta_{13.2} = \frac{r_{13} - r_{12}r_{23}}{1 - r_{23}^2} \tag{16.3b}$$

Similar equations apply, with a change of subscripts, when the independent variables are X_3 and X_4 instead of X_2 and X_3. In our example,

$$\beta_{13.4} = \frac{.583 - (.546)(.396)}{1 - (.396)^2} = .435$$

$$\beta_{14.3} = \frac{.546 - (.583)(.396)}{1 - (.396)^2} = .374$$

1 In this volume, Greek letters are generally used to indicate population parameters. The use of beta as a statistic computed from a sample is an exception, dictated by common usage.

We can now solve for the b coefficients by means of formulas (16.2*a*) and (16.2*b*):

$$b_{13.4} = \frac{9.1}{17.0}(.435) = .233$$

$$b_{14.3} = \frac{9.1}{19.4}(.374) = .175$$

For the complete regression equation, the a coefficient is still lacking. It is given by the general formula

$$a = \bar{X}_1 - b_{12.3}\bar{X}_2 - b_{13.2}\bar{X}_3 \tag{16.4}$$

Inserting the known values, we find

$$a = 73.8 - (.233)(49.5) - (.175)(61.1) = 51.58$$

The complete regression equation then reads

$$X_1' = 51.58 + .233X_3 + .175X_4$$

To interpret the equation, we may say that for every unit increase in X_3, X_1 increases .233 unit and that for every unit increase in X_4, X_1 increases .175 unit. To apply the equation to a particular student whose X_3 score is 25 and whose X_4 score is 32, we predict that the student's X_1 score will be

$$X_1' = 51.58 + 5.82 + 5.60 = 63.00$$

We use X_1' to stand for the predicted average freshman mark because the student has an actual average mark that we call X_1. Some other examples of individual students are presented in Table 16.2 to show how various combinations of values for X_3 and X_4 point to corresponding values of X_1.

MULTIPLE PREDICTIONS BY A GRAPHIC METHOD

A graphic method of making predictions of scores in X_1 from different combinations of scores in X_3 and X_4 is shown in Fig. 16.2. The chart is drawn to apply to the prediction of average freshman grades from scores in the analogies test and high-school averages. Diagonal lines are drawn in the figure, each representing locations of the same predicted value. These lines represent X_1' scores at intervals of five units. Note, for example, the line for $X_1' = 70$. A prediction of 70 may arise from many different combinations of X_3 and X_4. Choose several values, in turn, as possible scores in the analogies test—for example, 10, 30, 50, and 70. Corresponding values in high school average needed to yield predictions of 70 are 90, 65, 38, and 10, respectively. The chief use of the chart, however, is to find X_1' for two given values in X_3 and X_4. For an X_3 of 20 and an X_4 of 50, the prediction is 65. For an X_3 of 90 and an X_4 of 13, the prediction is 75.

When the prediction is not exactly on one of the diagonal lines,

FIGURE 16.2 Constant values in the dependent variable for different combinations of values in the two independent variables, each weighted as called for by the regression equation

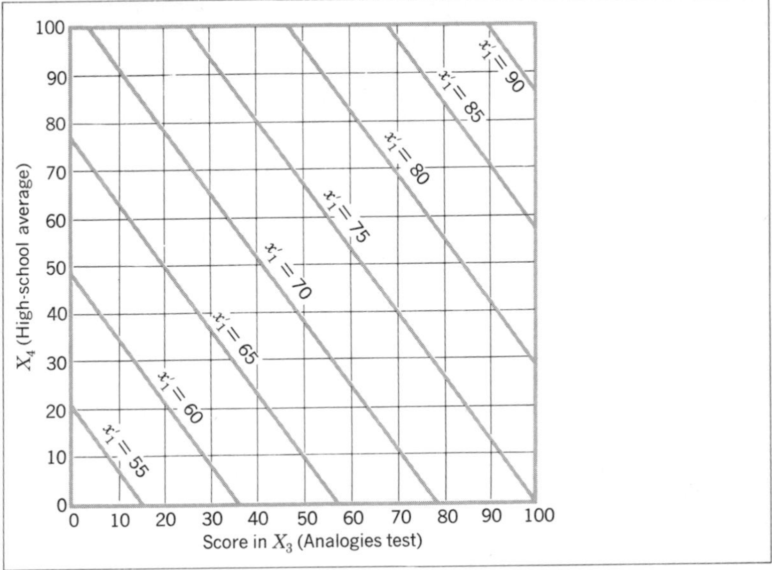

we interpolate, by inspection, between two lines. Thus, for $X_3 = 40$ and $X_4 = 70$, the most probable X_1 is 73. The proportion of the distance between two diagonal lines must be estimated by the perpendicular distance between them. The reader may obtain further practice in using the chart by verifying the predictions found by computation in Table 16.2.

CALCULATING THE MULTIPLE *R* FROM BETA COEFFICIENTS

If the beta coefficients are known, the shortest route to the multiple *R* is by way of the equation

$$R^2_{1.23} = \beta_{12.3}r_{12} + \beta_{13.2}r_{13} \tag{16.5}$$

Again, note that this gives R^2, from which the square root must be obtained. For the scholarship data and variables X_3 and X_4,

TABLE 16.2 Some predictions of scholarship mark from measures in two variables

	Student				
	A	*B*	*C*	*D*	*E*
X_3 analogies score	25	27	48	85	87
X_4 high-school average	32	61	65	90	52
$b_{13.4}X_3$	5.82	6.29	11.18	19.80	20.27
$b_{14.3}X_4$	5.60	10.68	11.38	15.75	9.10
X'_1 (predicted mark)	63.0	68.6	74.1	87.1	81.0

$$R^2_{1.34} = (.435)(.583) + (.374)(.546)$$
$$= .457809$$
$$R_{1.34} = .677$$

as was found by formula (16.1) previously.

INTERPRETATION OF A MULTIPLE *R*

Once computed, a multiple R is subject to the same kinds of interpretation, as to size and importance, that were described for a simple r. One kind of interpretation is in terms of R^2, which we call the *coefficient of multiple determination*. This tells us the proportion of variance in X_1 that is dependent upon, associated with, or predicted by X_3 and X_4 combined *with the regression weights used*. In this case, R^2 is .4578, and we can say that 45.78 percent of the variance in freshman marks is accounted for by whatever is measured by the analogies test and by high-school marks taken together. The remaining percentage of the variance, which is 54.22 $(1 - R^2)$, is still to be accounted for. This remainder is given the symbol K^2 and is known as the *coefficient of multiple nondetermination*. This is consistent with the fact that $R^2 + K^2 = 1.0$, just as $r^2 + k^2 = 1.0$ in the simple correlation problem.

THE STANDARD ERROR OF ESTIMATE FROM MULTIPLE PREDICTIONS

The standard error of estimate is again brought in to indicate about how far the predicted values would deviate from the obtained ones. The formula is the same as previously, except that the multiple R is substituted for r. It reads

$$S_{1.23} = S_1\sqrt{1 - R^2_{1.23}} \quad \text{(Standard error of multiple estimate)} \qquad (16.6)$$

In the illustrative problem,

$$S_{1.34} = 9.1\ \sqrt{1 - .457809} = 6.7$$

We can now say that two-thirds of the obtained X_1 values will lie within 6.7 points of the predicted X_1 values. The margin of error *with* knowledge of X_3 and X_4 is 73.6 percent as great as the margin of error would be without that knowledge. This is true because the radical term reduces to .736. These conclusions presuppose predictions made on the basis of the regression equation that was obtained.

The index of forecasting efficiency may also be used by way of interpretation and, because of its close relation to the standard error of estimate, may be mentioned at this point. The formula is the same as for a Pearson r [see formula (15.21)]. In the example of our three variables, $E = 26.4$ percent, which indicates that predictions using the equation have 26.4 percent less error than those made merely from a knowledge of the mean of the X_1 values.

MULTIPLE CORRELATION IN SMALL SAMPLES

For small samples—and for multiple-correlation problems this means anything less than an N of 100—degrees of freedom should be considered in dealing with questions of sampling. If the multiple R and the other statistics derived from it are to be used for estimating population parameters, there is even more bias than for a simple correlation problem.

It was stated earlier that the multiple R represents the maximum correlation between a dependent variable and a weighted combination of independent variables. The least-squares solution that is represented in computing the combined weights ensures this result; but it really ensures too much. It capitalizes upon any chance deviations that favor high multiple correlation. The multiple R is therefore an inflated value. It is a biased estimate of the multiple correlation in the population. If we were to apply the same regression weights in a new sample and to correlate predicted X_1 values with obtained X_1 values, we should probably find that the correlation would be smaller than R.

It is therefore desirable to find some way of estimating a parameter $\mathbf{P}$ which gives a more realistic picture of the general situation. A common way of "shrinking" R to a more probable population value is by the formula

$$_cR^2 = 1 - (1 - R^2)\left(\frac{N-1}{N-m}\right) \qquad \text{(Correction in } R \text{ for bias; an estimate of } \mathbf{P}) \qquad (16.7)$$

where N = number of cases in the sample correlated
m = number of variables correlated
$N - m$ = number of degrees of freedom, one degree being lost for each mean, there being one mean per variable

For the illustrative problem above, where $R = .677$, the corrected R^2 would be

$$_cR^2 = 1 - (1 - .4579)\left(\frac{174-1}{174-3}\right) = .4515$$

from which $_cR = .672$. The correction does not make much difference here because the sample was fairly large and the number of variables was small. There are problems in which the change would be very appreciable. It should follow that in computing a standard error of estimate that is without bias, one should use $_cR_{1.23}$ in formula (16.6).

SAMPLING ERRORS IN MULTIPLE-CORRELATION PROBLEMS

For a test of whether an obtained multiple R deviates significantly from a population $\mathbf{P}$ of zero, the following formula may be used:

$$s_R = \frac{1}{\sqrt{N-m}} \qquad \text{(Standard error of } R \text{ when the hypothesized population correlation is zero)} \qquad (16.8)$$

in which $N - m$ is the number of degrees of freedom. Unless N is very large, and much larger than m, this formula underestimates the amount of error. This SE is subject to the same limitations as s_r, and to a greater extent.

For testing the null hypothesis, Table R in the Appendix may also be used. It provides values of R at the .05 and .01 levels of significance for different degrees of freedom. In the illustrative problem, $N = 174$, $m = 3$, therefore the df is 171. From Table R we find that with 150 df (the next lower and nearest to 171), and with 3 variables, an R of .198 is significant at the .05 level and one of .244 at the .01 level. We should have little doubt that a multiple R different from zero exists in the population involved.

STANDARD ERROR OF A MULTIPLE-REGRESSION COEFFICIENT For the beta coefficient the standard error is estimated by the formula

$$s^2_{\beta_{12.34\ldots m}} = \frac{1 - R^2_{1.234\ldots m}}{(1 - R^2_{2.34\ldots m})(N - m)} \quad \text{(Variance error of a beta coefficient)} \tag{16.9a}$$

The new symbol here is $R_{2.34\ldots m}$, which is a multiple R with X_2 as the dependent variable and all other variables except X_1 as independent variables. There would be one of these standard errors for each of the independent variables in turn, each being substituted for X_2. For a three-variable problem, the R in the denominator reduces to r_{23}. Note that this formula gives the *variance error,* i.e., s^2.

For the b coefficient, the standard error is estimated by

$$s_{b_{12.34\ldots m}} = \frac{s_{1.234\ldots m}}{s_{2.34\ldots m}\sqrt{N - m}} \quad \text{(Standard error of a } b \text{ coefficient)} \tag{16.9b}$$

Needed in the denominator for each independent variable in turn is the standard error of estimate of that variable from all other independent variables. Beyond a three-variable problem this becomes quite laborious, but in the latter the denominator term reduces to s_{23}. Unlike the preceding formula, this gives the standard error *without* the need for extracting a square root after the equation is solved.

The chief use of these standard errors is to test the null hypothesis, to determine whether each independent variable has anything at all to contribute to prediction when its relation to other variables is taken into account. If the obtained beta or b is not significantly different from zero, that variable might well be dropped from the regression equation, and a new equation derived.

SIGNIFICANCE OF A DIFFERENCE BETWEEN MULTIPLE R's We sometimes want to know whether the multiple R with more independent variables included is significantly greater than the R with a smaller number of variables selected from the original set. There is available an F test for such a difference. The formula for computing F for this purpose reads

$$F = \frac{(R_1^2 - R_2^2)(N - m_1 - 1)}{(1 - R_1^2)(m_1 - m_2)} \tag{16.10}$$

where R_1 = multiple R with larger number of independent variables
R_2 = multiple R with one or more variables omitted
m_1 = larger number of independent variables
m_2 = smaller number of independent variables

In the use of the F tables, the df_1 degrees of freedom are given by $(m_1 - m_2)$, and the df_2 degrees of freedom by $(N - m_1 - 1)$.

Some principles of multiple correlation

Although multiple-correlation problems may be extended to any number of variables, before we consider the solution with more than three, it is desirable to examine some of the general principles which apply for any number of variables but which can be seen more clearly when there are only three.

The two main principles are that (1) a multiple correlation increases as the size of correlations between dependent and independent variables increases and (2) a multiple correlation increases as the size of intercorrelations of independent variables decreases. A maximum R will be obtained when the correlations with X_1 are large and when intercorrelations of $X_2, X_3, \ldots, X_m$ are small. In building a battery of tests to predict a criterion, test makers should try to maximize the validity of each test and to minimize the correlations between tests. There are limitations to the application of these objectives, however, and in practice they tend to conflict, as we shall see. There are also apparent exceptions to the rules, as examples will show. The whole story is not contained in the two principles as stated.

SOME TYPICAL COMBINATIONS OF r_{12}, r_{13}, AND r_{23}

Table 16.3 provides some examples of various combinations of correlations among three variables that enter into a multiple-correlation problem. The mathematically wise student will be able to predict the kind of outcome in each instance from a general inspection of formula (16.1). Repeated here for ready reference, it is

$$R_{1.23}^2 = \frac{r_{12}^2 + r_{13}^2 - 2r_{12}r_{13}r_{23}}{1 - r_{23}^2}$$

If the correlation r_{23} is zero, the third term in the numerator is zero, which has a tendency to make $R_{1.23}$ larger. On the other hand, there is a distinct advantage in having r_{23} very large, because of its role in the

TABLE 16.3
Examples of multiple correlations in a three-variable problem when intercorrelations vary

Example	r_{12}	r_{13}	r_{23}	$R^2_{1.23}$	$R_{1.23}$
1	.4	.4	.0	.3200	.57
2	.4	.4	.4	.2286	.48
3	.4	.4	.9	.1684	.41
4	.4	.2	.0	.2000	.45
5	.4	.2	.4	.1619	.40
6	.4	.2	.9	.2947	.54
7	.4	.0	.0	.1600	.40
8	.4	.0	.4	.1905	.44
9	.4	.0	.9	.8421	.92
10	.4	.2	−.4	.3143	.56
11	.4	−.4	−.4	.2286	.48

denominator. If r_{23} approaches 1.0, the denominator approaches zero. Even though the numerator may become small, under these conditions R can be quite large. A large R is thus favored by having r_{23} either very small or very large. This principle should be added to the two mentioned above. But it should also be said that a large r_{23} is more effective when the independent variables are unequally correlated with the dependent variable, and particularly when one of the correlations is very small.

Note the first example in Table 16.3, in which $r_{23} = .0$. For this event, formula (16.1) reduces to

$$R^2_{1.23} = r_{12}{}^2 + r_{13}{}^2 \qquad (16.11)$$

(Multiple R when intercorrelation of two independent variables is zero)

In other words, when independent variables have an intercorrelation of zero, the proportion of variance predicted by their combination is equal to the sum of the proportions of variance predicted by each separately. This holds for any number of independent variables whose intercorrelations are zero. A psychological interpretation of this is that when intercorrelations between predictive measures are zero, the total contribution of each to the prediction of a complex criterion containing all the things predicted is unique.

Note next the second and third examples and compare them with the first. In all three, the r_{12} and r_{13} correlations remain constant at .4, while r_{23} increases first to .4 and then to .9. As this happens, R goes from .57 to .48 to .41. In the last instance r_{23} is so high that there is practically no gain from combining the two variables X_2 and X_3. We shall see a modified result in the next three examples.

In examples 4 to 6, r_{12} remains constant at .4 and r_{13} remains constant at .2, while r_{23} varies from .0 to .9. In the first of these three we find formula (16.11) verified. The two variances sum to .2000, and R

is .45. As r_{23} increases to .4, R shrinks back to approximately .40. Thus we can conclude that if one test has a validity of .4, it may pay to add to it another with a validity of only .2, provided the two tests intercorrelate zero. But if there is any appreciable correlation between them, or only a moderate correlation, it would not pay.

What happens if we increase r_{23} still more? When it is as high as .9, R jumps to .54. This supports the third principle stated above: that r_{23} should be either very low or very high. One may ask why this principle does not appear to work in the first three examples. The answer is that it is obscured by the relation of r_{12} and r_{13}. In those examples r_{12} equals r_{13}, and in the next three examples these correlations are unequal. A better explanation is that one of them is very small. One may well ask what psychological meaning is involved in the increase in R when r_{23} is very large. This is best explained in connection with the next three examples.

In examples 7 to 9, r_{12} and r_{13} are still more uneven in size. They also have special interest because $r_{13} = .0$ in all three, while r_{23} varies from .0 to .4 to .9, as in the previous groups of three examples. It would seem, at first thought, that any test that correlates zero with a criterion would have no value in predicting that criterion. It is true that alone it has no value whatever for doing so. But it is not true if that test is combined with other tests with which it correlates. In example 7, the commonsense expectation is vindicated. The addition of an invalid test would offer no improvement. It would simply receive a regression weight of zero, which means it would not be included in the regression equation. But note that when r_{23} increases to .4, R becomes .44, and when r_{23} is .9, R becomes .92. Clearly a test with zero validity may add materially to prediction if it correlates substantially with another test that is valid.

SUPPRESSOR VARIABLES The psychological significance of this state of affairs is best explained by factor theory (see Chap. 18). Roughly, the explanation is that variable X_2, in spite of its positive correlation with X_1, has some variance in it that correlates zero, or perhaps even negatively, with the criterion. This same variance prevents X_2 from correlating as highly as it might with X_1. Variable X_2 correlates with X_3 because they have in common that variance not shared by X_1. In this kind of situation we find that X_3 acquires a *negative* regression weight, although it may correlate only zero, and not negatively, with the criterion. Such a variable is a *suppressor variable*. Its function in a regression equation is to suppress in other independent variables that variance which is not represented in the criterion but which may be in some variable that does otherwise correlate with the criterion.

An example of this came to the authors' attention in testing for pilot selection. It was a consistent finding that a vocabulary test, which is as pure a measure of the verbal-comprehension factor as we have, correlated zero or even slightly negatively with the criterion of success in pilot training (within the range of that ability among high-school graduates). The same kind of test correlated substantially with a reading-comprehension test, which also correlated positively with the pilot-training criterion. The reading test correlated *positively* with the criterion because it measured, besides verbal comprehension, such factors as mechanical experience and visualization, which were also component variances in the criterion. The combination of a vocabulary test with the reading test, with a negative weight for the vocabulary test, would have improved predictions over those possible with the reading test alone.

MULTIPLE-*R* PRINCIPLES IN LARGER BATTERIES

The principles illustrated above for the three-variable problems also apply in larger combinations of variables. The first two principles can be well illustrated by taking other hypothetical examples like those in Table 16.4. There we have a demonstration of how multiple *R*'s behave as the number of independent variables increases from 2 to 20 and as intercorrelations increase from .0 to .6.

Following Thorndike's choices, we shall assume that each variable correlates with a criterion to the extent of .3. This is a rather low validity coefficient and is about the lower limit of usefulness for a single test or other predictive device. We shall see, however, how valuable such instruments may be when combined in a battery, provided their intercorrelations are not too high.

In the second row of Table 16.4, when two such tests are combined, we see how the multiple R decreases from .42 when r_{23} is zero to .34 when r_{23} is .60. In each row the same expected phenomenon occurs: a decrease in R as intercorrelations increase. Inspection of the columns shows how R increases as we add more tests, having similar correlations, to the battery and how the gain in R continues up to a battery of 20, except for the case of zero intercorrelations, for which the limit of $R = 1.0$ was passed when the number of tests exceeded 11. In

TABLE 16.4
Multiple correlations from different numbers of independent variables, each correlating .30 with the dependent variable but with intercorrelations varying*

Number of independent variables	*Intercorrelations* .00	.10	.30	.60
1	(.30)	(.30)	(.30)	(.30)
2	.42	.40	.37	.34
4	.60	.53	.44	.36
9	.90	.67	.48	.37
20		.79	.52	.38

* Adapted from Thorndike, R. L. (Ed.) Research problems and techniques. *AAF Aviation Psychology Research Program Reports*, No. 3. Washington: GPO, 1947.

this situation (zero intercorrelations) the principle of formula (16.11) still applies. The proportion of predicted variance contributed by each test would be .09, and 11 tests would yield an R^2 of .99 and a multiple R of .995. In other columns the increases of R are less dramatic, but except in the last column, and perhaps in the one preceding, it would apparently pay to continue adding new tests until the 20 were included. Considerations of administrative effort would have to be balanced against gains in R.

Table 16.4 tells an even more important story. The value of having zero intercorrelation among tests in a battery is obvious. If one tries to achieve zero intercorrelations among tests, when each test measures a unique factor, however, one will often find that each test tends to correlate low with the criterion. This is because a practical criterion—e.g., training achievement or job performance—is usually a complex variable; it has a number of component variances, each component being a common factor (for a discussion of factor theory see Chap. 18). If one tries to increase the correlation of a single test with a criterion, the result is almost invariably an increase in the factorial complexity of the test; that is, more different factor variances are introduced. This automatically raises the correlation of this test with other tests because they have more factors in common. That is why in practice, the two principles mentioned first involve conflicting objectives.

Where there has to be a choice, it seems wisest to give less attention to the first principle (maximizing correlation of each test with the criterion) and greater attention to the second (minimizing intercorrelations). If there are 20 independent factors represented in a practical criterion, and if each is of equal importance, each would contribute .05 of the total variance. Each test, measuring only one of the factors, would need to correlate only $\sqrt{.05}$, which is .224, with the criterion. In this case, raising the correlation between any one test and the criterion would be of little use. There would be no objection to a higher correlation. Appropriate weighting would bring the test's contribution to prediction down to required proportions. Thus, it can be concluded that low correlations of tests with practical criteria can be tolerated, provided we can combine enough tests in a battery and provided their intercorrelations are near zero.[1]

Multiple correlation with more than three variables

With more than three variables, a good solution for a regression equation and for a multiple R may be carried out by what is known as the

[1] For a more detailed discussion of these problems, see Guilford, J. P. New standards for test evaluation. *Educational and Psychological Measurement,* 1946, **6,** 427–438.

TABLE 16.5 Solution of the regression coefficients for the multiple-regression equation

(1)	(2) β_{1k}	(3) r_{1k}	(4) $\beta_{1k}r_{1k}$	(5) S_1/S_k	(6) b_{1k}	(7) $\bar{X}_k$		(8) $(-\bar{X}_k)b_{1k}$
X_2	.1039	.465	.048314	1.750	.182	19.7		− 3.585
X_3	.3703	.583	.214885	.535	.198	49.5		− 9.801
X_4	.3022	.546	.165001	.469	.142	61.1		− 8.676
X_5	.1607	.365	.058655	2.459	.395	29.7		−11.732
			Σ .487855 = R^2				Σ	−33.794
			.698 = R				$\bar{X}_1$	73.800
							a =	40.006

Doolittle method. Because the computing operations are elaborate and because electronic computers are now more commonly available, this procedure will not be described here. A step-by-step description may be found in the earlier editions of this volume.[1] References to computer programs for dealing with multiple-regression analysis are also cited.[2]

When the Doolittle method was applied to all the data in Table 16.1, the following regression equation was found:

$$X_1' = 40.0 + .182X_2 + .198X_3 + .142X_4 + .395X_5$$

Using this equation and the information from scores in the four tests, we could compute a predicted average grade (X_1) for every student. The multiple R turned out to be .698, which is only a trifle greater than the R obtained earlier with only two predictors. That R equaled .677, with the two predictors X_3 and X_4.

Table 16.5 illustrates how the regression weights (b) and the multiple correlation may be obtained from the beta coefficients, when they are known, and other information. The betas are given in column 2. The corresponding correlations r_{1k} are given in column 3 (where k stands for variables X_2 to X_5 inclusive). The sum of the βr products equals R^2, as from an extension of formula (16.5). To obtain the b coefficients, we need ratios of S_1 to standard deviations S_2 to S_5 inclusive (see column 5) in order to apply extensions of formulas (16.2*a*) and (16.2*b*). The derivation of the constant a in the regression equation in-

1 Guilford, J. P. *Fundamental statistics in psychology and education.* (5th ed.) New York: McGraw-Hill, 1973.

2 Veldman, D. J. *FORTRAN programming for the behavioral sciences.* New York: Holt, 1967. Also Dixon, W. J. (Ed.) *BMD, biomedical computer programs.* Berkeley and Los Angeles: University of California Press, 1976. Nie, N. H., Hull, C. H., Jenkins, J. G., Steinbrenner, K., and Bent, D. H. *Statistical package for the social sciences.* (2d ed.) New York: McGraw-Hill, 1975.

volves application of an extended formula (16.4). Column 8 of the table takes care of these operations.

Interpretations of these results can be made as before. From the R^2 of .488, we can conclude that predictions from the obtained regression equation account for 48.8 percent of the variance in the predicted variable X_1'.

In general, when there are more than two predictor variables, it is difficult to untangle the variance accounted for in the dependent variable and to attribute portions of it to individual independent variables. One method which attempts to do this is commonality analysis.[1] An appreciation of the relative contributions of the independent variables in predicting the criterion variable is not readily grasped by simple inspection of the multiple-regression weights. The β^2's represent the independent contributions of the corresponding X's, but the joint contributions of the variables taken in all combinations must also be considered, and there is no unique way of doing it. Also, the interpretations are not general but apply to the set of predictors in the order in which they are listed.

OBTAINING A SHORT PREDICTIVE BATTERY

It may have been noted that the inclusion of two additional predictors in the regression equation improved accuracy of prediction only a little (48.8 percent, as compared with 45.8 percent of the variance in X_1 accounted for). This is a fairly frequent finding, but not a universal one by any means. In the earlier discussion of the principles of multiple correlation, it was noted that low intercorrelations of predictors are an important condition for high multiple R's. If certain variables add very little to prediction, it is likely that they measure one or more of the common factors that are covered by stronger predictors, in a redundant manner. The general question arises as to whether an added variable contributes enough to predictors already in the composite to justify its inclusion. There is a statistical answer to this problem.

The solution lies in what is called a *stepwise multiple-regression analysis*. In principle, the operation begins with selection of the test that by itself has the highest correlation with the dependent variable or criterion. In the illustrative problem represented in Table 16.1, the composite would be started with the analogies test X_3, which correlated .583 with X_1. The procedure then selects by computational steps the test that would make the largest gain in prediction. These operations are best left to a computer. At this point the computer would find the multiple R for the combination of the two best predictors, and it would make an F test to determine whether the new R is significantly

1 Kerlinger, F. N., and Pedhazur, E. J. *Multiple regression in behavioral research.* New York: Holt, Rinehart and Winston, 1973.

greater than the correlation without the last addition. Formula (16.10) provides the basis for the F test. The addition of variables would cease when the probability associated with the obtained F rose above an adopted alpha level.

In the prediction of a practical criterion, such as success in pilot training or achievement in algebra or geometry, instead of just trying out any standard tests that happen to be available for predictors, a more rational approach is recommended. First, it can be assumed that any such criterion is complex; it involves a number of abilities, which are known as *factors* and which are largely independent and therefore have low intercorrelations. One would then select for tryout some tests representing those hypothesized abilities. There would still be a use for the stepwise approach to multiple-regression analysis, for some predictor variables might not add enough prediction to achieve a significant F ratio. Even a variable that does add significantly might not contribute enough, as indicated by its βr product, to be useful, considering the cost of administering it. A decision to exclude it in that particular situation, however, would have to be made partly on other than statistical grounds.

Other combinations of measures

The regression equation is a means of combining different measures of the same object in order to derive a composite measure or score. The scores are summed, each weighted by its regression coefficient. There are other ways of combining scores to form a composite. For example, one might simply sum the raw scores for each person without applying differential weights. This is the common practice in deriving total scores of tests composed of subtests of different kinds, though in some cases there is some effort at weighting, for example, multiplying one score by 2, another by 3, and so on.

Actually, every test that is composed of items may be regarded as a *battery* of as many tests as there are items. The total score is usually an unweighted summation of the item scores. Rarely does a test maker resort to the determination of regression weights for test items, but the same principle that applies to test batteries could be adapted to single tests composed of parts. More often than not, even in the case of test batteries, there are so many parts, or they are used to predict in such a variety of situations, that there is not sufficient incentive to work out all the regression weights that would be required.

It is important to know some of the better substitute procedures for the multiple-regression equation and to be able to evaluate the effectiveness of a composite derived by any method. The multiple R applies only when the optimal regression weights are used; other weights will yield a composite that is likely to correlate less with the criterion. There are other problems connected with composite scores that call for attention, including that of what mean and what standard deviation will result when measures are combined each with a certain weight. These problems will be dealt with in the following paragraphs.

MEANS OF WEIGHTED COMPOSITES

When several measures of the same object are summed, each with its own weight, the mean of the same kind of composite for a sample of objects is given by the equation

$$\bar{X}_{ws} = \Sigma w_i \bar{X}_i \qquad \text{(Mean of a sum of weighted measures)} \qquad (16.12)$$

where w_i = weight applied to each variable X_i, when i varies from 1 to n in a list of n variables, and $\bar{X}_i$ = mean for the same sample of objects in variable X_i.

If we apply this formula with the b weights computed for the regression equation in the prediction of average freshman grades (see p. 384), the solution would be

$$\bar{X}_{ws} = (.182)(19.7) + (.198)(49.5) + (.142)(61.1) + (.395)(29.7)$$
$$= 33.8$$

Thus, the mean of the composite of four variables, including X_2 (arithmetic test), X_3 (analogies test), X_4 (high-school average), and X_5 (interest score), weighted with the coefficients .182, .198, .142, and .395, respectively, would be 33.8. This value is 40.0 units short of the mean for the criterion (freshman grades). By adding the difference (40.0), which is the a coefficient of the complete regression equation, we obtain a composite mean that coincides with that of the criterion. In other words, this discussion demonstrates the need for the a coefficient in the complete regression equation. If we were not interested in achieving that mean, we could drop the constant 40.0 and be left with a mean of 33.8.

STANDARD DEVIATIONS OF WEIGHTED COMPOSITES

We can likewise estimate the standard deviation of a composite measure when each component has a multiplier or weight. The computation of this statistic may be clearer, however, if we consider the standard deviation of a simple unweighted sum first.

THE STANDARD DEVIATION OF SUMS WHEN WEIGHTS ALL EQUAL 1 When scores from different tests are summed without applying differential weights, we may regard the weight for each test to be +1. When *two* scores are summed to make the composite, the variance of the composite scores is given by the equation

$$S_s^2 = S_1^2 + S_2^2 + 2r_{12}S_1S_2 \qquad \text{(Variance of a sum of two unweighted measures)} \qquad (16.13)$$

where S_1^2 and S_2^2 = variances of the components and r_{12} = coefficient of correlation between the two components.

The expression $r_{12}S_1S_2$ is the *covariance* of the two components. Its relation to correlation can be better shown by relating it to the Pearson formula, in which

$$r_{12} = \frac{\Sigma x_1 x_2}{NS_1S_2}$$

If we multiply both sides of this equation by S_1S_2, we have

$$r_{12}S_1S_2 = \frac{\Sigma x_1 x_2}{N}$$

The parallel between the term at the right and the expression for a variance should be obvious. A variance is of the form $\Sigma x_1^2/N$ or $\Sigma x_2^2/N$. A covariance is the mean of the cross products of deviations; a variance is a mean of the squares of deviations. With this new information as background, we may translate equation (16.13) into English by saying that the variance of a composite is equal to the sum of variances of the components plus twice the covariances of all pairs of those components. This is a general principle that is important to remember.

From equation (16.13) it follows, by taking square roots, that

$$S_s = \sqrt{S_1^2 + S_2^2 + 2r_{12}S_1S_2} \qquad \text{(Standard deviation of the sum of two unweighted measures)} \qquad (16.14)$$

A demonstration of how this works out in a particular sample is given in Table 16.6. Ten scores are given for the same individuals in X_a

TABLE 16.6 The variance and variability of a composite score that is the unweighted sum of two uncorrelated scores

Individual	X_a	x_a	x_a^2	X_b	x_b	x_b^2	X_c $(X_a + X_b)$	x_c	x_c^2
A	1	−4	16	6	0	0	7	−4	16
B	3	−2	4	7	+1	1	10	−1	1
C	4	−1	1	4	−2	4	8	−3	9
D	5	0	0	10	+4	16	15	+4	16
E	5	0	0	8	+2	4	13	+2	4
F	5	0	0	0	−6	36	5	−6	36
G	5	0	0	6	0	0	11	0	0
H	6	+1	1	8	+2	4	14	+3	9
I	7	+2	4	5	−1	1	12	+1	1
J	9	+4	16	6	0	0	15	+4	16
Σ	50	0	42	60	0	66	110	0	108
$\bar{X}$	5.0		4.2	6.0		6.6	11.0		10.8
S			2.05			2.57			3.29

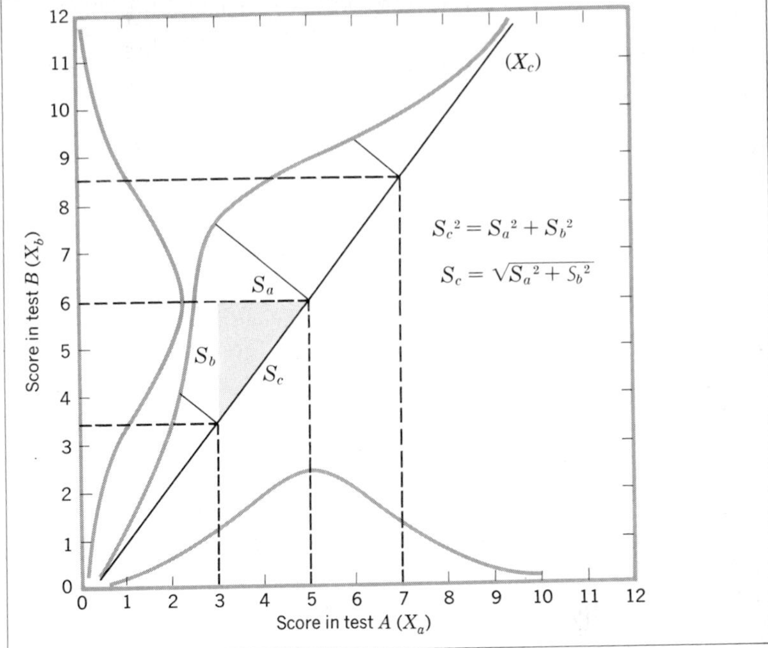

FIGURE 16.3 The way in which the standard deviation of an unweighted sum of two scores is related to the standard deviations of those two scores taken separately, when the two are uncorrelated

and in X_b between which the correlation r_{ab} equals zero. If $r = .0$, the third term in formula (16.13) drops out, and the variance of the composite is merely the sum of the variances of the components.

In the illustration in Table 16.6 the variances of the two components are 4.2 and 6.6 respectively. Their sum is 10.8, which checks with the mean square found from variable X_c. The way in which variances combine is also demonstrated in Fig. 16.3, which pictures hypothetical distributions for X_a, X_b, and their sum X_c. The position of the scale for X_c is determined by the juncture of the lines erected at distances of 1S from the means of X_a and X_b. The slanted scale of X_c is closer to that of X_b, consistent with the fact that X_b contributes more variance to it than X_a does and the fact that the composite correlates higher with X_b than with X_a. But these are incidental considerations here. The important demonstration is that when two variables like X_a and X_b are uncorrelated, we may regard the standard deviation of their composite X_c as the hypotenuse of a right triangle, of which S_a and S_b are the legs. The old, familiar Pythagorean theorem thus applies to the variance of the summation of two independent variables.

VARIANCE OF A COMPOSITE OF MORE THAN TWO COMPONENTS Equation (16.13) can be extended to include any number of unweighted components. For each component there would be a term for its variance, but there would be as many covariance terms to include as there are *pairs* of components. With three components there would be three covariance terms: $2r_{12}S_1S_2$, $2r_{13}S_1S_3$, and $2r_{23}S_2S_3$. Where there are n components, there are $n(n-1)/2$ pairs to consider. In terms of a general formula,

$$S_s^2 = \Sigma S_i^2 + 2\Sigma r_{ij}S_iS_j \quad \text{(Variance of a sum of any number of unweighted components)} \tag{16.15}$$

where S_i^2 = variance of any one component X_i
r_{ij} = correlation between any component X_i and any other component with a higher subscript number
S_i and S_j = standard deviations of the two components correlated

VARIANCE OF A COMPOSITE OF WEIGHTED COMPONENTS

When the components are weighted differently, the variance of the composite will reflect the weights. Let us begin with the special case of two components. If the summation equation is of the form

$$X_{ws} = w_1X_1 + w_2X_2$$

the variance of X_{ws} is given by the equation

$$S_{ws}^2 = w_1^2S_1^2 + w_2^2S_2^2 + 2r_{12}w_1S_1w_2S_2 \quad \text{(Variance of a composite of two weighted components)} \tag{16.16}$$

where w_1 and w_2 = weights applied to components X_1 and X_2, respectively.

As an example of this type of problem, let us use the data on X_4 and X_5 in Table 16.1. If these two variables are used in a composite to predict X_1, the least-squares solution gives b weights of .224 and .491, respectively, and a multiple R, based upon these weights, of .578. The predicted X values based upon the equation $X_1' = .224X_4 + .491X_5$ would be expected to have a standard deviation equal to $R_{1.45}$ times S_1 (since $R_{1.45} = S_{x'_1}/S_{x_1}$). This product is .578 × 9.1, which equals 5.26. Let us see whether formula (16.16) will lead to the same result. By substituting the appropriate values,

$$\begin{aligned} S_{ws}^2 &= (.224^2)(19.4^2) + (.491^2)(3.7^2) \\ &\quad + 2(.345)(.224)(19.4)(.491)(3.7) \\ &= 27.6319 \end{aligned}$$

from which

$$S_{ws} = 5.26$$

This agrees exactly with the expectation.

With weights of +1 for both X_4 and X_5, application of formula (16.13) would have given

$$\begin{aligned} S_s^2 &= 19.4^2 + 3.7^2 + 2(.345)(19.4)(3.7) \\ &= 439.5782 \end{aligned}$$

from which

$S_s = 21.0$

VARIANCE OF A COMPOSITE OF ANY NUMBER OF WEIGHTED COMPONENTS When there are more than two components, each weighted differently, the variance of the composite is given by the general formula

$$S_{ws}^2 = \Sigma w_i^2 S_i^2 + 2\Sigma r_{ij} w_i S_i w_j S_j \qquad \text{(Variance of a sum of any number of weighted components)} \qquad (16.17)$$

where w_i = weight assigned to variable X_i, where i takes on values 1 to n in turn, there being n predictors

r_{ij} = correlation between X_i and any other variable X_j, where j is a subscript greater than i

S_i and S_j = standard deviations of X_i and X_j, respectively

We could apply formula (16.17) to the four components of the regression equation predicting freshman grades with the appropriate b weight substituted for w in each case. We should find that the standard deviation is equal to R times S_1, which is .698 × 9.1 = 6.35. The inclusion of variables X_2 and X_3 in the regression equation raises the dispersion of the predicted grades from 5.26, which it would be with X_4 and X_5 only, to 6.35.

ACHIEVING ANY DESIRED STANDARD DEVIATION IN A COMPOSITE

In using regression equations, the dispersion of the predictions falls short of that of the obtained values because in the product RS_1, R is less than 1.0. This is all right and proper when we are interested in predicting an individual's most probable measure on the scale of obtained measures in X_1. The regression of predictions toward the general mean is a natural phenomenon of imperfect correlation, as was pointed out before (Chap. 15). There may be other uses of composites, however, that call for values other than those given by the regression equation. Suppose that we wanted predictions to spread just as much as the obtained values do, or suppose that we wanted them to be dispersed with some standard variability, for example, with an S of 10.0, as on a T scale, or an S of 2.0, as on a C scale (see Chap. 19). The way that such a goal can be achieved will now be explained.

Fortunately for the solution of this problem, it is not the absolute sizes of the weights that matter, but their ratios to one another. As long as they bear the same relations to one another, the correlation of the composite with some criterion will remain the same. Consequently, we could double, triple, or otherwise change the regression weights by some common multiple, without affecting the predictive value, if all we want is to predict individuals in the same relative positions in a distribution.

The S of the predictions is always related to the S of the obtained values by the extent of the correlation (when optimal weights are used). In a multiple-regression problem, the S of the predicted values equals R times the S of the obtained values. We can therefore make the S of the predictions equal the S of the obtained values by dividing each regression coefficient by R. An adjusted b coefficient, then, would be computed by the formula

$$b'_{12.34\ldots m} = \beta_{12.34\ldots m}\left(\frac{S_1}{S_2 R_{1.23\ldots m}}\right) \qquad \text{(16.18)}$$

(Regression coefficient adjusted to make the S of a composite equal S_1)

If the S desired in the composite is 10, 2, or any other chosen quantity, this could be achieved by substituting that quantity for S_1 in formula (16.18).

ACHIEVING ANY DESIRED MEAN FOR A COMPOSITE

In the optimally weighted regression equation, in order to make the mean of the predictions equal that of the obtained values, the a coefficient is introduced. The computation of a is given by formula (16.4). After one has determined any weights whatever to apply to the raw scores of the components of a composite measure, the same formula (properly extended) can be applied, putting in the place of $\bar{Y}$ any desired quantity. This is true because of the reasoning involved in the computation of the mean of a composite [see formula (16.12)]. Thus, if we had wanted the mean of the grades predicted by the regression equation to be 50, we would have substituted 50 for 73.8, the actual mean of the grades. The only practical restriction would be to choose a mean such that no composite measures would be negative. This means that any chosen mean should be at least 2.5 to 3.0 times the standard deviation of the composite.

SUBSTITUTES FOR THE OPTIMAL REGRESSION WEIGHTS

While regression weights derived from least-squares solutions, or weights proportional to them, yield the greatest accuracy of prediction from the variables available (in the particular sample), it is often expedient in the practical situation to deviate from the refined solution. It can be shown that we may substitute weights that approximate the regression coefficients, even very roughly at times, and still not affect the degree of correlation very much. Instead of applying weights to three decimal places, one significant digit will often suffice, in other words, simple integral weights.

In predicting freshman grades from high-school average and interest score combined, for example, the optimal weights were found to be .224 and .491. We might in practice round these to .2 and .5, respectively. It will be shown later[1] that the change in correlation between X'_1

[1] Methods for correlating composites or sums, either weighted or unweighted, will be described shortly.

and X_1 in the two cases is from .578, with the three-digit weights, to .577, with the one-digit weights. This loss is quite trivial. We could use weights of 2 and 5 if we so chose. Suppose we want even a simpler ratio of the two weights, such as ½ rather than ⅖. With weights of 1 and 2, the correlation of composites and grades would be .577. With equal weights the correlation would drop to .570. Even this much loss could be tolerated.

Before the reader draws the conclusion from this isolated example that all differential weighting is unnecessary, however, it is important to consider some points not yet brought out. There is no reason to believe that this is a typical example. Ordinarily, the larger the number of independent variables in a composite, the more one can depart from the weights demanded by least-squares solutions and yet maintain a high level of correlation between that composite and a criterion. This is why with a test composed of many items, we may forget to bother with differential weighting. In a two-variable composite, however, we have the minimum number of multiple predictors. We should therefore expect to find the validity of the composite to be sensitive to changes in weights.

Roughly, the explanation in this example is that X_4 (high-school average) has a beta weight about 2.4 times that for X_5 (interest score), and it has a standard deviation about five times as large as that for X_5. Even when X_4 and X_5 have the same weight in the composite, X_4 contributes to the composite in proportion to its standard deviation. This follows from equation (16.13), in which it is shown that *without* differential weights (and with zero intercorrelations) *each part's contribution to total variance is proportional to its own variance.* Without differential weighting of variables in the equation, then, X_4 is still weighted much more than X_5. This illustrates a fact that is not often realized. It is usually assumed that the act of summing several scores weights those scores equally. As a rule, it does not; *it weights them in proportion to their standard deviations.*

It should help to keep in mind that the optimal *b* coefficients are directly proportional to the beta coefficients and *inversely* related to the standard deviations of the predictor variables, as indicated in formulas (16.3*a* and *b*). Since the standard deviation of a variable is proportional to the range of scores, and since the latter is roughly proportional to the length of test, predictor weights should be inversely proportional to those two features of tests, also. Whatever weights might be adopted for other reasons might well be divided by the respective standard deviations.

Some investigators believe it important to consider reliabilities of measures in weighting them in combinations. By "reliability" is meant consistency of scores as indicated by some kind of a self-correlation. If regression weights have been computed, reliabilities have been automatically taken into account, and no modification of the weights for reliability would be necessary. But if some other method is used to arrive at weights and if the measures combined differ markedly in reliability, then some index of reliability should be considered. This tends to prevent "errors of measurement" in the less reliable instruments from being given too much weight. If reliability coefficients have been computed, the weight contributed from this source should be the square root of each reliability coefficient, rather than the reliability coefficient itself. The type of reliability coefficient should be one indicating internal consistency, i.e., an odd-even type or a Kuder-Richardson type (see Chap. 17).

THE CORRELATION OF COMPOSITE MEASURES WITH OTHER MEASURES

The multiple R is only one index of correlation between a composite measure and some other measure. To test the predictive value for composites with other than optimal weights, we have procedures called collectively the *correlation of sums*. The components may be unweighted (i.e., each weight is +1) or differentially weighted.

CORRELATION OF A COMPOSITE OF UNWEIGHTED MEASURES The simplest case is solved by the equation

$$r_{cs} = \frac{r_{c1}S_1 + r_{c2}S_2}{\sqrt{S_1^2 + S_2^2 + 2r_{12}S_1S_2}} \qquad \text{(Correlation of a sum of two unweighted components with a third variable)} \qquad (16.19)$$

where S_1 and S_2 = standard deviations of the two components and r_{c1} and r_{c2} = correlation of each component with the composite variable.

Let the illustrative summation equation be $X_s = X_4 + X_5$, where X_s stands for a sum of X_4 and X_5, which in recent illustrations have stood for high-school average and interest scores, respectively. What is the correlation of X_s with freshman grades, which here are symbolized by X_c? Applying formula (16.19),

$$r_{cs} = \frac{(.546)(19.4) + (.365)(3.7)}{\sqrt{19.4^2 + 3.7^2 + 2(.345)(19.4)(3.7)}}$$
$$= .570$$

When there are more than two components, the more general formula for the same kind of correlation is

$$r_{cs} = \frac{\Sigma r_{ci}S_i}{\sqrt{\Sigma S_i^2 + 2\Sigma r_{ij}S_iS_j}} \qquad \text{(Correlation between a sum of unweighted variables and another single variable)} \qquad (16.20)$$

where r_{ci} = correlation between any one component X_i and the outside single variable (i varies from 1 to n)

S_i = standard deviation of the same component

r_{ij} = correlation between X_i and any other component X_j, when j is a higher subscript number than i.[1]

CORRELATION OF A COMPOSITE OF WEIGHTED MEASURES When there are two components, each weighted differently, the correlation with a third measure is given by

$$r_{c\,ws)} = \frac{w_1 r_{c1} S_1 + w_2 r_{c2} S_2}{\sqrt{w_1{}^2 S_1{}^2 + w_2{}^2 S_2{}^2 + 2r_{12} w_1 S_1 w_2 S_2}} \tag{16.21}$$

(Correlation of a sum of two weighted measures with a third measure)

where w_1 and w_2 = weights attached to measures X_1 and X_2, respectively.

For the combination of high-school average and interest scores, let us assume weights of 2 and 5, respectively. These are closely proportional to the b coefficients of .224 and .491, respectively. Applying formula (16.21),

$$r_{c(ws)} = \frac{2(.546)(19.4) + 5(.365)(3.7)}{\sqrt{4(19.4^2) + 25(3.7^2) + 2(.345)(2)(19.4)(5)(3.7)}}$$
$$= .577$$

Thus, crude, integral weights of 2 and 5 would give as high a correlation of the combination of X_4 and X_5 with X_1 (freshman grades) as the three-digit b coefficients .224 and .491 would.

For the general case, with more than two components, the correlation with an outside variable is

$$r_{c(ws)} = \frac{\Sigma w_i r_{ci} S_i}{\sqrt{\Sigma w_i{}^2 S_i{}^2 + 2\Sigma r_{ij} w_i S_i w_j S_j}} \tag{16.22}$$

(Correlation of a weighted sum with an outside variable)

Alternative summarizing methods

Summative equations represent only one way in which several measures may be combined in order to make single predictions or arrive at single decisions. There are alternative methods, some of which are better than regression equations in certain situations. The two chief contenders are the multiple-cutoff method and the profile method. These will be described, and their variations discussed.

MULTIPLE-CUTOFF METHODS In a multiple-cutoff method, a minimum qualifying score or measure is adopted for each variable used in making a joint prediction. A good example of the method is the medical examination required for qualifi-

[1] Here, as in similar formulas, $r_{ij}S_iS_j$ implies covariances of all possible pairs of variables.

cation of individuals for military service, life insurance, or employment. Failure to meet the standard on any one test may disqualify the individual. A particularly good showing in one respect is not ordinarily allowed to compensate for a poor showing in another. The phenomenon of compensation, which the regression-equation approach allows, is the chief difference between the two methods, in principle.

MULTIPLE CUTOFFS CONTRASTED WITH MULTIPLE REGRESSION A geometric illustration of the difference between the two methods may be seen in Fig. 16.4. The two variables represented there (X_2 and X_3) are both independent variables, used jointly to predict some criterion X_1, which is not shown. A moderate correlation, of approximately .40, is assumed between X_2 and X_3, as represented by an elliptical distribution of the population. Let us assume a selection problem and also assume that we have the alternatives of applying two cutoff scores X_{2c} and X_{3c} or of applying a single cutoff score based upon a weighted sum of X_2 and X_3. Assume also that we reject the same proportion of the applicants by either method.

The use of two cutoff scores would reject all individuals to the left of the point X_{2c} and a vertical line erected at that point, as well as all individuals below the point X_{3c} and a horizontal line drawn at that level. Some individuals would be rejected on the basis of either variable alone, and some on the basis of failure to meet standards on both. The single cutoff on the weighted composite, however, would be represented by a slanted line. This is consistent with the slanted-line

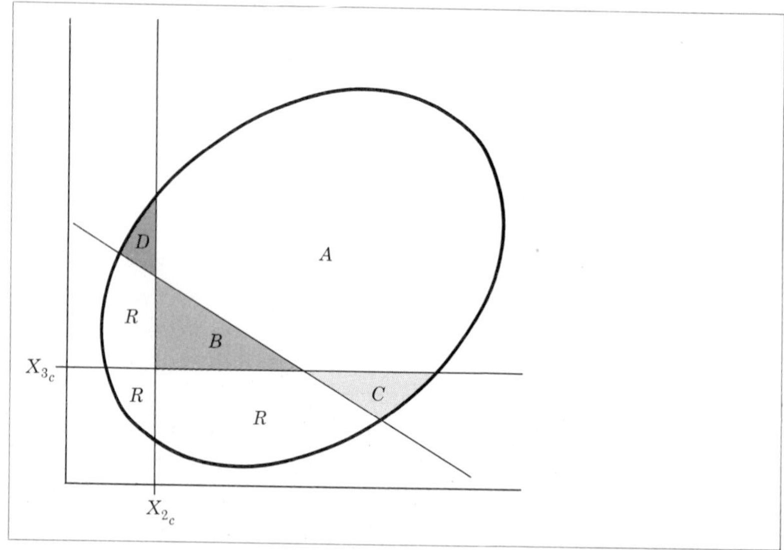

FIGURE 16.4 Geometric comparison of accepted and rejected personnel by the multiple-regression-equation method and the multiple-cutoff method, when approximately equal proportions are selected by either method. [After Thorndike, R. L. (Ed.) Research problems and techniques. *AAF Aviation Psychology Research Program Reports*, No. 3. Washington: GPO, 1947.]

system shown in Fig. 16.2. All individuals below and to the left of this slanted line would be rejected.

It is now possible to see which individuals would be accepted by the one method and rejected by the other and on which ones the two methods agree. The individuals in area *A* of the ellipse would be accepted by either method. The individuals in areas *R* would be rejected by either method. Individuals in area *B* would be rejected by the multiple-regression-equation method but would be accepted by the multiple-cutoff method. Individuals in areas *C* and *D* would be accepted by the regression method but rejected by the cutoff method—individuals in *C* for different reasons from those applying to individuals in *D*.

The crux of the comparison of values of the two methods lies in determining whether individuals in area *B* are any better in terms of the criterion than those in areas *C* and *D*. Individuals in area *B* are rejected by the one method because they combine low scores in X_2 and X_3. They just succeed in meeting minimum standards in both variables and so would be accepted by the other method. Individuals in areas *C* and *D*, although below standards in one variable, are allowed to present compensating strong scores in the other variable and hence to be accepted by the multiple-regression method. They are regarded as doubtful risks by the cutoff method.

It can be argued that not enough is known about compensatory effects in performances that serve as criteria, and that is quite true. There is a need for experimental studies of this kind. A vindication of the regression method, however, is found in the consistency with which composite scores continue to correlate as they do in line with multiple-correlation coefficients that forecast those correlations. If compensatory effects did not occur, there would probably be much more shrinkage in correlation of sums with criteria than there is.

AN EVALUATION OF THE MULTIPLE-CUTOFF METHOD If all regressions are linear, theoretically there should be no advantage in selection by multiple cutoffs over that by composites. This can be explained in general terms by the fact that in a linear regression there is a *continuous* improvement in criterion measures with increased score in an independent variable, and at a constant rate. Thus, as far as the relationship between the test and the criterion is concerned, there is no more reason for putting the cutoff at one point than there is for putting it at another. The cutoff would have to be established on the basis of some other determinants, such as success ratio or validity. In using a number of tests for selection for a single purpose, presumably it would be best to make the most rejections on the basis of the most valid test. When a regression is definitely curved, there is a real basis for using a cutoff on a single test. The cutoff would be established in line with the region of transition between low and high rates of increase in the criterion measure.

There are some practical difficulties in the administration of multiple cutoffs which make the method less appealing than a regression equation. There is the difficulty of establishing several different cutoff points which will take full advantage of the differences in validity between the tests and which will yield the appropriate numbers of qualified applicants. Once the minimum standards are established, however, the method is simple to apply. Failure to meet any one of the minimal scores automatically means rejection.

Rejection of an applicant on the basis of a single test is somewhat risky as compared with rejection on the basis of a composite score because of the fact that the reliability of a single test score is usually less than that of a composite. If the parts of a composite are positively intercorrelated, the total score is more reliable than the part scores.

PROFILE METHODS

For guidance work and clinical work in general there is usually preference for seeing an individual's scores represented in a pattern provided in a profile. A single summative score is unsuitable or may be unobtainable. A single composite score is unsuitable perhaps because the problem is not one of selection but of classification. In vocational guidance, clients are "sorted" into vocational categories. If there were single summative scores already established with satisfactory correlations with vocational criteria of many kinds, perhaps the profile method would be less important. Clinicians commonly express a desire to "see a personality in its totality," however, and a profile is one approach to this end.

There are several ways of using profiles. Some prefer to interpret an individual's profile intuitively, on the basis of a general impression of the plotted graph. Others prefer to match more definitely described job-requirement or adjustment-requirement patterns with individual trait patterns. It is possible, by means of careful research, to define certain adjustment requirements in terms of optimal scores in a number of different variables. This statement implies curved regressions, and that is precisely the condition which favors the choice of a profile method over a regression method.

Figure 16.5 demonstrates this kind of use of a profile. Through experience, it was found that female workers in a certain kind of routine task (pairing stockings in a factory) tended to be most suited to the job if they had scores in certain regions on the 13 traits scored in the Guilford-Martin personality inventories. Such workers were likely to be best if somewhat low in sociability, a little on the depressed and emotional side, less active than average (the task was sedentary), less dominating socially, somewhat beset with feelings of inferiority, somewhat subjective or hypersensitive, and perhaps none too agree-

C-Score	S	T	D	C	R	G	A	M	I	N	O	Ag	Co	C-Score
10	0	0 10	0 1	0 5	70+	24	35+	30 +	48+	41+	71+	56+	97 +	10
9	1 3	11 14	2 5	6 9	69 64	23 22	34 33	29 28	47 45	40 38	70 68	55 53	96 89	9
8	4 6	15 19	6 9	10 13	63 58	21 20	32 30	27 26	44 43	37 35	67 62	52 47	88 81	8
7	7 11	20 25	10 13	14 18	57 51	19 17	29 26	25 24	42 40	34 32	61 55	46 42	80 72	7
6	12 17	26 31	14 18	19 24	50 44	16 14	25 23	23 20	39 36	31 28	54 49	41 37	71 65	6
5	18 23	32 37	19 25	25 31	43 38	13 12	22 19	19 18	35 33	27 24	48 42	36 32	64 54	5
4	24 28	38 42	26 31	32 38	37 30	11 9	18 15	17 13	32 28	23 20	41 36	31 27	53 45	4
3	29 33	43 47	32 38	39 46	29 23	8 7	14 12	12 11	27 22	19 16	35 28	26 22	44 35	3
2	34 39	48 53	39 45	47 52	22 16	6 5	11 8	10 8	21 15	15 11	27 20	21 17	34 27	2
1	40 46	54 61	46 53	53 58	15 10	4 3	7 5	7 6	14 11	10 7	19 13	16 12	26 20	1
0	47+	62+	54+	59+	9 0	2 0	4 0	5 0	10 0	6 0	12 0	11 0	19 0	0
C-Score	S	T	D	C	R	G	A	M	I	N	O	Ag	Co	C-Score

FIGURE 16.5 The profile method of selection applied to personality-inventory scores. The clear portion of the chart represents what is believed to be the most favorable score ranges for personnel who are assigned to a certain routine kind of work. The scores of the worker whose profile is shown all fell within the clear region. (Courtesy of R. P. Kreuter, Hand Knit Hosiery Company, Sheboygan, Wis.)

able or cooperative. In most respects the tendencies listed would seem to present a generally "poor" personality picture. Low extremes were unfavorable, however; the general tendency was just average or slightly below in most traits. This is understandable in that such an individual is probably lacking in aspirations for positions that require the better qualities and is contented with routine work in which adjustments to social requirements are relatively easy. The profile is shown of a certain individual who was rated very high in performance at her task.

For selection purposes, a profile may be handled in various ways. That shown in Fig. 16.5 illustrates one procedure. The favorable zone is clear, and less favorable zones are shaded. The shading can be overprinted on the chart, or a plastic mask can be prepared to lay over individual charts. Decisions can be based upon the *number* of favorable scores or upon the trend of the individual's curve as compared with the trend of the optimal scores. If a single optimal score has been determined for every trait and an "ideal" profile has been drawn, the departure of a single profile from the ideal profile can be determined in various ways. The deviations of each person's scores from the ideal scores can be summarized in various ways. A way that meets common statistical principles would be to square the deviations, sum the squares, find a mean, and then find a square root. This would give a single summarizing index that has some statistical sanction.

CLASSIFICATION OF PERSONNEL Selection of personnel presupposes a supply of applicants and the possibility of rejecting a proportion of them. Attention is upon one kind of assignment to be filled. In the classification of personnel, there are two or more assignments that can be made, and one might even consider rejecting no applicants, provided proper assignments can be found for all. In some situations there is the double problem of selection and classification combined. The availability of more than one assignment, however, makes possible the utilization of many more applicants than would be the case if there were only one kind of place to fill, for, presumably, personnel who do not qualify for one place might well qualify for some other. The more different kinds of places there are to fill, the smaller the chance of any applicant's being completely rejected.

Classification, broadly defined, means assigning each individual to the most appropriate category. This would include the operations in educational and vocational guidance. In vocational guidance, the number of kinds of "assignments" is almost infinite, though the number of major categories is limited. In selection, we have an assignment with the need to find the person for it; in classification in general, we have a number of assignments with their requirements in terms of human resources, on the one hand, and a number of persons who have the resources to satisfy or not to satisfy each assignment, on the other. In vocational guidance, we have one individual, with a unique pattern of resources, on the one hand, and a large variety of possible occupations, on the other.

As demonstrated in this and in preceding chapters, we have solved many of the statistical problems involved in selection of personnel. These are bound up with the problems of prediction and with the evaluation of goodness of prediction. By contrast, the problems of classification have been solved more slowly. Assignment to alternative classes requires a *differential prediction,* rather than a prediction on a single variable. We have to predict how much better the individual will adjust or perform if assigned to one category than if assigned to another.

When only two assignments are being considered and two predictive indices, we attempt to predict a *difference* in the criterion variable (or between criterion variables) from a *difference* in the assessment variable (or between assessment variables). It is reasonable that the more independence there is between two criterion variables (the less they intercorrelate), the more easily we can make a differential prediction. The more easily, also, can we find relatively independent assessment variables. Lack of correlation between both the criterion measures and the two assessment measures seems to be very important for effective classification.[1]

CLASSIFICATION THROUGH SELECTION Regardless of the number of categories we have in which to place individuals, an approximate solution lies in the application of selection procedures. For each vocational category to be filled, we can derive a multiple-regression equation, where the criterion to be predicted is a measure of success in that vocation. The differences between composite scores would be the deciding factor in classification. If possible, each person would be assigned to that category for which he has the highest composite score. Profile methods could also be used. With an optimal profile developed for each category, and a method of comparing the extent to which an individual's profile approaches different criterion profiles, decisions could be reached.

USE OF THE DISCRIMINANT FUNCTION IN CLASSIFICATION A better procedure, one that introduces more directly the principle of differential prediction, is to use the *discriminant function*. The general principle is that the different scores or measures will be weighted in such a way as to maximize the difference between the means of two composites derived from two criterion groups, relative to the variance within those groups. Suppose that we have two groups of successful individuals in two vocations—selling life insurance and piloting airplanes. We also have scores from individuals in the two groups from several tests. We want to weight the tests (with the same weights applying to both groups) so that the two means of the composite scores will differ as much as possible. The overlapping of the two distributions of composite scores would then be as small as possible.

We can approach the problem from the point of view of correlation if we look at it in a different way. If we assign the criterion values of 1 and 0 to the two groups (which group is 1 and which is 0 does not matter) and if we treat the group differentiation as a genuine dichotomy, we have a multiple-point-biserial problem, as demonstrated by Wherry.[2] That is, the dichotomy is a criterion to be predicted by means of a multiple-regression equation, in which the components are optimally weighted. The information with which we start would be a point-biserial r between each measure and the criterion variable and a Pearson product-moment r (preferred) among the measures of assess-

1 These problems have been discussed at greater length by Thorndike, R. L. *Personnel selection*. New York: Wiley, 1949; and Brogden, H. E. An approach to the problem of differential prediction. *Psychometrika,* 1946, **11,** 139–154.

2 Wherry, R. J. Multiple bi-serial and multiple point bi-serial correlation. *Psychometrika,* 1947, **12,** 189–195.

ment. The procedure for determining the weights in the regression equation would be the same as illustrated in this chapter. The standard deviation of the criterion would be $\sqrt{pq}$, where $p =$ the proportion in one of the groups. A multiple-point-biserial R can also be computed to indicate the goodness of prediction afforded by this equation. A critical cutoff score could be found on the scale of X'.

When there are more than two classes to be predicted, the multiple-regression problem becomes quite complicated and more special computational methods are available.[1]

EXERCISES

In connection with each exercise, state your conclusions and interpretations.

1 Using information obtained from Data 16*A*, derive a regression equation involving X_1 (dependent variable) with X_2 and X_4. Compute the multiple R and its standard error.

DATA 16*A* Intercorrelations of scores from four examinations and marks received in freshman mathematics ($N = 100$)

Variable	X_1	X_2	X_3	X_4	X_5
X_1		.51	.51	.61	.39
X_2	.51		.70	.53	.39
X_3	.51	.70		.61	.29
X_4	.61	.53	.61		.28
X_5	.39	.39	.29	.28	
$\bar{X}$	5.70	4.10	5.44	5.37	4.95
S_x	2.42	1.92	1.84	2.26	2.14

X_1 = **grades in freshman mathematics**
X_2 = **Ohio State psychological examination**
X_3 = **English-usage examination**
X_4 = **algebra examination**
X_5 = **engineering-aptitude examination**

2 Do the same as in Exercise 1, substituting X_3 and X_5 as the independent variables.

3 If a computer is available, find a regression equation that includes all four of the independent variables in Data 16*A*, with a multiple R and its SE.

4 Two students, *A* and *B*, have the following scores:

	X_2	X_3	X_4	X_5
A	8	5	2	7
B	2	4	9	3

Estimate their most probable marks in freshman mathematics, using

[1] For treatments of multiple-discriminant-function analysis, see Bock, R D. *Multivariate statistical methods in behavioral research.* New York: McGraw-Hill, 1973; also Rozeboom, W. W. *Foundations of the theory of prediction.* Homewood, Ill.: Dorsey Press, 1966.

the regression equations derived in Exercises 1, 2, and 3 (see the answer to Exercise 3 for the third equation).

5 Compute the standard errors of multiple estimate, coefficients of multiple determination and multiple nondetermination, and indices of forecasting efficiency for the problems in Exercises 1 and 3.

6 Compute SE's of the regression coefficients in Exercise 1 and the *z* ratios.

7 Apply the shrinkage formulas to the multiple *R*'s and the SE's of estimate in connection with Exercises 1 and 3.

8 Estimate the means of the combinations of scores using the regression weights found in Exercises 1 and 3.

9 Estimate the standard deviation of:
a. An unweighted combination of scores X_2 and X_4 in Data 16*A*.
b. A weighted combination of the same scores, using the regression weights found in Exercise 1. Check by using the product $S_1R_{1.24}$.
c. A weighted combination of the same scores, using weights of 2 and 5, respectively.

10 Find the correlation of:
a. An unweighted combination of X_2 and X_4 with X_1.
b. A weighted combination of the same variables with X_1, using weights of 2 and 5, respectively.
Compare these correlations with the multiple $R_{1.24}$.

ANSWERS

1 $X_1' = .328X_2 + .505X_4 + 1.64$; $R_{1.24} = .649$; $s_R = .102$ (when $P = 0$).

2 $X_1' = .570X_3 + .299X_5 + 1.12$; $R_{1.35} = .569$; $s_R = .102$ (when $P = 0$).

3 $\beta_{12} = .150$; $\beta_{13} = .095$; $\beta_{13} = .422$; $\beta_{15} = .185$; $X_1' = .188X_2 + .124X_3 + .452X_4 + .209X_5 + .79$; $R_{1.2345} = .674$; $s_R = .103$ ($P = 0$).

4 X_1' (equation 1): 5.3, 6.8; X_1' (equation 2): 6.1, 4.3; X_1' (equation 3): 5.3, 6.4.

5 $S_{1.24} = 1.84$; $S_{1.2345} = 1.79$; $R^2_{1.24} = .421$; $R^2_{1.2345} = .454$; $K^2_{1.24} = .579$; $K^2_{1.2345} = .546$; $E_{1.24} = 23.9$; $E_{1.2345} = 26.1$.

6 $s_{\beta_{12.4}} = .091$; $s_{\beta_{14.2}} = .092$; $s_{b_{12.4}} = .115$; $s_{b_{14.2}} = .098$; $z_{12.4} = 2.85$; $z_{14.2} = 5.16$.

7 ${}_cS_{1.24} = 1.86$; ${}_cS_{1.2345} = 1.82$; ${}_cR_{1.24} = .639$; ${}_cR_{1.2345} = .656$.

8 $\bar{X}_{ws}$: 4.06, 4.91.

9 a. $S_s = 3.66$; b. $S_{ws} = 1.57$ (check: $S_1R_{1.24} = 1.57$); c. $S_{ws} = 13.73$.

10 a. $r_{cs} = .644$; b. $r_{c(ws)} = .645$.

FOUR

PSYCHOLOGICAL MEASUREMENTS

17 Reliability of measurements

THE IMPORTANCE OF RELIABILITY By a "perfectly reliable" measurement we mean one that is completely accurate or free from error. The same "yardstick" applied to the same individual or object in the same way should yield the same value from moment to moment, provided that the thing measured has itself not changed in the meantime.

There are times, both in theoretical investigations and in practical work, when reliability is very important. Although numbers, as such, are exact, the fact that we amass a series of numbers attached to individuals or observations is no assurance that those numbers mean what they seem to mean concerning the things measured.

There is no way of simply looking at numbers and telling whether they stand for any real values or whether they have been "pulled out of a hat." Some samples of measurements actually approach the latter, chance condition. Others are not exactly "chance" collections of numbers, but there are strong elements of chance involved in them. Conclusions derived from statistical results might differ considerably depending upon how reliable we know the measurements to be. Thus, the matter of reliability merits considerable attention.

Reliability theory

It is impossible to appreciate the many problems that arise in connection with reliability and the several meanings of the term itself without an understanding of some of the mathematical ideas underlying the concept. There exists a rigorous definition of reliability which makes it possible to understand many of the peculiarities of measurements, particularly those called test scores. There are also several operational conceptions of reliability, depending upon how it is estimated from empirical data—as by the internal-consistency, retest, and alternate-forms methods. Keeping in mind the several kinds of reliability and the fact that operational definitions do not coincide will aid in thinking about problems of reliability. We shall begin with some basic, theoretical conceptions of reliability.

A BASIC DEFINITION OF RELIABILITY

The reliability of a set of measurements is logically defined as the proportion of the variance that is true variance. Before elaborating upon the heart of this statement, which is the last part, attention should be called to the more incidental part. The statement begins with "the reliability of a set of measurements." Note that it is *measurements* that is said to have the property of reliability rather than the measuring instrument. That is because in psychological and educational measurement, and other behavioral and social measurements, reliability depends upon the population measured as well as upon the measuring instrument. It can rarely be said of any instrument, whether a test or some other device, that *the* reliability of the device is of a certain value, usually in the form of a coefficient of correlation. One should speak of the reliability of a certain instrument applied to a certain population under certain conditions.

The next comment on the definition, and a more important one, is about the definition of *true* variance. The total variance, in this context of tests, we shall call S_t^2. As usual, it is the mean of the squares of deviations of measurements from their mean. The idea of separating a total variance into components is not new. That idea was emphasized in the chapter on analysis of variance (Chap. 13) and the chapters on prediction of measurements (Chaps. 15 and 16). Here we make a new kind of segregation of variances. We think of the total variance of a set of measures as being composed of two sources of variance: *true* variance and *error* variance.[1] We think of each measurement as having two components, a true measure and an error. In terms of an equation,

$$X_t = X_\infty + X_e \qquad \text{(An obtained measure expressed as the sum of a true and an error component)} \qquad (17.1)$$

where X_t = obtained score or measurement
X_∞ = true score component[2]
X_e = error component

1 An important distinction should be made between error variance and sampling variance; the two are quite different. Error variance arises from imperfect measuring procedures. Sampling errors pertain to deviations from population values when samples are drawn. The development of test-measurement theory here ignores sampling errors.

2 The infinity-sign subscript indicates that X_∞ can be conceived as the mean of an infinitely large sample of repeated measurements of an individual.

Several assumptions are made in connection with this equation. The true value is assumed to be the genuine value of whatever is being measured, a value we should obtain if we had a perfect instrument applied under ideal conditions. An operational definition is that it is the mean we should obtain if we were to apply the measurement a very large number of times. Any obtained measurement at a particular time is determined in part by the true value and in part by conditions that bring about a departure from that value.

Another assumption regarding the equation is that in measuring a series of objects the error components occur independently and at random. They are thus independent of one another and of the true values. Their mean is zero; they are as often negative as positive. These conditions may not always be satisfied, but without evidence to the contrary we assume that they are. Knowledge of the nature of the instrument and of other conditions is often sufficient to support the application of these assumptions or to lead us to reject them in particular instances.

Reliability was defined as the proportion of the total variance that is true variance. The three variances—true, error, and total—are illustrated in Table 17.1. There we have a set of 10 hypothetical true measures whose mean is 25.0 and whose variance S_t^2 is 105.0. For each true measure there is a corresponding error component that is added to it to form a total score for each individual. The mean of these error components is zero, as assumed above. Their variance is 15.2.

TABLE 17.1 Dispersions of true measures, error components and of their sums, the total measures, with means, variances, and standard deviations

	True measures X_t	*Error components* X_e	*Total measures* X_t $(X_\infty + X_e)$
	5	−2	3
	15	+2	17
	20	−4	16
	25	−2	23
	25	+2	27
	25	0	25
	25	+10	35
	30	−4	26
	35	−2	33
	45	0	45
Σ	250	0.0	250
$\overline{X}$	25.0	0.0	25.0
Σx^2	1,050	152	1,202
S^2	105.0	15.2	120.2
S	10.2	3.9	11.0

The variance of the total measures can be estimated from the component variances by using formula (16.13). With zero correlation between the true and error components, the variance of the total scores is merely the sum of the two component variances. In terms of symbols,

$$S_t^2 = S_\infty^2 + S_e^2 \qquad \text{(A total variance as the sum of true and error variances)} \qquad (17.2)$$

The application of this equation in Table 17.1 gives a total variance of 120.2, which checks with that computed from the sum of the squares of the deviations x_t.

In order to satisfy the definition of reliability, we need to find the proportion of the total variance that is true variance. If we divide equation (17.2) through by S_t^2, we have the three proportions:

$$\frac{S_t^2}{S_t^2} = \frac{S_\infty^2}{S_t^2} + \frac{S_e^2}{S_t^2} = 1.00 \qquad \text{(Sum of proportions of true and error variances)} \qquad (17.3)$$

The reliability of these measurements is given by the ratio S_∞^2/S_t^2 or, in another form, by $1 - S_e^2/S_t^2$. Letting r_{tt} stand for the coefficient of reliability we thus have two alternative equations:

$$r_{tt} = \frac{S_\infty^2}{S_t^2} \qquad (17.4a)$$

$$r_{tt} = 1 - \frac{S_e^2}{S_t^2} \qquad (17.4b)$$

(Basic equations for the coefficient of reliability)

Applying these equations to the data in Table 17.1, we have

$$r_{tt} = \frac{105.0}{120.2} = .87, \text{ or } r_{tt} = 1 - \frac{15.2}{120.2} = .87$$

If we let e^2 stand for the proportion of error variance in the total, we have the equation

$$r_{tt} + e^2 = 1.00 \qquad \text{(Complementary nature of proportions of true and error variances)} \qquad (17.5)$$

The relationships just mentioned are demonstrated in Figs. 17.1 and 17.2. In Fig. 17.1 dispersions of true measures and of total measures are shown. Both have the same mean because the mean of the errors is zero. The standard deviation S_t is greater than S_∞. This is always true, unless by some very remote possibility they happen to be equal. The effect of errors of measurement is always to increase obtained dispersions, never to decrease them, unless they should happen to be correlated with the true measures or with one another.

Figure 17.2 presents the picture in a somewhat different manner. Here the summative properties of variances are apparent. Without the assumption of zero correlations for the errors, such a simple picture

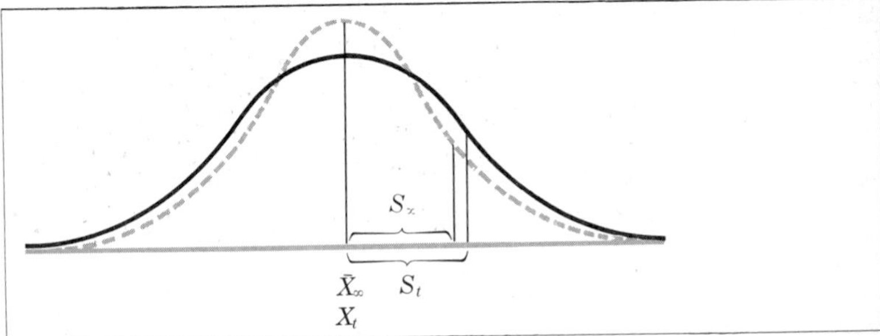

FIGURE 17.1 Distribution of obtained scores in a test (solid-line curve) and of the hypothetical true-score components (dashed-line curve). Means of obtained and true scores coincide on the assumption that the mean of the errors is zero. The standard deviation of the obtained scores is greater than that of the true scores

would not be possible. This kind of representation of variances, in tests particularly, will be encountered with increasing frequency in this and the next chapter.

THE INDEX OF RELIABILITY

The reliability coefficient for a test r_{tt}, as described thus far, is merely an abstract idea. Operationally, it is some kind of self-correlation of a test.

Before we go into the various operations for estimating r_{tt}, let us add more meaning to the fundamental idea of reliability. Let us think of the true score (X_∞) and the obtained score (X_t) as being two separate variables, the one dependent upon, or predictable from, the other. This is in spite of the fact that the one includes the other. Think of X_t as the dependent variable and of X_∞ as the independent variable. In a real sense, X_t is determined by, or dependent upon, X_∞. Figure 17.3 shows these two variables as coordinates and the line of regression of X_t on X_∞. The correlation between the two $r_{t\infty}$ is known as the *index of reliability*. It is, of course, a part-whole correlation, for X_∞ is a part of X_t, as shown in formula (17.1). The square of this correlation coefficient is an index of determination (see Chap. 14), and it indicates the proportion of variance in X_t that is determined by variance in X_∞. But this is

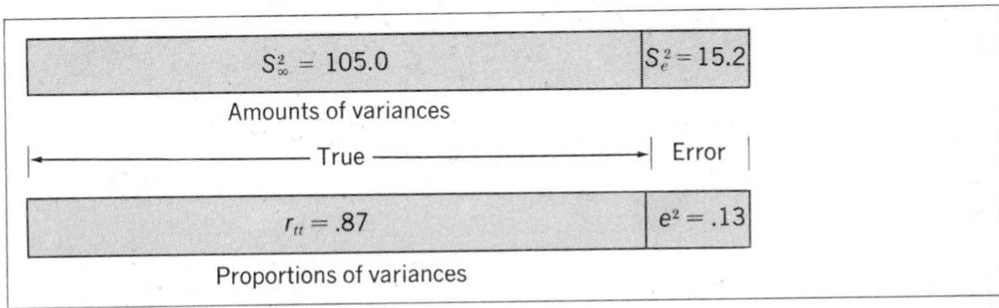

FIGURE 17.2 Amounts of true and error variance (first bar) in a test and proportions of true and error variance (second bar)

FIGURE 17.3 Regression of obtained scores on true scores, with parallel lines at vertical distances of one standard error ($S_{t\infty}$) from the regression line. (Compare this illustration with Fig. 15.6. The standard error of measurement is also a standard error of estimate when obtained score is "predicted" from true score.)

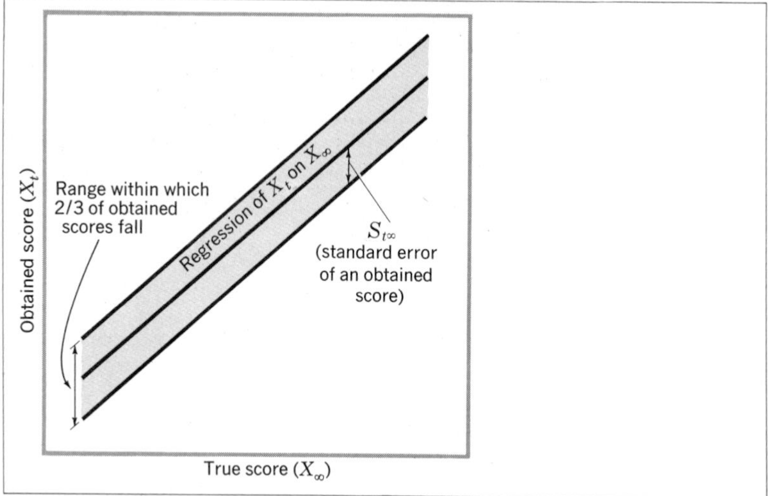

precisely what the reliability coefficient (r_{tt}) tells us. Consequently, we have shown that

$$r_{t\infty}^2 = r_{tt} \tag{17.6}$$

and (Relation of an index of reliability to a coefficient of reliability)

$$r_{t\infty} = \sqrt{r_{tt}} \tag{17.7}$$

When errors are entirely random, nothing can correlate with obtained scores higher than their correlation with corresponding true scores. The statistic $r_{t\infty}$, then, is often used as an indication of the upper limit of correlation of any variable with another. Since $r_{t\infty}$ is the square root of the reliability coefficient, it is always numerically larger than r_{tt}. Do not be surprised, then, to find that a test may correlate higher with another test than it correlates with itself. We cannot compute $r_{t\infty}$ directly from data, but it can be estimated from r_{tt} or from other information. It is a seldom-used statistic, but it has a definite meaning and could be used along with rt_{tt} or in place of it.[1]

THE STANDARD ERROR OF MEASUREMENT

Since we can estimate the correlation between obtained and true scores and can think in terms of prediction of one from the other, we can also inquire about the errors of prediction. We know the obtained

[1] Although theory of reliability has usually been developed in terms of true and error variances as explained here, Cronbach has approached the matter in a novel way, under the heading of "generalizability." In his view, a particular obtained score should be regarded as a representative of a "universe" of such scores that might be obtained from the same instrument under similar conditions. The question, then, is "How well does such a score represent its universe?" or "How far can we generalize from this score to all such scores?" The theory leads to much the same steps for estimating reliability or generalizability. (See Cronbach, L. J., and Gleser, G. C. *Dependability of behavioral measurement*. New York: Wiley, 1972.)

scores, and from them we could predict true scores (assuming any mean and standard deviation we please for the true-score scale). But there is nothing to be gained by so doing, for the predictions would be no more accurate than the scores from which they were obtained. Nothing would have happened except a change of unit and zero point.

Suppose that we think in terms of prediction in the other direction, from true scores to obtained scores. This is impossible, practically, since we do not know the true scores from which to make predictions. Let us think rather in terms of determination: of true scores *determining* obtained scores. But errors of measurement also help to determine obtained scores. We are interested in the extent of the discrepancies caused by these errors of measurement, in other words, in the size of distortions produced in the otherwise true-determined measurements. The average of these discrepancies is estimated by the formula

$$S_{t\infty} = S_t\sqrt{1 - r_{tt}} \qquad \text{(Standard error of measurement)} \qquad (17.8)$$

where S_t = standard deviation of the distribution of obtained scores and r_{tt} = reliability coefficient.

The standard error of measurement is a standard error of estimate and may be interpreted as such.[1] Figure 17.3 shows the limits marked off at distances of plus and minus 1 $S_{t\infty}$ from the regression line. In a certain test with a $S_{t\infty}$ equal to 2.0 units, we may say that two-thirds of the obtained scores are within 2.0 units of the true scores that determined them. If a certain individual's true score were 35, for example, the odds are 2 to 1 that his obtained score would not exceed 37 or fall below 33. Allowing a margin of 2*S*, we can say that the odds are 19 to 1 that his obtained score will not exceed 39 or fall below 31.

No obtained score tells us what the corresponding true score is, but with knowledge of the $S_{t\infty}$, and when it is small, we have a degree of confidence that the true score cannot be very far away. The same standard error gives us some basis for confidence as to whether the scores for two persons represent a real difference or whether we can tolerate the idea that they could have come from the same true score.

RELIABILITY AT DIFFERENT PARTS OF THE TEST SCALE Test users sometimes ask for the standard error of measurement rather than the reliability coefficient because it tells them more directly what they wish to know. It tells them whether they should be concerned about differences of 2, 4, 8, or 12 points or whether any or all of these differences are within the probable range that could have been produced by errors of measurement. One could set up confidence intervals here for single scores as is done for means and other statistics (see Chap. 8).

It may happen, however, that because of a peculiarity of the test itself, discriminations are better at one part of the scale than at other

1 This statistic is also called the *standard error of an obtained score.*

parts. The $S_{t\infty}$ statistic is a blanket index, implying approximately equal discriminating power all along the scale. If there is reason to suspect that discrimination is actually unequal along the scale, this can be examined by preparing a scatter diagram showing the relationship between two forms (or halves) of the same test. The standard deviations of the columns or rows at different score levels will indicate where predictions have the greatest accuracy. If the score distribution approaches normality and if obtained scores do not extend over the entire possible range, the standard error of measurement is probably uniform at all score levels.

Methods of estimating reliability

We leave theory for awhile and see how r_{tt} can be estimated from empirical data. There are many procedures, falling into three general categories: (1) internal-consistency reliability, or simply internal consistency; (2) alternate-forms reliability, comparable-forms reliability, or parallel-forms reliability; and (3) retest reliability, or test-retest reliability. Cronbach has proposed that we speak of the second and third types of estimate as coefficients of equivalence and of stability, respectively.[1] It would be convenient, also, to speak of the first type as a coefficient of consistency.

There is no one best way of estimating r_{tt}. The method employed will depend upon one's purposes and the meaning and use one wishes to attach to r_{tt}. A secondary consideration is availability of data in the proper form. Other considerations are testing conditions and the kind of test or other measure.

The various procedures differ most in the kinds of things that may be considered true variance and error variance. What may be regarded as true variance in computing one kind of r_{tt} may be regarded as error variance in computing one of the others. For the sake of clarity, let us look at some examples.

CONTRIBUTORS TO TRUE AND ERROR VARIANCES

On the whole, sources that contribute to an examinee's making the same score in "repeated" applications of a test are contributors to true variance in the obtained scores. The word "repeated" is in quotation marks here because the repetition is broadly defined so as to include alternate forms or two halves of the same test. On the whole, things that contribute to varying evaluations of performance of an individual in a test are contributors to error variance. The sources of true and error variances are numerous. Certain of them appear clearly and frequently enough to be recognized and named.

1 Cronbach, L. J. Test "reliability": its meaning and determination. *Psychometrika,* 1947, **12,** 1–16.

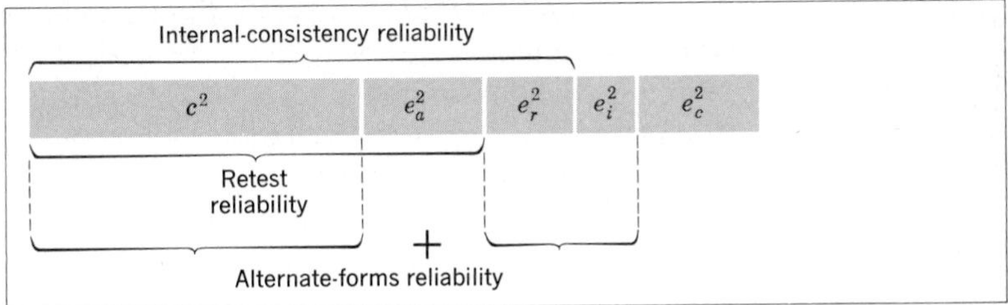

FIGURE 17.4 Proportions of the total-score variance that can be regarded as true variance or as error variance, depending upon which type of reliability estimate is made

Let the bar diagram in Fig. 17.4 represent the total variance in obtained scores of a test. Let c^2 be that proportion of the total variance that would be regarded as true variance no matter what method of estimating r_{tt} is employed. After all, such methods should have very much in common. Let e_a^2 be regarded as those sources of error variance which are unique to the alternate-forms method but are regarded as sources of true variance for the other methods. The relative sizes of these portions will vary from test to test. Actual examples of e_a^2 and of c^2 will be given shortly. Let e_i^2 be sources of error variance particularly when some internal-consistency method is used. This portion is also represented as providing sources of errors for the retest method. Finally, let e_r^2 be more distinctly the source of error when the retest method is applied, but a source of true variance for the other methods. The actual situation is probably not so simple as this, but it is hoped that this simplifying will contribute to clarity.

Now for some illustrations of actual determinants of the different kinds of variance. These determinants, it must be remembered, are to be thought of as contributing to individual differences between scores, either within a single application of a test or between applications or between forms. Among the determinants of individual differences that are consistent from time to time and from one form of a test to another is individual status in some enduring ability, skill, or other trait or traits. These are what we wish to measure. Incidental determinants that also belong under portion c^2 in the diagram (Fig. 17.4) are general skill in taking tests, skill in taking this particular kind of test (including the form of item used), and possibly the ability to understand test instructions. These additional sources of variance are only potential. For any given test, the task may require so little understanding or the type of item may be so well known to all examinees that they are practically on a par with respect to these determinants, and consequently the determinants would not contribute to individual differences in scores. If they do operate to affect variances, however, they would produce effects in the same directions in odd and even scores, and insofar as individuals do not change in these respects from one administration to another, they would contribute to true variance in all three types of reliability estimate.

Determinants that contribute to error variance in the retest method include temporary conditions, either of the examinee or of the testing environment, including the examiner. The examinee's state of health, feelings of fatigue or boredom, emotional condition, and the like may well change from one day to another. Environmental conditions can vary considerably without affecting scores materially, but insofar as they do, such factors as temperature, humidity, lighting, audibility of instructions or signals, ventilation, and the like may differ enough to contribute to error variance.

Probably more important changes occur in the examinees themselves. Having taken a certain test, they are not the same individuals when taking it again. Skills and knowledge acquired during the first administration and in the interval between administrations will have their effects upon the second performance. Memory of answers given on the first occasion may lead to repetitions of the same answers the second time and thus contribute to apparent true variance. Awareness of mistakes made in the first attempt, however, leads to changes in responses and hence to error variance. Besides possible improvement during the taking of the test the first time there is possible improvement resulting from transfer effects occurring during the interval between administrations. There are also possible maturational factors, particularly in young children. If learning and maturational effects were uniform for all individuals, or in proportion to their initial positions in the distribution, they would not contribute to error variance. But to the extent that learning and maturational effects differ from person to person, they do add much to error variance.

The longer the time interval between test administrations, the greater the error contributions. In some tests, continuous loss in reliability occurs as a function of time interval between test and retest. In some psychomotor tests, self-correlations of .90 to .96 may be found by the odd-even method (correlating two scores, one from odd-numbered items and one from even-numbered items), but test-retest correlations with a year interval between may give correlations of approximately .70. Results of this kind were found in testing aviation cadets in the AAF before training and again after air-crew training and perhaps some combat.

Error variance in the alternate-forms method is contributed chiefly by the change in content of the test. The knowledge and skill required for dealing with one particular set of items may vary somewhat from the knowledge and skill required for dealing with another set, and these variations differ from person to person. In addition, depending upon the time interval between administrations of the two forms, some of the causes of error variance just mentioned for the retest method may also apply to the alternate-forms method. An experiment in the

AAF[1] in which the two forms were given in immediate succession and also with 4 hours of other testing intervening showed no appreciable change in the size of the self-correlation. Longer periods might well be expected to have some effect.

If the odd-even technique is used in the split-half method (using two scores from two halves of the same test, in the same administration), the changes in conditions that may occur during a single administration of a test are rather uniformly distributed over all items in both halves, so that their effects would not show up as error variance. There are other ways of splitting tests into halves, however, which may allow more error variance to creep in. If the test is divided by blocks of items, as in odd and even half pages, odd and even 2-min trials, or first half against second half, there is room for systematic shifting of conditions. The effects of learning, of temporary changes in mental set (as for speed versus accuracy or as to mode of attack on the items), or of fatigue or motivation then might contribute to error variance. These are represented in section e_i^2 in Fig. 17.4.

The sources of error that would affect all methods of reliability estimate alike, represented by e_c^2, are such phenomena as fluctuations of attention or memory or motivation that occur from moment to moment or from item to item. In some tests, guessing is an important contributor to error variance. If a test is so difficult that everyone does a considerable amount of guessing (in the extreme case assume that every examinee guessed on every item), the total scores for all examinees approach chance distributions whose variances are very largely error variance. If guessing is a feature in any test, the more difficult the test is, the lower its reliability is likely to be. On the other hand, the easier the test, the lower the dispersion of scores and the lower the reliability. The smaller the number of alternative responses, the greater the importance of guessing. True-false tests of the same material and consisting of the same number of items are less reliable than four-choice tests, and these, in turn, are less reliable than tests of the completion form, other things being equal. The moral of this, of course, is to avoid items with too small a number of alternative responses or to compensate for the greater chance element by making the test longer.

WHEN DIFFERENT METHODS OF ESTIMATING r_{tt} ARE PREFERRED

Preference for one of the three types of reliability estimate depends mostly upon several considerations: the type of test, the meaning of the statistic, and the purpose for which the statistic will be used. These considerations will now be explained.

HOMOGENEOUS VERSUS HETEROGENEOUS TESTS Some psychological tests tend to be homogeneous, and others to be heterogeneous. A highly homogeneous test measures the same ability (or abilities) or

1 Guilford, J. P., and Lacey, J. I. (Eds.) Printed classification tests. *AAF Aviation Psychology Research Program Reports,* No. 5. Washington: GPO, 1947. Pp. 25–34.

some other trait (or traits) about equally well in all its parts. A heterogeneous test measures different psychological variables in its various parts. Homogeneity is a matter of internal consistency, the degree of which is to be indicated by an index of internal-consistency reliability. As we shall see later, however, special problems are created by a test in which speed of performance is an important feature. In such a test not all examinees attempt to answer all items before time is called. It is difficult to determine whether unattempted items are homogeneous with other items.

If a test is heterogeneous, so that different parts measure different traits, we should not expect a very high index of internal consistency. An example of such a test is a biographical-data inventory. This kind of test is composed of questions concerning the examinee's previous life and experiences. Each response to every item is usually validated by correlating it with some practical criterion, for example, success in pilot training. The reason one response is valid is not necessarily the same as the reason another is valid. They may both predict the criterion and yet correlate zero with each other. The parts of such a test, e.g., two randomly chosen halves, will probably not correlate very high with each other. The test has low internal consistency. An r_{tt} computed in this manner would not do justice to the test. Neither would an alternate-forms r_{tt}, if the forms were developed without regard for item intercorrelations.

The only meaningful estimate of reliability for a heterogeneous test is of the retest variety. If, by chance, a heterogeneous test were developed, each item of which correlated with a criterion and yet did not correlate with any other item, the internal-consistency reliability would be zero. Yet the retest reliability might be substantial or high. One biographical-data test of the type referred to above had a characteristic split-half reliability coefficient of about .35 and a retest reliability of about .65. Both these values are unusually low, but the test had a validity close to .40 for the selection of pilots and consequently was very useful.

It is clear from this discussion that the internal consistency and the stability of the same test need not agree very closely. There can be very low internal consistency and yet substantial or high retest reliability. It is probably not true, however, that there can be high internal consistency and at the same time low retest reliability, except after long time intervals. High internal-consistency reliability is in itself assurance that we are dealing with a homogeneous test, at least within the broad meaning of the term as defined above.

SPEED TESTS AND POWER TESTS Tests are sometimes roughly categorized as speed tests and power tests. There is no sharp line of demarcation. A genuine power test is one that all examinees have time to finish. It is expected that every examinee will attempt every item. Achievement examinations are in this category. Speed tests are those in which there is a time limit such that not all examinees can attempt all items. In this category are tests ranging all the way from those in which no one attempts all items to those in which 99 percent may do so. The latter are so close to the power type that many examiners would be inclined to place them in the power category. As a general (rough) criterion, we may say that a power test is one finished by at least 75 percent of the examinees.

It would be out of the question to use the odd-even method of self-correlation with a highly speeded test. If no examinee finished and if there were no errors, the correlation of halves would be +1.00, which would have no meaning except that the scorer had counted the numbers of reactions in the two halves correctly. If first and last halves were used, assuming that everyone finished the first half and that there were almost no errors, all scores for the first half would be about the same, and those for the last half would depend upon the rate of work. The correlation would be near zero, for lack of dispersion of the first-half scores.

In fact, any internal-consistency estimate of r_{tt} would be misapplied to a speed test. The errors just caricatured are present to some degree no matter which one of the internal-consistency methods we apply. A retest method will be adequate for many speed tests, except where there is identity of items and hence learning and memory are sources of variance, both true and error, in unknown proportions. For most speed tests, and this includes those in which any appreciable number of examinees fail to reach the last item, an alternate-forms type of reliability estimate is usually best.

A good device to use in the development of new tests is to prepare two equivalent halves and administer them in immediate succession as two separately timed tests. The correlation between the two halves, independently administered, can be treated as we treat the correlation of any other half scores by the Spearman-Brown formula (to be explained shortly) in order to estimate the reliability of the full-length test. The comparability of the halves can usually be accomplished by careful construction. Some check upon the adequacy of the efforts is in the comparability of means, standard deviations, and skewness of the two distributions.

MEANING AND USE OF THE INDICES OF RELIABILITY The retest method yields information about the stability of rank orders of individuals over a period of time. A high r_{tt} from this source indicates that persons change very little in status within their population from the first to the second testing and also that the test measures the same functions before and after the interval. A low r_{tt} of this type may mean that individuals have changed in different directions or in the same direction at different rates. Changes of means and of standard deviations will help to interpret the kinds of systematic changes taking place. Plots of scatter diagrams may show whether systematic changes are uniform over the range. These changes we call *function fluctuations of individuals.* If, after an interval, the test measures something different from what it measured before, we have a *function fluctuation of the test.* These changes can be examined by means of correlations of the test with other tests before and after the interval—or, better yet, by factor analysis (see Chap. 18).

There may be some practical reasons for wanting to know about the stability of scores over periods of time, and if so, the retest r_{tt} is the index to use. Usually, the length of time is a condition to be considered. The chief use of this information is in deciding whether to depend upon scores that were obtained in an earlier testing or to administer the same test or a new form to obtain some scores that better describe the individuals at the later time. As a general policy it would be desirable to establish the principles regarding what kinds of tests yield stable scores, with what kinds of populations, and over what periods of time, and what kinds of tests do not.

The meaning of internal consistency was covered in a superficial way in the discussion of homogeneous tests. We shall go more thoroughly into the matter shortly in treating the specific methods under this category. This concept probably comes closest to the basic idea of reliability. The methods make an estimate of reliability from a single administration of a single test form. The estimate is of an "on-the-spot" reliability. It tells us something of how closely the obtained score comes to the score the person would have made at this particular time if we had had a perfect measuring instrument. For some purposes this information will certainly not be sufficient. It is the kind of reliability that does have meaning in connection with factorial descriptions of tests. These descriptions (see Chap. 18) attempt to depict a test in terms of its component variances, some of which combine to make up its true variance. It tells us nothing about functional stability of persons or of tests.

The alternate-forms estimate of r_{tt} tells us something about functional stability in variations of the same test or in different items that have been designed to measure the same functions. It indicates how independent the measurements are of the particular items or content used. If the two forms happen to be two halves of the same test, then presumably both contain the same kind of items (verbal, numerical, pictorial, matching, multiple-choice, completion); only the specific problems change. The alternate-forms r_{tt} may tend to be slightly lower than the internal-consistency r_{tt}, but this may mean that it gives a more realistic picture of how accurately the test measures the general traits, ruling out whatever variance is dependent upon the particular content of one form of the test. The two estimates will be almost identical, probably, in power tests of very closely matched content. In power tests, then, the two methods could be used almost interchangeably. In speed tests, as indicated above, the alternate-forms method is the most justifiable approach to reliability estimate.

Internal-consistency reliability

There are several operations by which an internal-consistency estimate of reliability may be made, and there is so much basic test theory bound up with them that we need to give this approach special attention. First, we shall consider some theory.

THE STATISTICAL NATURE OF A TEST COMPOSED OF ITEMS

Most tests are composed of items. Most tests are scored by giving credit of +1 for each correct response to an item and a weight of 0 for each wrong answer or omission. The theory about to be explained assumes that kind of test. Furthermore, it applies best to a power test, in which both omissions and wrong answers probably mean inability to master the item. For the time being we shall not be concerned with the problem of chance success by guessing. We might assume completion items in which chance factors resulting from guessing are almost nil. The theory will probably apply to situations deviating appreciably from these specifications, enough so that the many conclusions to which it leads will have quite general application.

ITEM STATISTICS It is convenient to think of each item as a subtest in a larger composite. Each item, then, yields a distribution of scores, with a mean and a standard deviation. The mean of such a distribution, where the measures are either 0 or 1, is equal to p, the proportion of all attempting the item who get the right answer; the variance of the distribution is equal to pq, where $q = 1 - p$; and the standard deviation is $\sqrt{pq}$.

The total score on such a test is the sum of part scores. In equation form,

$$X_t = X_a + X_b + X_c + \cdot \cdot \cdot + X_i + \cdot \cdot \cdot + X_n \tag{17.9}$$

(The sum of item scores to make a total test score)

where $X_a, X_b, \ldots, X_n$ = scores in items $a, b, \ldots, n$, when there are n items in the test.

The variance of the total test score can be derived from the variances and covariances of the items, according to the principles brought out in the preceding chapter in connection with the variance of sums. Equation (16.16) applied to this particular use would therefore read

$$S_t^2 = p_a q_a + p_b q_b + p_c q_c + \cdots + p_i q_i + \cdots + p_n q_n + 2r_{ab}\sqrt{p_a q_a p_b q_b} + 2r_{ac}\sqrt{p_a q_a p_c q_c} + \cdots + 2r_{(n-1)n}\sqrt{(p_{(n-1)} q_{(n-1)} p_n q_n} \qquad (17.10)$$

(Total test variance as summation of item variances and covariances)

where $p_a, p_b, \ldots, p_n$ = proportion passing items $a, b, \ldots, n$
$q_a, q_b, \ldots, q_n = 1 - p_a, 1 - p_b, \ldots, 1 - p_n$
$r_{ab}, r_{ac}, \ldots, r_{(n-1)n}$ = intercorrelations of items

In abbreviated, summational form, the equation reads

$$S_t^2 = \Sigma p_i q_i + 2\Sigma r_{ij}\sqrt{p_i q_i p_j q_j} \qquad (17.11)$$

[Same as formula (17.10) in summation form]

where $p_i = p_a, p_b, \ldots, p_n$ in turn and r_{ij} = correlation between item i and item j, where subscript j is numerically greater than i.

AN EXAMPLE OF ITEM AND TEST STATISTICS As an example of what has just been presented in terms of equations, let us take an artificial test of eight items, which has been administered to ten fictitious examinees, with resulting data as shown in Table 17.2. The tabulation is known as an *item-score matrix* because of the rows and columns of item scores, a column for each item and a row for each examinee. The items are arranged in order of increasing difficulty from left to right, and the examinees are arranged in increasing order of ability from top to

TABLE 17.2 An item-score matrix, listing eight item scores (X_i) for each of ten examinees, with scores of 1 for right answers and 0 for wrong answers, along with odd and even scores and their differences

Examinees	*Items*										
	a	b	c	d	· · ·	i	· · ·	n	$\sum_{i=1}^{n} X_i = X_t$	X_o	X_e
1	0	0	0	0	0	0	0	0	0	0	0
2	1	0	0	0	0	0	0	0	1	1	0
3	1	0	1	0	0	0	0	0	2	2	0
4	1	1	0	0	1	0	0	0	3	2	1
5	0	1	0	1	0	0	1	0	3	1	2
·	1	1	1	0	1	0	1	0	5	4	1
·	1	1	1	1	1	1	0	0	6	3	3
j	1	1	1	1	1	1	0	0	6	3	3
·	1	1	1	1	0	1	0	1	6	2	4
N	1	1	1	1	1	1	1	1	8	4	4
$\sum_{j=1}^{N} X_i$	8	7	6	5	5	4	3	2	$40 = \Sigma X_t$	22	18
p_i	.8	.7	.6	.5	.5	.4	.3	.2	$4.0 = \bar{X}_t$	$S_t = 2.45$	
$p_i q_i$	.16	.21	.24	.25	.25	.24	.21	.16	$1.72 = \Sigma p_i q_i$	$S^2_t = 6.0$	

bottom, for the sake of convenience. The item score is 1 for a right answer and 0 for a wrong answer. All examinees attempted all items.

The sums of the rows of scores give total scores X_t for the individuals. The sums of the columns give the numbers of persons passing the various items. The sums of both sets yield the same overall sum, 40, which is ΣX_t. Dividing the sums of the columns by N, we obtain the means for the items, p_i, the proportion passing each item. The variance for each item is $p_i q_i$, giving values that appear in the last row of the table. The sum of the variances equals 1.72. The variance for the total scores X_t is equal to 6.0 (for which the computation is not shown). If we deduct the sum of the item variances from this quantity, we have $6.0 - 1.72$, which equals 4.28. From equation (17.11), it can be seen that this difference is the portion of the total-score variance that is contributed by the sum of the covariance terms doubled. With eight items scores summed, there are $8(8 - 1)/2$, or 28, different covariance terms, each added in twice. We are not concerned with those covariance terms, as such, here. Each one, of course, contains a term for the correlation between a pair of items, a phi coefficient. We shall use the covariance values later, along with other information from Table 17.2. For the moment, we return to further consideration of formula (17.11).

DEDUCTIONS DERIVED FROM THE ITEM-VARIANCE EQUATIONS

Many useful and enlightening inferences can be drawn from equation (17.11). We shall consider only the most important ones here.

RELATION OF VARIANCE TO ITEM DIFFICULTY The first thing to be noted is the relation of variance to item difficulty. Remembering that variance means individual differences and that the greater the variance, the more we have dispersed individuals in measurement, it can be stated that the item that will produce the greatest dispersion is of median difficulty. It is an item passed by half of the group and failed by half of the group. When $p = q = .5$, the pq product is at a maximum. As p approaches 0 or 1, the variance decreases toward the vanishing point. This principle has a commonsense explanation. Let us suppose an item that 1 person out of 100 can answer correctly. This item discriminates 1 person from each of 99, or makes 99 discriminations. Then, suppose an item that can be passed by 2 out of 100. This item makes 2×98 discriminations, or 196. Continue this to 50, and we get 2,500 discriminations, each one of the 50 who pass it from each one, in turn, of the 50 who fail it. Items of moderate difficulty, then, yield the maximum variance.

RELATION OF RELIABILITY TO ITEM INTERCORRELATIONS For the sake of internal consistency, however, large item variances by themselves would mean nothing. If equation (17.11) were limited to the item-variance terms alone, the test would have zero internal consistency, or zero reliability of the internal type. This kind of reliability comes entirely from the covariance terms, and these are composed of item intercorrelations as well as indices of dispersion. It is only by virtue of their entering into the covariance terms that the item variances contribute to internal consistency. The intercorrelations of the items are the essential sources of this kind of reliability. The larger the item intercorrelations, the greater the internal consistency.

EFFECT OF RANGE OF ITEM DIFFICULTY UPON RELIABILITY Reliability will be higher when the items are nearly equal in difficulty. A wide range of difficulty of items is not favorable to reliability. The reason is that the appropriate index of item intercorrelation is the r_ϕ coefficient. Operationally, with items scored as either 0 or +1, their distributions are best conceived as point distributions. If two items differ much in difficulty, the proportions passing the two differ, and r_ϕ is thus restricted in size. Only when the two items are equally difficult can the r_ϕ between them equal +1 as a maximum (see Chap. 14). Two items very far apart in difficulty might correlate less than .20 even when each measures the same thing and measures it well.

EFFECT OF ITEM INTERCORRELATIONS UPON TOTAL-SCORE DISTRIBUTIONS There is an interesting bearing of the internal consistency of a test upon the form of distribution of total scores on that test. Imagine a test of 10 items each of exactly median difficulty for the population ($p = q = .5$) and each correlated +1.0 with every other item. A person who passes one item would pass them all, and a person who fails one item would fail them all. There would be only two scores possible, 0 and 10. If 20 examinees took this test, the chances are good that their frequency distribution would be like the first diagram in Fig. 17.5. There would be perfect and maximal separation of the two groups. The form of the distribution would be U-shaped. Examples of U-shaped distributions can be found in Hull's book on hypnosis and suggestibility, though they are not so extreme as the one in Fig. 17.5.[1] It appears that some tests of suggestibility are such that if the examinee responds in the suggestible manner in one trial, he is likely to respond similarly in all trials.

If the item intercorrelations are not perfect but high, there will be some moderate scores, but there will be a distinct tendency toward bimodality. The second distribution in Fig. 17.5 shows this type of test. With still further reduction in item intercorrelation, the distribution

[1] Hull, C. L. *Hypnosis and suggestibility*. New York: Appleton-Century-Crofts, 1933. P. 68.

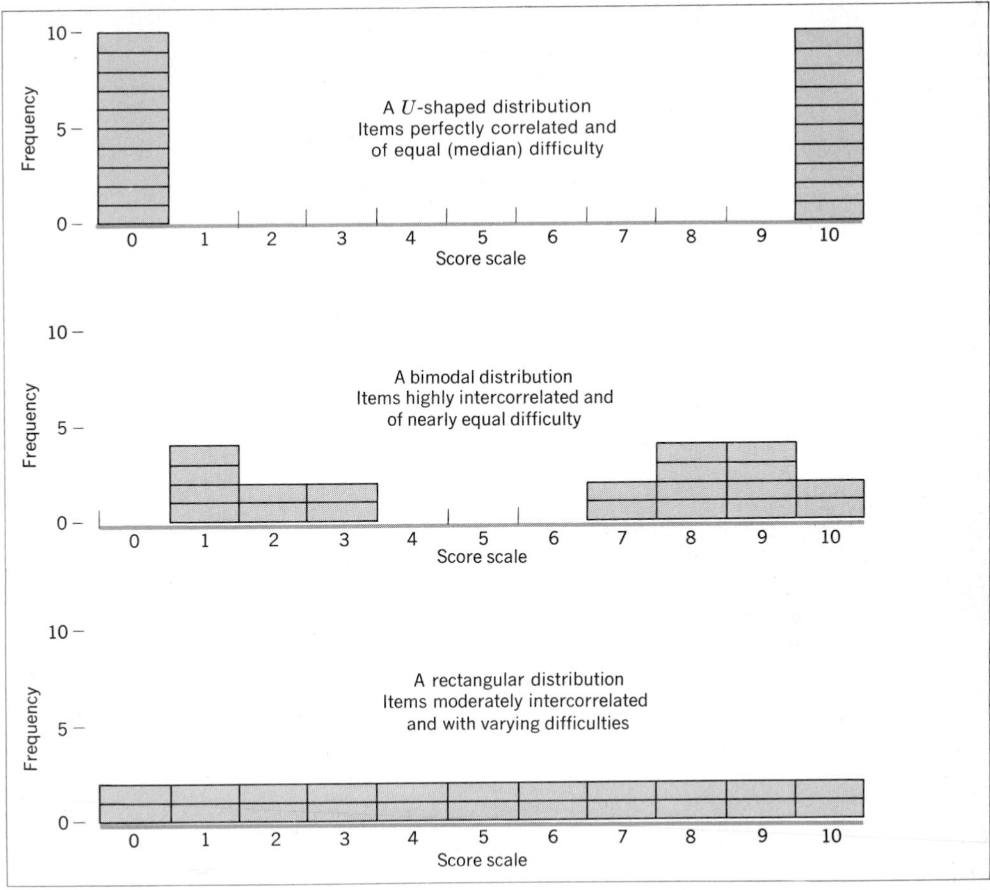

FIGURE 17.5 Illustration of the effects of item intercorrelation upon the form of the frequency distribution of total-test scores

approaches rectangular form, as in the third diagram in Fig. 17.5. With still further reduction in correlation, the distribution approaches normal form, but is somewhat platykurtic. A test of zero internal consistency, and with all items of median difficulty, would probably yield a normal distribution. It should not be concluded, however, that a normal distribution indicates zero reliability. It might do so if all items were of equal difficulty at the level of $p = .5$ and were uncorrelated. Rarely do tests conform to this condition.

THE SPEARMAN-BROWN FORMULA The Spearman-Brown (S-B) formula was designed to estimate the reliability of a test n times as long as the one for which we know a self-correlation. Many times a split-half correlation is known for a test, and the correlation of halves is an estimate of r_{tt} for the half test. The full-length test is not twice as reliable as the half test, but its reliability is greater and can be estimated by the special Spearman-Brown formula with $n = 2$. If we let r_{hh} stand for the correlation between halves of a test,

$$r_{tt} = \frac{2r_{hh}}{1 + r_{hh}} \quad \text{(Reliability of a total test estimated from reliability of one of its halves)} \tag{17.12}$$

AN ODD-EVEN ESTIMATE OF RELIABILITY To illustrate in a general way the application of this special case of the S-B formula, we use data from Table 17.2. For each examinee we have two split-half scores, one based upon item scores for odd-numbered items and the other based on item scores for even-numbered items. These two scores are listed under the headings of X_0 and X_e. The correlation between X_0 and X_e was found by means of ordinary correlation procedures to be .542. Applying formula (17.12),

$$r_{tt} = \frac{2(.542)}{1 + .542} = .70$$

The quantity .542 may be taken as an estimate of reliability of each of the two four-item tests, while .70 is the estimate of reliability of the total, eight-item test.

When this estimation formula is used, comparability of the halves must be assumed. Comparability is indicated to some degree by the similarity of means, standard deviations, item intercorrelations, skewness of distributions, and, of course, content. If comparability is lacking, the estimate of reliability of the total test will be somewhat in error. Since comparability is probably never perfect, an estimate by use of the Spearman-Brown formula is likely to be conservative; that is, it tends to be an underestimate.

Because the split-half method and the alternate-forms method in the form of two separately timed halves of the same test are so commonly used in practice, the chart in Fig. 17.6 is supplied as an aid in the use of formula (17.12). Since the estimates are approximate in any case, the graphic solution will probably serve for most purposes.

For the general case, in which n could be any ratio of test length to that for which r_{11} is known,

$$r_{nn} = \frac{nr_{11}}{1 + (n - 1)r_{11}} \quad \text{(Spearman-Brown formula for reliability of a test of length } n\text{)} \tag{17.13}$$

where r_{11} = reliability of a test of unit length.

The ratio n in equation (17.13) could be fractional as well as integral. If we knew the self-correlation for a test of 50 items, we could estimate reliabilities for tests of 75 items ($n = 1.5$) or of 25 items ($n = 0.5$).

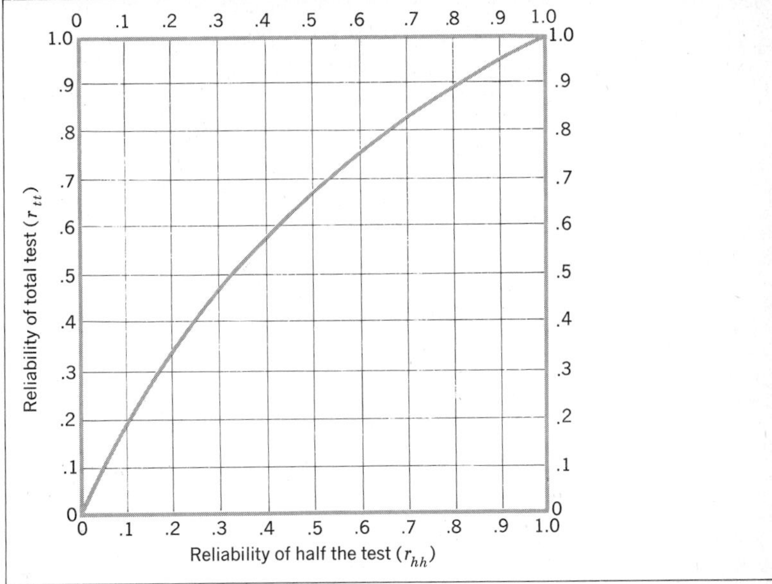

FIGURE 17.6 Reliability of a total-test score as a function of known reliability of a half-test score when the Spearman-Brown formula is applied

THE KUDER-RICHARDSON ESTIMATES OF RELIABILITY

In accordance with item theory, most of the Kuder-Richardson (K-R) formulas for estimating r_{tt} depend upon item statistics. They were developed because of dissatisfaction with split-half methods. A test can be split into halves in a great many ways, and each split might yield a somewhat different estimate of r_{tt}. The use of item statistics gets away from such biases as may arise from arbitrary splitting into halves.

The Kuder-Richardson methods make the same assumptions that are involved in the use of the Spearman-Brown formula. Those assumptions call for items of equal, or nearly equal, difficulty and equal intercorrelation.

The most accurate of the practical Kuder-Richardson formulas (known as their *formula 20*) is[1]

$$r_{tt} = \left(\frac{n}{n-1}\right)\left(\frac{S_t^2 - \Sigma pq}{S_t^2}\right) \quad \text{(General Kuder-Richardson formula for estimating reliability)} \qquad (17.14)$$

where n = number of items in the test

p = proportion passing an item (or responding in some specified manner)

$q = 1 - p$

It will be recognized, in comparing this formula with equation (17.11), that the numerator term $(S_t^2 - \Sigma pq)$ is the sum of the co-

[1] Richardson, M. W., and Kuder, G. F. The calculation of test reliability coefficients based upon the method of rational equivalence. *Journal of Educational Psychology*, 1939, **30,** 681–687.

variance terms in the summation of item variances and covariances used to express the total-test variance. The expression Σpq is the sum of the variances of all items. Deducting this quantity from the total test variance, we have left the sum of the covariances. It is in these covariances that the source of the *true* variance lies. The ratio of this quantity $(S_t^2 - \Sigma pq)$ to the total-test variance would thus seem to satisfy the basic definition of reliability given in the first part of this chapter. The factor $n(n-1)$ provides a correction to help take into account the fact that the numerator of the other fraction could never reach the quantity S_t^2 and hence that fraction could never equal 1.0, for Σpq would never be zero.

Having information on item variances in the data of Table 17.2, we can apply the Kuder-Richardson formula 20 to the eight-item test. Substituting the known values from that source,

$$\begin{aligned} r_{tt} &= \left(\frac{8}{7}\right)\left(\frac{6.00 - 1.72}{6.00}\right) \\ &= \frac{8(4.28)}{42} \\ &= .81 \end{aligned}$$

This estimate is higher than that from the odd-even approach, but this should not be surprising. The split-half approach would yield different estimates for the various ways in which the two halves are chosen, the odd-even division being only one of them, and some estimates will be smaller than others and possibly smaller than that by the K-R 20 approach.

The K-R formula is appropriately applied only when items are scored either 0 or 1. Cronbach has provided a more general formula that can be applied when items have a greater number of possible scores.[1] His coefficient α (alpha) is computed by means of a formula just like K-R 20 except that the term ΣV_i is substituted for the Σpq, where V_i is the variance of each item.

AN APPROXIMATION K-R FORMULA If we are justified in assuming that all items in the test have approximately the same level of difficulty, we may use a formula that requires less information. It reads

$$r_{tt} = \left(\frac{n}{n-1}\right)\left(\frac{S_t^2 - n\bar{p}\bar{q}}{S_t^2}\right). \qquad \text{(An approximation formula for Kuder-Richardson reliability)} \qquad (17.15)$$

where $\bar{p}$ and $\bar{q}$ = average proportion of passing and failing examinees in all items. The equation is known as the *Kuder-Richardson formula 21*.

Applying the K-R formula 21 to the data of Table 17.1, where the mean $\bar{p}$ is .5 and therefore $\overline{pq}$ is .25, we have

1 Cronbach, L. J. Coefficient alpha and the internal structure of tests. *Psychometrika*, 1951, **16**, 297–334.

$$r_{tt} = \left(\frac{8}{7}\right)\left(\frac{6.00 - 8(.25)}{6.00}\right)$$
$$= .76$$

This estimate is lower than that from the K-R formula 20 but higher than the odd-even estimate. The estimate from K-R 21 is expected to be generally lower than that from K-R 20.

One advantage of the use of K-R 21 is that we need not even make item-score counts, for the average $\bar{p}$ is equal to the mean of the total scores divided by n, and $\bar{q} = 1 - \bar{p}$. From these facts, the formula can be simplified to

$$r_{tt} = \frac{nS_t^2 - \overline{RW}}{(n-1)S_t^2} \qquad \text{[Alternate to formula (17.15)]} \qquad (17.16)$$

where $\bar{R}$ = average number of right responses and $\overline{W}$ = average number of wrong responses (or $n - \bar{R}$). $\bar{R}$ is of course, the mean of the total scores, where the total score is the number of right answers. Thus, for $\bar{R}$ and $\overline{W}$ we may substitute $\bar{X}$ and $n - \bar{X}$, respectively.

It should be said that all the Kuder-Richardson formulas, indeed all the internal-consistency formulas that depend upon a single administration of a test, tend to underestimate the reliability of a test, formula (17.16) most of all. Even formula (17.14) gives an underestimate when there is a wide dispersion of item difficulties.

HORST'S MODIFICATION Horst has suggested a modification of formula (17.14) which allows for variation of item difficulties.[1] The modification takes into account the extent to which the test approaches the maximum variance that a test with the same distribution of difficulties could have. That maximum variance is estimated by the formula

$$S_m^2 = 2\Sigma R_i p_i - \bar{X}\,(1 + \bar{X}) \qquad \text{(Maximum variance a test could achieve with its given distribution of item difficulties)} \qquad (17.17)$$

where R_i is the rank position of an item in the test, where items are ranked for difficulty, the easiest item being ranked 1; p_i is the item mean, or proportion passing the item; and $\bar{X}$ is the total-test mean. For the test represented in Table 17.2, the $R_i p_i$ products are summed as follows:

$$1(.8) + 2(.7) + 3(.6) + 4(.5) + 5(.5)\ 6(.4) + 7(.3) + 8(.2) = 14.6$$

Applying formula (17.17), we find

$$S_m^2 = 2(14.6) - 4(5)$$
$$= 9.2$$

1 Horst, P. Correcting the Kuder-Richardson reliability for dispersion of item difficulties. *Psychological Bulletin*, 1953, **50,** 371–374.

This value is to be compared with an obtained S_t^2 of 6.0. Horst's modified K-R formula is designed to tell us how nearly the obtained variance approaches the maximum variance possible. The modified formula reads

$$r_{tt} = \left(\frac{S_t^2 - \Sigma pq}{S_m^2 - \Sigma pq}\right)\left(\frac{S_m^2}{S_t^2}\right) \qquad \text{(Horst's modified Kuder-Richardson formula)} \tag{17.18}$$

Substituting the appropriate values, we have

$$\begin{aligned} r_{tt} &= \left(\frac{6 - 1.72}{9.2 - 1.72}\right)\left(\frac{9.2}{6}\right) \\ &= \left(\frac{4.28}{7.48}\right)\left(\frac{9.2}{6}\right) \\ &= .88 \end{aligned}$$

Assuming that the Horst formula gives the most nearly correct estimate of r_{tt}, we can see how much the other formulas fall short, in this particular instance. It is likely that in tests containing larger numbers of items and less dispersion of item difficulty, the variations between such estimates would be smaller.

It should be emphasized that the K-R formulas were designed for power tests, in which every examinee has a chance to attempt every item. They are entirely precluded for speed tests and for many others that depart very far from the power-test condition.

A SUMMARY OF INTERNAL CONSISTENCY RELIABILITY Internal-consistency reliability is most appropriately applied to homogeneous tests, i.e., tests composed of equivalent units—equivalent in several respects. The parts (usually items) all measure the same trait, or traits, to about the same degree. The total variance of a test can be conceived as a sum of the variances and covariances of its parts. The true variance of a test is contributed by its covariances, to which both the item variance and item intercorrelations are important contributors. Internal-consistency reliability is the greatest when:

1 The item intercorrelations are greatest.
2 The variance of items is greatest. This occurs when the proportion passing an item is .50.
3 The items are of equal difficulty. Then the item intercorrelations can be at a maximum.

Some special problems in reliability

Like all coefficients of correlation, r_{tt}, however estimated, must be interpreted in a relativistic manner. Its size depends upon many conditions under which it is obtained experimentally. Some of the more important conditions and considerations will be mentioned in what follows.

RELIABILITY IN DIFFERENT RANGES OF MEASUREMENT

Like intercorrelations of different variables, self-correlations are affected by the range of ability or of a trait present in the population samples. The narrower the range, the smaller r_{tt} tends to be. This can be seen mathematically if one examines formula (17.4*b*), where r_{tt} is given as equal to $1 - S_{t\infty}^2/S_t^2$, substituting $S_{t\infty}^2$ for its equivalent S_e^2. If the standard error of measurement $S_{t\infty}$ remains constant and S_t decreases, the denominator S_t^2 decreases, the ratio $S_{t\infty}/S_t^2$ increases and r_{tt} decreases. This is why occasional test users prefer to know $S_{t\infty}$ rather than r_{tt} concerning a test, since it is probably more stable from population to population. Figure 17.7 illustrates how in a restricted sample (small square) the same scatter of points gives a relatively wider spread and hence a lower correlation. Restriction is ordinarily not so clear-cut or so severe as this in practice, but the principle is the same.

If we wish to estimate the reliability coefficient in one range from the known reliability in another range, the following formula may be used. It assumes the same standard error of measurement in both ranges.

$$r_{nn} = 1 - \frac{S_o^2(1 - r_{oo})}{S_n^2} \qquad (17.19)$$

(Estimation of r_{tt} in a population of one dispersion from that in another similar population of different dispersion)

where S_o = standard deviation of the distribution for which the reliability coefficient is known; S_n = the standard deviation in the dis-

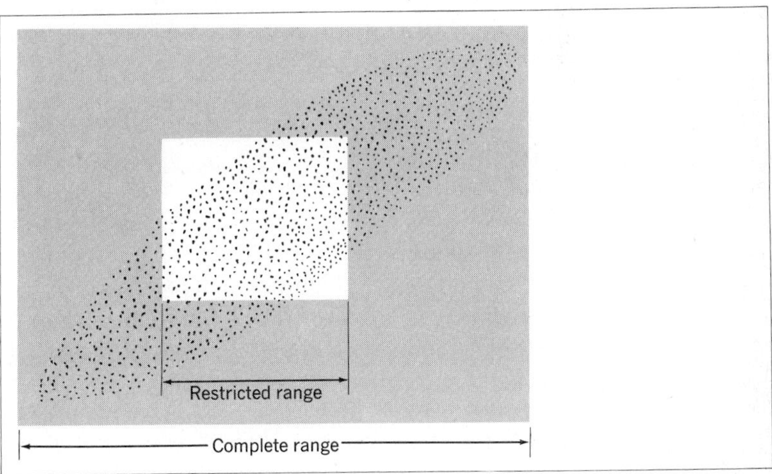

FIGURE 17.7 An extreme instance of curtailment of range of scores. The correlation for the cases within the small rectangle is much smaller than that for the cases within the larger rectangle

tribution in which the reliability is not known; and r_{oo} and r_{nn} = reliabilities in the two respective distributions.

If we know that a more limited group has a standard deviation of 8.0 and a reliability coefficient of .85 for a test, what should be the reliability coefficient in a more variable group whose S is 10.0? Applying formula (17.19),

$$r_{nn} = 1 - \frac{8^2(1 - .85)}{10^2} = .904$$

RELIABILITY AND LENGTH OF TEST It was indicated in connection with the split-half method that the whole test is more reliable than either half and that in general terms there is an increase in reliability with an increase in the length of the test. *This is true if the additional items added to a test are homogeneous with the ones to which they are added.* By "homogeneous" we mean that they have about the same intercorrelation with the items already in the test as those items have among themselves and that they possess about the same level of difficulty. If a test is lengthened to n times its present length under these conditions, we have a right to expect a change in reliability in accordance with the Spearman-Brown equation, which was given previously [formula (17.13)].

LENGTHENING A TEST TO ATTAIN A CERTAIN DESIRED RELIABILITY We can use the Spearman-Brown formula in reverse. If we know that the reliability of a short test is .75, we can ask how long the test would have to be to attain a reliability of .90. If we solve the equation of the Spearman-Brown formula to find n, it becomes

$$n = \frac{r_{nn}(1 - r_{11})}{r_{11}(1 - r_{nn})} \qquad \text{(Estimation of length of test required for a given reliability)} \qquad (17.20)$$

Substituting the known values in this equation, we have

$$n = \frac{.90(1 - .75)}{.75(1 - .90)} = 3.0$$

The test with $r_{11} = .75$ would have to be three times as long to attain a reliability of .90.

Any other level of reliability, larger *or smaller,* in which we are interested can serve as r_{nn}, and the necessary n ratio can then be computed. Experience will show that some tests of low reliability cannot reach some desired high reliability without being made so long as to be impractical. Others will exhibit promising improvements in reliability with a moderate amount of extension. The formula is useful in the respect that it is helpful in making decisions about rejection or extension of tests and is useful in cases in which a test is already too long for comfort and we need to decide whether shortening it by a desired amount would sacrifice too much in reliability.

RELIABILITY OF RATINGS AND OTHER JUDGMENTS

Many of the statistics described in connection with test scores also apply fairly well to human judgments of various kinds. The judgments may be in the form of rank order, rating-scale evaluations, pair-comparison scaling, judgments in equal-appearing intervals, and the like. We can correlate the same observer's judgments obtained at two different times, or we can assume that similar judges are interchangeable and intercorrelate their evaluations (see the discussion of intraclass correlation in Chap. 13). We can pool judgments for two comparable groups of observers and correlate them as long as they apply to the same objects or persons.

Experience has shown that, with the appropriate cautions exercised, these applications may be made with meaningful results. Every coefficient must, as usual, be interpreted in the light of the manner in which it was obtained. Even the Spearman-Brown formula has been shown to apply, as, for example, in the pooling of judgments from two observers which yields increased reliability in a manner found for the doubling of a test in length. For example, on page 270 the reliability of one observer's ratings was estimated to be .33. For a pool of three raters' ratings the S-B formula tells us that the reliability should be .60, which agrees exactly with the estimate from the intraclass *r*. The comparability of judges must exist here, just as the comparability of items must exist in applying this formula to the change in length of test.[1]

EXERCISES

1 The following reliability coefficients were presented for a certain test:

Split half	.96	Retest after 1 month	.91
Alternate form	.94	Retest after 2 years	.86

Are these coefficients reasonable? Explain.

2 In six tests, the following correlations were found between halves composed of comparable items: .43, .55, .66, .74, .86, .94. Determine estimates of reliability for the full-length tests.

3 In a test of 55 items, the SD of the total scores was 7.5. The sum of the variances of the items was 9.8327. Estimate the reliability of the total scores.

4 Another test of 150 items has an SD of 24.4 and a mean of 94.2. Estimate the reliability of the scores, assuming that the items are approximately equal in difficulty and intercorrelation.

5 In four tests, the reliability coefficients were .65, .76, .87, and .94. Determine $r_{t\infty}$ and $S_{t\infty}$ in each case, assuming an SD of 10.0.

6 Determine the r_{nn} values lacking in Data 17*A*.

1 A special application of the Kuder-Richardson formula has been proposed for estimating the reliability of peer ratings in the special case where each person in the group rates every other person on a trait (Gordon, L. V. Estimation of the reliability of peer ratings. *Educational and Psychological Measurement,* 1969, **29,** 305–313).

DATA 17*A*
Reliability coefficients as *n* varies, for different values of r_{11}

r_{11}	n					
	1.5	*2*	*4*	*6*	*10*	*20*
.30	**.39**			**.72**		**.90**
.70		**.82**	**.90**	**.93**		**.98**
.90	**.93**	**.95**			**.99**	

DATA 17*B*
Lengths of tests needed to achieve certain levels of reliability

r_{11}	r_{nn}			
	.65	*.75*	*.84*	*.95*
.30	**4.33**		**13.22**	
.50		**3.00**		**19.00**
.70	**0.80**		**2.43**	
.90		**0.33**		**2.11**

7 For the coefficients in the completed table in Data 17*A*, plot on graph paper the increase in r_{nn} (on the ordinate) as n (on the abscissa) increases, for each value of r_{11}. State some general conclusions.

8 Complete the table for Data 17*B*, computing the necessary *n*'s.

9 A test has an SD of 7.2 and $r_{tt} = .86$. In another group the SD is 6.0. Assuming equal standard errors of measurement in the two samples, what should be the reliability in the second sample? In still another group, the SD is 9.0. What reliability should be expected in the third group?

10 As a mathematical exercise, assuming a constant standard error of estimate, derive formula (17.19), starting with the equations $r_{oo} = 1 - S_e^2/S_o^2$ and $r_{nn} = 1 - S_e^2/S_n^2$, and noting that S_e^2 is an identical value in the two equations.

11 Solving the Spearman-Brown equation for n, derive formula (17.20).

ANSWERS

2 r_{tt}: .60, .71, .80, .85, .92, .97.

3 $r_{tt} = .84$, by formula (17.14).

4 $r_{tt} = .95$, by formula (17.16).

5 $r_{t\infty}$: .81, .87, .93, .97; $S_{t\infty}$: 5.9, 4.9, 3.6, 2.4.

6 When $r_{11} = .30$, $r_{nn} = .46$, .63, .81; when $r_{ii} = .70$, $r_{nn} = .78$, .96; when $r_{11} = .90$, $r_{nn} = .97$, .98, .99.

8 When $r_{11} = .30$, $n = 7.00$, 44.33; when $r_{11} = .50$, $n = 1.86$, 5.25; when $r_{11} = .70$, $n = 1.28$, 8.14; when $r_{11} = .90$, $n = 0.21$, 0.58.

9 $r_{nn} = .80$; .91.

18 Validity of measurements

Although most of the comments in this chapter will be about validity of tests, the problem of validity arises in connection with all kinds of measurements. Most of what is said about the validity of tests applies, in a general way, to other kinds of evaluations or measurements as well.

Problems of validity

It is often easy enough to apply a measuring instrument and to obtain some numerical data. In the physical sciences the meaning of numbers that are used to describe phenomena is usually well established. The values stand for degrees of electrical resistance, pressure of gas, or mass of a particle, for example. In the behavioral sciences, however, the connection between a number and the thing or things for which it stands is not nearly so obvious.

This chapter will maintain that validity is a highly relative concept. If the question is asked about any particular test, "Is it valid?" the answer should be in the form of another question, "Is it valid *for what?*" Furthermore, just as we found in the preceding chapter that we cannot, strictly speaking, state that any particular figure indicates *the reliability* of a test, so we cannot give a single number to indicate *the validity* of a test.

There was a time, unfortunately still not entirely in the past, when investigators would conceive of a hypothetical variable of personality and proceed to construct a test for it, assuming that the trait as they had conceived it existed and that the operations involved in the test gave measurements of that variable. It might have been a test of "intelligence," "introversion," or "neurotic tendency." The designed measuring scales became operational definitions of those variables, without solid evidence as to the psychological nature of the concepts involved.

CONSTRUCT VALIDITY Over the years there has been an increasing call for evidence that an instrument purported to measure trait *X* really measures it. The most common approach has been to see that different instruments that are supposed to measure the same trait intercorrelate highly with one another. But even after demonstrating high intercorrelations, there are remaining doubts as to the nature of the variable that the instruments measure in common. And the high intercorrelations may well indicate that there is not just one trait variable in common but several.

Particularly since Thurstone developed his multiple-factor theory and his multiple-factor methods of analysis, such historical concepts as "intelligence," "introversion," and "neurotic tendency" have been broken down into much more limited variables in personality, sometimes called *primary traits.* They are primary in the sense of resisting further reduction into combinations of still narrower traits. Nearly a hundred primary intellectual abilities and nearly a score of psychomotor abilities have been demonstrated by factor analysis,[1] and the number of factor traits of temperament, needs, interests, and attitudes that have been suggested probably exceeds 60.[2] The validity of a test for measuring a particular factorial trait is indicated by the *loading* of the test on the factor that it represents. The concept of factor loading will be explained shortly.

It is recognized by many of those who factor-analyze that scarcely any test is an unadulterated measure of any one factorial trait. Not only is it diluted by errors of measurement, as seen in the preceding chapter, but it is also adulterated with variances in other factorial traits. This situation can be overcome to some extent by carefully combining tests, an exacting procedure that we cannot go into here.[3] It is the authors' belief that the best answer to the question, "What does this test measure?" is in the form of a list of factors with which it correlates, or its loadings on those factors. This kind of information is *factorial validity*. For many individuals concerned with construct validity, factorial validity provides the best answers. The con-

1 See Guilford, J. P., and Hoepfner, R. *The analysis of intelligence*. New York: McGraw-Hill, 1971; Guilford, J. P. A system of psychomotor abilities. *The American Journal of Psychology*, 1958, **71,** 164–174.

2 See, for example, Guilford, J. P. *Personality*. New York: McGraw-Hill, 1959.

3 See Guilford, J. P., and Michael, W. B. Approaches to univocal factor scores. *Psychometrika*, 1948, **13,** 1–22.

cept will be explained in some detail, and it will be shown that it is basic to an understanding of other kinds of validity, as well as correlations in general.

A distinction has been made between two types of construct validity.[1] One type is known as *convergent* validity and the other as *discriminant* validity. The distinction is concerned with the operations by which validity is determined, but more than that, it calls for finding what a test does not measure as well as what it does.

There is said to be convergent validity when a test or other measure of a proposed trait correlates strongly with instruments of other kinds designed to measure the same trait or that are thought to measure it. There is a call for evidence of the measurement of the construct by different methods, e.g., ratings (by self and by others) and performance tests as well as by printed tests, for example. This is not to say that any trait must be equally measurable by all methods, for, as a matter of fact, many traits are not well assessed by all methods. The behavioral signs of those traits are not equally observable from all approaches.

Discriminant validity is shown by the fact that the test correlates little or not at all with measures of other traits, whether by the same method or by other methods. There is some danger that measures of traits by the same method may correlate higher than they should because of the method in common. In ratings by outside observers, for example, there is the well-known "halo error," which is seen in the fact that the rater is inclined to rate the observed person similarly in all traits. The rater has a general conception of the person as being at a certain level.

The search for high correlations among different methods of measuring the same trait and low correlations between measures of different traits would seem to be only a step in the direction of the more refined conception of factorial validity. Tests very valid for measuring the same factor *are* usually strongly intercorrelated. That is the way in which factors are discovered. Tests correlate very little with instruments that measure only other factors because the factors themselves have little or no correlation with one another.

PREDICTIVE VALIDITY The vocational counselor and the personnel manager face a different kind of problem when they are concerned about the validity of a test. They want to know whether the test provides the information that they need for predicting outcomes in specific kinds of tasks and assignments—clerical performance, academic achievement, salesmanship, and the like. A test is valid in

1 Campbell, D. T., and Fiske, D. W. Convergent and discriminant validation by the multitrait-multimethod matrix. *Psychological Bulletin,* 1959, **56,** 81–105.

predicting outcomes in clerical work if it correlates with an assessment of clerical proficiency. Another test is valid for salesmanship if it predicts the extent of sales under standard conditions. From this point of view a test is valid for any sphere of behavior if it enables us to forecast performance within that sphere, regardless of the name of the test or of the trait or traits that it is said to measure. For example, the same test may predict about equally well who can learn to fly an airplane and who can excel on the pistol range. From a practical standpoint, the validity of a test is its forecasting proficiency in connection with any measurable aspect of daily living. It should be noted, however, that, as pointed out in Chap. 15 (p. 358), the *utility* of a test in selection and classification in general depends upon circumstances other than its correlation with a criterion.

CRITERIA OF PERFORMANCE One of the most difficult aspects of the *predictive*-validity problem is that of obtaining adequate criteria of what represents success. The factor-analytic approach has a good solution when it is a question of what traits we wish to measure. But practical criteria needed in the operation of determining predictive validity are often very difficult to obtain and to measure adequately.

As an example for the last statement, let us consider a criterion for academic achievement. It has often been assumed that academic achievement, like intelligence, is a unitary attribute of each individual—one variable. But this is far from true. Although commonly there are positive correlations between indices of achievement in different courses, there is sufficient disagreement to permit some students to receive marks all the way from A to F in different subjects. It is best, therefore, to examine the validity of each test used for guidance purposes in connection with every school subject by itself. The writers have known data showing correlations all the way from .37 to .74 between the Ohio State Psychological Examination, a college-aptitude test, and marks in freshman courses at a certain university, for example.

The point is that success in almost any sphere of life is highly complex, being determined by a number of psychological factors in the individual. If we single out certain aspects of performance and measure them to use as criteria, we are checking on the validity of the test or tests for predicting those particular aspects. We should, of course, single out the most significant aspects as criteria, not some inconsequential aspects because they are just more observable and measurable.

Having selected some aspects, we have the further problem of securing dependable measurements and perhaps of combining them and weighting them. The criterion measurements themselves should be reliable and valid. A criterion is valid if it measures important aspects of performance. We can allow statistically for the unreliability of criteria when we know coefficients of reliability for them, as we shall see later, but we cannot so easily know or allow for lack of validity of criteria.

A brief introduction to factor theory

Because so many of the facts of validity are explainable in terms of factor theory, it is desirable for us to examine the basic features of that theory in order to gain a better grasp of the problems and methods involved. There is not space in this volume to describe the procedures for making a factor analysis of tests. Properly described, these procedures would take up a volume in themselves.[1]

BASIC ASSUMPTIONS IN FACTOR THEORY

It is best to begin with basic theorems, two of which will provide the foundation needed for the logic of validity.

THEOREM I The total variance of a test may be regarded as the sum of three kinds of component variances: (1) that contributed by one or more common factors, *common* because they appear in more than one test; (2) that unique to the test itself and to its equivalent forms; and (3) error variance.

We are now ready to partition what was called *true* variance in the preceding chapter into component variances. Both the common-factor variances and the specific variance in a test contribute to its internal-consistency reliability and to its equivalent-forms reliability. It is not necessary to assume that the common factors are all independent or uncorrelated, but to do so relieves us of having to deal with covariance terms, thus simplifying the picture.[2]

Theorem I may be stated in the form of the equation

$$S_t^2 = S_a^2 + S_b^2 + \cdot \cdot \cdot + S_q^2 + S_s^2 + S_e^2 \tag{18.1}$$

where S_t^2 = total variance of a test

$S_a^2\ S_b^2, \cdot \cdot \cdot, S_q^2$ = variances in common factors $A, B, \ldots, Q$

S_s^2 = variance specific to the test

S_e^2 = error variance

If we divide equation (18.1) through by S_t^2, we have

1 For treatments of factor analysis see Comrey, A. L. *A first course in factor analysis*. New York: Academic Press, 1973. Fruchter, B. *Introduction to factor analysis*. New Delhi: Affiliated East-West Press, Pvt. Ltd., 1967. Gorsuch, R. L. *Factor analysis*. Philadelphia: W. B. Saunders, 1974.

2 This theorem and the second one follow from the basic postulate that an obtained test score (in standard form) is a linear summation of components from the sources indicated in the theorem.

$$\frac{S_t^2}{S_t^2} = \frac{S_a^2}{S_t^2} + \frac{S_b^2}{S_t^2} + \cdots + \frac{S_q^2}{S_t^2} + \frac{S_s^2}{S_t^2} + \frac{S_e^2}{S_t^2} = 1.00 \qquad (18.2)$$

Substituting new symbols for these ratios, which are proportions, we have

$$1.00 = a_x^2 + b_x^2 + \cdots + q_x^2 + s_x^2 + e_x^2 \qquad \text{(Proportions of factor variances in a test)} \qquad (18.3)$$

where $a_x^2, b_x^2, \ldots, q_x^2$ = proportions of total variance contributed to test *X* by factors *A, B, . . . , Q* respectively
s_x^2 = proportion of specific variance in test *X*
e_x^2 = proportion of error variance in text *X*

In the same notation, the reliability of test *X* can be written as

$$r_{tt} = 1 - e_x^2 = a_x^2 + b_x^2 + \cdots + q_x^2 + s_x^2 \qquad (18.4)$$

(Reliability as a sum of proportions of nonerror variance)

This equation will be useful in discussions of the relation of validity to reliability later on.

COMMUNALITY A new concept that should be pointed out here, although we shall not have occasion to do much with it in this chapter, is the *communality* of a test. The communality of a test is the sum of the proportions of common-factor variances. In equation form,

$$h_x^2 = a_x^2 + b_x^2 + \cdots + q_x^2 \qquad \text{(Communality of a test)} \qquad (18.5)$$

The communality of a test contains all the nonerror variance except the specific variance. Communality is what gives any test the chance of correlating with other tests and with practical criteria. If there were no communality in a test, it could be quite reliable and still not correlate with anything else. On the other hand, a test could have relatively low reliability, and yet if all its nonerror variance were in common with variance in other variables, its correlations with other things could be rather substantial; hence its validity could be good.

A NUMERICAL EXAMPLE OF COMPONENT VARIANCES As an example, let us consider three tests and a practical criterion. Five common factors are represented in these four variables. Table 18.1 lists the proportions of common-factor, specific, and error variance for each variable. Test 1 has 36 percent of its variance accounted for by factor *A*, and 36 percent by factor C. The sum of these two components equals 72 percent, which represents the communality of this test. Add the 10 percent specific variance, and we have 82 percent, which represents the test's true variance and a reliability of .82. The remaining 18 percent is error

TABLE 18.1 Proportions of common-factor, specific, and error variance in three tests and a practical criterion of proficiency

Variable	*Common factors*					*Specific,* S	*Error,* E	*Commu-nality,* h^2	*Relia-bility,* r_{xx}
	A	*B*	*C*	*D*	*F*				
Test 1	**.36**	**.00**	**.36**	**.00**	**.00**	.10	.18	.72	.82
Test 2	**.16**	**.00**	**.12**	**.00**	**.64**	.00	.08	.92	.92
Test 3	**.00**	**.49**	**.00**	**.25**	**.00**	.09	.17	.74	.83
Criterion *J*	**.16**	**.09**	**.16**	**.25**	**.00**	.14	.20	.66	.80

variance. The other tests and criterion *J* can be interpreted in a similar manner. Figure 18.1 shows the component variances for these same four variables, each as a segment of a bar diagram.

FACTOR LOADINGS The proportion of a total variance contributed by one component may be regarded as a coefficient of determination of the total by the part. The square root of each proportion of variance contributed by a common factor may therefore be regarded as the correlation between the total variable and the factor. These square roots are correlation coefficients and are known as *factor loadings*. For the three tests and criterion *J*, the common-factor loadings are given in Table 18.2. Test 2 correlates .40 with factor *A*, .35 with factor *C*, and .80 with factor *F*. Factor *F* has no correlations with other variables in this list, but in order to be regarded as a common factor, it must have some correlation with other variables not in this list.

The square roots of specific variance are not listed because it is not certain what the specific variances represent. A certain specific variance may indeed be unique to its own test, but it may be a composite of some kind, in which case each component of the specific variance would have its own correlation with the total. On the other

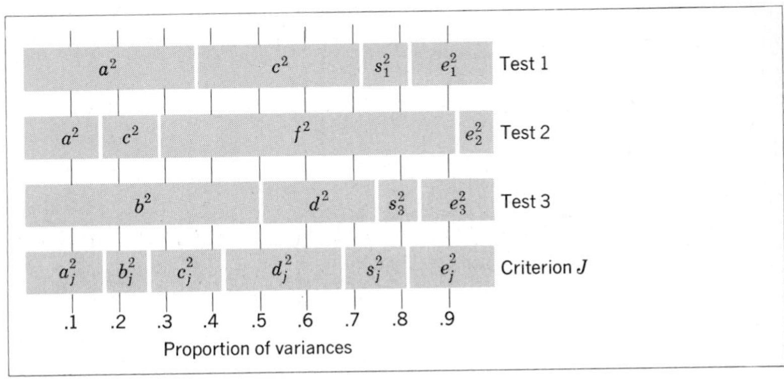

FIGURE 18.1 Proportions of common-factor, specific, and error variances in three hypothetical tests and a criterion

TABLE 18.2 Factor loadings (correlations of common factors with experimental variables) for the three tests and a criterion

Variables	*Common factors*				
	A	*B*	*C*	*D*	*F*
Test 1	**.60**	**.00**	**.60**	**.00**	**.00**
Test 2	**.40**	**.00**	**.35**	**.00**	**.80**
Test 3	**.00**	**.70**	**.00**	**.50**	**.00**
Criterion *J*	**.40**	**.30**	**.40**	**.50**	**.00**

hand, some specific variances might turn out in later analyses to be one or more unrecognized common-factor variances. Certain tests have been known to lack any specific variance at all, the entire true variance being composed of common-factor components and the communality equaling the reliability of the test.

THEOREM II The second major theorem of factor analysis is that the correlation between two experimental variables (such as tests and criteria) is equal to the sum of the cross products of their common-factor loadings. In equation form,

$$r_{jx} = a_j a_x + b_j b_x + \cdot \cdot \cdot + q_j q_x \quad \text{(A correlation as a sum of factor-loading products)} \tag{18.6}$$

where a_j and a_x = loadings of factor *A* in criterion *J* and test *X*, b_j and b_x = loadings of factor *B* in criterion *J* and test *X*, etc.

HOW FACTOR THEORY EXPLAINS PREDICTIVE VALIDITY

Applied to the loadings given in Table 18.2, the correlation between tests 1 and 2 would be

$$r_{12} = (.6)(.4) + (.0)(.0) + (.6)(.35) + (.0)(.0) + (.0)(.8) = .45$$

The correlation between test 1 and criterion *J* (its validity for predicting criterion *J*) would be

$$r_{j1} = (.4)(.6) + (.3)(.0) + (.4)(.6) + (.5)(.0) + (.0)(.0) = .48$$

The other intercorrelations and validity coefficients found in a similar manner are listed in Table 18.3. In experimental practice we do not

TABLE 18.3 Intercorrelations of tests and criterion *J* derived from their common-factor loadings

Variables	*Tests*			*Criterion*
	1	*2*	*3*	*J*
Test 1		**.45**	**.00**	**.48**
Test 2	**.45**		**.00**	**.30**
Test 3	**.00**	**.00**		**.46**
Criterion *J*	**.48**	**.30**	**.46**	

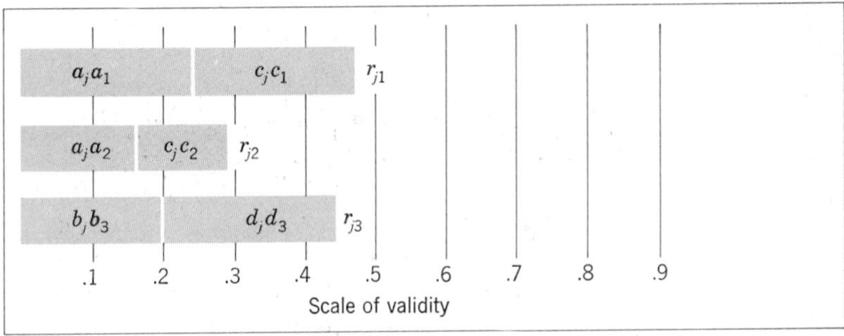

FIGURE 18.2 Segments of three intercorrelations of tests, each with a criterion, that are contributed by different common factors

know the factor loadings first and derive from them the intercorrelations; we know the intercorrelations and by factor analysis arrive at the factor loadings. We have assumed that the factor loadings are known here for the sake of illustration.

Examination of the three validity coefficients in Table 18.3 shows that they are .48, .30, and .46, for tests 1, 2, and 3, respectively. The three validity coefficients are represented graphically in Fig. 18.2. The reasons for the validity of tests 1 and 2 are the same; their common ground with the criterion is in factors *A* and *C*. The reason test 3 is valid, however, is totally different from this. Test 3 is valid because it has factors *B* and *D* in common with the criterion. Test 2 has the lowest validity for predicting criterion *J*, but its unusually large loading in factor *F* offers strong possibilities for its validity in predicting some other criterion that has a substantial loading in factor *F*.

HOW FACTOR THEORY EXPLAINS MULTIPLE-CORRELATION PRINCIPLES The multiple correlations of some of these tests with criterion *J* can be nicely explained by the various factor loadings. The multiple correlation $R_{j.12} = .49$, which is only .01 higher than the correlation r_{j1}. Adding test 2 to test 1 in a battery to predict *J* is of little or no value because both bring to the composite a coverage of the same common factors in *J*, and test 1 does this much better. The multiple $R_{j.13}$, however, is equal to .66. Adding test 3 to test 1 to make a joint prediction of *J* is very effective because the two tests cover totally different components in *J*. The multiple $R_{j.23}$ is less than $R_{j.13}$, being .55. The reason for this is that test 2 does not cover factors *A* and *C* nearly so well as test 1 does.

OPTIMAL WEIGHTING OF FACTORS IN COMPOSITES We might well raise the question at this point as to whether tests 1 and 3, optimally weighted, with their multiple *R* of .66, have yielded the maximum amount of validity possible for a weighted composite that contains factors *A*, *B*, *C*, and *D*. Reference to equation (18.6) will show that the correlations r_{j1} and r_{j3} could have been higher if the tests' factor loadings a_1, c_1, b_3, and d_3 had been larger. The only limits to those factor loadings would be that the communalities should not exceed 1.0.

This is not the whole story, however. We could make those loadings as large as the communalities would allow, and they would still not yield the maximal correlation with criterion *J* unless they were in the right proportions. The right proportions would have to take into consideration the proportions of loadings a_j, b_j, c_j, and d_j in the criterion. With sufficient loadings of the four factors in the tests and with proper weightings, the maximum validity for the composite in predicting criterion *J* would be equal to the square root of the communality of that criterion. The square root of .66 is .81. This principle is reminiscent of the one mentioned in the preceding chapter regarding the index of reliability, which is the square root of the reliability coefficient. It gives the maximum possible correlation of anything with the variable in question. In this statement, however, is latent the assumption that all the true variance is common-factor variance—that $h^2 = r_{tt}$.

It is doubtful whether tests 1 and 3 could ever be weighted appropriately to yield a validity for their composite equal to the maximum .81 with criterion *J*, even though their common-factor loadings were as large as possible. The reason is that factors *A* and *C* are tied together in the same test and factors *B* and *D* are tied together in the other test. Since factors *A* and *C* have equal loadings in criterion *J* and also in test 1, as long as they keep the same ratio in test 1 they would be properly weighted in a regression equation. This is merely a coincidence in this particular problem. Factors *B* and *D*, however, are weighted in reverse order in test 3 and criterion *J*. For optimal prediction of *J*, the loading d_3 should be greater than the loading b_3, to correspond with the fact that the loading d_j is greater than b_j.

The moral of this is that for the freedom to weight each factor in a composite as it should be weighted to get the maximal prediction of a criterion, it is best to use unique, or univocal, tests, i.e., each test with but one common factor. In practice, a regression weight has to be applied to the test as a whole, and all factors in it are weighted the same, insofar as external weights are applied.

INCREASING VALIDITY BY ADDING FACTORS We have just seen that increasing the predictive validity of a composite depends upon large factor loadings for factors represented in the criterion and an *optimal weighting of the individual factors*. There is another important way of increasing the validity of a composite, and that is to bring in a new test that covers a common factor in the criterion that is not already covered. Criterion *J* was reported to have 14 percent of its variance devoted to specific sources. It is possible that this portion of the variance in *J* is really contributed by unrecognized common factors, which could be revealed by further factor analysis. Suppose that it were found that only one additional common factor was involved, a factor *G*. In order to contribute all the remaining true variance in *J*, the loading g_j would be approximately .37 (the square root of .14). With an additional test in the composite predictor to measure *G*, the multiple *R* could be increased materially.

Conditions upon which validity depends

RELATION OF VALIDITY TO RELIABILITY It has been a common belief that the predictive validity of a test is directly proportional to its reliability—the more reliable the test, the more valid it is. Factor theory supports this principle to some extent, and under certain conditions. As the proportion of true variance in a test is increased, as by lengthening it or by applying item analysis, other things being equal, the communality should increase and the loadings on common factors. And if one or more of those common factors are shared by the criterion predicted, predictive validity should also increase. If test and criterion share no common factors, no amount of increase in reliability will have any effect upon validity.

One kind of exception to the principle that high reliability means high validity is seen in connection with homogeneous and heterogeneous tests. Homogeneous tests have high internal-consistency reliability, but they can have no validity for predicting variations in certain criteria, for lack of factors in common, as indicated above. Heterogeneous tests have relatively low reliability, and yet some have been known to show relatively high predictive validity. A heterogeneous test measures different factors within its content and thus has a better chance of sharing common factors with a factorially complex criterion. The reasons will be more fully brought out in the discussion that follows. One qualifying statement needs to be made regarding homogeneous tests. By definition, all parts of such a test measure the same variables. But every part, even each item, could be a little factorially complex in the same way, as when it assesses verbal comprehension and also reasoning by analogy, two different factors. In this situation, as well as in a heterogeneous test, the number of factors represented is a favorable condition for validity.

GOALS OF VALIDITY AND RELIABILITY SOMETIMES INCOMPATIBLE When we seek to make a single test both highly reliable (internally) and also highly valid, we are often working at cross purposes. The two goals are incompatible in some respects. In aiming for one goal, we may defeat efforts toward the other.

Maximal reliability requires high interrelations between items; maximal predictive validity requires low intercorrelations. Maximal reliability requires items of equal difficulty; maximal predictive validity requires items differing in difficulty. This point needs some explanation. Tucker has demonstrated this fact mathematically, but there is a simpler, commonsense rationale.[1] A range of difficulty is very desirable, of course, in order to obtain graded measures of individuals. It was shown in Chap. 17 how with perfect intercorrelation of items (which could occur with r_ϕ coefficients only when items are of equal difficulty), there were only two scores—perfect scores and zeros. For spacing individuals in fine-enough gradations for measurement purposes, it is necessary to have a continuous distribution, not a U-shaped one. It would be ideal, for fine measurements, to space items, each discriminating well between all those above a certain point on the scale and those below, rather evenly all along the range of ability in the population. With such spacings, intercorrelations could not be perfect, and some would indeed be very low.

There must be some compromising of aims; both reliability and validity cannot be maximal. Fortunately, the kind of moderate item intercorrelations usually obtained for well-constructed items are of the size that, according to Tucker's conclusions, will yield good validities. They will also yield satisfactory reliabilities, but those reliabilities will not often be above .90. To be more specific, the item-test correlations for well-constructed items range between .30 and .80, which means item intercorrelations approximately between .10 and .60. Items within these ranges of correlation should provide tests of both satisfactory reliability and validity. There is probably better reason for going below these limits than above them in constructing items. To do so would probably be to err on the side of validity, which, for some purposes, is the more important.

HOMOGENEOUS TESTS; HETEROGENEOUS BATTERIES The relation of heterogeneity to validity deserves more attention. One way to make a test more valid for predictive purposes is to make it more heterogeneous. In factorial language this means adding new factors. If we succeeded in getting into the scores of the single test all the factors that are also in the practical criterion, and if we weighted them properly, we could achieve maximal accuracy of predictions from the single test.

Recall, in this connection, the principles of the multiple-regression equation. Maximal multiple correlation is achieved by minimizing the intercorrelations of the independent variables. If we apply this to test items, as separate variables, the principle still holds. The ideal test, from this point of view, would be one in which each item measured a different factor (and measured it consistently). This would mean a test of low internal reliability. It would also mean a test which,

[1] Tucker, L. R. Maximum validity of a test with equivalent items. *Psychometrika,* 1946, **11,** 1–13.

though correlating well with the criterion, would make very crude discriminations for each factor. Each item would ordinarily differentiate only two categories—those who pass it and those who fail it—for each trait measured. If we brought in a number of items to measure each factor, with differences in difficulty to overcome this defect, we should have virtually a battery of tests within a single test.

The solution to the incompatibility of goals of reliability and predictive validity is precisely what has just been suggested: to use a battery of tests rather than single tests. Reliability should be the goal emphasized for each test, and predictive validity the goal emphasized for the battery. Even in single tests some reliability should be sacrificed for the sake of well-graded measurements. It is strongly urged that, if possible, each test be designed to measure one common factor. It should be univocal, its contribution unique. In this way minimal intercorrelation of tests is ensured, which satisfies one of the major principles in multiple regression. It was also shown that when tests are univocal, the various factors can be weighted in the best way to make each prediction. The univocal test will correlate less with a practical criterion than a heterogeneous test will, but what we lose in validity for the single test will be more than made up by forming batteries which cover the factors to be predicted and in a more manageable manner. For the sake of meaningful profiles also, a battery of univocal tests has no equal. The use of single-test scores in a profile, however, calls for high reliability for each test.

RELIABILITIES AND TEST BATTERIES If a composite score from a battery is to be used and not part scores from the components, as in a profile, it is likely that not much will be gained by achieving reliabilities for single tests higher than .60 or by having tests longer than 30 items each.[1] The reliability of the composite score of *independent* tests will be approximately a weighted average of the reliabilities of the components.[2] This means that if the components have a generally low reliability, in such a battery the reliability of the composite will be low. This need not be disturbing, provided the validity of the composite is high. To the extent that the components are intercorrelated, the reliability of the composite will exceed the average reliability of the components. In general, if there is a choice between lengthening tests in a predictive battery to make them more reliable and adding more tests of different kinds that contribute unique valid variances, the decision should certainly go to the second alternative. If single-test scores are to be used separately, however, attention must also be given to reliability of components.

1 Dailey, J. T. Determination of optimal test reliability in a battery of aptitude tests. Technical memorandum No. 10, Lackland Air Force Base, 1948.

2 Mosier, C. I. On the reliability of a weighted composite. *Psychometrika,* 1943, **8,** 161–168.

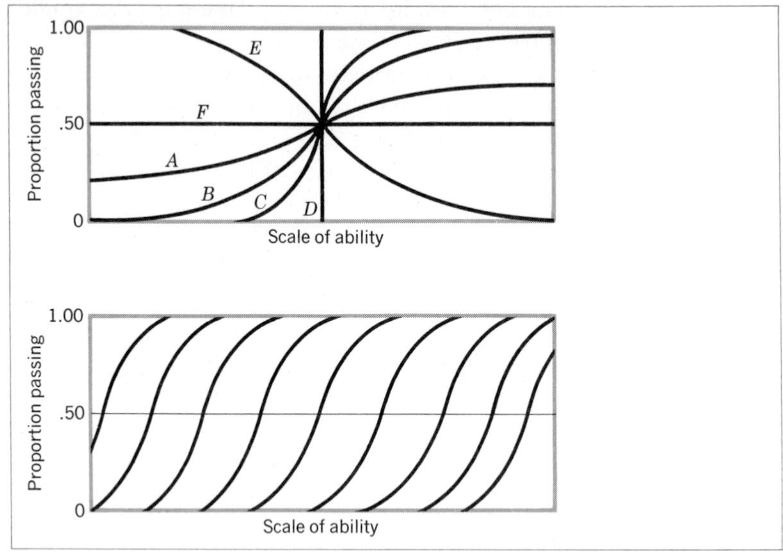

FIGURE 18.3 Proportion passing an item (responding correctly) as a function of ability level on the scale of the kind or kinds of ability to pass the item

DISCRIMINATION VALUES OF ITEMS

Some of the points just discussed may be made a little clearer if we approach the item theory from a different aspect. Figure 18.3 is used to illustrate this approach. Imagine a scale of ability or of any other trait that we attempt to measure by means of a test. We want each item to correlate with that variable, to indicate the status of individuals with respect to the variable, and to discriminate between individuals.

Suppose we already know the positions of large numbers of individuals on this scale. We apply to them an item that we will call item C. The item is of median difficulty, for of the entire group 50 percent respond in the acceptable manner and 50 percent do not. According to the requirements of good reliability, this knowledge about the difficulty of item C is promising but not sufficient evidence that the item would contribute to a reliable test. We do not yet know whether it is at all related to the variable we want to measure. It could be of median difficulty and still be uncorrelated with other items in the test. Let us subdivide the large sample into subsamples grouped in class intervals as if for known values along the scale. We are now interested in seeing whether those groups higher on the scale have any greater probability of passing the item than those lower on the scale. Theory states, and experimental evidence supports the idea, that the increase in the probability of passing the item follows the normal cumulative frequency curve. The regression of proportions passing the item upon ability is the S-shaped or ogive form. For item C, not very far below average ability, we find a point below which none pass the item. Above a point just as far above the mean we find that all pass the item. The interval between is sometimes called the *transition zone,* a concept borrowed from psychophysics.[1]

Other items may have the same difficulty level as item *C*, but like items *B* and *A* in the diagram (Fig. 18.3) they have different degrees of discriminating power. Both *B* and *A* have much wider transition zones (they both actually go beyond the range of the given horizontal scale), and their curves have slopes that are less steep than that for *C*. The steepness of the slope is known as the curve's *precision*. The term applies well here because the steeper the precision of the curve, the greater the precision of discrimination. A perfectly discriminating item is *D*, whose slope is infinite. A nondiscriminating item is *F*, whose slope is zero. There is a mathematical relationship between the precision of an ogive like these and the correlation between the item and a good measure of the trait.[2] Item *E* would have a negative correlation with the variable to be measured. This would be an unusual event and would probably mean that the item was keyed wrong in scoring. Items like *D* would seem to be ideal; they are perfectly discriminating. But it can be seen how only one such item used alone would be almost futile, for it discriminates only at one point. A set of such items, equally spaced as to difficulty, however, would make an ideal test.

The second diagram is more realistic and yet pictures a somewhat ideal situation. It shows a series of items about equally spaced as to difficulty and all with excellent discriminating power. With the extensive range of difficulty level, there could not be as high internal reliability as some might desire. But the possibility of accurately grading individuals on a continuous scale is greater because of that dispersion. To appreciate the full value of the items that depart from median difficulty, one would need either to use a biserial *r* or a tetrachoric *r* in correlating item with total score or to make allowance for the effect of divergencies in difficulty upon the phi coefficient.

VALIDITY AND LENGTH OF TEST

Since the homogeneous lengthening of a homogeneous test increases its reliability, in accordance with the Spearman-Brown formula, it will also increase its validity, either factorial or predictive. If the change in length is by some ratio *n* (the new length divided by the old), the new validity of the test is estimated by the formula

$$r_{y(nx)} = \frac{r_{yx}}{\sqrt{\dfrac{1 - r_{xx}}{n} + r_{xx}}} \qquad \text{(18.7)}$$

(Validity of a homogeneous test increased in length *n* times)

[1] Woodworth, R. S. *Experimental psychology*. New York: Holt, 1938. p. 401.

[2] For proof of this, see Richardson, M. W. Relation between the difficulty and the differential validity of a test. *Psychometrika*, 1936, **1**, 33–49.

where r_{yx} = validity coefficient for predicting criterion Y from test X and r_{xx} = reliability of test X.

A certain line-drawing test developed to predict creative abilities of students in a course in designing had a reliability of .57 and a correlation with teacher's ratings of .65.[1] If this test were made twice as long, what validity could be expected? Applying formula (18.7),

$$r_{y(2x)} = \frac{.65}{\sqrt{\frac{1-.57}{2} + .57}} = .73$$

It would thus definitely pay to make this test longer and more reliable in order to improve its validity.

If we wanted to know how much homogeneous lengthening is needed in order to achieve a desired level of validity, we could do this by solving formula (18.7) for n, which gives[2]

$$n = \frac{1 - r_{xx}}{\frac{r_{yx}^2}{r_{y(nx)}^2} - r_{xx}} \qquad \text{(Ratio of new length of test for a required validity)} \qquad (18.8)$$

If we wanted a validity of .80 for the line-drawing test, the ratio of the revised length to the former length would have to be

$$n = \frac{1 - .57}{\frac{.4225}{.64} - .57} = 4.8$$

Whether it would be practical to devote nearly five times as much effort to this test is a question of policy that goes beyond statistical answers.

RELATION OF VALIDITY COEFFICIENTS TO ERRORS OF MEASUREMENT

When two measured variables are correlated, the errors of measurement, if uncorrelated among themselves, serve to lower the coefficient of correlation as compared with what it would have been had the two measures been perfectly reliable. We say that the degree of correlation has been attenuated. If we want to know what the correlation would have been if the two variables had been perfectly measured, we must resort to the *correction for attenuation,* for which we have a formula:

$$r_{\infty\omega} = \frac{r_{xy}}{\sqrt{r_{xx}r_{yy}}} \qquad \text{(A correlation coefficient corrected for attenuation)} \qquad (18.9)$$

1 Guilford, J. P., and Guilford, R. B. A prognostic test for students in design. *Journal of Applied Psychology,* 1931, **15,** 335–345.

2 If n should turn out to be negative, we ignore the algebraic sign.

where r_{xx} and r_{yy} = reliability coefficients of the two tests. The subscripts ∞ and ω indicate true scores in X and Y.

The correlation obtained between a figure-classification test and a form-perception test was .36. The reliability coefficients for the two tests were .60 and .94, respectively. Applying formula (18.9),

$$r_{\infty\omega} = \frac{.36}{\sqrt{(.60)(.94)}} = .48$$

We should therefore expect the correlation between true scores in these two tests to be .48 rather than the obtained one of .36.

Rather commonly, when making this correction for attenuation in two fallible tests, we are dealing with two forms of the same test for purposes of finding reliability. There is a possibility of determining four intercorrelations between the two tests, i.e., each form of the one correlated with the two forms of the other. In this case, it is well to use all the information available concerning the intercorrelation of the two tests by computing the four coefficients and using their arithmetic mean as a better estimate of the numerator of the fraction in formula (18.9).

FACTORIAL EXPLANATION OF ATTENUATION AND ITS CORRECTION It may not be clear to the reader why errors of measurement always lower intercorrelations, and why, when the corrective formula is applied, correlations should not be perfect. The answers to both these questions can best be given by reference to factor theory.

Consider test 1 and criterion J of the illustration used earlier when factor theory was introduced. Error variance made up 18 percent of the total variance of test 1 and 20 percent of criterion J. Let us suppose that we could rid each variable of all errors of measurement, all error variance. In doing so, let us further suppose that the remaining true variance is expanded with all its components in proportion to their original amounts. Figure 18.4 demonstrates what happens when the error components are "squeezed out" of variables and the true-variance components expand to take their places. Proportions of variances that were .36 and .36 for factors A and C in test 1 before correction become .439 and .439 after correction. The new factor loadings are .663 on each factor. In the criterion the corresponding loadings become .447 in place of .40. By equation (18.6), the new correlation r_{j1} becomes .59, whereas it was .48. The use of formula (18.9) applied to the original r_{j1} gives

$$r_{\infty\omega} = \frac{.48}{\sqrt{(.82)(.80)}} = .59$$

The change in validity from .48 to .59 is shown graphically in Fig. 18.4.

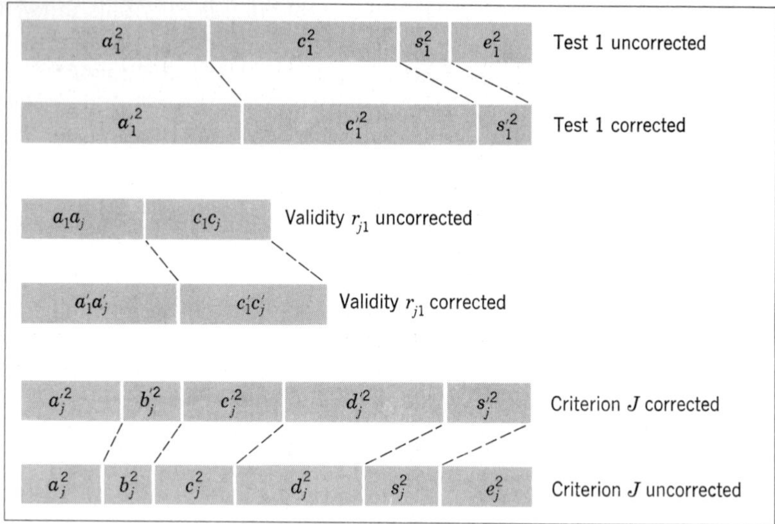

FIGURE 18.4 Proportions of certain component variances in a test and a criterion after correction for attenuation (elimination of error variance statistically), also the contributions of factors to the validity coefficient before and after correction

CORRECTION FOR ATTENTUATION IN THE CRITERION ONLY

The correction procedure of formula (18.9) has limited application except in theoretical problems. In practice, we are compelled to deal with fallible tests. If the tests from which we wish to predict something else are not perfect, that fact must be faced, and our predictions are reduced in accuracy accordingly. But we should hardly expect to be asked to overlook the fallibility of the criterion we are trying to predict. If it measures success inaccurately, this lack of accuracy should not be permitted to make it appear that the test is less valid than it really is. It is desirable, therefore, to correct predictive-validity coefficients for attenuation in the criterion measurements but not in the test scores. This one-sided correction is made by the formula

$$r_{x\omega} = \frac{r_{xy}}{\sqrt{r_{yy}}} \qquad \text{(Validity coefficient corrected for attenuation in the criterion only)} \qquad (18.10)$$

As an application of this formula, we cite the line-drawing test previously mentioned that correlated with a teacher's rank-order judgments of creative ability of students in design to the extent of .65. The reliability of the teacher's ratings (combined from two rank orders a month apart) was found to be .82. Had the teacher's ratings been perfectly reliable measures of the thing she was judging, the correlation with test scores would have been $.65/\sqrt{.82} = .72$. The correlation of .72 is accordingly taken as the genuine validity of the test, unless we are concerned about predicting the teacher's judgments, contaminated by flaws as they obviously are, rather than genuine ability as evidenced by those ratings.

Many a validity coefficient reported in the literature is of very uncertain meaning because errors of measurement in the criterion were not taken into account. The reliability of ratings made by a single observer, even of the better ones, is characteristically about .60. With such criteria, validity coefficients are about 25 percent underestimated. Too often the reliability problem of a criterion is entirely ignored. The writers have known of purported criteria of a performance criterion (bombing errors of bombardiers in training) which at best had reliabilities of only approximately .30. What is even more important, but incidental to the discussion here, is the *construct validity* of the criterion. That is, does the criterion measure represent relevant and desirable aspects of performance and measure it well? Investigators who hope to develop successful selective instruments are often beaten before they start if they do not first ensure reliable and valid criteria or if they do not estimate these features and make allowances for their shortcomings.

LIMITATIONS TO THE USE OF CORRECTION FOR ATTENUATION The correction of a correlation for attenuation requires that we have a rather accurate estimate of reliability for each variable that enters into the situation. If either r_{yy} or r_{xx} is underestimated, the corrected r_{yx} will be overestimated. If either reliability coefficient is overestimated, the corrected r_{yx} will be underestimated. It is probably best, if one wishes to be on the conservative side, for a reliability estimate to be too large when used for this purpose. On the other hand, it is likely that most estimates of internal-consistency reliability are too low, which is in the wrong direction for conservatism.

There is also the question as to which of the three main types of reliability coefficient is desirable in correcting for attenuation. There are proponents for the use of each type in this connection. It is best to decide what kind of errors of measurement should be ruled out in the particular situation or particular use of r_{tt}. Once this decision is made, the type of reliability will be selected accordingly, since it was shown in the preceding chapter that each type emphasizes certain sources of variance as error. The tendency of underestimation of r_{tt} by internal-consistency methods works against their use where there is a reasonably good alternative. In general, the alternate-forms approach is probably best.

VALIDITY OF RIGHT AND WRONG RESPONSES

Many tests are scored with a formula score in which the wrong responses are given a negative fractional weight, and the right responses a weight of +1.

A PRIORI SCORING FORMULAS One of the reasons behind such scoring formulas is a priori reasoning about chance success and the need for correcting for it. In a true-false test we have a two-alternative situation, and the assumption is that an examinee who does not know an answer will guess at random. When the examinee guesses, the probability of getting the right answer is .5. When there are three alternatives, the theoretical proportion of right answers in guessing is .33; in a four-choice item the probability is .25; and so on. This has led to the stock scoring formula of the form

$$S = R - \frac{W}{k-1} \quad \text{(A test score with a priori correction for guessing)} \tag{18.11}$$

where R = number of right responses
W = number of wrong responses (not including omitted items)
k = number of alternative responses to each item

In a true-false test this reduces to the familiar $R - W$. In a five-choice-item test it becomes $R - W/4$. Incidentally, a similar correction could be made by the general formula

$$S = R + \frac{O}{k} \quad \text{(Alternative scoring formula with correction for guessing)} \tag{18.12}$$

where O = number of omissions (including items not attempted). This score is not numerically identical to that from formula (18.11), but is perfectly correlated with it.

It should be emphasized that neither of these formulas will tend to reduce the error variance introduced by guessing unless there is an appreciable number of omissions or failures to attempt items. If every examinee attempts all items, the correlation between R and W will be a perfect -1.0, which offers no freedom for improvement by scoring formula. The formula scores would then correlate $+1$ with R, and the correction operation would be of no value for the purposes of measurement. In a speed test, however, and in a power test in which the examinees voluntarily omit many items, such a scoring formula may help to eliminate some of the error variance and thus promote better reliability and validity. The more difficult the test, the more important it is to apply the correction formula, for as difficulty increases, the amount of guessing increases.

If a scoring formula of this type is to be used in a test, and particularly if it is a power test, there should be explicit instructions to the examinees that there will be a deduction of a fraction of a point for each wrong answer (or a bonus of a fraction of a point for an omission). The second formula is naturally more palatable to examinees. But there can be better scoring formulas than those based upon a priori reasoning about guessing, as we shall see next.

It might be pointed out, incidentally, that when examinees do not know the answer to an item, their habits of taking tests are such that they do not choose among the alternatives entirely at random. Certain positions in a list of five responses may be favored by habits of reading or of attention. This is probably not sufficiently important in itself to overthrow the usefulness of "chance" scoring formulas. In the long run, if the position of the right answer is randomized, the correction may work well enough. More serious, however, is the fact that many test writers, in preparing four- or five-choice items, do not provide "misleads" or "distractors" that are equally attractive. It is easy, perhaps, for the test writer to think of one good wrong answer to an item, but to think of more than one and to make all equally attractive is a trying art. Many a four- or five-choice item reduces virtually to a three- or two-choice item because of this fact. The a priori scoring formula as given above then undercorrects; we do not know by how much.

EMPIRICAL WEIGHTING OF RIGHT AND WRONG ANSWERS When R and W scores are not too highly intercorrelated and when there is a practical criterion, it often pays to treat the two as if they were two different variables, as if they had arisen from two different tests. One then applies multiple-regression procedures and derives optimal weights which will maximize the correlation of a weighted combination of R and W scores and the criterion. Since, as was pointed out before (Chap. 16), it is the *relative* sizes of the weights that are important and we do not care whether the formula scores have the same mean as the criterion or represent predictions in proper sizes, we can let the R score have a weight of +1 and find what weight the W score must then have. We should expect it to have a fractional negative weight, though it might differ markedly from the weight given by formula (18.11). For this purpose, Thurstone has given the following equation to determine the weight, v, for the W score:[1]

$$v = \frac{\sigma_r(r_{cr}r_{wr} - r_{cw})}{\sigma_w(r_{cw}r_{wr} - r_{cr})} \qquad \text{(18.13)}$$

(Optimal weight for error scores when weight for rights scores is +1)

where the subscripts c, r, and w stand for criterion, rights, and wrongs scores, respectively. The correlation between these formula scores and the criterion is given by the usual multiple-R formula for three variables. In symbols that apply here,

$$R^2_{c.rw} = \frac{r_{cr}^2 + r_{cw}^2 - 2r_{cr}r_{cw}r_{wr}}{1 - r_{wr}^2} \qquad \text{(18.14)}$$

(Correlation of optimally weighted formula score with a criterion)

Note that this gives R^2.

The application of these formulas sometimes leads to surprising

1 Thurstone, L. L. *The reliability and validity of tests.* Ann Arbor, Mich.: Edwards, 1931. P. 80.

results. A two-choice numerical-operations test, a fairly simple and unique measure of the ability of facility with numbers, should have had a scoring formula of $R - 3W$ to yield maximal validity for the selection of navigators in the AAF. Another, five-choice numerical-operations test should have had a weight of -2 for wrong answers. Thus the importance of accuracy was much greater than the a priori weights would have provided for. For the selection of bombardier students, the weight for wrong responses should have been about $-.5$ for the two-choice items and about zero for the five-choice items, for maximal validity of the test. For the bombardier criterion, accuracy was of relatively less importance than for the navigator.

For still other tests, there were results deviating from a priori weighting; for example, one test involving estimations of lengths or distances on a map seemed to require a *positive* weight for wrong answers, for maximal validity for pilots, indicating that speed was of great importance in this test, even at the expense of accuracy.

On the whole, the experience with scoring formulas tended to show that empirical formulas give validities slightly better than a priori weighting of wrong responses, with gains of the order of .02 to .03 being typical. On the whole, optimal weighting of wrongs gives increases of the order of .03 to .06 over validities for the rights scores used alone. There are some instances in which the optimal weight for W is zero.

FACTORIAL VALIDITY OF RIGHTS AND WRONGS SCORES The procedure for maximizing predictive validity for a test by using the proper scoring weights can also be applied to maximizing the correlation of a test with a factor—in other words, in increasing its loading on a factor. Experience shows that error scores might well be given much attention as sources of certain kinds of variance that it is worth our while to measure. Some AAF findings indicated that a trait of carefulness was quite measurable by using wrongs scores in several tests, whereas the number of right responses usually failed to measure it.[1]

Fruchter has found by factor-analyzing rights scores and wrongs scores in the same tests that while the two scores in the same test may measure the same factors (in reverse), they do so to different degrees. He also found that some factors are more measurable by wrongs scores than others. In fact, it is possible that a certain kind of reasoning should be measured by errors rather than by correct solutions. These results are suggestive of the rich possibilities that may exist in the fuller use and weighting of wrong responses.

1 Guilford, J. P., and Lacey, J. I. (Eds.) Printed classification tests. *AAF Aviation Psychology Research Program Reports,* No. 5. Washington: GPO, 1947. Chap. 25.

2 Fruchter, B. Differences in factor content of rights and wrongs scores. *Psychometrika,* 1953, **18,** 257–265.

Item analysis

Many of the statistical operations involved in dealing with tests, particularly during test construction, have to do with item analysis, wherever tests are composed of items or other small parts. The major goals of item analysis are the improvement of total-score reliability or total-score validity (or both) and the achievement of better item sequences and types of score distributions. We want to be sure that all items in a test are functioning—that they do something for us in the way of measurement or at least make some contribution toward that end. Item-analysis procedures, including appropriate statistical methods, enable us to differentiate between the better and the poorer items.

RATIONAL AND EMPIRICAL APPROACHES TO TEST DEVELOPMENT

The basic philosophies of test makers differ considerably. On the one hand, there are those who prefer to develop tests that measure recognized, basic psychological traits or variables. In other words, there is much concern with construct validity. This school of thought is generally divided between those who regard factor analysis, properly used, as the best means of demonstrating construct validity and those who do not. Supporters of factor analysis point out that it both isolates the trait involved and ascertains which tests measure it best. Their opponents depend upon intercorrelations of tests and other variables for evidence of construct validity, but balk at going further in the way of factor analysis.

The empirical school of thought is commonly represented among those who face practical problems of measurement. Their immediate concern is to make testing instruments that will help to solve pressing everyday problems. For example, if they wish to be able to identify individuals with high aptitude for creative production in science or in management, test makers may adopt the specific approach of predicting a criterion of success in creative performance from a very large list of items containing biographical information. Any item that is found to be correlated significantly with the criterion is retained in the test; there may be little or no concern regarding the personality traits involved in what is measured by the total-test scores. Of course, this does not mean that those who adopt primarily the empirical view are behaving irrationally or that they necessarily have no interest in basic concepts. There is quite a range of dispositions toward the rational school on the part of the empiricists. Both groups make much use of empirical steps of item analysis in test development.

COMMON ITEM STATISTICS

Of the descriptive statistics commonly used in item analysis, three kinds stand out. For one thing, we want to know about the difficulty level of an item, if the test is one of ability, or about the "popularity" level of an item, if it is designed to measure a nonaptitude trait. In this context, "popularity" does not mean "social desirability"; it refers to the extent to which the individuals of a population answer the item in the keyed direction. In both aptitude and nonaptitude tests, the basic index of difficulty and of popularity is the proportion of the individuals passing the item (answering in the keyed direction). The proportion "passing" item I, p_i, is the item mean, as was pointed out in the preceding chapter. We shall see that there are more meaningful indices of difficulty or popularity, but they are based upon p_i.

The other two statistics of common interest in item analysis are r_{it}, the correlation of an item with the total score from the test of which it is a part, and r_{ic}, the correlation of the item with some outside criterion. The criterion may be a recognized measure of a trait, when construct validity is of primary interest, or some measure of success in everyday life, when predictive validity is of primary interest. We shall see that these two kinds of correlation may take different forms and that statistics have been proposed as substitutes for them.

ITEM DIFFICULTY AND INDICES OF DIFFICULTY

When the term *difficulty* is used in an item-analysis context, it also covers the concept of "popularity," mentioned above in connection with difficulty. There are two things wrong with using the item mean as an index of difficulty for an item. One is that the larger the p_i, the *easier* the item; i.e., the scale of difficulty is reversed. The other is that the scale of proportions is not an interval scale. Neither of these objections is fatal. A correct rank ordering of items can be achieved, and items having the same p_i may be assumed to be of equal difficulty for the same population. One can also get used to a reversal of direction of the scale. But for certain purposes a better scale is desirable.

A RATIONAL SCALE FOR ITEM DIFFICULTY

A rational scaling of difficulty of items is achieved by making a transformation from proportions to corresponding standard-score values z_i. To illustrate, let us use some data on six words taken from a vocabulary test, as represented in Table 18.4. A test including many other words had been given to 360 elderly people, many of them in various stages of senility, so that a wide range of ability was involved. The total sample was divided into upper and

TABLE 18.4 Means and standard-score values for six vocabulary-test items*

	p_u	p_l	p_i	z_u	z_l	z_i
Hat	.993	.723	.858	−2.46	−0.59	−1.07
Forest	.903	.500	.701	−1.30	0.00	−0.53
Middle	.747	.333	.540	−0.66	+0.60	−0.10
Prefer	.550	.173	.362	−0.13	+0.94	+0.35
Adept	.267	.060	.164	+0.62	+1.55	+0.98
Coherent	.050	.027	.038	+1.64	+1.93	+1.77

* Adapted from item-analysis data provided by Dr. Oscar J. Kaplan, with his permission.

lower halves on the basis of a standardized interview that asked questions designed to assess immediate memory, remote memory, knowledge of current events, and other common knowledge. Six words, representing a wide range of difficulty, were selected from the vocabulary test for this illustration. For each word we have the two proportions giving fully satisfactory answers in the upper and lower groups (designated by p_u and p_l) and in the two groups combined (p_i).

The conventional rational scaling procedure assumes that the proportions represent areas under the normal-distribution curve. A given proportion p_i is taken to represent the proportion of the area under the normal distribution *above* a certain corresponding z_i value on the base line of the curve. Thus, the greater the proportion, the lower the z_i value. Items with proportions above .50 receive negative z_i values, and proportions below .50 receive positive z_i values (see Fig. 18.5). The relationship between p_i and z_i in a coordinate system is that of a descending cumulative frequency curve. In Fig. 18.5 the curve is shown with a point on the curve for each of the six items. Two of the words are shown with special information. The z_i values seen in Table 18.4 are found from Table C of the Appendix.

It will be noted that the z values for the upper and lower groups are numerically different. Naturally, the same items should be more difficult for the lower group than for the upper one, so we should expect systematically higher z values for the lower group. There is also another difference in the ranges of the two sets of z values: Those for the upper group have a greater range. The difference in range is due to a difference in actual range of ability in the two groups.

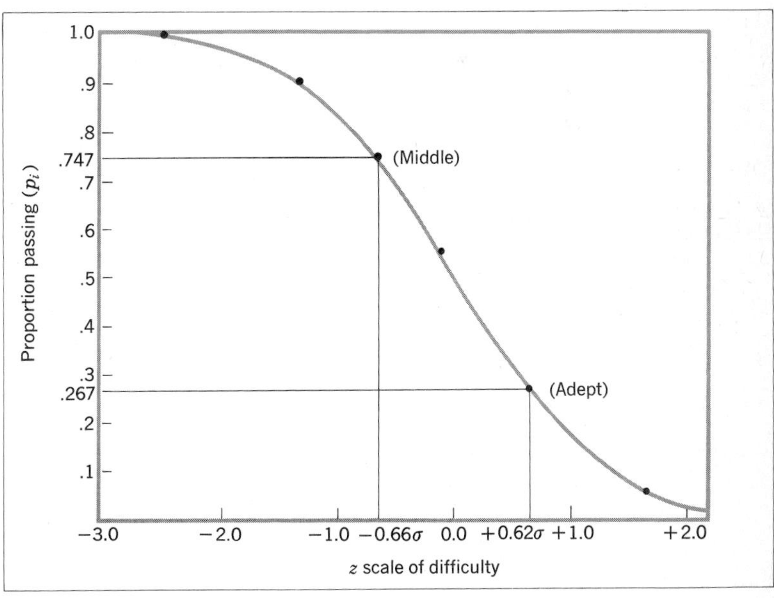

FIGURE 18.5 The determination of a z-scale value for each of six vocabulary-test items from their proportions of correct responses in a certain population

If we plot the two sets of z values z_u and z_l on a pair of coordinates, we find that there is a linear relationship with near-perfect correlation. There is merely a difference in unit and mean in the two sets. By making a linear transformation (explained in Chap. 19), it is possible to modify the scale values in the one set to make them coincide in mean and standard deviation with those in the other set. The two transformation equations are

$$z_u' = 1.42z_l - 1.48$$
$$z_l' = .714z_u + 1.05$$

depending upon whether we wish to adopt the upper-group scale or the lower-group scale as the common measuring stick for difficulty of these items. Although the zero point on such a z scale is a meaningful quantity, being the point of median difficulty for the group, if we wished to be rid of negative signs in the z values, we could add a constant of 5.00 to each of them.

Scaling items for difficulty on z scales is, of course, not always necessary. Doing so, however, makes feasible a number of things that would otherwise not be possible. Having equal-unit, or interval-scale, values for items, we find it possible to talk about functional relationships between difficulty of the items and some other properties. For example, if the items are sounds to be judged for differences in pitch or in loudness, stimulus properties can be related to item difficulty. Relations to chronological age and other variables also become possible in quantitative terms.

ITEM DIFFICULTY WHEN THERE IS CHANCE SUCCESS Many frequently used tests are composed of items to each of which one response is to be chosen from several (commonly two to five) possible answers. In such cases, the proportion of right answers is inflated by an increment due to chance success. The smaller the number of alternatives, the greater the chance contribution. A formula has been provided for correcting for this chance element, giving a proportion that would probably have been obtained had there been no possibility for the examinee to guess.[1] The formula reads

$$cp = \frac{kp - 1}{k - 1} \quad \text{(Proportion of correct responses to an item, with chance success eliminated)} \qquad (18.15)$$

where cp = corrected proportion of correct answers
p = obtained proportion
k = number of alternative answers to each item

The all-important assumption is that the examinee who does not know the right answer makes guesses completely at random from

[1] Guilford, J. P. The determination of item difficulty when chance success is a factor. *Psychometrika* 1936, **1,** 259–264.

among the alternatives.[1] It is probable that in some items the examinee knows that one or two alternatives are incorrect, and so guessing actually lies among a number of alternatives smaller than k. Thus the formula undercorrects where there is a bias.

To illustrate how much difference the correction for guessing makes, let us assume an obtained p of .75. With 5 alternatives, cp becomes .69; with 4 alternatives, .67; with 3 alternatives, .625; and with 2 alternatives, .50. The corresponding z values would be changed, ranging after correction from a z of $-.67$ when p is .75 to a z of 0.00 when p is .50. The use of corrected proportions is especially desirable when difficulties of items having different numbers of possible responses are being compared.[2]

ITEM CORRELATIONS

More important than the matter of item difficulty are the questions concerning whether a test item discriminates individuals in line with other items in the test, whether responses to the item predict some criterion, and whether the criterion is the total score on the test of which it is a part or some outside evaluation of individuals. The problem is often known as that of *item validity:* a case of construct validity when the criterion is the total score and a case of predictive validity when the criterion is an outside measure, particularly when it is a practical variable of some kind. Of course, raising the correlations of items with total score has the objective of increasing the internal-consistency reliability of the test. But, as just stated, it is also a matter of construct validity, i.e., ensuring that the psychological variable, or variables, measured will be more uniform for all items.

DISCRIMINATION VALUE OF ITEMS One principle of item validation, less commonly utilized than others, is the discrimination value of items. In this connection, the question is, "How sharply does the item segregate persons higher on the scale of the criterion from those lower on the scale?" Figure 18.3 shows some ogive functions, each presenting the increase in proportion passing an item as a function of the ability that it measures. The more steeply the curve rises, the better the discrimination and the fewer false positives and false negatives predicted by the item. An index indicating the degree of steepness of the curve would serve as a measure of the discrimination value of the item. The discrimination principle is a good one, but the procedures for applying the principle require considerable computational effort, and hence they have lost out in popularity to correlation methods, which will be discussed next. It can be readily inferred that the higher the discrimi-

1 The principle for correction here is the same as that applied to total scores, mentioned earlier (p. 454).

2 Procedures have recently been developed for determining when examinees are choosing randomly among wrong alternatives and when they are not, and procedures for making allowances for the fact. See Weitzman, R. A. Ideal multiple-choice items. *Journal of the American Statistical Association,* 1970, **65,** 71–89.

nation value of an item, the greater its correlation with the criterion; hence the one approach is a real alternative to the other, and both yield about the same conclusions regarding the goodness of an item.

ITEM CORRELATIONS WITH AN EXTERNAL CRITERION For an illustration of item-criterion correlations r_{ic}, let us use data from the same source as those represented in Table 18.4, the six words from a vocabulary test. We should consider, first, what particular kind of correlation coefficient is most appropriate. If we have the appropriate information regarding results with an item and the criterion, a number of coefficients can be computed—biserial r, point-biserial r, tetrachoric r, and the phi coefficient. All are possible and meaningful.

The choice of correlation depends upon what kind of question we want to answer. If we want to know whether the attribute or attributes measured by the criterion are also measured by the item and the extent to which the item measures them, we should use a biserial r or a tetrachoric r. If we want to know how much predictive power the item has and how it would contribute to predictions, we should use a point-biserial r. With the item giving information in only one of two categories, when scored 0 to 1, its predictive power is accordingly limited. The point-biserial r takes this into consideration. If the prediction to be made is in one of two categories only—for example, predicting whether an individual will succeed or fail—a phi coefficient is to be preferred.

Actually, if the only use to be made of the correlation coefficients is in selecting and rejecting items in making up a test, one coefficient is about as good as another. They are generally interrelated by ratios such that for the same items, the coefficients of different kinds would come in about the same rank order. Other considerations, then, have some bearing: the amount of information needed for computing the coefficient, the amount of labor (and availability of laborsaving aids), and whether or not there is a good statistical test of whether the obtained coefficient differs significantly from zero.

If one were to apply the most convenient formulas for the two biserial coefficients, reference to Chap. 14 will show that the information required includes $\bar{X}_p$, the mean total-test score for those individuals who passed the item (which changes from item to item); X_t, the mean of the entire sample (which is the same for all items); p, the proportion passing the item, and other normal-curve values derived from it; and S_t, the standard deviation of the total-score distribution.

Fortunately, all four kinds of coefficients can be estimated with much less information required. In each case the crucial information includes p_u and p_l, when the division of cases on the criterion variable is at $p' = q' = .50$. For example, the formula for the phi coefficient then reduces to

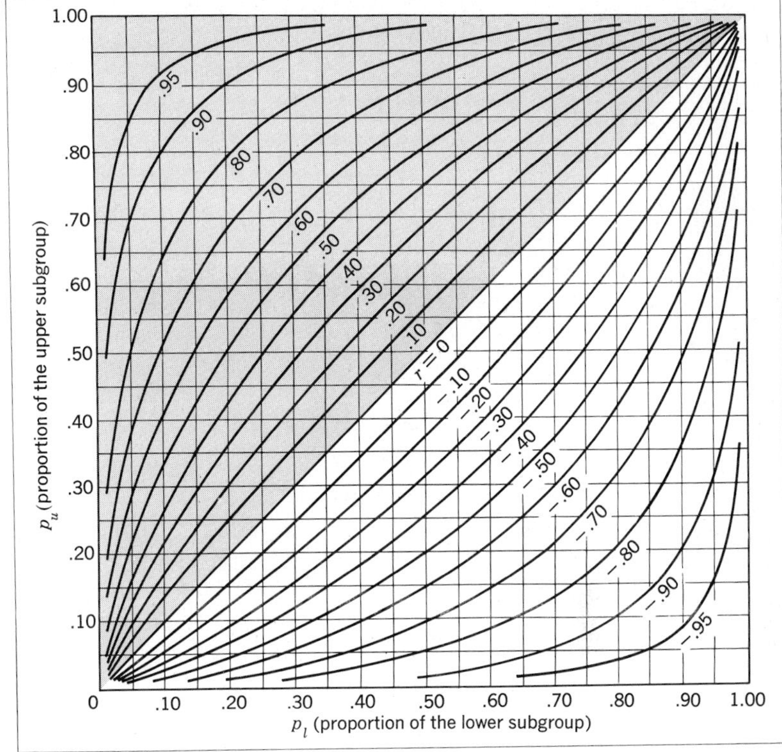

FIGURE 18.6 An abac for graphic estimates of the tetrachoric *r* when one variable has been dichotomized at the median of the distribution. (Prepared from the Pearson tables of r_t by Harvey F. Dingman.)

$$r_\phi = \frac{p_u - p_l}{2\sqrt{pq}} \quad \text{(Phi coefficient for an item with a variable on which the mean is .5)} \qquad (18.16)$$

where p_u and p_l = proportions of individuals passing the item in upper and lower criterion groups, respectively

p = proportion of total sample passing the item

With an equal division of cases, $p = (p_u + p_l)/2$. The terms p_u and p_l are readily obtained in the process of counting the number of passing individuals.

To determine whether an obtained r_ϕ is significantly divergent from zero, we make use of the relation of r_ϕ to chi square: $Nr_\phi{}^2 = \chi^2$. Phi coefficients significant at the .05 and .01 levels can be estimated by the equations

$$r_{\phi.05} = \frac{1.960}{\sqrt{N}} \qquad r_{\phi.01} = \frac{2.576}{\sqrt{N}} \quad \text{(Phi coefficients significant at the .05 and .01 levels, respectively)} \qquad (18.17)$$

where N is the number of cases in the two groups combined.

For convenience in estimation of item-criterion correlations,

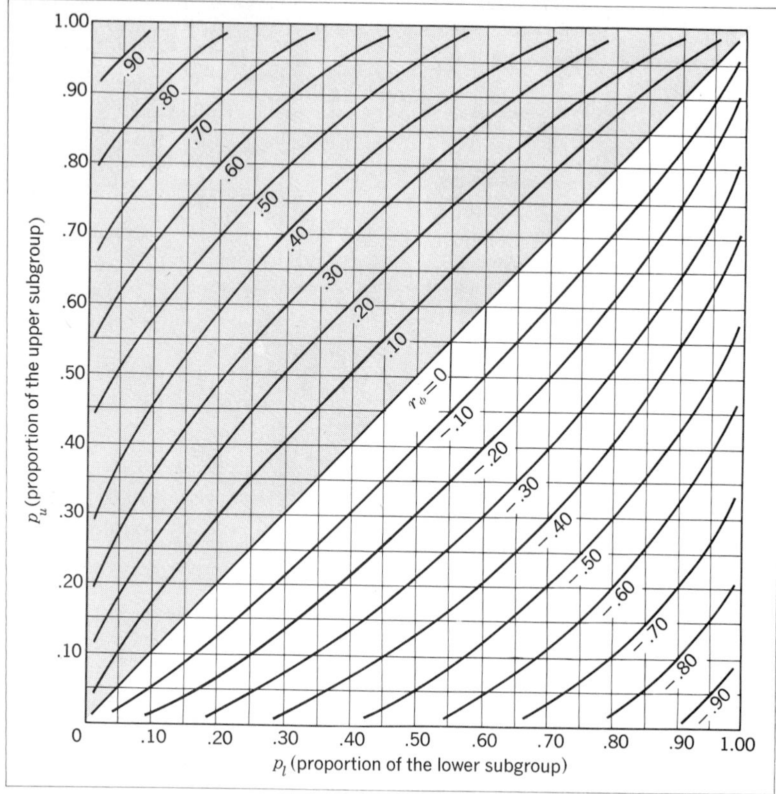

FIGURE 18.7
An abac for graphic estimate of the phi coefficient when one variable has been dichotomized at the median

abacs like those in Figs. 18.6 and 18.7 have been developed for all four kinds of coefficients. Only those for r_t and r_ϕ are given here, since they represent all four directly or indirectly.[1] The biserial r would presumably be very similar to r_t in value, and one can be taken as an estimate of the other. The point-biserial r is related to both the biserial r and r_ϕ by certain ratios, provided the proportion p is not extreme:

$$r_{pbi} = r_b \frac{y}{\sqrt{pq}} \qquad (18.18)$$

(Estimation of r_{pbi} from r_b and from r_ϕ)

$$r_{pbi} = r_\phi \frac{\sqrt{pq}}{y} \qquad (18.19)$$

We are now ready to examine the data in Table 18.5. For each item the values p_u and p_l are given. Using the abacs, we find the r_t and r_ϕ estimates given in the fourth and sixth columns of numbers. The r_{pbi} values have been estimated from the r_t values by means of formula (18.18). By checking, the reader may see how close the r_{pbi} values would come to those estimated from the r_ϕ coefficient, using formula

[1] For the other abacs and for further details on item-analysis methods, see Guilford, *Psychometric methods.* (2d ed.) New York: McGraw-Hill, 1954.

TABLE 18.5 Item-criterion correlations for six vocabulary-test items, derived from proportions of correct answers in two criterion groups of elderly subjects*

Item	p_u	p_l	p_i	r_t	r_{pbi}	ϕ
Hat	.993	.723	.858	.80	.52	.40
Forest	.903	.500	.701	.70	.52	.44
Middle	.747	.333	.540	.62	.49	.41
Prefer	.550	.173	.362	.59	.46	.39
Adept	.267	.060	.164	.55	.39	.29
Coherent	.050	.027	.038	.20	.09	.05

* Adapted from item-analysis data provided by Dr. Oscar J. Kaplan, with his permission.

(18.19). With $N = 360$, use of formula (18.17) tells us that it takes an r_ϕ of .103 to be significant at the .05 level and an r_ϕ of .135 to be significant at the .01 level. Five of the six r_ϕ's are significant beyond the .01 level. Using formulas for the standard errors of r_t and r_{pbi} (the SE for the latter being essentially the same as for the Pearson r), with $\rho = 0$ [see formula (8.12), p. 142], we could also determine which of those coefficients are significantly different from zero.

ITEM-TOTAL-SCORE CORRELATIONS For an illustration of correlation of items with total scores, let us use data from Table 17.2 (p. 422), where we have item scores of ten fictitious examinees on eight hypothetical items. Dividing the ten cases into the highest five and lowest five on total score, we can determine the proportions in each group who pass each item. The p_u and p_l values are given in Table 18.6. Such information would be the basis for entering the abac in Fig. 18.6 or 18.7. But it will be noted that so many of the proportions are 0 to 1.0 that the utility of graphic estimates of correlations is often very dubious for these data. On the other hand, enough information is available in Table 17.2 to compute either a biserial r or a point-biserial r. Let us compute the latter. We can do so, for we know the mean $\bar{X}_t$ to be 4.0 and the standard deviation to be 2.45, and $\bar{X}_p$ can be readily computed for each item, as well as p, the mean of p_u and p_l. The point-biserial r's, thus computed, are listed in the third row of Table 18.6. We could also have computed the phi coefficients, preferably by means of formula (18.16). It would be unwise to estimate tetrachoric r's from these data by computation because in so many of the 2×2 tables there are zero frequencies, which lead to estimates of $+1.0$ and -1.0 for r_t, as pointed out in Chap. 14.

CORRECTION OF AN ITEM-TOTAL CORRELATION FOR OVERLAP An item-total correlation is a part-whole correlation and is thus spuriously high because the item's specific and error variances contribute to the correlation as well as to its common-factor variance, where only the latter should be tolerated for complete accuracy. The smaller the number of items in a test, the more serious the inflation of r from this source. As pointed out in Chap. 14, where there are n components of a total, all with equal variance and mutually independent (uncorrelated), each component would correlate with the total to the extent of $1/\sqrt{n}$. In this eight-item test, such a value would be $1/\sqrt{8}$, or .35. It will be noted that one of the obtained r's in Table 18.6 is about that size; all are larger than the actual relationships would warrant.

One solution to this problem would be to adapt formula (14.35) (p. 331), which gives the general equation for the correlation of a part (in this case a single item) with the sum of the remaining part of the composite. In adapting the formula for this purpose, the standard deviation of the item is $\sqrt{p_i q_i}$, and its variance is $p_i q_i$, where p_i is the proportion passing the item and $q_i = 1 - p_i$.

But this formula overcorrects to some extent and is mentioned here only to call attention to the principle involved. One difficulty with it is that each remainder is a different composite, thus the criterion differs slightly for every item. It overcorrects, since each item is correlated with $n - 1$ items rather than n items. The shorter the test, the more appreciable such discrepancies become. Henrysson has developed a formula that is said to overcome these irregularities.[1] It reads

$$_c r_{pbi} = \left(\sqrt{\frac{n}{n-1}}\right)\left(\frac{r_{pbi}S_t - \sqrt{p_i q_i}}{\sqrt{S_t^2 - \Sigma p_i q_i}}\right) \quad \text{(Point-biserial } r \text{ between a test item and the total-test score, corrected for part-whole overlap)} \tag{18.20}$$

where n = number of items in the test
r_{pbi} = obtained item-total correlation
p_i = proportion passing the item
$q_i = 1 - p_i$

As an example, let us apply formula (18.20) to the data in Table 17.2 (p. 422). The obtained point-biserial r's are given in the second row, and those corrected by formula (18.20) in the third row of Table 18.6. It will be noted that for high obtained correlations, the corrections are small, but for lower r's they may be considerable. For only one item (g), however, would the decision be to reject the item for lack of homogeneity with the total score.

The usefulness of these corrections for part-whole overlap is most apparent when obtained correlations are small. Such (uncorrected) r's might well be due almost entirely to the spurious part-whole component. The statistical significance of small, uncorrected item-total correlations should be questioned. If they do not reach the required level of significance after correction, they are not to be regarded as significant before correction.

1 Henrysson, S. Correction of item-total correlations in item analysis. *Psychometrika*, 1963, **28**, 211–218.

TABLE 18.6 Correlations of eight items with total scores derived from those items, with corrections for part-whole overlap

	a	b	c	d	e	f	g	h
p_i	.8	.7	.6	.5	.5	.4	.3	.2
r_{pbi}	.51	.78	.75	.73	.65	.83	.36	.61
$_c r_{pbi}$	.44	.75	.70	.67	.56	.80	.22	.57

Assuming statistical significance, uncorrected r's are still useful, for they are probably in approximately correct order as to size. Correction would probably not change the order materially. Correction is usually of little importance when tests exceed 20 items in length, but in deciding whether or not to apply a correction, the investigator who is concerned about the statistical significance of the r_{it} correlations should consider whether the quantity $1/\sqrt{n}$ is of about the same order of magnitude as the r required for significant departure from zero.

SOME GENERAL CONSIDERATIONS ON ITEM CORRELATIONS For most purposes of item correlations, it does not matter much which kind of coefficient is used, as shown by experience.[1] A minor point in favor of phi is that it does not require an assumption of normal distribution, as is true of r_b and r_t. Although r_ϕ does not apply logically to the case of a continuous criterion measure, where upper and lower halves of the cases are used in a dichotomy, it does apply logically when a certain proportion of the middle scores are omitted—the middle half, 46 percent (when the two extreme 27 percentages of the cases form the criterion groups), or any arbitrarily chosen portion.

On the practice of eliminating central cases, the writers are rather negative. On the one hand, this procedure does not use all the available data; it throws away information. In addition to the loss of power of statistical tests that is involved, one loses the opportunity to determine whether there are nonlinear relationships between items and criterion. In aptitude items it is probably safe to assume linear relationships, in which case the loss of middle cases is immaterial from this point of view. But in the case of nonaptitude variables, instances of nonlinear relationships have been known. The middle third of the cases might behave much like the upper group or like the lower group, and not lie between them in the proportion passing. There are even cases in which the two extremes are alike in responding to an item, both different from the middle. Such information would be missed if only extreme groups were utilized.

One weakness of the phi coefficient in item correlations, a weakness shared to some extent with the point-biserial r, is the fact that it favors items of median difficulty. Recall that phi can be maximum when the means p and p' are equal. When p' is arbitrarily fixed at .5, as

1 As demonstrated in Guilford and Lacey, op. cit., pp. 28–33.

is usually the case, items with p near .5 can correlate higher with the criterion. Items with extreme means have little chance of significant correlations. In one sense this is not a fault but a virtue, in that items of median difficulty mean greater internal-consistency reliability. But a test composed of items all of near-median difficulty would fail to be a good discriminating measure of cases near the extremes of the range. For discriminating among the very highest and lowest on the score scale, we need items of high and low difficulties.

If we are looking for valid items at extreme levels of difficulty, there is one thing we can do. A second item analysis could be done in which the cases are subdivided into the highest one-fifth against the second one-fifth (or even the lowest four-fifths). Phi coefficients could be computed with this new division of criterion cases. A similar item analysis could be done with the lowest one-fifth of the cases in a criterion group. Items at means of .8 and .2 would then be favored by the use of a phi coefficient in these two analyses, respectively.

ITEM COMPOSITES FOR PREDICTIVE VALIDITY If the test maker's objective is to produce a test instrument with maximum validity for predicting some practical criterion, the item-analysis procedures differ somewhat from those described. A practical criterion is likely to be factorially complex, with a number of common factors involved. Different items will be needed to represent each of the common factors, except where the items themselves are factorially complex in favorable ways, i.e., where they have the same combinations of factors represented as in the criterion. Multiple-regression principles apply to this situation, which calls for items that correlate strongly with the criterion but low with one another. Thus, in item analysis for constructing a test of this nature, we need correlations of items both with the outside criterion and with one another. The alternative to information on item intercorrelations is in item-total correlations. Selecting items with higher r_{ic}'s and lower r_{it}'s should result in the kind of heterogeneous test that is required for maximum validity, but such a procedure is not recommended for combining items (see pp. 417–418).

WEIGHTING ITEMS FOR RELIABILITY The mention of weighting items differentially, which means other than the usual 0 or 1 weights, to achieve maximum validity, raises the question of whether the same kind of practice might not also be applied when the construct to be measured is a fundamental trait. After all, items do not correlate equally with a total-score criterion. It seems reasonable to suppose that items correlating higher with this kind of criterion should be given more voice in the composite score. Methods of weighting items in such scales have been devised and used.[1] But the general experience is

[1] For methods of weighting items, see Guilford, *Psychometric methods*, 1954.

that if the test contains many items scored for a trait, differential weights contribute very little to increasing its reliability. When a test is very short, containing 10 to 15 items, let us say, differential weighting may contribute enough to reliability to make the practice worthwhile.

A NEED FOR CROSS VALIDATION Where methods are used to achieve good differential weights for items, as in applying multiple-regression principles to derive optimal weights, a cross-validation study becomes very important in order to check on the applicability of the weights in new samples, even from the same population. Cross validation means checking, by applying weights or regression equations found in one sample to predictions in another. Since this procedure requires two different samples, a double cross-validation study is commonly made, in which each sample serves in turn for deriving weights and for testing weights.

As pointed out in connection with multiple-regression procedures in Chap. 16, the solution for optimal weights capitalizes upon all favorable chance effects available. A cross validation would involve applying the same weights in scoring in a new sample and determining the correlation between predicted criterion values and obtained criterion values. Some shrinkage of validity is ordinarily expected. Indeed, the mere selection of the more valid items, without differential weighting, should also be followed by a cross validation in a new sample, where possible.

Give your conclusions and interpretations in connection with the solution to each of the following problems:

DATA 18*A* Loadings for four uncorrelated common factors and reliability coefficients in two tests and a criterion

Variable	*Factors*				r_{tt}
	A	*B*	*C*	*D*	
Test 1	.10	.60	.40	.00	.80
Test 2	.20	.30	.50	.70	.87
Criterion *J*	.20	.50	.10	.00	.65

1 For the data in Data 18*A*, compute (*a*) communalities, (*b*) proportions of specific variance, and (*c*) intercorrelations.

2 Test *X* has a reliability coefficient of .92, and a criterion *Y* has a reliability of .65. Assume that the coefficient for predictive validity in connection with four different samples has values of .35, .48, .61, and .72.

a. Determine the probable correlation between the "true" test scores and the "true" criterion measures in each of the four situations.

b. Determine the validity of the fallible test for predicting the "true" criterion in each situation.

3 Using the information given in Data 18*B*:
a. Estimate the validity coefficient in each case, assuming that each test is doubled homogeneously in length.
b. Do the same, assuming that each test is made five times as long.
c. Do the same, assuming that each test is made half as long.

DATA 18*B*
Reliability coefficients and validity coefficients for four tests

Coefficient	*Test*			
	X_1	X_2	X_3	X_4
r_{xx}	.80	.80	.60	.80
r_{yx}	.70	.50	.50	.30

4 How long (in ratio to original lengths) would it be necessary to make tests X_1 and X_2 in Exercise 3 in order to achieve a validity coefficient of .60?

5 Assume the following data for a certain test:

$S_r = 10.0 \qquad S_w = 4.0 \qquad r_{cr} = .3 \qquad r_{cw} = -.2 \qquad r_{wr} = -.4$

(where the subscripts stand for "right," "wrong," and "criterion" scores, respectively).
a. Compute the optimal weight for the wrong-responses score (*W*), when the right-responses score (*R*) is weighted +1.
b. Compute the correlation of formula scores obtained by use of these weights with the criterion (*C*).
c. Assume, in turn, arbitrary weights of −2.0 and +1.0 for the wrong responses (with a weight of +1 for the right responses) and estimate the correlation with *C* for such weighted combinations.

6 For the items in Data 18C, determine the *z*-scale values for difficulty of the items.
a. What can you conclude concerning the relative levels of ability of the two groups and concerning their dispersions in ability measured by the test?
b. Plot points representing the z_{I} values and the corresponding z_{II} values and fit a straight line to the points by inspection.

DATA 18C
Proportion of examinees passing each of five items in two groups

Group	*Item*				
	A	*B*	*C*	*D*	*E*
I	.02	.10	.33	.71	.96
II	.09	.31	.54	.84	.97

7 For the items represented in Data 18*D*, give the difficulty values *z* for the corrected and uncorrected proportions, taking into account the number of alternative responses in each case.

DATA 18*D*
Items differing in number of alternative responses and in proportions of examinees passing them

	Alternatives					
	2	2	3	3	5	5
Proportion passing	.85	.55	.85	.55	.85	.55

8 For the information in Data 18*E*:
a. Compute the phi coefficients by formula; also look them up in the abac given in the chapter.
b. Look up the tetrachoric correlations for the items in the abac.
c. Estimate the corresponding point-biserial *r*'s from both r_ϕ and r_t and compare them.
d. Determine which r_ϕ and r_{pbi} coefficients are significantly different from 0 at the .05 level.

DATA 18*E*
Proportions of examinees from upper and lower halves on a total-test score who passed each of four items ($N = 100$)

Group	*Item*			
	G	*H*	*I*	*J*
Upper	.75	.95	.40	.57
Lower	.35	.80	.05	.43

ANSWERS

1 h^2: .53, .87, .30; s^2: .27, .00, .35; $r_{12} = .40$; $r_{1J} = .36$; $r_{2J} = .24$.
2 a. $r_{\infty\omega}$: .45, .62, .79, .93.
r_ω : .43, .59, .76, .89.
3 a. .74, .53, .56, .32.
b. .76, .55, .61, .33.
c. .64, .46, .42, .27.
4 *n*: 0.36; 1.89 (an obtained negative sign for *n* should be disregarded).
5 a. $v = -0.91$; *b*. $R = .31$; c. r_{cs}: .30, .24.
6 z_{I}: +2.05, +1.28, +0.44, −0.55, −1.75.
z_{II}: +1.34, +0.50, −0.10, −0.99, −1.88.
7 z (without correction): −1.04, −0.13, −1.04, −0.13, −1.04, −0.13.
z (with correction): −0.52, +1.28, −0.76, +0.45, −0.89, +0.16.
8 Computed r_ϕ: .40, .22, .42, .14.
r_t: .60, .48, .72, .20.
Estimated r_{pbi}: .48, .30, .52, .16.
Only item *J* fails to achieve significant correlations.

19 Test scales and norms

In this chapter we consider in some detail the problems of measurement by means of test scores. In previous chapters where test scores played a role, it was usually assumed that they approximate scales with equal units and that equal increments of numbers correspond to equal increments of psychological quantity. Such an assumption is necessary for the meaningful application of most statistical operations. When a test is composed of many items and when it is of an appropriate level of difficulty for the population examined, this assumption is fairly sound.

In the following pages we shall consider some ways of transforming raw-score scales into other scales for various reasons. One objective is to effect a more reasonable scale of measurement. Another important objective is to derive comparable scales for different tests. The raw scores from each test yield numbers that have no necessary comparability with the same numbers from another test. There are many occasions for wanting not only comparable values from different tests but also values that have some standard meaning. These are the problems of test norms and test standards.

WHY COMMON SCALES ARE NECESSARY

Aside from a few tests that yield scores in terms of physical-stimulus values (such as tests of sensory acuity) or of response values (such as time, distance, or energy values), most tests yield numerical values that have no unique significance.

If modern psychology and education have taught anything about measurement, they have amply demonstrated the fact that there are few, if any, *absolute* measures of human behavior. The search for absolute measures has given way to an emphasis upon the concept of individual differences. The mean of the population has become the reference point, and out of the differences between individuals has come the basis for scale units. Even when the test happens to yield such objective scores as those in time, space, or energy units, it is sometimes doubted that such units, although unquestionably equal in a physical sense, really represent equal psychological increments along scales of ability or performance. These considerations, among others, send us in search of more rational and meaningful scales of measurements for behavioral events.

There is also the practical consideration that scales for different tests should be comparable. The most obvious need for comparable scales is seen in educational and vocational guidance, particularly when profiles of scores are used. A profile is intended to provide a somewhat extensive picture of an individual. The comparison of trait positions for a person depends upon having scores that are comparable numerically.

For such a purpose, conversion of raw scores to values on some common scale is essential. Centile ranks were mentioned and illustrated earlier (Chap. 3). They do make possible some comparable values for different tests; they do make use of the mean or median as the main point of reference, and they are rather easily understood by the layman. But they have some limitations that make them less than fully useful to those who are accustomed to asking more of measurements. Centile values are ranks only, and do not represent equal units of individual differences. It is possible to have scales that possess approximately equal units as well as comparability of means, dispersions, and form of distribution.

SOME COMMON DERIVED SCALES The chief interest in what follows will be in such scales—those with comparability of means, dispersions, and form of distribution. The four kinds of scales to be discussed here are the standard-score scale, the *T* scale, the *C* scale, and the stanine scale. Their application to derivation of test norms and profile charts will be demonstrated. The treatment will be kept at a rather elementary level, emphasizing basic concepts. For a more advanced treatment of some of these problems the reader is referred to a discussion by Angoff.[1]

Standard scores

AN EXAMPLE OF THE NEED FOR COMPARABLE SCORES A concrete example will illustrate some of the ideas expressed above. A student earns scores of 195 in an English examination, 20 in a reading test, 39 in an information test, 139 in a general academic-aptitude test, and 41 in a nonverbal psychological test. Is the student

1 Angoff, W. H. Scales, norms and equivalent scores. In R. L. Thorndike (Ed.), *Educational Measurement* (2nd ed.), Washington D.C. American Council on Education, 1971, Chap. 15.

therefore best in English and poorest in reading? Could the student perhaps be equally good in all the tests? From the raw scores alone, we can answer neither of these questions, nor can we answer many others that could be legitimately asked. This student's five scores are listed in column 4 of Table 19.1 (student I).

Knowing the means of students in the five tests helps to some extent, since they serve as comparable reference points. The means are listed in column 2. We now see that the student is well above average in English and in academic aptitude and is somewhat below average in reading and information, just as the numbers seem to indicate at their face value. The second student, whose raw scores are also in column 4, is numerically highest in the same two and lowest in the same three. When we consider the averages again, however, we find that student II is only about average in English, in academic aptitude, and in the psychological test but is above average in reading and in the information test.

When a student is above the mean in two tests, in which one is he actually superior? Student I is 39.3 points above the mean in English and 16.2 points above the mean in the psychological test (see column 5 of Table 19.1). Is the student's superiority in English really greater than that in the psychological test? Student II is 20.3 points above the mean in reading and 17.5 points above the mean in information. Is this student about equally superior in the two tests?

And how do the two students compare? The superiority of student I is apparent in three tests (English, academic aptitude, and psychological), and that of student II in the other two tests. This we can tell from the raw scores. But suppose the two were competing for a scholarship at a university; which one, if there is to be a choice between them,

TABLE 19.1 A comparison of standard scores and raw scores earned by two students in five examinations

(1) Examination	*(2)* Mean	*(3)* Standard deviation	*(4)* X Raw scores		*(5)* x Deviations		*(6)* z Standard scores	
			I	*II*	*I*	*II*	*I*	*II*
English	155.7	26.4	195	162	+39.3	+ 6.3	+1.49	+0.24
Reading	33.7	8.2	20	54	−13.7	+20.3	−1.67	+2.48
Information	54.5	9.3	39	72	−15.5	+17.5	−1.67	+1.88
Academic aptitude	87.1	25.8	139	84	+51.9	− 3.1	+2.01	−0.12
Psychological	24.8	6.8	41	25	+16.2	+ 0.2	+2.38	+0.03
Sums			434	397			+2.54	+4.51
Means							+0.51	+0.90

should win? The totals of the five scores are 434 and 397, in favor of student I. Assuming that the five different abilities are equally important, have we done justice by comparing sums of raw scores? Are we justified in finding a sum or an average of each student's five raw scores?

Suppose that we are interested in determining which student is the more consistent in abilities, as shown by these five tests, and which one has the greater variability? Would a standard deviation of a student's raw scores give the answer? As the reader has probably already concluded, the reply to most of these questions is in the negative. We are extremely limited in making direct comparisons in terms of raw scores for the reason that raw-score scales are arbitrary and unique. We need a common scale before comparisons of the kind mentioned can be made. Standard scores furnish one such common scale.

THE NATURE OF THE STANDARD-SCORE SCALE

A standard-score scale has a mean of zero and a standard deviation of 1.0. An illustration of the conversion of a raw-score scale into a standard scale is shown in Fig. 19.1, distributions *A, B,* and *C*. Distribution *A* is based upon the obtained or raw scores. The mean is 80.0 and the SD is 14.0. The distribution is obviously somewhat skewed.

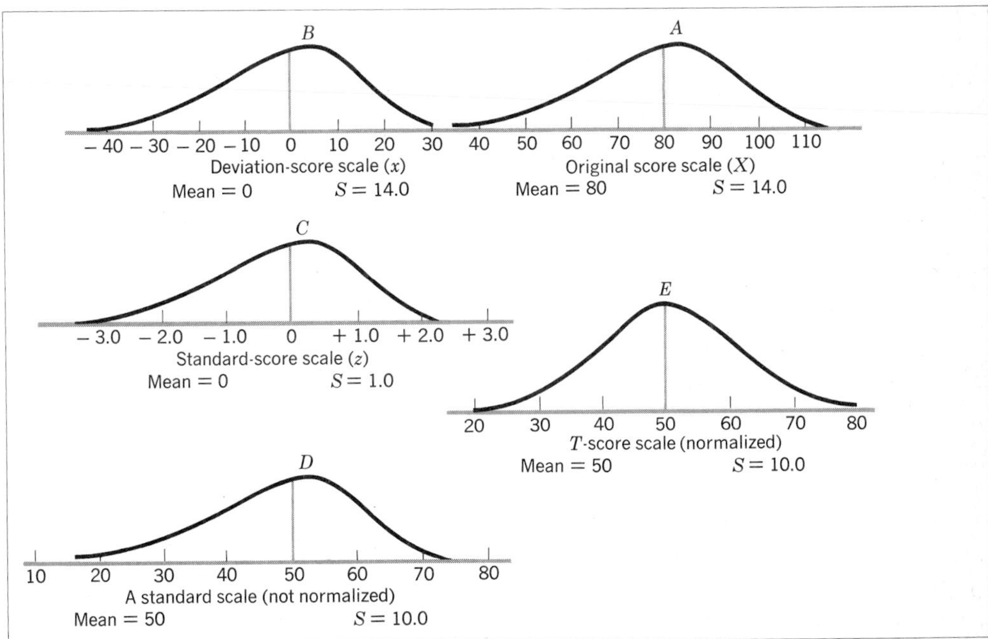

FIGURE 19.1 Distributions before and after conversion from a raw-score scale to a standardized-score scale with a desired mean and standard deviation, with and without normalizing the distribution

As we have previously seen, a standard score is obtained from a raw score by means of the formula

$$z = \frac{X - \bar{X}}{S} = \frac{x}{S} \qquad \text{(Standard score } z \text{ corresponding to a raw score } X \text{ and a deviation } x) \qquad (19.1)$$

An intermediate step between the raw-score scale and the standard-score scale is to find the deviation $X - \bar{X}$, or x. The step is illustrated in Fig. 19.1*B*. Deducting the mean from every raw score has the effect of shifting the entire distribution down the score scale so that the mean becomes zero. The final step, arriving at the z scale, is shown in Fig. 19.1*C*. Distribution *C* is drawn so that the mean is directly beneath that in distribution *B*, both at zero, so that deviations of 14 units on the original scale correspond with a deviation of 1*S* on the standard scale. Note especially that the form of distribution has not changed; it is still skewed exactly as it was originally. This procedure *does not* normalize the distribution as some other scaling procedures do.

APPLICATION TO COMPARISONS OF SCORES

The two students represented in Table 19.1 will now be compared in terms of their standard scores. Before we take these comparisons very seriously, however, we must consider two possible limitations to this procedure. Applying formula (19.1), we arrive at the standard scores in column 6 of Table 19.1. In order to make accurate comparisons between different tests, two necessary conditions must be satisfied. The population of students from which the distributions of scores arose must be assumed to have equal means and dispersions in all the abilities measured by the different tests, and the form of distribution, in terms of skewness and kurtosis, must be very similar from one ability to another.

Unfortunately, we have no ideal scales common to all these tests, with measurements which would tell us about these population parameters. Certain selective features might have brought about a higher mean, a narrower dispersion, and a negatively skewed distribution on the actual continuum of ability measured by one test, and a lower mean, a wider dispersion, and a symmetrical distribution on the continuum of another ability represented by another test. Since we can never know definitely about these features for any given population, in common scaling we often have to proceed on the assumption that actual means, standard deviations, and form of distribution are uniform for all abilities measured. In spite of these limitations, it is almost certain that derived scales provide more nearly comparable scales than do raw scores.

Returning to Table 19.1, with the standard scores that we have for the two students, we can now give more satisfactory answers to the questions raised above regarding them. Student I is most superior in the psychological test, next in academic aptitude, and third in

English. In terms of student I's deviation scores we should have decided otherwise. In terms of standard scores we find that this student is equally deficient in reading ability and information, whereas the deviations would place the student in the reverse order. Student II's five standard scores come in about the same rank order as the deviation scores but certainly not in the same order as the raw scores for this student.

Comparing the two students in terms of raw scores, we should conclude that student I has the advantage in academic aptitude. In terms of deviations this ranking would be the same, but in terms of standard scores, it is in the psychological test that the advantage is greatest. Student II has about the same superiority in the reading and information tests in terms of raw scores and deviations, but has a decidedly greater advantage in reading ability in terms of standard scores. Whereas the raw-score total gives student I the distinct advantage of 17 points, or an average superiority of about 7 points, the standard-score average reverses the order, giving student II a 0.39 lead. Student II might thus win the scholarship contest.

DISADVANTAGES OF STANDARD SCALES

Although standard scales will do all that has been stated and more, under the proper conditions, there are several features that make them less useful than some other scales. One shortcoming is the fact that half the scores are negative in sign, which makes computations awkward. Another disadvantage is the very large unit, which is one whole standard deviation.

We could, of course, overcome the first shortcoming by adding a constant to all the scores to make them all positive, and we could multiply them by another constant, preferably by 10, to make the unit smaller and the range in total units greater. If we did both of these, we could achieve almost any mean and standard deviation we wanted, depending upon the choice of constants. If we wanted a mean of 50 and a standard deviation of 10, we would multiply every standard score by 10 and add 50.

DIRECT SCALING TO A DESIRED MEAN AND STANDARD DEVIATION

This brings us to a more general procedure. If we knew from the time we had acquired the distribution of raw scores that we were to convert them to a common scale with a certain mean and standard deviation, we should not go to the trouble of converting first to standard scores and then to the new scale. We can do the operation in one step by the equation

$$X_s = \left(\frac{S_s}{S_o}\right) X_o - \left[\left(\frac{S_s}{S_o}\right) \bar{X}_o - \bar{X}_s\right] \tag{19.2}$$

(Conversion of scores in one scale directly to comparable scores in another scale; a linear transformation)

where X_s = a score on the standard scale, corresponding to X_o
X_o = a score on the obtained scale; a raw score
$\bar{X}_o$ and $\bar{X}_s$ = means of X_o and X_s, respectively
S_o and S_s = standard deviations of X_o and X_s, respectively

If the desired mean is 50 and the desired standard deviation is 10, with these substitutions the equation becomes

$$X_s = \left(\frac{10}{S_o}\right) X_o - \left[\left(\frac{10}{S_o}\right) \bar{X}_o - 50\right]$$

Knowing S_o and $\bar{X}_o$ from the particular distribution of raw scores, we can reduce the equation to the very simple form describing a straight line. Taking the illustration of Fig. 19.1, where $\bar{X}_o = 80$ and $S_o = 14.0$,

$$X_s = \left(\frac{10}{14}\right) X_o - \left[\left(\frac{10}{14}\right) 80 - 50\right]$$
$$= .714X_o - 7.12$$

By this formula a raw score of 100 would become a scaled (integral) score of 64. A raw score of 50 would become a scaled score of 29. We can see a graphic exhibition of this transformation by relating distributions *A* and *D* in Fig. 19.1. A score of 100 in *A* is in a position comparable to a score of 64 in *D*, and a score of 50 in *A* is in a position similar to a score of 29 in *D*.

Scaling by this procedure, as by the standard-score method, assumes that the obtained form of distribution is the same as the genuine population distribution would be on a scale of equal units. If this assumption is correct, the derived scales are interval scales. As far as improving equality of units is concerned, then, nothing was gained, nor was anything to be gained. We know, however, that there can be distortions in obtained-score scales, for various reasons: difficulty level of the test relative to the population's status on the traits measured, intercorrelations of the items, and variations in difficulty and in intercorrelations of the items (see p. 424). Therefore, we should not feel compelled to retain the same form of distribution in scaled scores as in the obtained scores. If there is a real discrepancy between the genuine population distribution and the sample distribution, there is room for improvement of the scale in terms of equality of units. The next methods to be described have the probable advantage that by normalizing distributions, they also achieve better metric scales.

The T scale and T scaling of tests

The well-known *T* scale overcomes the objections raised against standard scores and adds besides an advantage peculiar to itself. It adopts as its unit one-tenth of a standard deviation, so that an ordinary dis-

TABLE 19.2 The calculation of *T* scores for a distribution of English-examination scores

(1) Scores	(2) Upper limit of interval	(3) Frequency	(4) Cumulative frequency	(5) Cumulative proportion	(6) T score (from Table 19.3)
225–229	229.5	1	83	1.000	
220–224	224.5	0	82	.988	72.6
215–219	219.5	1	82	.988	72.6
210–214	214.5	5	81	.976	69.8
205–209	209.5	5	76	.916	63.8
200–204	204.5	7	71	.855	60.6
195–199	199.5	6	64	.771	57.4
190–194	194.5	6	58	.700	55.2
185–189	189.5	6	52	.627	53.2
180–184	184.5	11	46	.554	51.4
175–179	179.5	9	35	.422	48.0
170–174	174.5	5	26	.313	45.1
165–169	169.5	5	21	.253	43.3
160–164	164.5	6	16	.193	41.3
155–159	159.5	5	10	.120	38.2
150–154	154.5	2	5	.060	34.5
145–149	149.5	1	3	.036	32.0
140–144	144.5	1	2	.024	30.2
135–139	139.5	0	1	.012	27.4
130–134	134.5	1	1	.012	27.4

tribution of substantial N, with a range of 5 to 6S on its base line, yields 50 to 60 integral *T*-scale scores. Its peculiar advantage has been in its extensions beyond the range of distribution for a single population. By using samples from populations with both higher and lower means, it is possible to extend the scale to cover up to 100 units. The units of the scales from different populations are made equivalent, in order to achieve a common scale. We shall be concerned here only with scaling within only one population.

HOW TO DERIVE *T*-SCALE EQUIVALENTS FOR RAW SCORES

A college or university or a single school system may wish to use the *T*-scale idea as its common yardstick for all its tests. The freshmen entering a large university, for example, may be taken as the standard group for this purpose. As an illustration, let us use the data in Table 19.2. Here is a distribution of 83 scores obtained by freshmen in an English examination of the objectively scored type. The procedure will be described step by step:

step 1 List the class intervals as usual. Here a large number of class intervals is desirable.

step 2 List the exact upper limits of class intervals; cumulative frequencies are to be used.

step 3 List the frequencies.

step 4 List the cumulative frequencies (see Chap. 3 for instructions, if needed).

step 5 Find the cumulative proportions for the class intervals.

step 6 Find the corresponding T scores from Table 19.3. These are then listed in the last column of Table 19.2, given to one decimal place. We usually want finally a ready means of reading directly the T score corresponding to any integral raw score. It is recommended that the remaining steps be taken to satisfy this objective.

TABLE 19.3 An aid in the calculation of T scores

Proportion below the point	*T score*	*Proportion below the point*	*T score*	*Proportion below the point*	*T score*
.0005	17.1	.100	37.2	.900	62.8
.0007	18.1	.120	38.3	.910	63.4
.0010	19.1	.140	39.2	.920	64.1
.0015	20.3	.160	40.1	.930	64.8
.0020	21.2	.180	40.8	.940	65.5
.0025	21.9	.200	41.6	.950	66.4
.0030	22.5	.220	42.3	.960	67.5
.0040	23.5	.250	43.3	.965	68.1
.0050	24.2	.300	44.8	.970	68.8
.0070	25.4	.350	46.1	.975	69.6
.010	26.7	.400	47.5	.980	70.5
.015	28.3	.450	48.7	.985	71.7
.020	29.5	.500	50.0	.990	73.3
.025	30.4	.550	51.3	.993	74.6
.030	31.2	.600	52.5	.995	75.8
.035	31.9	.650	53.9	.9960	76.5
.040	32.5	.700	55.2	.9970	77.5
.050	33.6	.750	56.7	.9975	78.1
.060	34.5	.780	57.7	.9980	78.7
.070	35.2	.800	58.4	.9985	79.7
.080	35.9	.820	59.2	.9990	80.9
.090	36.6	.840	59.9	.9993	81.9
		.860	60.8	.9995	82.9
		.880	61.7		

step 7 Plot a point to represent each *T* score in Table 19.2 corresponding to the upper limit of the class interval, as in Fig. 19.2. If the original distribution of raw scores is normal, the points should fall rather close to a straight line. The reason why they are not perfectly in line is that there are some irregularities in the original data. With a ruler, draw through the points a line that will come as close to all the points as seems possible. Among those which do not touch the line, as many of them should be above it as below it. The line may be extended beyond the ends of the points at both ends. If the raw-score distribution is skewed, the trend in the points will show some curvature. It is best, then, to attempt to follow the curvature, but with a smooth trend. If the curvature is not followed, the distribution on the scaled scores will not be normalized.

step 8 For any integral raw-score point, find the corresponding *T*-score points. For example, in Fig. 19.2 a raw score of 220 corresponds to a *T* score of 70, and a raw score of 150 corresponds to a *T* score of 33. In this we favor integral *T* scores, but at times have to resort to half points when we cannot decide upon the nearest unit.

step 9 Prepare a table in which every integral raw score, or every second, third, or fifth one, appears in one column and the corresponding *T* scores in the other, such as Table 19.4. It will serve for all future purposes of translation where the original tested group remains the standard.

A NORMAL GRAPHIC PROCEDURE FOR *T* SCALING

It is possible to accomplish *T* scaling graphically by the use of normal-probability paper. This graph paper is especially designed with spacing for cumulative proportions along one axis in a manner consistent with the cumulative normal-curve function. Figure 19.3 shows how the English-examination data can be so treated. Using the cumulative proportions appearing in Table 19.2, column 5, we plot each one

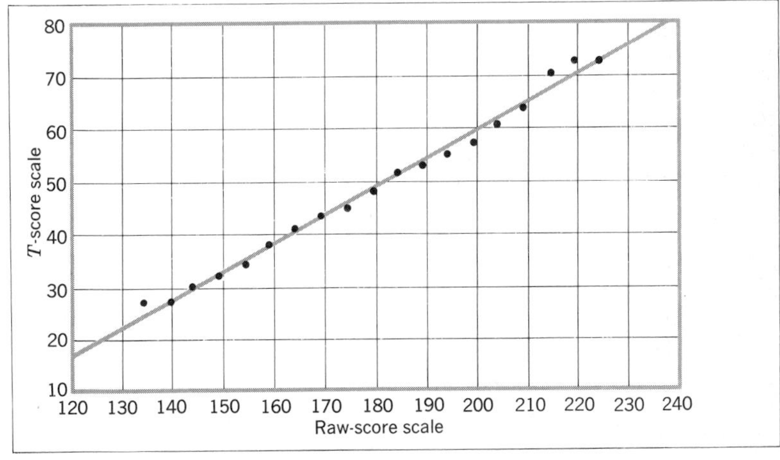

FIGURE 19.2 A smoothing process applied in deriving *T*-scale equivalents for English-examination scores (see Table 19.2)

TABLE 19.4 Rectified scaling with T scores for the distribution of English-examination scores

Examination score	*T score*	*Examination score*	*T score*	*Examination score*	*T score*
240	81	195	57	155	35.5
235	78	190	54	150	33
230	75.5	185	51.5	145	30
225	73	180	49	140	27.5
220	70	175	46	135	25
215	67.5	170	43.5	130	22
210	65	165	41	125	20.5
205	62	160	38	120	17
200	59.5				

FIGURE 19.3 A graphic solution to scaling, which utilizes normal-probability graph paper

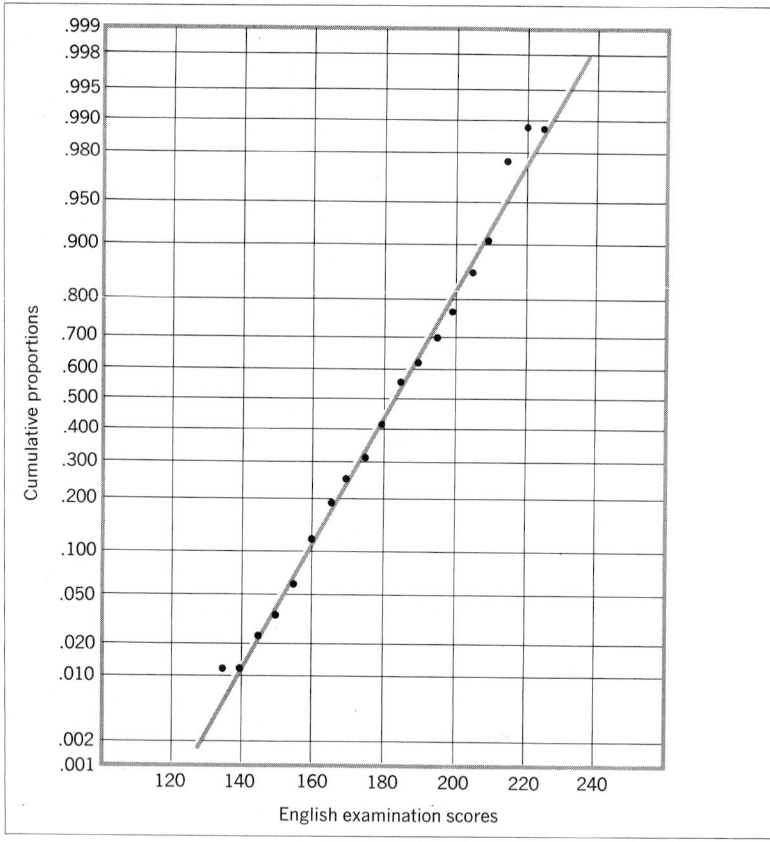

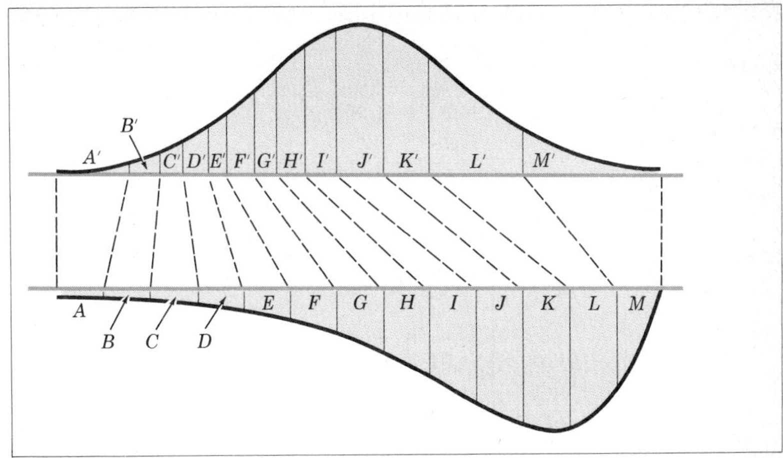

FIGURE 19.4
A graphic illustration of what happens in scaling so as to normalize a distribution. Intervals are matched so as to equate corresponding *areas* under the curves

against its corresponding raw-score value given in column 2. The trend of the points will be in a straight line if the distribution of raw scores is normal. If that distribution is skewed, there will be some curvature in the trend which one should try to follow in smoothing. To find the *T* equivalent for any raw score, we find that raw score on the base line, follow it up to the line drawn through the points, locate the equivalent proportion, and then go to Table 19.3 for the corresponding *T*.

AN EVALUATION OF THE *T*-SCALE PROCEDURE

The *T* scale is one of the most widely used derived scales.[1] Its advantages are many, its disadvantages few. When the scaling is carried out, as described, the procedure normalizes distributions. This effect is pictured in Fig. 19.1. Compare distributions *D* and *E* in that illustration. Both have a mean of 50 and an *S* of 10. The one is skewed like the original distribution, and the other is normal. The normalizing process comes about through the conversion to centile ranks and then to corresponding deviations from the mean in a normal distribution. Table 19.3 is based upon the normal curve. For a given proportion (area below a given point) a *T*-score equivalent is given instead of a standard-score equivalent.

The normalizing process may be pictured as in Fig. 19.4. In that diagram an obtained distribution, seriously skewed, is given below, and the normalized distribution on the derived scale is given above. The process ensures that the *areas A, B, C,* . . . , *M* correspond, in the proportions that they occupy, to *areas A', B', C',* . . . , *M'*. The correspondences of scale distances are also shown, by connecting dashed lines. If the units on the derived scale (not shown) represent genuinely equal increments of the measured variable, then obviously those on the original scale do not. We may not know that the population is nor-

1 Some college-aptitude tests are commonly scaled to a variant of the *T* scale, with mean and SD 10 times as great (500 and 100).

mally distributed on a trait, but by normalizing distributions, where there is no inhibiting information to the contrary, we achieve more common and meaningful scores.

Other advantages of the *T* scale have been mentioned—the possibility of extending it beyond limited populations; its convenient mean, unit, and standard deviation; and its general applicability. It has some limitations which should be pointed out. In much practical use of tests, as fine a unit as .1*S* may be a considerable overrefinement. Much coarser discriminations may be all that is necessary. Furthermore, the unit may give quite a false sense of accuracy of the measurement that is actually being made. If the original scores had a standard deviation much smaller than 10—for example, one of five score units—then the substitution of a unit of .1*S* is in a sense "hairsplitting." Two whole units on the *T* scale are then as fine a distinction as we could actually make between individuals.

Furthermore, every test, even the best of them, has an error of measurement whose size is indicated by its "standard error of measurement" (see Chap. 17). This stems from the fact that the test is not perfectly reliable. If the error of measurement is as much as two units on the raw-score scale, it might be even larger on the *T* scale. If the error is such that the best practical discriminations we can make between individuals are of the order of one-half *S*, it is rather presumptuous to apply a scale that pretends to distinguish to one-tenth *S*. For this reason, particularly, and because many test users require less refinement than the *T* scale offers, the *C* scale, which will be described next, was proposed.

The C scale and stanines

THE C-SCALE SYSTEM

The principles of the *C* scale and the derivation of *C*-scale equivalents for raw scores are illustrated in Table 19.5. The *C* scale is so arranged that the mean will be exactly at 5.0, with the two limiting classes being 0 and 10. Column 2 gives the exact limits of the 11 units in terms of standard scores. The corresponding centile-rank limits (derived from Table B in the Appendix) are given in column 3. The percentage of cases within each unit is found by subtracting within neighboring pairs of centile-rank limits. Thus, the middle unit has the difference $59.9 - 40.1 = 19.8$ percent, each of the two tail units has 1.2 percent (when all remaining cases in the extremes of the tails are included), and so on. Since it is more convenient to think in terms of whole numbers, the approximate percentages of the cases falling in the different classes are given to the nearest whole numbers, in column 5. These values can be used either as a guide in thinking of the makeup of the standard distribution when scores are based on a *C* scale or in the operations of subdividing lists of obtained scores of individuals when

TABLE 19.5 The 11-point scaled-score system and its application to some memory-test data

(1) C-scale score	*(2)* Standard-score limits	*(3)* Centile-rank limits	*(4)* Percentage within each interval	*(5)* Percentage in whole numbers	*(6)* Corresponding score points in memory test	*(7)* Memory-test scores in each scaled-score interval
	+2.75	99.7				
10			0.9	1		41+
	+2.25	98.8			40.5	
9			2.8	3		38 – 40
	+1.75	96.0			37.6	
8			6.6	7		35 – 37
	+1.25	89.4			34.6	
7			12.1	12		31 – 34
	+0.75	77.3			30.8	
6			17.4	17		28 – 30
	+0.25	59.9			27.8	
5			19.8	20		25 – 27
	−0.25	40.1			24.4	
4			17.4	17		21 – 24
	−0.75	22.7			20.8	
3			12.1	12		18 – 20
	−1.25	10.6			17.7	
2			6.6	7		15 – 17
	−1.75	4.0			14.5	
1			2.8	3		12 – 14
	−2.25	1.2			11.8	
0			0.9	1		0 – 11
	−2.75	0.3				

they are arranged in rank order. Thus, if we had 100 persons lined up in rank order in a test, the highest person would be given a score of 10, the next three a score of 9, the next seven a score of 8, etc., until the last in line was given a score of 0.

STEPS IN DERIVING A *C* SCALE The operations for deriving a *C* scale are much the same as those for deriving a *T* scale. There are some differences in the steps to be recommended, however, and so all the steps will be listed here:

step 1 List the class intervals.

step 2 List the exact upper limits of the intervals.

step 3 List the frequencies.

step 4 List the cumulative frequencies.

step 5 Find the cumulative proportions for the intervals.

step 6 From here on the steps differ from those for *T* scaling. Next, plot the cumulative proportions on the ordinate corresponding to *X* values (exact upper limits) on the abscissa of ordinary coordinate paper.

step 7 Draw by inspection a smooth S-shaped curve through the trend of the points. If the distribution is obviously skewed and one tail of the S is short, or even if it vanishes, follow the general trend of the points anyway. At this stage one sees the advantage of having a liberal number of classes.

step 8 Look for each of the centile limits (from column 3 of Table 19.5) on the ordinate, find the intersection of that centile-rank level with the curve, and drop down to the abscissa to locate the corresponding raw-score point. Avoid arriving at a point exactly at integers, so that it is clear whether each integral raw score goes above or below the division point. The values thus obtained are like those in column 6 of Table 19.5.

step 9 Determine within which *C* intervals the various integral-score values lie and write the limiting scores as in column 7 of Table 19.5.

The normal-graphic procedure described in connection with *T* scaling can also be applied here; in fact, it is even more convenient in this connection and is to be recommended in preference to steps 6 and 7. Since the centile ranks are marked on probability paper (see Fig. 19.3), one would locate the centile-rank limits (column 3 of Table 19.5) and from the plot, usually a straight line, find the corresponding raw-score division points.

AN EVALUATION OF THE *C* SCALE

The *C* scale has many of the advantages of the *T* scale. It refers obtained scores to a common scale that is related to the normal distribution. If the population distribution on a measured trait is normal, then the distribution of *C* scores properly represents that population, and the units of measurement may be regarded as equal. It lacks the refinement of a small unit such as that provided by the *T* scale. On the other hand, it probably more nearly represents the accuracy of discrimination actually made by means of tests, and its broader categories will do for guidance purposes.

There is a handicap in selection of personnel in that a change of minimum qualifying score of only one *C*-scale unit may result in quite a difference in percentage of cases selected. For example, if the cutoff score were changed from 5 to 6, 20 percent more rejections would have to be made. For selection purposes, however, raw-score cutoffs would be just as feasible as derived scores. The reference of any chosen raw-score cutoff to equivalent *C*-score limits or centiles would add meaning to that particular value.

For guidance and counseling purposes, the use of a zero *C* score may be unwise. Unless more sophisticated than most people, a counselee would hardly relish being told that he or she earned a score of zero. To meet this contingency, one could let the scores range from 1 through 11 instead of 0 through 10, or one could resort to a condensed scale, to be described next.

THE STANINE SCALE There are several reasons for condensing the *C* scale to some extent by giving it a nine-unit range. This is usually done by combining the two categories at either end, with 4 percent of the distribution in categories 1 and 9. Such a scale was standard for the Army Air Force Aviation Psychology Program during World War II. All test scores and composites were eventually scaled to this system, called *stanine*, as a contraction of "standard nine." The mean of such a norm distribution is 5.0, as in the *C* scale, but the standard deviation is a trifle lower—1.96—because of the contractions at the tails of the curve.

Perhaps the chief practical benefit to be derived from nine units rather than eleven is that such scores occupy only one column on computer punched-card records. For research purposes, however, a slight grouping error is thus introduced. In guidance work, many counselors might not like to have the rare one person in a hundred at either extreme submerged with the other adjacent 3 percent. There is presumably as much discrimination between the hundredth person and the next 3 percent as there is between any other neighboring categories. This loss of discrimination in the stanine scale might not be tolerable and is unnecessary in the use of profiles in guidance.

Some norm and profile suggestions

Suggestions were made in Chap. 3 concerning the derivation of centile points. Here we shall find use for centile values. We see a profile chart, in which raw scores can be interpreted in terms of the *C* scale, *T* scale, and centile rank.

A PROFILE CHART WITH THREE INTERPRETIVE SCALES Figure 19.5 shows an example of a profile chart by means of which raw scores on several tests may be readily translated into *C*-scale, *T*-scale, or centile equivalents. The seven tests are the parts of the Guilford-Zimmerman Aptitude Survey. The score on each part measures a different aptitude factor.

The operation of *C* scaling determined the ranges of obtained scores falling in each *C*-scale interval. The spacing of the obtained-score integers in the chart has been made accordingly, giving as many integral scores as possible without crowding. Equal units are assumed for the *C* scale, which also means for the *T* scale. The centile-rank val-

Norms for college men
Parts of the survey

Centile	T score	C score	I VC	II GR	III NO	IV PS	V SO	VI SV	VII MK
99.7, 99	75	10	67, 65, 63	26, 25	128, 123, 119	72, 71, 69	50, 47, 44	60, 59, 57	53, 52, 51
95	70	9	60, 57	24, 23, 22	114, 109	67, 65, 63	42, 40	54, (51)	(50), 49
90	65	8	54, 51	21, 20	105, 101, (98)	61, 59	38, 36, 34	48, 45	48, 45
80	60	7	48, 45, 42	19, 18, 17	93, 88	57, 55, 53	(32), 30	42, 39	42, 39
70, 60	55	6	39, 36	16, (15), 14	82, 77	51, (49)	28, 26, 24	36, 33	36, 33
50, 40	50	5	(33), 30	13, 12, 11	71, 66	47, 45	22, 20, 18	30, 27, 24	30, 27
30	45	4	27, 24	10, 9	63, 61, 58	43, 41, 39	16, 14	21, 18	24, 21
20	40	3	21, 18	8, 7, 6	54, 49, 45	37, 35, 33	12, 10, 8	15, 12	18, 15
10, 5	35	2	15, 14, 13	5, 4, 3	42, 38	31, 29	6, 4	9, 6	12, 10
	30	1	12, 11, 9	2, 1	35, 31, 27	27, 25	2, 0	3, 1	8, 6
1, 0.3	25	0	7, 6, 5	0, −1	23, 19, 15	23, 20, 17	−3, −5	0, −2	5, 4

FIGURE 19.5 A profile chart for the seven parts of the Guilford-Zimmerman Aptitude Survey, based upon scores for college men. The code: VC = Verbal Comprehension; GR = General Reasoning; NO = Numerical Operations; PS = Perceptual Speed; SO = Spatial Orientation; SV = Spatial Visualization; and MK = Mechanical Knowledge

ues are aligned with the *C* and *T* scales. The norms were derived from entering freshmen at a West Coast state university.

The chart in Fig. 19.5 shows the profile for a particular student. His obtained scores have been encircled, and the circles have been connected to show trends. As a very brief interpretation of this particular profile, it may be noted that the student is above the mean in all except verbal comprehension, which measures a factor long known by the same name. The student might therefore be expected to do average work in a variety of curricula. But because of his special superiority in spatial visualization and mechanical knowledge, it looks as if he would stand the best chance of success in engineering. His above-average status in numerical operations and in the two spatial-ability tests suggests some talent for mathematics. Being relatively lower in the verbal abilities represented in the first two tests, he might do less well in the humanities and social sciences. Decisions should, of course, depend also upon expressed or measured interests and on any other relevant information available. The example discussed is intended only to show one of the many kinds of useful consequences of measurement with tests and to show the role of scaling of scores in working toward certain goals.

EXERCISES

1 a. Determine the standard scores for the two hypothetical students in Data 19*A*.
   b. Give a rank order to each student in the five tests, first in terms of raw scores and then in terms of standard scores. Explain discrepancies in rank order.

2 a. Derive a linear conversion equation for transforming scores in the syllogism test into a scale that would give a mean of 50 and an SD of 10.
   b. Using the equation, determine the scores for students *A* and *B* on the new scale.

3 Determine the equivalent *T* scores for the upper-category limits of the form-perception scores in Data 19*B*.

4 By a graphic smoothing process, find a modified set of equivalent *T* scores for the same category limits.

5 Using the results of Exercise 4, find equivalent *T* scores for the following raw scores in the form-perception test: 8, 12, 16, 22, 37, 42.

6 Determine for the form-perception test the exact score limits (to one decimal place) corresponding to the *C*-score categories. Use a smoothing process, on regular or probability graph paper.

DATA 19*A* Means and standard deviations in five parts of an engineering-aptitude examination and scores of two students

Test	*Figure classification*	*Cube visualizing*	*Syllogism*	*Paper folding*	*Form perception*
Mean	22	15	28	33	26
SD	4	6	8	5	7
Student *A*	28	26	30	17	35
Student *B*	15	32	15	32	41

DATA 19*B*
Frequency distribution of scores for engineering freshmen in the form-perception test

Scores	*Frequencies*
40–44	2
35–39	16
30–34	42
25–29	52
20–24	55
15–19	26
10–14	13
5–9	1
	Σ 207

7 Determine *C*-score equivalents for the six raw scores listed in Exercise 5.

8 Through the relationship of either *T* scores or *C* scores to centiles, determine the centile equivalents to the raw scores listed in Exercise 5.

ANSWERS

1 a. *A*: +1.50, +1.83, +0.25, −3.20, +1.29.
B: −1.75, +2.83, −1.62, −0.20, +2.14.

2 a. $X_s = 1.25X_o + 15$.
b. X_s: 52.5, 33.75.

3 *T*: 73.3, 63.6, 55.5, 48.9, 41.3, 35.1, 24.2.

4 *T*: (79), 72, 64, 56, 49, 41, 34, 26.

5 *T* scores: 23, 30, 36, 45, 68, 75.

6 *C*-score limits: 39.9, 36.2, 32.8, 29.6, 26.4, 23.2, 19.9, 16.7, 13.3, 9.7.

7 *C* scores: 0, 1, 2, 4, 9, 10.

8 Centiles: 0.5, 2.5, 9.0, 33.5, 97.0, 99.6.

Appendix

A list of brief titles

- *A* Squares, and square roots and their reciprocals, for numbers 1 to 1,000
- *B* Proportions of the area under the normal distribution curve
- *C* Standard scores and ordinates corresponding to areas under the normal curve
- *D* Significance levels of the *t*-ratio
- *E* Chi square
- *F* Significant *F* ratios
- *G* Functions of *p*, *q*, *z*, and *y*
- *H* Fisher's *Z* for values of *r*
- *I* Trigonometric functions
- *J* Four-place logarithms of numbers
- *K* Significance levels for the Spearman rank-difference coefficient of correlation
- *L* Values for the estimation of the cosine-pi coefficient of correlation
- *M* Significant chi squares in small samples
- *N* Probabilities in tails of binomial distributions
- *O* Significant *T* values for ranked differences
- *P* Significant *R* values for sums of ranks
- *Q* Significant Pearson-*r* coefficients
- *R* Significant multiple-correlation coefficients
- *S* Critical values in the Mann-Whitney *U* test
- *T* Coefficients of orthogonal polynomials

TABLE A Squares, and square roots and their reciprocals, for numbers 1 to 1,000*

N	N^2	$\sqrt{N}$	$1/\sqrt{N}$	N	N^2	$\sqrt{N}$	$1/\sqrt{N}$
1	1	1.0000	1.0000	51	26 01	7.1414	.1400
2	4	1.4142	.7071	52	27 04	7.2111	.1387
3	9	1.7321	.5774	53	28 09	7.2801	.1374
4	16	2.0000	.5000	54	29 16	7.3485	.1361
5	25	2.2361	.4472	55	30 25	7.4162	.1348
6	36	2.4495	.4082	56	31 36	7.4833	.1336
7	49	2.6458	.3780	57	32 49	7.5498	.1325
8	64	2.8284	.3536	58	33 64	7.6158	.1313
9	81	3.0000	.3333	59	34 81	7.6811	.1302
10	1 00	3.1623	.3162	60	36 00	7.7460	.1291
11	1 21	3.3166	.3015	61	37 21	7.8102	.1280
12	1 44	3.4641	.2887	62	38 44	7.8740	.1270
13	1 69	3.6056	.2774	63	39 69	7.9373	.1260
14	1 96	3.7417	.2673	64	40 96	8.0000	.1250
15	2 25	3.8730	.2582	65	42 25	8.0623	.1240
16	2 56	4.0000	.2500	66	43 56	8.1240	.1231
17	2 89	4.1231	.2425	67	44 89	8.1854	.1222
18	3 24	4.2426	.2357	68	46 24	8.2462	.1213
19	3 61	4.3589	.2294	69	47 61	8.3066	.1204
20	4 00	4.4721	.2236	70	49 00	8.3666	.1195
21	4 41	4.5826	.2182	71	50 41	8.4261	.1187
22	4 84	4.6904	.2132	72	51 84	8.4853	.1179
23	5 29	4.7958	.2085	73	53 29	8.5440	.1170
24	5 76	4.8990	.2041	74	54 76	8.6023	.1162
25	6 25	5.0000	.2000	75	56 25	8.6603	.1155
26	6 76	5.0990	.1961	76	57 76	8.7178	.1147
27	7 29	5.1962	.1925	77	59 29	8.7750	.1140
28	7 84	5.2915	.1890	78	60 84	8.8318	.1132
29	8 41	5.3852	.1857	79	62 41	8.8882	.1125
30	9 00	5.4772	.1826	80	64 00	8.9443	.1118
31	9 61	5.5678	.1796	81	65 61	9.0000	.1111
32	10 24	5.6569	.1768	82	67 24	9.0554	.1104
33	10 89	5.7446	.1741	83	68 89	9.1104	.1098
34	11 56	5.8310	.1715	84	70 56	9.1652	.1091
35	12 25	5.9161	.1690	85	72 25	9.2195	.1085
36	12 96	6.0000	.1667	86	73 96	9.2736	.1078
37	13 69	6.0828	.1644	87	75 69	9.3274	.1072
38	14 44	6.1644	.1622	88	77 44	9.3808	.1066
39	15 21	6.2450	.1601	89	79 21	9.4340	.1060
40	16 00	6.3246	.1581	90	81 00	9.4868	.1054
41	16 81	6.4031	.1562	91	82 81	9.5394	.1048
42	17 64	6.4807	.1543	92	84 64	9.5917	.1043
43	18 49	6.5574	.1525	93	86 49	9.6437	.1037
44	19 36	6.6332	.1508	94	88 36	9.6954	.1031
45	20 25	6.7082	.1491	95	90 25	9.7468	.1026
46	21 16	6.7823	.1474	96	92 16	9.7980	.1021
47	22 09	6.8557	.1459	97	94 09	9.8489	.1015
48	23 04	6.9282	.1443	98	96 04	9.8995	.1010
49	24 01	7.0000	.1429	99	98 01	9.9499	.1005
50	25 00	7.0711	.1414	100	1 00 00	10.0000	.1000

* See page 126 for a suggestion on the use of $1/\sqrt{N}$.

TABLE A Squares, and square roots and their reciprocals, for numbers 1 to 1,000 (continued)

N	N^2	$\sqrt{N}$	$1/\sqrt{N}$	N	N^2	$\sqrt{N}$	$1/\sqrt{N}$
101	1 02 01	10.0499	.0995	151	2 28 01	12.2882	.0814
102	1 04 04	10.0995	.0990	152	2 31 04	12.3288	.0811
103	1 06 09	10.1489	.0985	153	2 34 09	12.3693	.0808
104	1 08 16	10.1980	.0981	154	2 37 16	12.4097	.0806
105	1 10 25	10.2470	.0976	155	2 40 25	12.4499	.0803
106	1 12 36	10.2956	.0971	156	2 43 36	12.4900	.0801
107	1 14 49	10.3441	.0967	157	2 46 49	12.5300	.0798
108	1 16 64	10.3923	.0962	158	2 49 64	12.5698	.0796
109	1 18 81	10.4403	.0958	159	2 52 81	12.6095	.0793
110	1 21 00	10.4881	.0953	160	2 56 00	12.6491	.0791
111	1 23 21	10.5357	.0949	161	2 59 21	12.6886	.0788
112	1 25 44	10.5830	.0945	162	2 62 44	12.7279	.0786
113	1 27 69	10.6301	.0941	163	2 65 69	12.7671	.0783
114	1 29 96	10.6771	.0937	164	2 68 96	12.8062	.0781
115	1 32 25	10.7238	.0933	165	2 72 25	12.8452	.0778
116	1 34 56	10.7703	.0928	166	2 75 56	12.8841	.0776
117	1 36 89	10.8167	.0925	167	2 78 89	12.9228	.0774
118	1 39 24	10.8628	.0921	168	2 82 24	12.9615	.0772
119	1 41 61	10.9087	.0917	169	2 85 61	13.0000	.0769
120	1 44 00	10.9545	.0913	170	2 89 00	13.0384	.0767
121	1 46 41	11.0000	.0909	171	2 92 41	13.0767	.0765
122	1 48 84	11.0454	.0905	172	2 95 84	13.1149	.0762
123	1 51 29	11.0905	.0902	173	2 99 29	13.1529	.0760
124	1 53 76	11.1355	.0898	174	3 02 76	13.1909	.0758
125	1 56 25	11.1803	.0894	175	3 06 25	13.2288	.0756
126	1 58 76	11.2250	.0891	176	3 09 76	13.2665	.0754
127	1 61 29	11.2694	.0887	177	3 13 29	13.3041	.0752
128	1 63 84	11.3137	.0884	178	3 16 84	13.3417	.0750
129	1 66 41	11.3578	.0880	179	3 20 41	13.3791	.0747
130	1 69 00	11.4018	.0877	180	3 24 00	13.4164	.0745
131	1 71 61	11.4455	.0874	181	3 27 61	13.4536	.0743
132	1 74 24	11.4891	.0870	182	3 31 24	13.4907	.0741
133	1 76 89	11.5326	.0867	183	3 34 89	13.5277	.0739
134	1 79 56	11.5758	.0864	184	3 38 56	13.5647	.0737
135	1 82 25	11.6190	.0861	185	3 42 25	13.6015	.0735
136	1 84 69	11.6619	.0857	186	3 45 96	13.6382	.0733
137	1 87 69	11.7047	.0854	187	3 49 69	13.6748	.0731
138	1 90 44	11.7473	.0851	188	3 53 44	13.7113	.0729
139	1 93 21	11.7898	.0848	189	3 57 21	13.7477	.0727
140	1 96 00	11.8322	.0845	190	3 61 00	13.7840	.0725
141	1 98 81	11.8743	.0842	191	3 64 81	13.8203	.0724
142	2 01 64	11.9164	.0839	192	3 68 64	13.8564	.0722
143	2 04 49	11.9583	.0836	193	3 72 49	13.8924	.0720
144	2 07 36	12.0000	.0833	194	3 76 36	13.9284	.0718
145	2 10 25	12.0416	.0830	195	3 80 25	13.9642	.0716
146	2 13 16	12.0830	.0828	196	3 84 16	14.0000	.0714
147	2 16 09	12.1244	.0825	197	3 88 09	14.0357	.0712
148	2 19 04	12.1655	.0822	198	3 92 04	14.0712	.0711
149	2 22 01	12.2066	.0819	199	3 96 01	14.1067	.0709
150	2 25 00	12.2474	.0816	200	4 00 00	14.1421	.0707

TABLE A Squares, and square roots and their reciprocals, for numbers 1 to 1,000 (continued)

N	N^2	$\sqrt{N}$	$1/\sqrt{N}$
201	4 04 01	14.1774	.0705
202	4 08 04	14.2127	.0704
203	4 12 09	14.2478	.0702
204	4 16 16	14.2829	.0700
205	4 20 25	14.3178	.0698
206	4 24 36	14.3527	.0697
207	4 28 49	14.3875	.0695
208	4 32 64	14.4222	.0693
209	4 36 81	14.4568	.0692
210	4 41 00	14.4914	.0690
211	4 45 21	14.5258	.0688
212	4 49 44	14.5602	.0687
213	4 53 69	14.5945	.0685
214	4 57 96	14.6287	.0684
215	4 62 25	14.6629	.0682
216	4 66 56	14.6969	.0680
217	4 70 89	14.7309	.0679
218	4 75 24	14.7648	.0677
219	4 79 61	14.7986	.0676
220	4 84 00	14.8324	.0674
221	4 88 41	14.8661	.0673
222	4 92 84	14.8997	.0671
223	4 97 29	14.9332	.0670
224	5 01 76	14.9666	.0668
225	5 06 25	15.0000	.0667
226	5 10 76	15.0333	.0665
227	5 15 29	15.0665	.0664
228	5 19 84	15.0997	.0662
229	5 24 41	15.1327	.0661
230	5 29 00	15.1658	.0659
231	5 33 61	15.1987	.0658
232	5 38 24	15.2315	.0657
233	5 42 89	15.2643	.0655
234	5 47 56	15.2971	.0654
235	5 52 25	15.3297	.0652
236	5 56 96	15.3623	.0651
237	5 61 69	15.3948	.0650
238	5 66 44	15.4272	.0648
239	5 71 21	15.4596	.0647
240	5 76 00	15.4919	.0645
241	5 80 81	15.5242	.0644
242	5 85 64	15.5563	.0643
243	5 90 49	15.5885	.0642
244	5 95 36	15.6205	.0640
245	6 00 25	15.6525	.0639
246	6 05 16	15.6844	.0638
247	6 10 09	15.7162	.0636
248	6 15 04	15.7480	.0635
249	6 20 01	15.7797	.0634
250	6 25 00	15.8114	.0632

N	N^2	$\sqrt{N}$	$1/\sqrt{N}$
251	6 30 01	15.8430	.0631
252	6 35 04	15.8745	.0630
253	6 40 09	15.9060	.0629
254	6 45 16	15.9374	.0627
255	6 50 25	15.9687	.0626
256	6 55 36	16.0000	.0625
257	6 60 49	16.0312	.0624
258	6 65 64	16.0624	.0623
259	6 70 81	16.0935	.0621
260	6 76 00	16.1245	.0620
261	6 81 21	16.1555	.0619
262	6 86 44	16.1864	.0618
263	6 91 69	16.2173	.0617
264	6 96 96	16.2481	.0615
265	7 02 25	16.2788	.0614
266	7 07 56	16.3095	.0613
267	7 12 89	16.3401	.0612
268	7 18 24	16.3707	.0611
269	7 23 61	16.4012	.0610
270	7 29 00	16.4317	.0609
271	7 34 41	16.4621	.0607
272	7 39 84	16.4924	.0606
273	7 45 29	16.5227	.0605
274	7 50 76	16.5529	.0604
275	7 56 25	16.5831	.0603
276	7 61 76	16.6132	.0602
277	7 67 29	16.6433	.0601
278	7 72 84	16.6733	.0600
279	7 78 41	16.7033	.0599
280	7 84 00	16.7332	.0598
281	7 89 61	16.7631	.0597
282	7 95 24	16.7929	.0595
283	8 00 89	16.8226	.0594
284	8 06 56	16.8523	.0593
285	8 12 25	16.8819	.0592
286	8 17 96	16.9115	.0591
287	8 23 69	16.9411	.0590
288	8 29 44	16.9706	.0589
289	8 35 21	17.0000	.0588
290	8 41 00	17.0294	.0587
291	8 46 81	17.0587	.0586
292	8 52 64	17.0880	.0585
293	8 58 49	17.1172	.0584
294	8 64 36	17.1464	.0583
295	8 70 25	17.1756	.0582
296	8 76 16	17.2047	.0581
297	8 82 09	17.2337	.0580
298	8 88 04	17.2627	.0579
299	8 94 01	17.2916	.0578
300	9 00 00	17.3205	.0577

TABLE A Squares, and square roots and their reciprocals, for numbers 1 to 1,000 (continued)

N	N^2	$\sqrt{N}$	$1/\sqrt{N}$
301	9 06 01	17.3494	.0576
302	9 12 04	17.3781	.0575
303	9 18 09	17.4069	.0574
304	9 24 16	17.4356	.0574
305	9 30 25	17.4642	.0573
306	9 36 36	17.4929	.0572
307	9 42 49	17.5214	.0571
308	9 48 64	17.5499	.0570
309	9 54 81	17.5784	.0569
310	9 61 00	17.6068	.0568
311	9 67 21	17.6352	.0567
312	9 73 44	17.6635	.0566
313	9 79 69	17.6918	.0565
314	9 85 96	17.7200	.0564
315	9 92 25	17.7482	.0563
316	9 98 56	17.7764	.0563
317	10 04 89	17.8045	.0562
318	10 11 24	17.8326	.0561
319	10 17 61	17.8606	.0560
320	10 24 00	17.8885	.0559
321	10 30 41	17.9165	.0558
322	10 36 84	17.9444	.0557
323	10 43 29	17.9722	.0556
324	10 49 76	18.0000	.0556
325	10 56 25	18.0278	.0555
326	10 62 76	18.0555	.0554
327	10 69 29	18.0831	.0553
328	10 75 84	18.1108	.0552
329	10 82 41	18.1384	.0551
330	10 89 00	18.1659	.0550
331	10 95 61	18.1934	.0550
332	11 02 24	18.2209	.0549
333	11 08 89	18.2483	.0548
334	11 15 56	18.2757	.0547
335	11 22 25	18.3030	.0546
336	11 28 96	18.3303	.0546
337	11 35 69	18.3576	.0545
338	11 42 44	18.3848	.0544
339	11 49 21	18.4120	.0543
340	11 56 00	18.4391	.0542
341	11 62 81	18.4662	.0542
342	11 69 64	18.4932	.0541
343	11 76 49	18.5203	.0540
344	11 83 36	18.5472	.0539
345	11 90 25	18.5742	.0538
346	11 97 16	18.6011	.0538
347	12 04 09	18.6279	.0537
348	12 11 04	18.6548	.0536
349	12 18 01	18.6815	.0535
350	12 25 00	18.7083	.0535
351	12 32 01	18.7350	.0534
352	12 39 04	18.7617	.0533
353	12 46 09	18.7883	.0532
354	12 53 16	18.8149	.0531
355	12 60 25	18.8414	.0531
356	12 67 36	18.8680	.0530
357	12 74 49	18.8944	.0529
358	12 81 64	18.9209	.0529
359	12 88 81	18.9473	.0528
360	12 96 00	18.9737	.0527
361	13 03 21	19.0000	.0526
362	13 10 44	19.0263	.0526
363	13 17 69	19.0526	.0525
364	13 24 96	19.0788	.0524
365	13 32 25	19.1050	.0523
366	13 39 56	19.1311	.0523
367	13 46 89	19.1572	.0522
368	13 54 24	19.1833	.0521
369	13 61 61	19.2094	.0521
370	13 69 00	19.2354	.0520
371	13 76 41	19.2614	.0519
372	13 83 84	19.2873	.0518
373	13 91 29	19.3132	.0518
374	13 98 76	19.3391	.0517
375	14 06 25	19.3649	.0516
376	14 13 76	19.3907	.0516
377	14 21 29	19.4165	.0515
378	14 28 84	19.4422	.0514
379	14 36 41	19.4679	.0514
380	14 44 00	19.4936	.0513
381	14 51 61	19.5192	.0512
382	14 59 24	19.5448	.0512
383	14 66 89	19.5704	.0511
384	14 74 56	19.5959	.0510
385	14 82 25	19.6214	.0510
386	14 89 96	19.6469	.0509
387	14 97 69	19.6723	.0508
388	15 05 44	19.6977	.0508
389	15 13 21	19.7231	.0507
390	15 21 00	19.7484	.0506
391	15 28 81	19.7737	.0506
392	15 36 64	19.7990	.0505
393	15 44 49	19.8242	.0504
394	15 52 36	19.8494	.0504
395	15 60 25	19.8746	.0503
396	15 68 16	19.8997	.0503
397	15 76 09	19.9249	.0502
398	15 84 04	19.9499	.0501
399	15 92 01	19.9750	.0501
400	16 00 00	20.0000	.0500

TABLE A Squares, and square roots and their reciprocals, for numbers 1 to 1,000 (continued)

N	N^2	$\sqrt{N}$	$1/\sqrt{N}$	N	N^2	$\sqrt{N}$	$1/\sqrt{N}$
401	16 08 01	20.0250	.0499	451	20 34 01	21.2368	.0471
402	16 16 04	20.0499	.0499	452	20 43 04	21.2603	.0470
403	16 24 09	20.0749	.0498	453	20 52 09	21.2838	.0470
404	16 32 16	20.0998	.0498	454	20 61 16	21.3073	.0469
405	16 40 25	20.1246	.0497	455	20 70 25	21.3307	.0469
406	16 48 36	20.1494	.0496	456	20 79 36	21.3542	.0468
407	16 56 49	20.1742	.0496	457	20 88 49	21.3776	.0468
408	16 64 64	20.1990	.0495	458	20 97 64	21.4009	.0467
409	16 72 81	20.2237	.0494	459	21 06 81	21.4243	.0467
410	16 81 00	20.2485	.0494	460	21 16 00	21.4476	.0466
411	16 89 21	20.2731	.0493	461	21 25 21	21.4709	.0466
412	16 97 44	20.2978	.0493	462	21 34 44	21.4942	.0465
413	17 05 69	20.3224	.0492	463	21 43 69	21.5174	.0465
414	17 13 96	20.3470	.0491	464	21 52 96	21.5407	.0464
415	17 22 25	20.3715	.0491	465	21 62 25	21.5639	.0464
416	17 30 56	20.3961	.0490	466	21 71 56	21.5870	.0463
417	17 38 89	20.4206	.0490	467	21 80 89	21.6102	.0463
418	17 47 24	20.4450	.0489	468	21 90 24	21.6333	.0462
419	17 55 61	20.4695	.0489	469	21 99 61	21.6564	.0462
420	17 64 00	20.4939	.0488	470	22 09 00	21.6795	.0461
421	17 72 41	20.5183	.0487	471	22 18 41	21.7025	.0461
422	17 80 84	20.5426	.0487	472	22 27 84	21.7256	.0460
423	17 89 29	20.5670	.0486	473	22 37 29	21.7486	.0460
424	17 97 76	20.5913	.0486	474	22 46 76	21.7715	.0459
425	18 06 25	20.6155	.0485	475	22 56 25	21.7945	.0459
426	18 14 76	20.6398	.0485	476	22 65 76	21.8174	.0458
427	18 23 29	20.6640	.0484	477	22 75 29	21.8403	.0458
428	18 31 84	20.6882	.0483	478	22 84 84	21.8632	.0457
429	18 40 41	20.7123	.0483	479	22 94 41	21.8861	.0457
430	18 49 00	20.7364	.0482	480	23 04 00	21.9089	.0456
431	18 57 61	20.7605	.0482	481	23 13 61	21.9317	.0456
432	18 66 24	20.7846	.0481	482	23 23 24	21.9545	.0455
433	18 74 89	20.8087	.0481	483	23 32 89	21.9773	.0455
434	18 83 56	20.8327	.0480	484	23 42 56	22.0000	.0455
435	18 92 25	20.8567	.0479	485	23 52 25	22.0227	.0454
436	19 00 06	20.8806	.0479	486	23 61 96	22.0454	.0454
437	19 09 69	20.9045	.0478	487	23 71 69	22.0681	.0453
438	19 18 44	20.9284	.0478	488	23 81 44	22.0907	.0453
439	19 27 21	20.9523	.0477	489	23 91 21	22.1133	.0452
440	19 36 00	20.9762	.0477	490	24 01 00	22.1359	.0452
441	19 44 81	21.0000	.0476	491	24 10 81	22.1585	.0451
442	19 53 64	21.0238	.0476	492	24 20 64	22.1811	.0451
443	19 62 49	21.0476	.0475	493	24 30 49	22.2036	.0450
444	19 71 36	21.0713	.0475	494	24 40 36	22.2261	.0450
445	19 80 25	21.0950	.0474	495	24 50 25	22.2486	.0449
446	19 89 16	21.1187	.0474	496	24 60 16	22.2711	.0449
447	19 98 09	21.1424	.0473	497	24 70 09	22.2935	.0449
448	20 07 04	21.1660	.0472	498	24 80 04	22.3159	.0448
449	20 16 01	21.1896	.0472	499	24 90 01	22.3383	.0448
450	20 25 00	21.2132	.0471	500	25 00 00	22.3607	.0447

TABLE A Squares, and square roots and their reciprocals, for numbers 1 to 1,000 (continued)

N	N^2	$\sqrt{N}$	$1/\sqrt{N}$
501	25 10 01	22.3830	.0447
502	25 20 04	22.4054	.0446
503	25 30 09	22.4277	.0446
504	25 40 16	22.4499	.0445
505	25 50 25	22.4722	.0445
506	25 60 36	22.4944	.0445
507	25 70 49	22.5167	.0444
508	25 80 64	22.5389	.0444
509	25 90 81	22.5610	.0443
510	26 01 00	22.5832	.0443
511	26 11 21	22.6053	.0442
512	26 21 44	22.6274	.0442
513	26 31 69	22.6495	.0442
514	26 41 96	22.6716	.0441
515	26 52 25	22.6936	.0441
516	26 62 56	22.7156	.0440
517	26 72 89	22.7376	.0440
518	26 83 24	22.7596	.0439
519	26 93 61	22.7816	.0439
520	27 04 00	22.8035	.0439
521	27 14 41	22.8254	.0438
522	27 24 84	22.8473	.0438
523	27 35 29	22.8692	.0437
524	27 45 76	22.8910	.0437
525	27 56 25	22.9129	.0436
526	27 66 76	22.9347	.0436
527	27 77 29	22.9565	.0436
528	27 87 84	22.9783	.0435
529	27 98 41	23.0000	.0435
530	28 09 00	23.0217	.0434
531	28 19 61	23.0434	.0434
532	28 30 24	23.0651	.0434
533	28 40 89	23.0868	.0433
534	28 51 56	23.1084	.0433
535	28 62 25	23.1301	.0432
536	28 72 96	23.1517	.0432
537	28 83 69	23.1733	.0432
538	28 94 44	23.1948	.0431
539	29 05 21	23.2164	.0431
540	29 16 00	23.2379	.0430
541	29 26 81	23.2594	.0430
542	29 37 64	23.2809	.0430
543	29 48 49	23.3024	.0429
544	29 59 36	23.3238	.0429
545	29 70 25	23.3452	.0428
546	29 81 16	23.3666	.0428
547	29 92 09	23.3880	.0428
548	30 03 04	23.4094	.0427
549	30 14 01	23.4307	.0427
550	30 25 00	23.4521	.0426

N	N^2	$\sqrt{N}$	$1/\sqrt{N}$
551	30 36 01	23.4734	.0426
552	30 47 04	23.4947	.0426
553	30 58 09	23.5160	.0425
554	30 69 16	23.5372	.0425
555	30 80 25	23.5584	.0424
556	30 91 36	23.5797	.0424
557	31 02 49	23.6008	.0424
558	31 13 64	23.6220	.0423
559	31 24 81	23.6432	.0423
560	31 36 00	23.6643	.0423
561	31 47 21	23.6854	.0422
562	31 58 44	23.7065	.0422
563	31 69 69	23.7276	.0421
564	31 80 96	23.7487	.0421
565	31 92 25	23.7697	.0421
566	32 03 56	23.7908	.0420
567	32 14 89	23.8118	.0420
568	32 26 24	23.8328	.0420
569	32 37 61	23.8537	.0419
570	32 49 00	23.8747	.0419
571	32 60 41	23.8956	.0418
572	32 71 84	23.9165	.0418
573	32 83 29	23.9374	.0418
574	32 94 76	23.9583	.0417
575	33 06 25	23.9792	.0417
576	33 17 76	24.0000	.0417
577	33 29 29	24.0208	.0416
578	33 40 84	24.0416	.0416
579	33 52 41	24.0624	.0416
580	33 64 00	24.0832	.0415
581	33 75 61	24.1039	.0415
582	33 87 24	24.1247	.0415
583	33 98 89	24.1454	.0414
584	34 10 56	24.1661	.0414
585	34 22 25	24.1868	.0413
586	34 33 96	24.2074	.0413
587	34 45 69	24.2281	.0413
588	34 57 44	24.2487	.0412
589	34 69 21	24.2693	.0412
590	34 81 00	24.2899	.0412
591	34 92 81	24.3105	.0411
592	35 04 64	24.3311	.0411
593	35 16 49	24.3516	.0411
594	35 28 36	24.3721	.0410
595	35 40 25	24.3926	.0410
596	35 52 16	24.4131	.0410
597	35 64 09	24.4336	.0409
598	35 76 04	24.4540	.0409
599	35 88 01	24.4745	.0409
600	36 00 00	24.4949	.0408

TABLE A Squares, and square roots and their reciprocals, for numbers 1 to 1,000 (continued)

N	N^2	$\sqrt{N}$	$1/\sqrt{N}$	N	N^2	$\sqrt{N}$	$1/\sqrt{N}$
601	36 12 01	24.5153	.0408	651	42 38 01	25.5147	.0392
602	36 24 04	24.5357	.0408	652	42 51 04	25.5343	.0392
603	36 36 09	24.5561	.0407	653	42 64 09	25.5539	.0391
604	36 48 16	24.5764	.0407	654	42 77 16	25.5734	.0391
605	36 60 25	24.5967	.0407	655	42 90 25	25.5930	.0391
606	36 72 36	24.6171	.0406	656	43 03 36	25.6125	.0390
607	36 84 49	24.6374	.0406	657	43 16 49	25.6320	.0390
608	36 96 64	24.6577	.0406	658	43 29 64	25.6515	.0390
609	37 08 81	24.6779	.0405	659	43 42 81	25.6710	.0390
610	37 21 00	24.6982	.0405	660	43 56 00	25.6905	.0389
611	37 33 21	24.7184	.0405	661	43 69 21	25.7099	.0389
612	37 45 44	24.7386	.0404	662	43 82 44	25.7294	.0389
613	37 57 69	24.7588	.0404	663	43 95 69	25.7488	.0388
614	37 69 96	24.7790	.0404	664	44 08 96	25.7682	.0388
615	37 82 25	24.7992	.0403	665	44 22 25	25.7876	.0388
616	37 94 56	24.8193	.0403	666	44 35 56	25.8070	.0387
617	38 06 89	24.8395	.0403	667	44 48 89	25.8263	.0387
618	38 19 24	24.8596	.0402	668	44 62 24	25.8457	.0387
619	38 31 61	24.8797	.0402	669	44 75 61	25.8650	.0387
620	38 44 00	24.8998	.0402	670	44 89 00	25.8844	.0386
621	38 56 41	24.9199	.0401	671	45 02 41	25.9037	.0386
622	38 68 84	24.9399	.0401	672	45 15 84	25.9230	.0386
623	38 81 29	24.9600	.0401	673	45 29 29	25.9422	.0385
624	38 93 76	24.9800	.0400	674	45 42 76	25.9615	.0385
625	39 06 25	25.0000	.0400	675	45 56 25	25.9808	.0385
626	39 18 76	25.0200	.0400	676	45 69 76	26.0000	.0385
627	39 31 29	25.0400	.0399	677	45 83 29	26.0192	.0384
628	39 43 84	25.0599	.0399	678	45 96 84	26.0384	.0384
629	39 56 41	25.0799	.0399	679	46 10 41	26.0576	.0384
630	39 69 00	25.0998	.0398	680	46 24 00	26.0768	.0383
631	39 81 61	25.1197	.0398	681	46 37 61	26.0960	.0383
632	39 94 24	25.1396	.0398	682	46 51 24	26.1151	.0383
633	40 06 89	25.1595	.0397	683	46 64 89	26.1343	.0383
634	40 19 56	25.1794	.0397	684	46 78 56	26.1534	.0382
635	40 32 25	25.1992	.0397	685	46 92 25	26.1725	.0382
636	40 44 96	25.2190	.0397	686	47 05 96	26.1916	.0382
637	40 57 69	25.2389	.0396	687	47 19 69	26.2107	.0382
638	40 70 44	25.2587	.0396	688	47 33 44	26.2298	.0381
639	40 83 21	25.2784	.0396	689	47 47 21	26.2488	.0381
640	40 96 00	25.2982	.0395	690	47 61 00	26.2679	.0381
641	41 08 81	25.3180	.0395	691	47 74 81	26.2869	.0380
642	41 21 64	25.3377	.0395	692	47 88 64	26.3059	.0380
643	41 34 49	25.3574	.0394	693	48 02 49	26.3249	.0380
644	41 47 36	25.3772	.0394	694	48 16 36	26.3439	.0380
645	41 60 25	25.3969	.0394	695	48 30 25	26.3629	.0379
646	41 73 16	25.4165	.0393	696	48 44 16	26.3818	.0379
647	41 86 09	25.4362	.0393	697	48 58 09	26.4008	.0379
648	41 99 04	25.4558	.0393	698	48 72 04	26.4197	.0379
649	42 12 01	25.4775	.0393	699	48 86 01	26.4386	.0378
650	42 25 00	25.4951	.0392	700	49 00 00	26.4575	.0378

TABLE A Squares, and square roots and their reciprocals, for numbers 1 to 1,000 (continued)

N	N^2	$\sqrt{N}$	$1/\sqrt{N}$
701	49 14 01	26.4764	.0378
702	49 28 04	26.4953	.0377
703	49 42 09	26.5141	.0377
704	49 56 16	26.5330	.0377
705	49 70 25	26.5518	.0377
706	49 84 36	26.5707	.0376
707	49 98 49	26.5895	.0376
708	50 12 64	26.6083	.0376
709	50 26 81	26.6271	.0376
710	50 41 00	26.6458	.0375
711	50 55 21	26.6646	.0375
712	50 69 44	26.6833	.0375
713	50 83 69	26.7021	.0375
714	50 97 96	26.7208	.0374
715	51 12 25	26.7395	.0374
716	51 26 56	26.7582	.0374
717	51 40 89	26.7769	.0373
718	51 55 24	26.7955	.0373
719	51 69 61	26.8142	.0373
720	51 84 00	26.8328	.0373
721	51 98 41	26.8514	.0372
722	52 12 84	26.8701	.0372
723	52 27 29	26.8887	.0372
724	52 41 76	26.9072	.0372
725	52 56 25	26.9258	.0371
726	52 70 76	26.9444	.0371
727	52 85 29	26.9629	.0371
728	52 99 84	26.9815	.0371
729	53 14 41	27.0000	.0370
730	53 29 00	27.0185	.0370
731	53 43 61	27.0370	.0370
732	53 58 24	27.0555	.0370
733	53 72 89	27.0740	.0369
734	53 87 56	27.0924	.0369
735	54 02 25	27.1109	.0369
736	54 16 96	27.1293	.0369
737	54 31 69	27.1477	.0368
738	54 46 44	27.1662	.0368
739	54 61 21	27.1846	.0368
740	54 76 00	27.2029	.0368
741	54 90 81	27.2213	.0367
742	55 05 64	27.2397	.0367
743	55 20 49	27.2580	.0367
744	55 35 36	27.2764	.0367
745	55 50 25	27.2947	.0366
746	55 65 16	27.3130	.0366
747	55 80 09	27.3313	.0366
748	55 95 04	27.3496	.0366
749	56 10 01	27.3679	.0365
750	56 25 00	27.3861	.0365

N	N^2	$\sqrt{N}$	$1/\sqrt{N}$
751	56 40 01	27.4044	.0365
752	56 55 04	27.4226	.0365
753	56 70 09	27.4408	.0364
754	56 85 16	27.4591	.0364
755	57 00 25	27.4773	.0364
756	57 15 36	27.4955	.0364
757	57 30 49	27.5136	.0363
758	57 45 64	27.5318	.0363
759	57 60 81	27.5500	.0363
760	57 76 00	27.5681	.0363
761	57 91 21	27.5862	.0362
762	58 06 44	27.6043	.0362
763	58 21 69	27.6225	.0362
764	58 36 96	27.6405	.0362
765	58 52 25	27.6586	.0362
766	58 67 56	27.6767	.0361
767	58 82 89	27.6948	.0361
768	58 98 24	27.7128	.0361
769	59 13 61	27.7308	.0361
770	59 29 00	27.7489	.0360
771	59 44 41	27.7669	.0360
772	59 59 84	27.7849	.0360
773	59 75 29	27.8029	.0360
774	59 90 76	27.8209	.0359
775	60 06 25	27.8388	.0359
776	60 21 76	27.8568	.0359
777	60 37 29	27.8747	.0359
778	60 52 84	27.8927	.0359
779	60 68 41	27.9106	.0358
780	60 84 00	27.9285	.0358
781	60 99 61	27.9464	.0358
782	61 15 24	27.9643	.0358
783	61 30 89	27.9821	.0357
784	61 46 56	28.0000	.0357
785	61 62 25	28.0179	.0357
786	61 77 96	28.0357	.0357
787	61 93 69	28.0535	.0356
788	62 09 44	28.0713	.0356
789	62 25 21	28.0891	.0356
790	62 41 00	28.1069	.0356
791	62 56 81	28.1247	.0356
792	62 72 64	28.1425	.0355
793	62 88 49	28.1603	.0355
794	63 04 36	28.1780	.0355
795	63 20 25	28.1957	.0355
796	63 36 16	28.2135	.0354
797	63 52 09	28.2312	.0354
798	63 68 04	28.2489	.0354
799	63 84 01	28.2666	.0354
800	64 00 00	28.2843	.0354

TABLE A Squares, and square roots and their reciprocals, for numbers 1 to 1,000 (continued)

N	N^2	$\sqrt{N}$	$1/\sqrt{N}$	N	N^2	$\sqrt{N}$	$1/\sqrt{N}$
801	64 16 01	28.3019	.0353	851	72 42 01	29.1719	.0343
802	64 32 04	28.3196	.0353	852	72 59 04	29.1890	.0343
803	64 48 09	28.3373	.0353	853	72 76 09	29.2062	.0342
804	64 64 16	28.3549	.0353	854	72 93 16	29.2233	.0342
805	64 80 25	28.3725	.0352	855	73 10 25	29.2404	.0342
806	64 96 36	28.3901	.0352	856	73 27 36	29.2575	.0342
807	65 12 49	28.4077	.0352	857	73 44 49	29.2746	.0342
808	65 28 64	28.4253	.0352	858	73 61 64	29.2916	.0341
809	65 44 81	28.4429	.0352	859	73 78 81	29.3087	.0341
810	65 61 00	28.4605	.0351	860	73 96 00	29.3258	.0341
811	65 77 21	28.4781	.0351	861	74 13 21	29.3428	.0341
812	65 93 44	28.4956	.0351	862	74 30 44	29.3598	.0341
813	66 09 69	28.5132	.0351	863	74 47 69	29.3769	.0340
814	66 25 96	28.5307	.0350	864	74 64 96	29.3939	.0340
815	66 42 25	28.5482	.0350	865	74 82 25	29.4109	.0340
816	66 58 56	28.5657	.0350	866	74 99 56	29.4279	.0340
817	66 74 89	28.5832	.0350	867	75 16 89	29.4449	.0340
818	66 91 24	28.6007	.0350	868	75 34 24	29.4618	.0339
819	67 07 61	28.6182	.0349	869	75 51 61	29.4788	.0339
820	67 24 00	28.6356	.0349	870	75 69 00	29.4958	.0339
821	67 40 41	28.6531	.0349	871	75 86 41	29.5127	.0339
822	67 56 84	28.6705	.0349	872	76 03 84	29.5296	.0339
823	67 73 29	28.6880	.0349	873	76 21 29	29.5466	.0338
824	67 89 76	28.7054	.0348	874	76 38 76	29.5635	.0338
825	68 06 25	28.7228	.0348	875	76 56 25	29.5804	.0338
826	68 22 76	28.7402	.0348	876	76 73 76	29.5973	.0338
827	68 39 29	28.7576	.0348	877	76 91 29	29.6142	.0338
828	68 55 84	28.7750	.0348	878	77 08 84	29.6311	.0337
829	68 72 41	28.7924	.0347	879	77 26 41	29.6479	.0337
830	68 89 00	28.8097	.0347	880	77 44 00	29.6648	.0337
831	69 05 61	28.8271	.0347	881	77 61 61	29.6816	.0337
832	69 22 24	28.8444	.0347	882	77 79 24	29.6985	.0337
833	69 38 89	28.8617	.0346	883	77 96 89	29.7153	.0337
834	69 55 56	28.8791	.0346	884	78 14 56	29.7321	.0336
835	69 72 25	28.8964	.0346	885	78 32 25	29.7489	.0336
836	69 88 96	28.9137	.0346	886	78 49 96	29.7658	.0336
837	70 05 69	28.9310	.0346	887	78 67 69	29.7825	.0336
838	70 22 44	28.9482	.0345	888	78 85 44	29.7993	.0336
839	70 39 21	28.9655	.0345	889	79 03 21	29.8161	.0335
840	70 56 00	28.9828	.0345	890	79 21 00	29.8329	.0335
841	70 72 81	29.0000	.0345	891	79 38 81	29.8496	.0335
842	70 89 64	29.0172	.0345	892	79 56 64	29.8664	.0335
843	71 06 49	29.0345	.0344	893	79 74 49	29.8831	.0335
844	71 23 36	29.0517	.0344	894	79 92 36	29.8998	.0334
845	71 40 25	29.0689	.0344	895	80 10 25	29.9166	.0334
846	71 57 16	29.0861	.0344	896	80 28 16	29.9333	.0334
847	71 74 09	29.1033	.0344	897	80 46 09	29.9500	.0334
848	71 91 04	29.1204	.0343	898	80 64 04	29.9666	.0334
849	72 08 01	29.1376	.0343	899	80 82 01	29.9833	.0334
850	72 25 00	29.1548	.0343	900	81 00 00	30.0000	.0333

N	N^2	$\sqrt{N}$	$1/\sqrt{N}$	N	N^2	$\sqrt{N}$	$1/\sqrt{N}$
901	81 18 01	30.0167	.0333	951	90 44 01	30.8383	.0324
902	81 36 04	30.0333	.0333	952	90 63 04	30.8545	.0324
903	81 54 09	30.0500	.0333	953	90 82 09	30.8707	.0324
904	81 72 16	30.0666	.0333	954	91 01 16	30.8869	.0324
905	81 90 25	30.0832	.0332	955	91 20 25	30.9031	.0324
906	82 08 36	30.0998	.0332	956	91 39 36	30.9192	.0323
907	82 26 49	30.1164	.0332	957	91 58 49	30.9354	.0323
908	82 44 64	30.1330	.0332	958	91 77 64	30.9516	.0323
909	82 62 81	30.1496	.0332	959	91 96 81	30.9677	.0323
910	82 81 00	30.1662	.0331	960	92 16 00	30.9839	.0323
911	82 99 21	30.1828	.0331	961	92 35 21	31.0000	.0323
912	83 17 44	30.1993	.0331	962	92 54 44	31.0161	.0322
913	83 35 69	30.2159	.0331	963	92 73 69	31.0322	.0322
914	83 53 96	30.2324	.0331	964	92 92 96	31.0483	.0322
915	83 72 25	30.2490	.0331	965	93 12 25	31.0644	.0322
916	83 90 56	30.2655	.0330	966	93 31 56	31.0805	.0322
917	84 08 89	30.2820	.0330	967	93 50 89	31.0966	.0322
918	84 27 24	30.2985	.0330	968	93 70 24	31.1127	.0321
919	84 45 61	30.3150	.0330	969	93 89 61	31.1288	.0321
920	84 64 00	30.3315	.0330	970	94 09 00	31.1448	.0321
921	84 82 41	30.3480	.0330	971	94 28 41	31.1609	.0321
922	85 00 84	30.3645	.0329	972	94 47 84	31.1769	.0321
923	85 19 29	30.3809	.0329	973	94 67 29	31.1929	.0321
924	85 37 76	30.3974	.0329	974	94 86 76	31.2090	.0320
925	85 56 25	30.4138	.0329	975	95 06 25	31.2250	.0320
926	85 74 76	30.4302	.0329	976	95 25 76	31.2410	.0320
927	85 93 29	30.4467	.0328	977	95 45 29	31.2570	.0320
928	86 11 84	30.4631	.0328	978	95 64 84	31.2730	.0320
929	86 30 41	30.4795	.0328	979	95 84 41	31.2890	.0320
930	86 49 00	30.4959	.0328	980	96 04 00	31.3050	.0319
931	86 67 61	30.5123	.0328	981	96 23 61	31.3209	.0319
932	86 86 24	30.5287	.0328	982	96 43 24	31.3369	.0319
933	87 04 89	30.5450	.0327	983	96 62 89	31.3528	.0319
934	87 23 56	30.5614	.0327	984	96 82 56	31.3688	.0319
935	87 42 25	30.5778	.0327	985	97 02 25	31.3847	.0319
936	87 60 96	30.5941	.0327	986	97 21 96	31.4006	.0318
937	87 79 69	30.6105	.0327	987	97 41 69	31.4166	.0318
938	87 98 44	30.6268	.0327	988	97 61 44	31.4325	.0318
939	88 17 21	30.6431	.0326	989	97 81 21	31.4484	.0318
940	88 36 00	30.6594	.0326	990	98 01 00	31.4643	.0318
941	88 54 81	30.6757	.0326	991	98 20 81	31.4802	.0318
942	88 73 64	30.6920	.0326	992	98 40 64	31.4960	.0318
943	88 92 49	30.7083	.0326	993	98 60 49	31.5119	.0317
944	89 11 36	30.7246	.0325	994	98 80 36	31.5278	.0317
945	89 30 25	30.7409	.0325	995	99 00 25	31.5436	.0317
946	89 49 16	30.7571	.0325	996	99 20 16	31.5595	.0317
947	89 68 09	30.7734	.0325	997	99 40 09	31.5753	.0317
948	89 87 04	30.7896	.0325	998	99 60 04	31.5911	.0317
949	90 06 01	30.8058	.0325	999	99 80 01	31.6070	.0316
950	90 25 00	30.8221	.0324	1,000	1 00 00 00	31.6228	.0316

THE USE OF TABLES B AND C

Tables B and C following assume a normal distribution whose standard deviation is equal to 1.00 and whose total area under the curve also equals 1.00. Under these conditions, there are fixed mathematical relationships between values on the base line (as measured in σ units) and areas under the curve (A, B, and C) and also ordinate values (y).

The use of Tables B and C is explained in Chap. 7. Figures *J*.1, *J*.2, *K*.1, and *K*.2 below should help to relate the symbols to the normal curve.

Table B is best used when we know z and want to find a corresponding A, B, or C area, or the ordinate y. Table C is best used when we know any one of the areas A, B, or C and want to find the corresponding z or y. In case any one of these areas is known, it can be readily used to find a corresponding area of the other kinds by means of the following relationships:

$A = B - .50$
$A = .50 - C$ (since $A + C = .50$)
$B = A + .50$
$B = 1.00 - C$ (since $B + C = 1.00$)
$C = 1.00 - B$

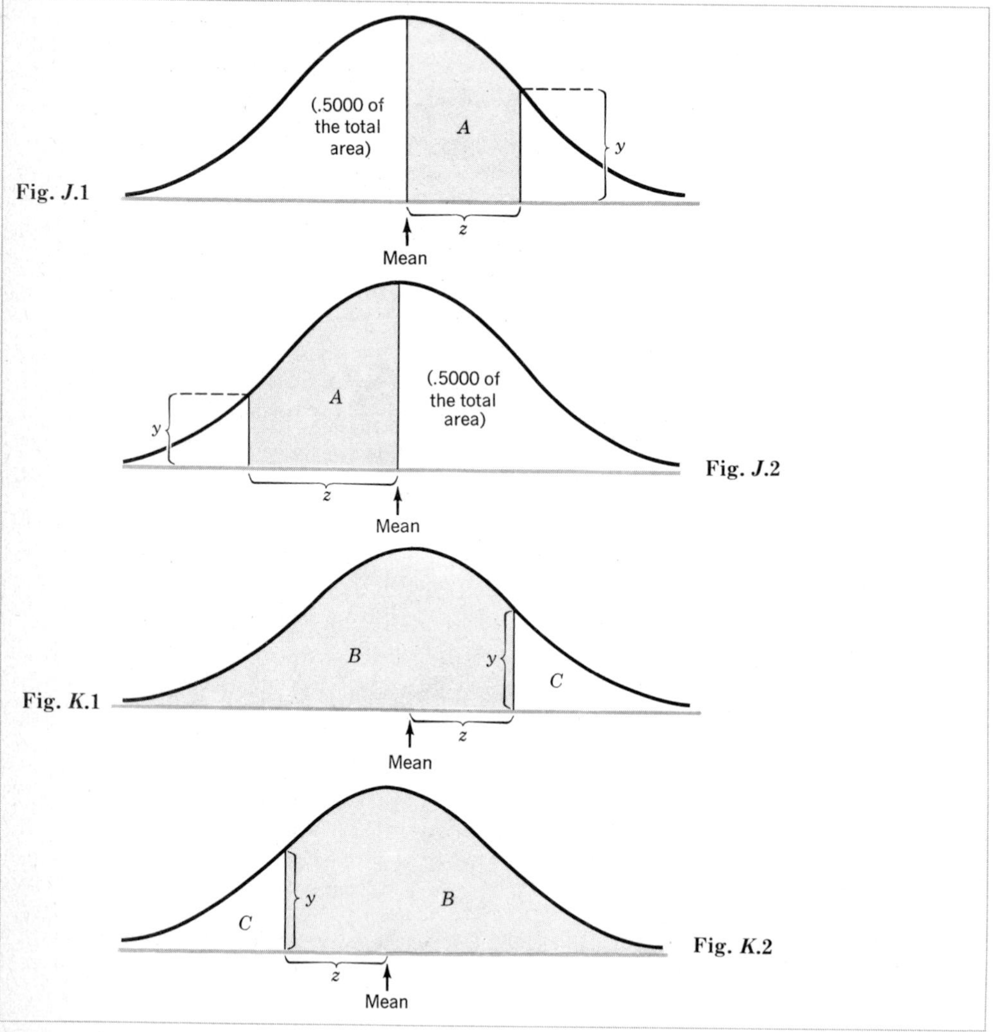

TABLE B Areas and ordinates of the normal curve in terms of x/σ

(1) z Standard score $\left(\frac{x}{\sigma}\right)$	(2) A Area from mean to $\frac{x}{\sigma}$	(3) B Area in larger portion	(4) C Area in smaller portion	(5) y Ordinate at $\frac{x}{\sigma}$
0.00	.0000	.5000	.5000	.3989
0.01	.0040	.5040	.4960	.3989
0.02	.0080	.5080	.4920	.3989
0.03	.0120	.5120	.4880	.3988
0.04	.0160	.5160	.4840	.3986
0.05	.0199	.5199	.4801	.3984
0.06	.0239	.5239	.4761	.3982
0.07	.0279	.5279	.4721	.3980
0.08	.0319	.5319	.4681	.3977
0.09	.0359	.5359	.4641	.3973
0.10	.0398	.5398	.4602	.3970
0.11	.0438	.5438	.4562	.3965
0.12	.0478	.5478	.4522	.3961
0.13	.0517	.5517	.4483	.3956
0.14	.0557	.5557	.4443	.3951
0.15	.0596	.5596	.4404	.3945
0.16	.0636	.5636	.4364	.3939
0.17	.0675	.5675	.4325	.3932
0.18	.0714	.5714	.4286	.3925
0.19	.0753	.5753	.4247	.3918
0.20	.0793	.5793	.4207	.3910
0.21	.0832	.5832	.4168	.3902
0.22	.0871	.5871	.4129	.3894
0.23	.0910	.5910	.4090	.3885
0.24	.0948	.5948	.4052	.3876
0.25	.0987	.5987	.4013	.3867
0.26	.1026	.6026	.3974	.3857
0.27	.1064	.6064	.3936	.3847
0.28	.1103	.6103	.3897	.3836
0.29	.1141	.6141	.3859	.3825
0.30	.1179	.6179	.3821	.3814
0.31	.1217	.6217	.3783	.3802
0.32	.1255	.6255	.3745	.3790
0.33	.1293	.6293	.3707	.3778
0.34	.1331	.6331	.3669	.3765
0.35	.1368	.6368	.3632	.3752
0.36	.1406	.6406	.3594	.3739
0.37	.1443	.6443	.3557	.3725
0.38	.1480	.6480	.3520	.3712
0.39	.1517	.6517	.3483	.3697
0.40	.1554	.6554	.3446	.3683

Source: Edwards, A. L. *Statistical Methods for the Behavioral Sciences*. New York: Holt, 1954. Reprinted by permission of author and publisher.

TABLE B Areas and ordinates of the normal curve in terms of x/σ (continued)

(1) z Standard score $\left(\frac{x}{\sigma}\right)$	(2) A Area from mean to $\frac{x}{\sigma}$	(3) B Area in larger portion	(4) C Area in smaller portion	(5) y Ordinate at $\frac{x}{\sigma}$
0.41	.1591	.6591	.3409	.3668
0.42	.1628	.6628	.3372	.3653
0.43	.1664	.6664	.3336	.3637
0.44	.1700	.6700	.3300	.3621
0.45	.1736	.6736	.3264	.3605
0.46	.1772	.6772	.3228	.3589
0.47	.1808	.6808	.3192	.3572
0.48	.1844	.6844	.3156	.3555
0.49	.1879	.6879	.3121	.3538
0.50	.1915	.6915	.3085	.3521
0.51	.1950	.6950	.3050	.3503
0.52	.1985	.6985	.3015	.3485
0.53	.2019	.7019	.2981	.3467
0.54	.2054	.7054	.2946	.3448
0.55	.2088	.7088	.2912	.3429
0.56	.2123	.7123	.2877	.3410
0.57	.2157	.7157	.2843	.3391
0.58	.2190	.7190	.2810	.3372
0.59	.2224	.7224	.2776	.3352
0.60	.2257	.7257	.2743	.3332
0.61	.2291	.7291	.2709	.3312
0.62	.2324	.7324	.2676	.3292
0.63	.2357	.7357	.2643	.3271
0.64	.2389	.7389	.2611	.3251
0.65	.2422	.7422	.2578	.3230
0.66	.2454	.7454	.2546	.3209
0.67	.2486	.7486	.2514	.3187
0.68	.2517	.7517	.2483	.3166
0.69	.2549	.7549	.2451	.3144
0.70	.2580	.7580	.2420	.3123
0.71	.2611	.7611	.2389	.3101
0.72	.2642	.7642	.2358	.3079
0.73	.2673	.7673	.2327	.3056
0.74	.2704	.7704	.2296	.3034
0.75	.2734	.7734	.2266	.3011
0.76	.2764	.7764	.2236	.2989
0.77	.2794	.7794	.2206	.2966
0.78	.2823	.7823	.2177	.2943
0.79	.2852	.7852	.2148	.2920
0.80	.2881	.7881	.2119	.2897
0.81	.2910	.7910	.2090	.2874
0.82	.2939	.7939	.2061	.2850
0.83	.2967	.7967	.2033	.2827

TABLE B Areas and ordinates of the normal curve in terms of x/σ (continued)

(1) z Standard score $\left(\frac{x}{\sigma}\right)$	(2) A Area from mean to $\frac{x}{\sigma}$	(3) B Area in larger portion	(4) C Area in smaller portion	(5) y Ordinate at $\frac{x}{\sigma}$
0.84	.2995	.7995	.2005	.2803
0.85	.3023	.8023	.1977	.2780
0.86	.3051	.8051	.1949	.2756
0.87	.3078	.8078	.1922	.2732
0.88	.3106	.8106	.1894	.2709
0.89	.3133	.8133	.1867	.2685
0.90	.3159	.8159	.1841	.2661
0.91	.3186	.8186	.1814	.2637
0.92	.3212	.8212	.1788	.2613
0.93	.3238	.8238	.1762	.2589
0.94	.3264	.8264	.1736	.2565
0.95	.3289	.8289	.1711	.2541
0.96	.3315	.8315	.1685	.2516
0.97	.3340	.8340	.1660	.2492
0.98	.3365	.8365	.1635	.2468
0.99	.3389	.8389	.1611	.2444
1.00	.3413	.8413	.1587	.2420
1.01	.3438	.8438	.1562	.2396
1.02	.3461	.8461	.1539	.2371
1.03	.3485	.8485	.1515	.2347
1.04	.3508	.8508	.1492	.2323
1.05	.3531	.8531	.1469	.2299
1.06	.3554	.8554	.1446	.2275
1.07	.3577	.8577	.1423	.2251
1.08	.3599	.8599	.1401	.2227
1.09	.3621	.8621	.1379	.2203
1.10	.3643	.8643	.1357	.2179
1.11	.3665	.8665	.1335	.2155
1.12	.3686	.8686	.1314	.2131
1.13	.3708	.8708	.1292	.2107
1.14	.3729	.8729	.1271	.2083
1.15	.3749	.8749	.1251	.2059
1.16	.3770	.8770	.1230	.2036
1.17	.3790	.8790	.1210	.2012
1.18	.3810	.8810	.1190	.1989
1.19	.3830	.8830	.1170	.1965
1.20	.3849	.8849	.1151	.1942
1.21	.3869	.8869	.1131	.1919
1.22	.3888	.8888	.1112	.1895
1.23	.3907	.8907	.1093	.1872
1.24	.3925	.8925	.1075	.1849
1.25	.3944	.8944	.1056	.1826
1.26	.3962	.8962	.1038	.1804

TABLE B Areas and ordinates of the normal curve in terms of x/σ (continued)

(1) z Standard score $\left(\frac{x}{\sigma}\right)$	(2) A Area from mean to $\frac{x}{\sigma}$	(3) B Area in larger portion	(4) C Area in smaller portion	(5) y Ordinate at $\frac{x}{\sigma}$
1.27	.3980	.8980	.1020	.1781
1.28	.3997	.8997	.1003	.1758
1.29	.4015	.9015	.0985	.1736
1.30	.4032	.9032	.0968	.1714
1.31	.4049	.9049	.0951	.1691
1.32	.4066	.9066	.0934	.1669
1.33	.4082	.9082	.0918	.1647
1.34	.4099	.9099	.0901	.1626
1.35	.4115	.9115	.0885	.1604
1.36	.4131	.9131	.0869	.1582
1.37	.4147	.9147	.0853	.1561
1.38	.4162	.9162	.0838	.1539
1.39	.4177	.9177	.0823	.1518
1.40	.4192	.9192	.0808	.1497
1.41	.4207	.9207	.0793	.1476
1.42	.4222	.9222	.0778	.1456
1.43	.4236	.9236	.0764	.1435
1.44	.4251	.9251	.0749	.1415
1.45	.4265	.9265	.0735	.1394
1.46	.4279	.9279	.0721	.1374
1.47	.4292	.9292	.0708	.1354
1.48	.4306	.9306	.0694	.1334
1.49	.4319	.9319	.0681	.1315
1.50	.4332	.9332	.0668	.1295
1.51	.4345	.9345	.0655	.1276
1.52	.4357	.9357	.0643	.1257
1.53	.4370	.9370	.0630	.1238
1.54	.4382	.9382	.0618	.1219
1.55	.4394	.9394	.0606	.1200
1.56	.4406	.9406	.0594	.1182
1.57	.4418	.9418	.0582	.1163
1.58	.4429	.9429	.0571	.1145
1.59	.4441	.9441	.0559	.1127
1.60	.4452	.9452	.0548	.1109
1.61	.4463	.9463	.0537	.1092
1.62	.4474	.9474	.0526	.1074
1.63	.4484	.9484	.0516	.1057
1.64	.4495	.9495	.0505	.1040
1.65	.4505	.9505	.0495	.1023
1.66	.4515	.9515	.0485	.1006
1.67	.4525	.9525	.0475	.0989
1.68	.4535	.9535	.0465	.0973
1.69	.4545	.9545	.0455	.0957

TABLE B Areas and ordinates of the normal curve in terms of x/σ (continued)

(1) z Standard score $\left(\frac{x}{\sigma}\right)$	(2) A Area from mean to $\frac{x}{\sigma}$	(3) B Area in larger portion	(4) C Area in smaller portion	(5) y Ordinate at $\frac{x}{\sigma}$
1.70	.4554	.9554	.0446	.0940
1.71	.4564	.9564	.0436	.0925
1.72	.4573	.9573	.0427	.0909
1.73	.4582	.9582	.0418	.0893
1.74	.4591	.9591	.0409	.0878
1.75	.4599	.9599	.0401	.0863
1.76	.4608	.9608	.0392	.0848
1.77	.4616	.9616	.0384	.0833
1.78	.4625	.9625	.0375	.0818
1.79	.4633	.9633	.0367	.0804
1.80	.4641	.9641	.0359	.0790
1.81	.4649	.9649	.0351	.0775
1.82	.4656	.9656	.0344	.0761
1.83	.4664	.9664	.0336	.0748
1.84	.4671	.9671	.0329	.0734
1.85	.4678	.9678	.0322	.0721
1.86	.4686	.9686	.0314	.0707
1.87	.4693	.9693	.0307	.0694
1.88	.4699	.9699	.0301	.0681
1.89	.4706	.9706	.0294	.0669
1.90	.4713	.9713	.0287	.0656
1.91	.4719	.9719	.0281	.0644
1.92	.4726	.9726	.0274	.0632
1.93	.4732	.9732	.0268	.0620
1.94	.4738	.9738	.0262	.0608
1.95	.4744	.9744	.0256	.0596
1.96	.4750	.9750	.0250	.0584
1.97	.4756	.9756	.0244	.0573
1.98	.4761	.9761	.0239	.0562
1.99	.4767	.9767	.0233	.0551
2.00	.4772	.9772	.0228	.0540
2.01	.4778	.9778	.0222	.0529
2.02	.4783	.9783	.0217	.0519
2.03	.4788	.9788	.0212	.0508
2.04	.4793	.9793	.0207	.0498
2.05	.4798	.9798	.0202	.0488
2.06	.4803	.9803	.0197	.0478
2.07	.4808	.9808	.0192	.0468
2.08	.4812	.9812	.0188	.0459
2.09	.4817	.9817	.0183	.0449
2.10	.4821	.9821	.0179	.0440
2.11	.4826	.9826	.0174	.0431
2.12	.4830	.9830	.0170	.0422

TABLE B Areas and ordinates of the normal curve in terms of x/σ (continued)

(1) z Standard score $\left(\frac{x}{\sigma}\right)$	(2) A Area from mean to $\frac{x}{\sigma}$	(3) B Area in larger portion	(4) C Area in smaller portion	(5) y Ordinate at $\frac{x}{\sigma}$
2.13	.4834	.9834	.0166	.0413
2.14	.4838	.9838	.0162	.0404
2.15	.4842	.9842	.0158	.0396
2.16	.4846	.9846	.0154	.0387
2.17	.4850	.9850	.0150	.0379
2.18	.4854	.9854	.0146	.0371
2.19	.4857	.9857	.0143	.0363
2.20	.4861	.9861	.0139	.0355
2.21	.4864	.9864	.0136	.0347
2.22	.4868	.9868	.0132	.0339
2.23	.4871	.9871	.0129	.0332
2.24	.4875	.9875	.0125	.0325
2.25	.4878	.9878	.0122	.0317
2.26	.4881	.9881	.0119	.0310
2.27	.4884	.9884	.0116	.0303
2.28	.4887	.9887	.0113	.0297
2.29	.4890	.9890	.0110	.0290
2.30	.4893	.9893	.0107	.0283
2.31	.4896	.9896	.0104	.0277
2.32	.4898	.9898	.0102	.0270
2.33	.4901	.9901	.0099	.0264
2.34	.4904	.9904	.0096	.0258
2.35	.4906	.9906	.0094	.0252
2.36	.4909	.9909	.0091	.0246
2.37	.4911	.9911	.0089	.0241
2.38	.4913	.9913	.0087	.0235
2.39	.4916	.9916	.0084	.0229
2.40	.4918	.9918	.0082	.0224
2.41	.4920	.9920	.0080	.0219
2.42	.4922	.9922	.0078	.0213
2.43	.4925	.9925	.0075	.0208
2.44	.4927	.9927	.0073	.0203
2.45	.4929	.9929	.0071	.0198
2.46	.4931	.9931	.0069	.0194
2.47	.4932	.9932	.0068	.0189
2.48	.4934	.9934	.0066	.0184
2.49	.4936	.9936	.0064	.0180
2.50	.4938	.9938	.0062	.0175
2.51	.4940	.9940	.0060	.0171
2.52	.4941	.9941	.0059	.0167
2.53	.4943	.9943	.0057	.0163
2.54	.4945	.9945	.0055	.0158
2.55	.4946	.9946	.0054	.0154

TABLE B Areas and ordinates of the normal curve in terms of x/σ (continued)

(1) z Standard score $\left(\frac{x}{\sigma}\right)$	(2) A Area from mean to $\frac{x}{\sigma}$	(3) B Area in larger portion	(4) C Area in smaller portion	(5) y Ordinate at $\frac{x}{\sigma}$
2.56	.4948	.9948	.0052	.0151
2.57	.4949	.9949	.0051	.0147
2.58	.4951	.9951	.0049	.0143
2.59	.4952	.9952	.0048	.0139
2.60	.4953	.9953	.0047	.0136
2.61	.4955	.9955	.0045	.0132
2.62	.4956	.9956	.0044	.0129
2.63	.4957	.9957	.0043	.0126
2.64	.4959	.9959	.0041	.0122
2.65	.4960	.9960	.0040	.0119
2.66	.4961	.9961	.0039	.0116
2.67	.4962	.9962	.0038	.0113
2.68	.4963	.9963	.0037	.0110
2.69	.4964	.9964	.0036	.0107
2.70	.4965	.9965	.0035	.0104
2.71	.4966	.9966	.0034	.0101
2.72	.4967	.9967	.0033	.0099
2.73	.4968	.9968	.0032	.0096
2.74	.4969	.9969	.0031	.0093
2.75	.4970	.9970	.0030	.0091
2.76	.4971	.9971	.0029	.0088
2.77	.4972	.9972	.0028	.0086
2.78	.4973	.9973	.0027	.0084
2.79	.4974	.9974	.0026	.0081
2.80	.4974	.9974	.0026	.0079
2.81	.4975	.9975	.0025	.0077
2.82	.4976	.9976	.0024	.0075
2.83	.4977	.9977	.0023	.0073
2.84	.4977	.9977	.0023	.0071
2.85	.4978	.9978	.0022	.0069
2.86	.4979	.9979	.0021	.0067
2.87	.4979	.9979	.0021	.0065
2.88	.4980	.9980	.0020	.0063
2.89	.4981	.9981	.0019	.0061
2.90	.4981	.9981	.0019	.0060
2.91	.4982	.9982	.0018	.0058
2.92	.4982	.9982	.0018	.0056
2.93	.4983	.9983	.0017	.0055
2.94	.4984	.9984	.0016	.0053
2.95	.4984	.9984	.0016	.0051
2.96	.4985	.9985	.0015	.0050
2.97	.4985	.9985	.0015	.0048
2.98	.4986	.9986	.0014	.0047
2.99	.4986	.9986	.0014	.0046

TABLE B Areas and ordinates of the normal curve in terms of x/σ (continued)

(1) z Standard score $\left(\frac{x}{\sigma}\right)$	(2) A Area from mean to $\frac{x}{\sigma}$	(3) B Area in larger portion	(4) C Area in smaller portion	(5) y Ordinate at $\frac{x}{\sigma}$
3.00	.4987	.9987	.0013	.0044
3.01	.4987	.9987	.0013	.0043
3.02	.4987	.9987	.0013	.0042
3.03	.4988	.9988	.0012	.0040
3.04	.4988	.9988	.0012	.0039
3.05	.4989	.9989	.0011	.0038
3.06	.4989	.9989	.0011	.0037
3.07	.4989	.9989	.0011	.0036
3.08	.4990	.9990	.0010	.0035
3.09	.4990	.9990	.0010	.0034
3.10	.4990	.9990	.0010	.0033
3.11	.4991	.9991	.0009	.0032
3.12	.4991	.9991	.0009	.0031
3.13	.4991	.9991	.0009	.0030
3.14	.4992	.9992	.0008	.0029
3.15	.4992	.9992	.0008	.0028
3.16	.4992	.9992	.0008	.0027
3.17	.4992	.9992	.0008	.0026
3.18	.4993	.9993	.0007	.0025
3.19	.4993	.9993	.0007	.0025
3.20	.4993	.9993	.0007	.0024
3.21	.4993	.9993	.0007	.0023
3.22	.4994	.9994	.0006	.0022
3.23	.4994	.9994	.0006	.0022
3.24	.4994	.9994	.0006	.0021
3.30	.4995	.9995	.0005	.0017
3.40	.4997	.9997	.0003	.0012
3.50	.4998	.9998	.0002	.0009
3.60	.4998	.9998	.0002	.0006
3.70	.4999	.9999	.0001	.0004

TABLE C Standard scores (or deviates) and ordinates corresponding to divisions of the area under the normal curve into a larger proportion (B) and a smaller proportion (C); also the value $\sqrt{BC}$

B The larger area	z Standard score	y Ordinate	$\sqrt{BC}$	C The smaller area
.500	.0000	.3989	.5000	.500
.505	.0125	.3989	.5000	.495
.510	.0251	.3988	.4999	.490
.515	.0376	.3987	.4998	.485
.520	.0502	.3984	.4996	.480
.525	.0627	.3982	.4994	.475
.530	.0753	.3978	.4991	.470
.535	.0878	.3974	.4988	.465
.540	.1004	.3969	.4984	.460
.545	.1130	.3964	.4980	.455
.550	.1257	.3958	.4975	.450
.555	.1383	.3951	.4970	.445
.560	.1510	.3944	.4964	.440
.565	.1637	.3936	.4958	.435
.570	.1764	.3928	.4951	.430
.575	.1891	.3919	.4943	.425
.580	.2019	.3909	.4936	.420
.585	.2147	.3899	.4927	.415
.590	.2275	.3887	.4918	.410
.595	.2404	.3876	.4909	.405
.600	.2533	.3863	.4899	.400
.605	.2663	.3850	.4889	.395
.610	.2793	.3837	.4877	.390
.615	.2924	.3822	.4867	.385
.620	.3055	.3808	.4854	.380
.625	.3186	.3792	.4841	.375
.630	.3319	.3776	.4828	.370
.635	.3451	.3759	.4814	.365
.640	.3585	.3741	.4800	.360
.645	.3719	.3723	.4785	.355
.650	.3853	.3704	.4770	.350
.655	.3989	.3684	.4754	.345
.660	.4125	.3664	.4737	.340
.665	.4261	.3643	.4720	.335
.670	.4399	.3621	.4702	.330
.675	.4538	.3599	.4684	.325
.680	.4677	.3576	.4665	.320
.685	.4817	.3552	.4645	.315
.690	.4959	.3528	.4625	.310
.695	.5101	.3503	.4604	.305
.700	.5244	.3477	.4583	.300
.705	.5388	.3450	.4560	.295
.710	.5534	.3423	.4538	.290
.715	.5681	.3395	.4514	.285
.720	.5828	.3366	.4490	.280

TABLE C Standard scores (or deviates) and ordinates corresponding to divisions of the area under the normal curve into a larger proportion (B) and a smaller proportion (C); also the value $\sqrt{BC}$ (continued)

B *The larger area*	z *Standard score*	y *Ordinate*	$\sqrt{BC}$	C *The smaller area*
.725	.5978	.3337	.4465	.275
.730	.6128	.3306	.4440	.270
.735	.6280	.3275	.4413	.265
.740	.6433	.3244	.4386	.260
.745	.6588	.3211	.4359	.255
.750	.6745	.3178	.4330	.250
.755	.6903	.3144	.4301	.245
.760	.7063	.3109	.4271	.240
.765	.7225	.3073	.4240	.235
.770	.7388	.3036	.4208	.230
.775	.7554	.2999	.4176	.225
.780	.7722	.2961	.4142	.220
.785	.7892	.2922	.4108	.215
.790	.8064	.2882	.4073	.210
.795	.8239	.2841	.4037	.205
.800	.8416	.2800	.4000	.200
.805	.8596	.2757	.3962	.195
.810	.8779	.2714	.3923	.190
.815	.8965	.2669	.3883	.185
.820	.9154	.2624	.3842	.180
.825	.9346	.2578	.3800	.175
.830	.9542	.2531	.3756	.170
.835	.9741	.2482	.3712	.165
.840	.9945	.2433	.3666	.160
.845	1.0152	.2383	.3619	.155
.850	1.0364	.2332	.3571	.150
.855	1.0581	.2279	.3521	.145
.860	1.0803	.2226	.3470	.140
.865	1.1031	.2171	.3417	.135
.870	1.1264	.2115	.3363	.130
.875	1.1503	.2059	.3307	.125
.880	1.1750	.2000	.3250	.120
.885	1.2004	.1941	.3190	.115
.890	1.2265	.1880	.3129	.110
.895	1.2536	.1818	.3066	.105
.900	1.2816	.1755	.3000	.100
.905	1.3106	.1690	.2932	.095
.910	1.3408	.1624	.2862	.090
.915	1.3722	.1556	.2789	.085
.920	1.4051	.1487	.2713	.080
.925	1.4395	.1416	.2634	.075
.930	1.4757	.1343	.2551	.070
.935	1.5141	.1268	.2465	.065
.940	1.5548	.1191	.2375	.060
.945	1.5982	.1112	.2280	.055

TABLE C Standard scores (or deviates) and ordinates corresponding to divisions of the area under the normal curve into a larger proportion (B) and a smaller proportion (C); also the value $\sqrt{BC}$ (continued)

B The larger area	z Standard score	y Ordinate	$\sqrt{BC}$	C The smaller area
.950	1.6449	.1031	.2179	.050
.955	1.6954	.0948	.2073	.045
.960	1.7507	.0862	.1960	.040
.965	1.8119	.0773	.1838	.035
.970	1.8808	.0680	.1706	.030
.975	1.9600	.0584	.1561	.025
.980	2.0537	.0484	.1400	.020
.985	2.1701	.0379	.1226	.015
.990	2.3263	.0267	.0995	.010
.995	2.5758	.0145	.0705	.005
.996	2.6521	.0118	.0631	.004
.997	2.7478	.0091	.0547	.003
.998	2.8782	.0063	.0447	.002
.999	3.0902	.0034	.0316	.001
.9995	3.2905	.0018	.0224	.0005

TABLE D Significance levels of the t-ratio*

	Level of significance for two-tail test				
df	*.10*	*.05*	*.02*	*.01*	*.001*
1	6.314	12.706	31.821	63.657	636.619
2	2.920	4.303	6.965	9.925	31.598
3	2.353	3.182	4.541	5.841	12.941
4	2.132	2.776	3.747	4.604	8.610
5	2.015	2.571	3.365	4.032	6.859
6	1.943	2.447	3.143	3.707	5.959
7	1.895	2.365	2.998	3.499	5.405
8	1.860	2.306	2.896	3.355	5.041
9	1.833	2.262	2.821	3.250	4.781
10	1.812	2.228	2.764	3.169	4.587
11	1.796	2.201	2.718	3.106	4.437
12	1.782	2.179	2.681	3.055	4.318
13	1.771	2.160	2.650	3.012	4.221
14	1.761	2.145	2.624	2.977	4.140
15	1.753	2.131	2.602	2.947	4.073
16	1.746	2.120	2.583	2.921	4.015
17	1.740	2.110	2.567	2.898	3.965
18	1.734	2.101	2.552	2.878	3.922
19	1.729	2.093	2.539	2.861	3.883
20	1.725	2.086	2.528	2.845	3.850
21	1.721	2.080	2.518	2.831	3.819
22	1.717	2.074	2.508	2.819	3.792
23	1.714	2.069	2.500	2.807	3.767
24	1.711	2.064	2.492	2.797	3.745
25	1.708	2.060	2.485	2.787	3.725
26	1.706	2.056	2.479	2.779	3.707
27	1.703	2.052	2.473	2.771	3.690
28	1.701	2.048	2.467	2.763	3.674
29	1.699	2.045	2.462	2.756	3.659
30	1.697	2.042	2.457	2.750	3.646
40	1.684	2.021	2.423	2.704	3.551
60	1.671	2.000	2.390	2.660	3.460
120	1.658	1.980	2.358	2.617	3.373
∞	1.645	1.960	2.326	2.576	3.291
	.05	*.025*	*.01*	*.005*	*.0005*
	Level of significance for one-tail test				

* Abridged from Fisher, R. A. and F. Yates. *Statistical tables for biological, agricultural, and medical research.* Published by Longman Group Ltd., London (previously published by Oliver & Boyd, Edinborough). Used with permission of the authors and publishers.

TABLE E Table of chi square*

df	P = .99	.98	.95	.90	.80	.70	.50	.30	.20	.10	.05	.02	.01	.001
1	.000157	.000628	.00393	.0158	.0642	.148	.455	1.074	1.642	2.706	3.841	5.412	6.635	10.827
2	.0201	.0404	.103	.211	.446	.713	1.386	2.408	3.219	4.605	5.991	7.824	9.210	13.815
3	.115	.185	.352	.584	1.005	1.424	2.366	3.665	4.642	6.251	7.815	9.837	11.341	16.268
4	.297	.429	.711	1.064	1.649	2.195	3.357	4.878	5.989	7.779	9.488	11.668	13.277	18.465
5	.554	.752	1.145	1.610	2.343	3.000	4.351	6.064	7.289	9.236	11.070	13.388	15.086	20.517
6	.872	1.134	1.635	2.204	3.070	3.828	5.348	7.231	8.558	10.6645	12.592	15.033	16.812	22.457
7	1.239	1.564	2.167	2.833	3.822	4.671	6.346	8.383	9.803	12.017	14.067	16.622	18.475	24.322
8	1.645	2.032	2.733	3.490	4.594	5.527	7.344	9.524	11.030	13.362	15.507	18.168	20.090	26.125
9	2.088	2.532	3.325	4.168	5.380	6.393	8.343	10.656	12.242	14.684	16.919	19.679	21.666	27.877
10	2.558	3.059	3.940	4.865	6.179	7.267	9.342	11.781	13.442	15.987	18.307	21.161	23.209	29.588
11	3.053	3.609	4.575	5.578	6.989	8.148	10.341	12.899	14.631	17.275	19.675	22.618	24.725	31.264
12	3.571	4.178	5.226	6.304	7.807	9.034	11.340	14.011	15.812	18.549	21.026	24.054	26.217	32.909
13	4.107	4.765	5.892	7.042	8.634	9.926	12.340	15.119	16.985	19.812	22.362	25.472	27.688	34.528
14	4.660	5.368	6.571	7.790	9.467	10.821	13.339	16.222	18.151	21.064	23.685	26.873	29.141	36.123
15	5.229	5.985	7.261	8.547	10.307	11.721	14.339	17.322	19.311	22.307	24.996	28.259	30.578	37.697
16	5.812	6.614	7.962	9.312	11.152	12.624	15.338	18.418	20.465	23.542	26.296	29.633	32.000	39.252
17	6.408	7.255	8.672	10.085	12.002	13.531	16.338	19.511	21.615	24.769	27.587	30.995	33.409	40.790
18	7.015	7.906	9.390	10.865	12.857	14.440	17.338	20.601	22.760	25.989	28.869	32.346	34.805	42.312
19	7.633	8.567	10.117	11.651	13.716	15.352	18.338	21.689	23.900	27.204	30.144	33.687	36.191	43.820
20	8.260	9.237	10.851	12.443	14.578	16.266	19.337	22.775	25.038	28.412	31.410	35.020	37.566	45.315
21	8.897	9.915	11.591	13.240	15.445	17.182	20.337	23.858	26.171	29.615	32.671	36.343	38.932	46.797
22	9.542	10.600	12.338	14.041	16.314	18.101	21.337	24.939	27.301	30.813	33.924	37.659	40.289	48.268
23	10.196	11.293	13.091	14.848	17.187	19.021	22.337	26.018	28.429	32.007	35.172	38.968	41.638	49.728
24	10.856	11.992	13.848	15.659	18.062	19.943	23.337	27.096	29.553	33.196	36.415	40.270	42.980	51.179
25	11.524	12.697	14.611	16.473	18.940	20.867	24.337	28.172	30.675	34.382	37.652	41.566	44.314	52.620
26	12.198	13.409	15.379	17.292	19.820	21.792	25.336	29.246	31.795	35.563	38.885	42.856	45.642	54.052
27	12.879	14.125	16.151	18.114	20.703	22.719	26.336	30.319	32.912	36.741	40.113	44.140	46.963	55.476
28	13.565	14.847	16.928	18.939	21.588	23.647	27.336	31.391	34.027	37.916	41.337	45.419	48.278	56.893
29	14.256	15.574	17.708	19.768	22.475	24.577	28.336	32.461	35.139	39.087	42.557	46.693	49.588	58.302
30	14.953	16.306	18.493	20.599	23.364	25.508	29.336	33.530	36.250	40.256	43.773	47.962	50.892	59.703

* Table E is reprinted from Table III of Fisher's *Statistical methods for research workers.* Edinburgh: Oliver & Boyd, 1932, by kind permission of the author and publishers. For df larger than 30, the value from the expression $\sqrt{2x^2} - \sqrt{2df - 1}$ may be interpreted as a *t* ratio. For instructions on the use of this table, see p. 195.

TABLE F .05 (lightface type) and .01 (boldface type) points for the distribution of F*

df_2	df_1 degrees of freedom (for greater mean square)																								df_2
	1	2	3	4	5	6	7	8	9	10	11	12	14	16	20	24	30	40	50	75	100	200	500	∞	
1	161	200	216	225	230	234	237	239	241	242	243	244	245	246	248	249	250	251	252	253	253	254	254	254	
	4,052	**4,999**	**5,403**	**5,625**	**5,764**	**5,859**	**5,928**	**5,981**	**6,022**	**6,056**	**6,082**	**6,106**	**6,142**	**6,169**	**6,208**	**6,234**	**6,258**	**6,286**	**6,302**	**6,323**	**6,334**	**6,352**	**6,361**	**6,366**	
2	18.51	19.00	19.16	19.25	19.30	19.33	19.36	19.37	19.38	19.39	19.40	19.41	19.42	19.43	19.44	19.45	19.46	19.47	19.47	19.48	19.49	19.49	19.50	19.50	
	98.49	**99.00**	**99.17**	**99.25**	**99.30**	**99.33**	**99.34**	**99.36**	**99.38**	**99.40**	**99.41**	**99.42**	**99.43**	**99.44**	**99.45**	**99.46**	**99.47**	**99.48**	**99.48**	**99.49**	**99.49**	**99.49**	**99.50**	**99.50**	
3	10.13	9.55	9.28	9.12	9.01	8.94	8.88	8.84	8.81	8.78	8.76	8.74	8.71	8.69	8.66	8.64	8.62	8.60	8.58	8.57	8.56	8.54	8.54	8.53	3
	34.12	**30.82**	**29.46**	**28.71**	**28.24**	**27.91**	**27.67**	**27.49**	**27.34**	**27.23**	**27.13**	**27.05**	**26.92**	**26.83**	**26.69**	**26.60**	**26.50**	**26.41**	**26.35**	**26.27**	**26.23**	**26.18**	**26.14**	**26.12**	
4	7.71	6.94	6.59	6.39	6.26	6.16	6.09	6.04	6.00	5.96	5.93	5.91	5.87	5.84	5.80	5.77	5.74	5.71	5.70	5.68	5.66	5.65	5.64	5.63	4
	21.20	**18.00**	**16.69**	**15.98**	**15.52**	**15.21**	**14.98**	**14.80**	**14.66**	**14.54**	**14.45**	**14.37**	**14.24**	**14.15**	**14.02**	**13.93**	**13.83**	**13.74**	**13.69**	**13.61**	**13.57**	**13.52**	**13.48**	**13.46**	
5	6.61	5.79	5.41	5.19	5.05	4.95	4.88	4.82	4.78	4.74	4.70	4.68	4.64	4.60	4.56	4.53	4.50	4.46	4.44	4.42	4.40	4.38	4.37	4.36	5
	16.26	**13.27**	**12.06**	**11.39**	**10.97**	**10.67**	**10.45**	**10.27**	**10.15**	**10.05**	**9.96**	**9.89**	**9.77**	**9.68**	**9.55**	**9.47**	**9.38**	**9.29**	**9.24**	**9.17**	**9.13**	**9.07**	**9.04**	**9.02**	
6	5.99	5.14	4.76	4.53	4.39	4.28	4.21	4.15	4.10	4.06	4.03	4.00	3.96	3.92	3.87	3.84	3.81	3.77	3.75	3.72	3.71	3.69	3.68	3.67	6
	13.74	**10.92**	**9.78**	**9.15**	**8.75**	**8.47**	**8.26**	**8.10**	**7.98**	**7.87**	**7.79**	**7.72**	**7.60**	**7.52**	**7.39**	**7.31**	**7.23**	**7.14**	**7.09**	**7.02**	**6.99**	**6.94**	**6.90**	**6.88**	
7	5.59	4.74	4.35	4.12	3.97	3.87	3.79	3.73	3.68	3.63	3.60	3.57	3.52	3.49	3.44	3.41	3.38	3.34	3.32	3.29	3.28	3.25	3.24	3.23	7
	12.25	**9.55**	**8.45**	**7.85**	**7.46**	**7.19**	**7.00**	**6.84**	**6.71**	**6.62**	**6.54**	**6.47**	**6.35**	**6.27**	**6.15**	**6.07**	**5.98**	**5.90**	**5.85**	**5.78**	**5.75**	**5.70**	**5.67**	**5.65**	
8	5.32	4.46	4.07	3.84	3.69	3.58	3.50	3.44	3.39	3.34	3.31	3.28	3.23	3.20	3.15	3.12	3.08	3.05	3.03	3.00	2.98	2.96	2.94	2.93	8
	11.26	**8.65**	**7.59**	**7.01**	**6.63**	**6.37**	**6.19**	**6.03**	**5.91**	**5.82**	**5.74**	**5.67**	**5.56**	**5.48**	**5.36**	**5.28**	**5.20**	**5.11**	**5.06**	**5.00**	**4.96**	**4.91**	**4.88**	**4.86**	
9	5.12	4.26	3.86	3.63	3.48	3.37	3.29	3.23	3.18	3.13	3.10	3.07	3.02	2.98	2.93	2.90	2.86	2.82	2.80	2.77	2.76	2.73	2.72	2.71	9
	10.56	**8.02**	**6.99**	**6.42**	**6.06**	**5.80**	**5.62**	**5.47**	**5.35**	**5.26**	**5.18**	**5.11**	**5.00**	**4.92**	**4.80**	**4.73**	**4.64**	**4.56**	**4.51**	**4.45**	**4.41**	**4.36**	**4.33**	**4.31**	
10	4.96	4.10	3.71	3.48	3.33	3.22	3.14	3.07	3.02	2.97	2.94	2.91	2.86	2.82	2.77	2.74	2.70	2.67	2.64	2.61	2.59	2.56	2.55	2.54	10
	10.04	**7.56**	**6.55**	**5.99**	**5.64**	**5.39**	**5.21**	**5.06**	**4.95**	**4.85**	**4.78**	**4.71**	**4.60**	**4.52**	**4.41**	**4.33**	**4.25**	**4.17**	**4.12**	**4.05**	**4.01**	**3.96**	**3.93**	**3.91**	
11	4.84	3.98	3.59	3.36	3.20	3.09	3.01	2.95	2.90	2.86	2.82	2.79	2.74	2.70	2.65	2.61	2.57	2.53	2.50	2.47	2.45	2.42	2.41	2.40	11
	9.65	**7.20**	**6.22**	**5.67**	**5.32**	**5.07**	**4.88**	**4.74**	**4.63**	**4.54**	**4.46**	**4.40**	**4.29**	**4.21**	**4.10**	**4.02**	**3.94**	**3.86**	**3.80**	**3.74**	**3.70**	**3.66**	**3.62**	**3.60**	
12	4.75	3.88	3.49	3.26	3.11	3.00	2.92	2.85	2.80	2.76	2.72	2.69	2.64	2.60	2.54	2.50	2.46	2.42	2.40	2.36	2.35	2.32	2.31	2.30	12
	9.33	**6.93**	**5.95**	**5.41**	**5.06**	**4.82**	**4.65**	**4.50**	**4.39**	**4.30**	**4.22**	**4.16**	**4.05**	**3.98**	**3.86**	**3.78**	**3.70**	**3.61**	**3.56**	**3.49**	**3.46**	**3.41**	**3.38**	**3.36**	
13	4.67	3.80	3.41	3.18	3.02	2.92	2.84	2.77	2.72	2.67	2.63	2.60	2.55	2.51	2.46	2.42	2.38	2.34	2.32	2.28	2.26	2.24	2.22	2.21	13
	9.07	**6.70**	**5.74**	**5.20**	**4.86**	**4.62**	**4.44**	**4.30**	**4.19**	**4.10**	**4.02**	**3.96**	**3.85**	**3.78**	**3.67**	**3.59**	**3.51**	**3.42**	**3.37**	**3.30**	**3.27**	**3.21**	**3.18**	**3.16**	

* Reproduced from Snedecor, C. W., *Statistical methods*. Ames, Iowa: Collegiate, 1937. By permission of the author. For instructions on the use of this table, see p. 166.

TABLE F .05 (lightface type) and .01 (boldface type) points for the distribution of F (continued)

df_2	df_1 degrees of freedom (for greater mean square)																							df_2
	1	2	3	4	5	6	7	8	9	10	11	12	14	16	20	24	30	40	50	75	100	200	500	∞	
14	4.60	3.74	3.34	3.11	2.96	2.85	2.77	2.70	2.65	2.60	2.56	2.53	2.48	2.44	2.39	2.35	2.31	2.27	2.24	2.21	2.19	2.16	2.14	2.13	14
	8.86	**6.51**	**5.56**	**5.03**	**4.69**	**4.46**	**4.28**	**4.14**	**4.03**	**3.94**	**3.86**	**3.80**	**3.70**	**3.62**	**3.51**	**3.43**	**3.34**	**3.26**	**3.21**	**2.14**	**3.11**	**3.06**	**3.02**	**3.00**	
15	4.54	3.68	3.29	3.06	2.90	2.79	2.70	2.64	2.59	2.55	2.51	2.48	2.43	2.39	2.33	2.29	2.25	2.21	2.18	2.15	2.12	2.10	2.08	2.07	15
	8.68	**6.36**	**5.42**	**4.89**	**4.56**	**4.32**	**4.14**	**4.00**	**3.89**	**3.80**	**3.73**	**3.67**	**3.56**	**3.48**	**3.36**	**3.29**	**3.20**	**3.12**	**3.07**	**3.00**	**2.97**	**2.92**	**2.89**	**2.87**	
16	4.49	3.63	3.24	3.01	2.85	2.74	2.66	2.59	2.54	2.49	2.45	2.42	2.37	2.33	2.28	2.24	2.20	2.16	2.13	2.09	2.07	2.04	2.02	2.01	16
	8.53	**6.23**	**5.29**	**4.77**	**4.44**	**4.20**	**4.03**	**3.89**	**3.78**	**3.69**	**3.61**	**3.55**	**3.45**	**3.37**	**3.25**	**3.18**	**3.10**	**3.01**	**2.96**	**2.89**	**2.86**	**2.80**	**2.77**	**2.75**	
17	4.45	3.59	3.20	2.96	2.81	2.70	2.62	2.55	2.50	2.45	2.41	2.38	2.33	2.29	2.23	2.19	2.15	2.11	2.08	2.04	2.02	1.99	1.97	1.96	17
	8.40	**6.11**	**5.18**	**4.67**	**4.34**	**4.10**	**3.93**	**3.79**	**3.68**	**3.59**	**3.52**	**3.45**	**3.35**	**3.27**	**3.16**	**3.08**	**3.00**	**2.92**	**2.86**	**2.79**	**2.76**	**2.70**	**2.67**	**2.65**	
18	4.41	3.55	3.16	2.93	2.77	2.66	2.58	2.51	2.46	2.41	2.37	2.34	2.29	2.25	2.19	2.15	2.11	2.07	2.04	2.00	1.98	1.95	1.93	1.92	18
	8.28	**6.01**	**5.09**	**4.58**	**4.25**	**4.01**	**3.85**	**3.71**	**3.60**	**3.51**	**3.44**	**3.37**	**3.27**	**3.19**	**3.07**	**3.00**	**2.91**	**2.83**	**2.78**	**2.71**	**2.68**	**2.62**	**2.59**	**2.57**	
19	4.38	3.52	3.13	2.90	2.74	2.63	2.55	2.48	2.43	2.38	2.34	2.31	2.26	2.21	2.15	2.11	2.07	2.02	2.00	1.96	1.94	1.91	1.90	1.88	19
	8.18	**5.93**	**5.01**	**4.50**	**4.17**	**3.94**	**3.77**	**3.63**	**3.52**	**3.43**	**3.36**	**3.30**	**3.19**	**3.12**	**3.00**	**2.92**	**2.84**	**2.76**	**2.70**	**2.63**	**2.60**	**2.54**	**2.51**	**2.49**	
20	4.35	3.49	3.10	2.87	2.71	2.60	2.52	2.45	2.40	2.35	2.31	2.28	2.23	2.18	2.12	2.08	2.04	1.99	1.96	1.92	1.90	1.87	1.85	1.84	20
	8.10	**5.85**	**4.94**	**4.43**	**4.10**	**3.87**	**3.71**	**3.56**	**3.45**	**3.37**	**3.30**	**3.23**	**3.13**	**3.05**	**2.94**	**2.86**	**2.77**	**2.69**	**2.63**	**2.56**	**2.53**	**2.47**	**2.44**	**2.42**	
21	4.32	3.47	3.07	2.84	2.68	2.57	2.49	2.42	2.37	2.32	2.28	2.25	2.20	2.15	2.09	2.05	2.00	1.96	1.93	1.89	1.87	1.84	1.82	1.81	21
	8.02	**5.78**	**4.87**	**4.37**	**4.04**	**3.81**	**3.65**	**3.51**	**3.40**	**3.31**	**3.24**	**3.17**	**3.07**	**2.99**	**2.88**	**2.80**	**2.72**	**2.63**	**2.58**	**2.51**	**2.47**	**2.42**	**2.38**	**2.36**	
22	4.30	3.44	3.05	2.82	2.66	2.55	2.47	2.40	2.35	2.30	2.26	2.23	2.18	2.13	2.07	2.03	1.98	1.93	1.91	1.87	1.84	1.81	1.80	1.78	22
	7.94	**5.72**	**4.82**	**4.31**	**3.99**	**3.76**	**3.59**	**3.45**	**3.35**	**3.26**	**3.18**	**3.12**	**3.02**	**2.94**	**2.83**	**2.75**	**2.67**	**2.58**	**2.53**	**2.46**	**2.42**	**2.37**	**2.33**	**2.31**	
23	4.28	3.42	3.03	2.80	2.64	2.53	2.45	2.38	2.32	2.28	2.24	2.20	2.14	2.10	2.04	2.00	1.96	1.91	1.88	1.84	1.82	1.79	1.77	1.76	23
	7.88	**5.66**	**4.76**	**4.26**	**3.94**	**3.71**	**3.54**	**3.41**	**3.30**	**3.21**	**3.14**	**3.07**	**2.97**	**2.89**	**2.78**	**2.70**	**2.62**	**2.53**	**2.48**	**2.41**	**2.37**	**2.32**	**2.28**	**2.26**	
24	4.26	3.40	3.01	2.78	2.62	2.51	2.43	2.36	2.30	2.26	2.22	2.18	2.13	2.09	2.02	1.98	1.94	1.89	1.86	1.82	1.80	1.76	1.74	1.73	24
	7.82	**5.61**	**4.72**	**4.22**	**3.90**	**3.67**	**3.50**	**3.36**	**3.25**	**3.17**	**3.09**	**3.03**	**2.93**	**2.85**	**2.74**	**2.66**	**2.58**	**2.49**	**2.44**	**2.36**	**2.33**	**2.27**	**2.23**	**2.21**	
25	4.24	3.38	2.99	2.76	2.60	2.49	2.41	2.34	2.28	2.24	2.20	2.16	2.11	2.06	2.00	1.96	1.92	1.87	1.84	1.80	1.77	1.74	1.72	1.71	25
	7.77	**5.57**	**4.63**	**4.13**	**3.86**	**3.63**	**3.46**	**3.32**	**3.21**	**3.13**	**3.05**	**2.99**	**2.89**	**2.81**	**2.70**	**2.62**	**2.54**	**2.45**	**2.40**	**2.32**	**2.29**	**2.23**	**2.19**	**2.17**	
26	4.22	3.37	2.98	2.74	2.59	2.47	2.39	2.32	2.27	2.22	2.18	2.15	2.10	2.05	1.99	1.95	1.90	1.85	1.82	1.78	1.76	1.72	1.70	1.69	26
	7.72	**5.53**	**4.64**	**4.14**	**3.82**	**3.59**	**3.42**	**3.29**	**3.17**	**3.09**	**3.02**	**2.96**	**2.86**	**2.77**	**2.66**	**2.58**	**2.50**	**2.50**	**2.36**	**2.28**	**2.25**	**2.19**	**2.15**	**2.13**	

TABLE F .05 (lightface type) and .01 (boldface type) points for the distribution of F (continued)

df_2	df_1 degrees of freedom (for greater mean square)																								df_2
	1	2	3	4	5	6	7	8	9	10	11	12	14	16	20	24	30	40	50	75	100	200	500	∞	
27	4.21	3.35	2.96	2.73	2.57	2.46	2.37	2.30	2.25	2.20	2.16	2.13	2.08	2.03	1.97	1.93	1.88	1.84	1.80	1.76	1.74	1.71	1.68	1.67	27
	7.68	**5.49**	**4.60**	**4.11**	**3.79**	**3.56**	**3.39**	**3.26**	**3.14**	**3.06**	**2.98**	**2.93**	**2.83**	**2.74**	**2.63**	**2.55**	**2.47**	**2.38**	**2.33**	**2.25**	**2.21**	**2.16**	**2.12**	**2.10**	
28	4.20	3.34	2.95	2.71	2.56	2.44	2.36	2.29	2.24	2.19	2.15	2.12	2.06	2.02	1.96	1.91	1.87	1.81	1.78	1.75	1.72	1.69	1.67	1.65	28
	7.64	**5.45**	**4.57**	**4.07**	**3.76**	**3.53**	**3.36**	**3.23**	**3.11**	**3.03**	**2.95**	**2.90**	**2.80**	**2.71**	**2.60**	**2.52**	**2.44**	**2.35**	**2.30**	**2.22**	**2.18**	**2.13**	**2.09**	**2.06**	
29	4.18	3.33	2.93	2.70	2.54	2.43	2.35	2.28	2.22	2.18	2.14	2.10	2.05	2.00	1.94	1.90	1.85	1.80	1.77	1.73	1.71	1.68	1.65	1.64	29
	7.60	**5.42**	**4.54**	**4.04**	**3.73**	**3.50**	**3.33**	**3.20**	**3.08**	**3.00**	**2.92**	**2.87**	**2.77**	**2.68**	**2.57**	**2.49**	**2.41**	**2.32**	**2.27**	**2.19**	**2.15**	**2.10**	**2.06**	**2.03**	
30	4.17	3.32	2.92	2.69	2.53	2.42	2.34	2.27	2.21	2.16	2.12	2.09	2.04	1.99	1.93	1.89	1.84	1.79	1.76	1.72	1.69	1.66	1.64	1.62	30
	7.56	**5.39**	**4.51**	**4.02**	**3.70**	**3.47**	**3.30**	**3.17**	**3.06**	**2.98**	**2.90**	**2.84**	**2.74**	**2.66**	**2.55**	**2.47**	**2.38**	**2.29**	**2.24**	**2.16**	**2.13**	**2.07**	**2.03**	**2.01**	
32	4.15	3.30	2.90	2.67	2.51	2.40	2.32	2.25	2.19	2.14	2.10	2.07	2.02	1.97	1.91	1.86	1.82	1.76	1.74	1.69	1.67	1.64	1.61	1.59	32
	7.50	**5.34**	**4.46**	**3.97**	**3.66**	**3.42**	**3.25**	**3.12**	**3.01**	**2.94**	**2.86**	**2.80**	**2.70**	**2.62**	**2.51**	**2.42**	**2.34**	**2.25**	**2.20**	**2.12**	**2.08**	**2.02**	**1.98**	**1.96**	
34	4.13	3.28	2.88	2.65	2.49	2.38	2.30	2.23	2.17	2.12	2.08	2.05	2.00	1.95	1.89	1.84	1.80	1.74	1.71	1.67	1.64	1.61	1.59	1.57	34
	7.44	**5.29**	**4.42**	**3.93**	**3.61**	**3.38**	**3.21**	**3.08**	**2.97**	**2.89**	**2.82**	**2.76**	**2.66**	**2.58**	**2.47**	**2.38**	**2.30**	**2.21**	**2.15**	**2.08**	**2.04**	**1.98**	**1.94**	**1.91**	
36	4.11	3.26	2.86	2.63	2.48	2.36	2.28	2.21	2.15	2.10	2.06	2.03	1.98	1.93	1.87	1.82	1.78	1.72	1.69	1.65	1.62	1.59	1.56	1.55	36
	7.39	**5.25**	**4.38**	**3.89**	**3.58**	**3.35**	**3.18**	**3.04**	**2.94**	**2.86**	**2.78**	**2.72**	**2.62**	**2.54**	**2.43**	**2.35**	**2.26**	**2.17**	**2.12**	**2.04**	**2.00**	**1.94**	**1.90**	**1.87**	
38	4.10	3.25	2.85	2.62	2.46	2.35	2.26	2.19	2.14	2.09	2.05	2.02	1.96	1.92	1.85	1.80	1.76	1.71	1.67	1.63	1.60	1.57	1.54	1.53	38
	7.35	**5.21**	**4.34**	**3.86**	**3.54**	**3.32**	**3.15**	**3.02**	**2.91**	**2.82**	**2.75**	**2.69**	**2.59**	**2.51**	**2.40**	**2.32**	**2.22**	**2.14**	**2.08**	**2.00**	**1.97**	**1.90**	**1.86**	**1.84**	
40	4.08	3.23	2.84	2.61	2.45	2.34	2.25	2.18	2.12	2.07	2.04	2.00	1.95	1.90	1.84	1.79	1.74	1.69	1.66	1.61	1.59	1.55	1.53	1.51	40
	7.31	**5.18**	**4.31**	**3.83**	**3.51**	**3.29**	**3.12**	**2.99**	**2.88**	**2.80**	**2.73**	**2.66**	**2.56**	**2.49**	**2.37**	**2.29**	**2.20**	**2.11**	**2.05**	**1.97**	**1.94**	**1.88**	**1.84**	**1.81**	
42	4.07	3.22	2.83	2.59	2.44	2.32	2.24	2.17	2.11	2.06	2.02	1.99	1.94	1.89	1.82	1.78	1.73	1.68	1.64	1.60	1.57	1.54	1.51	1.49	42
	7.27	**5.15**	**4.29**	**3.80**	**3.49**	**3.26**	**3.10**	**2.96**	**2.86**	**2.77**	**2.70**	**2.64**	**2.54**	**2.46**	**2.35**	**2.26**	**2.17**	**2.08**	**2.02**	**1.94**	**1.91**	**1.85**	**1.80**	**1.78**	
44	4.06	3.21	2.82	2.58	2.43	2.31	2.23	2.16	2.10	2.05	2.01	1.98	1.92	1.88	1.81	1.76	1.72	1.66	1.63	1.58	1.56	1.52	1.50	1.48	44
	7.24	**5.12**	**4.26**	**3.78**	**3.46**	**3.24**	**3.07**	**2.94**	**2.84**	**2.75**	**2.68**	**2.62**	**2.52**	**2.44**	**2.32**	**2.24**	**2.15**	**2.06**	**2.00**	**1.92**	**1.88**	**1.82**	**1.78**	**1.75**	
46	4.05	3.20	2.81	2.57	2.42	2.30	2.22	2.14	2.09	2.04	2.00	1.97	1.91	1.87	1.80	1.75	1.71	1.65	1.62	1.57	1.54	1.51	1.48	1.46	46
	7.21	**5.10**	**4.24**	**3.76**	**3.44**	**3.22**	**3.05**	**2.92**	**2.82**	**2.73**	**2.66**	**2.60**	**2.50**	**2.42**	**2.30**	**2.22**	**2.13**	**2.04**	**1.98**	**1.90**	**1.86**	**1.80**	**1.76**	**1.72**	
48	4.04	3.19	2.80	2.56	2.41	2.30	2.21	2.14	2.08	2.03	1.99	1.96	1.90	1.86	1.79	1.74	1.70	1.64	1.61	1.56	1.53	1.50	1.47	1.45	48
	7.19	**5.08**	**4.22**	**3.74**	**3.42**	**3.20**	**3.04**	**2.90**	**2.80**	**2.71**	**2.64**	**2.58**	**2.48**	**2.40**	**2.28**	**2.20**	**2.11**	**2.02**	**1.96**	**1.88**	**1.84**	**1.78**	**1.73**	**1.70**	

TABLE F .05 (lightface type) and .01 (boldface type) points for the distribution of F (continued)

df_2	df_1 degrees of freedom (for greater mean square)																							df_2
	1	2	3	4	5	6	7	8	9	10	11	12	14	16	20	24	30	40	50	75	100	200	500	∞	
50	4.03	3.18	2.79	2.56	2.40	2.29	2.20	2.13	2.07	2.02	1.98	1.95	1.90	1.85	1.78	1.74	1.69	1.63	1.60	1.55	1.52	1.48	1.46	1.44	50
	7.17	**5.06**	**4.20**	**3.72**	**3.41**	**3.18**	**3.02**	**2.88**	**2.78**	**2.70**	**2.62**	**2.56**	**2.46**	**2.39**	**2.26**	**2.18**	**2.10**	**2.00**	**1.94**	**1.86**	**1.82**	**1.76**	**1.71**	**1.68**	
55	4.02	3.17	2.78	2.54	2.38	2.27	2.18	2.11	2.05	2.00	1.97	1.93	1.88	1.83	1.76	1.72	1.67	1.61	1.58	1.52	1.50	1.46	1.43	1.41	55
	7.12	**5.01**	**4.16**	**3.68**	**3.37**	**3.15**	**2.98**	**2.85**	**2.75**	**2.66**	**2.59**	**2.53**	**2.43**	**2.35**	**2.23**	**2.15**	**2.06**	**1.96**	**1.90**	**1.82**	**1.78**	**1.71**	**1.66**	**1.64**	
60	4.00	3.15	2.76	2.52	2.37	2.25	2.17	2.10	2.04	1.99	1.95	1.92	1.86	1.81	1.75	1.70	1.65	1.59	1.56	1.50	1.48	1.44	1.41	1.39	60
	7.08	**4.98**	**4.13**	**3.65**	**3.34**	**3.12**	**2.95**	**2.82**	**2.72**	**2.63**	**2.56**	**2.50**	**2.40**	**2.32**	**2.20**	**2.12**	**2.03**	**1.93**	**1.87**	**1.79**	**1.74**	**1.68**	**1.63**	**1.60**	
65	3.99	3.14	2.75	2.51	2.36	2.24	2.15	2.08	2.02	1.98	1.94	1.90	1.85	1.80	1.73	1.68	1.63	1.57	1.54	1.49	1.46	1.42	1.39	1.37	65
	7.04	**4.95**	**4.10**	**3.62**	**3.31**	**3.09**	**2.93**	**2.79**	**2.70**	**2.61**	**2.54**	**2.47**	**2.37**	**2.30**	**2.18**	**2.09**	**2.00**	**1.90**	**1.84**	**1.76**	**1.71**	**1.64**	**1.60**	**1.56**	
70	3.98	3.13	2.74	2.50	2.35	2.23	2.14	2.07	2.01	1.97	1.93	1.89	1.84	1.79	1.72	1.67	1.62	1.56	1.53	1.47	1.45	1.40	1.37	1.35	70
	7.01	**4.92**	**4.08**	**3.60**	**3.29**	**3.07**	**2.91**	**2.77**	**2.67**	**2.59**	**2.51**	**2.45**	**2.35**	**2.28**	**2.15**	**2.07**	**1.98**	**1.88**	**1.82**	**1.74**	**1.69**	**1.62**	**1.56**	**1.53**	
80	3.96	3.11	2.72	2.48	2.33	2.21	2.12	2.05	1.99	1.95	1.91	1.88	1.82	1.77	1.70	1.65	1.60	1.54	1.51	1.45	1.42	1.38	1.35	1.32	80
	6.96	**4.88**	**4.04**	**3.56**	**3.25**	**3.04**	**2.87**	**2.74**	**2.64**	**2.55**	**2.48**	**2.41**	**2.32**	**2.24**	**2.11**	**2.03**	**1.94**	**1.84**	**1.78**	**1.70**	**1.65**	**1.57**	**1.52**	**1.49**	
100	3.94	3.09	2.70	2.46	2.30	2.19	2.10	2.03	1.97	1.92	1.88	1.85	1.79	1.75	1.68	1.63	1.57	1.51	1.48	1.42	1.39	1.34	1.30	1.28	100
	6.90	**4.82**	**3.98**	**3.51**	**3.20**	**2.99**	**2.82**	**2.69**	**2.59**	**2.51**	**2.43**	**2.36**	**2.26**	**2.19**	**2.06**	**1.98**	**1.89**	**1.79**	**1.73**	**1.64**	**1.59**	**1.51**	**1.46**	**1.43**	
125	3.92	3.07	2.68	2.44	2.29	2.17	2.08	2.01	1.95	1.90	1.86	1.83	1.77	1.72	1.65	1.60	1.55	1.49	1.45	1.39	1.36	1.31	1.27	1.25	125
	6.84	**4.78**	**3.94**	**3.47**	**3.17**	**2.95**	**2.79**	**2.65**	**2.56**	**2.47**	**2.40**	**2.33**	**2.23**	**2.15**	**2.03**	**1.94**	**1.85**	**1.75**	**1.68**	**1.59**	**1.54**	**1.46**	**1.40**	**1.37**	
150	3.91	3.06	2.67	2.43	2.27	2.16	2.07	2.00	1.94	1.89	1.85	1.82	1.76	1.71	1.64	1.59	1.54	1.47	1.44	1.37	1.34	1.29	1.25	1.22	150
	6.81	**4.75**	**3.91**	**3.44**	**3.14**	**2.92**	**2.76**	**2.62**	**2.53**	**2.44**	**2.37**	**2.30**	**2.20**	**2.12**	**2.00**	**1.91**	**1.83**	**1.72**	**1.66**	**1.56**	**1.51**	**1.43**	**1.37**	**1.33**	
200	3.89	3.04	2.65	2.41	2.26	2.14	2.05	1.98	1.92	1.87	1.83	1.80	1.74	1.69	1.62	1.57	1.52	1.45	1.42	1.35	1.32	1.26	1.22	1.19	200
	6.76	**4.71**	**3.88**	**3.41**	**3.11**	**2.90**	**2.73**	**2.60**	**2.50**	**2.41**	**2.34**	**2.28**	**2.17**	**2.09**	**1.97**	**1.88**	**1.79**	**1.69**	**1.62**	**1.53**	**1.48**	**1.39**	**1.33**	**1.28**	
400	3.86	3.02	2.62	2.39	2.23	2.12	2.03	1.96	1.90	1.85	1.81	1.78	1.72	1.67	1.60	1.54	1.49	1.42	1.38	1.32	1.28	1.22	1.16	1.13	400
	6.70	**4.66**	**3.83**	**3.36**	**3.06**	**2.85**	**2.69**	**2.55**	**2.46**	**2.37**	**2.29**	**2.23**	**2.12**	**2.04**	**1.92**	**1.84**	**1.74**	**1.64**	**1.57**	**1.47**	**1.42**	**1.32**	**1.24**	**1.19**	
1,000	3.85	3.00	2.61	2.38	2.22	2.10	2.02	1.95	1.89	1.84	1.80	1.76	1.70	1.65	1.58	1.53	1.47	1.41	1.36	1.30	1.26	1.19	1.13	1.08	1,000
	6.66	**4.62**	**3.80**	**3.34**	**3.04**	**2.82**	**2.66**	**2.53**	**2.43**	**2.34**	**2.26**	**2.20**	**2.09**	**2.01**	**1.89**	**1.81**	**1.71**	**1.61**	**1.54**	**1.44**	**1.38**	**1.28**	**1.19**	**1.11**	
∞	3.84	2.99	2.60	2.37	2.21	2.09	2.01	1.94	1.88	1.83	1.79	1.75	1.69	1.64	1.57	1.52	1.46	1.40	1.35	1.28	1.24	1.17	1.11	1.00	∞
	6.64	**4.60**	**3.78**	**3.32**	**3.02**	**2.80**	**2.64**	**2.51**	**2.41**	**2.32**	**2.24**	**2.18**	**2.07**	**1.99**	**1.87**	**1.79**	**1.69**	**1.59**	**1.52**	**1.41**	**1.36**	**1.25**	**1.15**	**1.00**	

TABLE G Functions of p, q, z, and y, where p and q are proportions ($p + q = 1.00$) and z and y are constants of the unit normal-distribution curve*

	A	B	C	D	E	F	G	H	I	J	K	L	M	
p	pq	$\sqrt{pq}$	pq/y	$\sqrt{pq}/y$	p/y	y/p	zy/p	y	zy/q	y/q	q/y	$\sqrt{p/q}$	$\sqrt{q/p}$	p
.99	.0099	.0995 –	.3715	3.733	37.15 –	.02692	–.06262	.02665	6.2002	2.665	.3752	9.950	.1005	.01
.98	.0196	.1400	.4048	2.892	20.24	.04941	–.1015	.04842	4.9719	2.421	.4131	7.000	.1429	.02
.97	.0291	.1706	.4277	2.507	14.26	.07015	–.1319	.06804	4.2657	2.268	.4409	5.686	.1759	.03
.96	.0384	.1960	.4456	2.274	11.14	.08976	–.1571	.08617	3.7717	2.154	.4642	4.899	.2041	.04
.95	.0475	.2179	.4605	2.113	9.211	.1086	–.1786	.1031	3.3928	2.063	.4848	4.359	.2294	.05
.94	.0564	.2375 –	.4735	1.994	7.891	.1267	–.1970	.1191	3.0868	1.985	.5037	3.958	.2526	.06
.93	.0651	.2551	.4848	1.900	6.926	.1444	–.2131	.1343	2.8307	1.918	.5213	3.645	.2743	.07
.92	.0736	.2713	.4951	1.825	6.188	.1616	–.2271	.1487	2.6110	1.858	.5381	3.391	.2949	.08
.91	.0819	.2862	.5043	1.762	5.604	.1785	–.2393	.1624	2.4191	1.804	.5542	3.180	.3145	.09
.90	.0900	.3000	.5128	1.709	5.128	.1950	–.2499	.1755	2.2491	1.755	.5698	3.000	.3333	.10
.89	.0979	.3129	.5206	1.664	4.733	.2113	–.2591	.1880	2.0966	1.709	.5850	2.844	.3516	.11
.88	.1056	.3250	.5279	1.625	4.399	.2273	–.2671	.2000	1.9587	1.667	.5999	2.708	.3693	.12
.87	.1131	.3363	.5346	1.590	4.112	.2432	–.2739	.2115	1.8330	1.627	.6145	2.587	.3865	.13
.86	.1204	.3470	.5409	1.559	3.864	.2588	–.2796	.2226	1.7175	1.590	.6290	2.478	.4035	.14
.85	.1275	.3571	.5468	1.532	3.646	.2743	–.2843	.2332	1.6110	1.554	.6433	2.380	.4201	.15
.84	.1344	.3666	.5524	1.507	3.452	.2896	–.2880	.2433	1.5123	1.521	.6576	2.291	.4365	.16
.83	.1411	.3756	.5576	1.484	3.280	.3049	–.2909	.2531	1.4203	1.489	.6718	2.210	.4525	.17
.82	.1476	.3842	.5625	1.464	3.125	.3200	–.2929	.2624	1.3344	1.458	.6860	2.134	.4685	.18
.81	.1539	.3923	.5671	1.446	2.985	.3350	–.2941	.2714	1.2538	1.428	.7002	2.065	.4844	.19
.80	.1600	.4000	.5715	1.429	2.858	.3500	–.2946	.2800	1.1781	1.400	.7144	2.000	.5000	.20
.79	.1659	.4073	.5756	1.413	2.741	.3648	–.2942	.2882	1.1067	1.372	.7287	1.940	.5156	.21
.78	.1716	.4142	.5796	1.399	2.634	.3796	–.2931	.2961	1.0393	1.346	.7430	1.883	.5311	.22
.77	.1771	.4208	.5832	1.386	2.536	.3943	–.2913	.3036	.9754	1.320	.7575	1.830	.5465	.23
.76	.1824	.4271	.5867	1.374	2.445	.4090	–.2889	.3109	.9149	1.295	.7720	1.780	.5620	.24
.75	.1875	.4330	.5900	1.363	2.360	.4237	–.2858	.3178	.8573	1.271	.7867	1.732	.5774	.25

* When p is less than .50, look for it in the last column.

TABLE G Functions of p, q, z, and y, where p and q are proportions ($p + q = 1.00$) and z and y are constants of the unit normal-distribution curve (continued)

p	A	B	C	D	E	F	G	H	I	J	K	L	M	p
	pq	$\sqrt{pq}$	pq/y	$\sqrt{pq}/y$	p/y	y/p	zy/p	y	zy/q	y/q	q/y	$\sqrt{p/q}$	$\sqrt{q/p}$	
.74	.1924	.4386	.5931	1.352	2.281	.4384	−.2820	.3244	.8026	1.248	.8016	1.687	.5928	.26
.73	.1971	.4440	.5961	1.343	2.208	.4529	−.2775	.3306	.7504	1.225	.8166	1.644	.6082	.27
.72	.2016	.4490	.5989	1.334	2.139	.4675	−.2725	.3366	.7006	1.202	.8318	1.604	.6236	.28
.71	.2059	.4538	.6015	1.326	2.074	.4822	−.2668	.3423	.6532	1.180	.8472	1.565	.6391	.29
.70	.2100	.4583	.6040	1.318	2.013	.4967	−.2605	.3477	.6078	1.159	.8628	1.528	.6547	.30
.69	.2139	.4625−	.6063	1.311	1.956	.5113	−.2535	.3528	.5643	1.138	.8787	1.492	.6703	.31
.68	.2176	.4665−	.6085	1.304	1.902	.5259	−.2460	.3576	.5227	1.118	.8949	1.458	.6860	.32
.67	.2211	.4702	.6106	1.298	1.850	.5405	−.2378	.3621	.4828	1.097	.9112	1.425	.7018	.33
.66	.2244	.4737	.6124	1.293	1.801	.5552	−.2290	.3664	.4445	1.078	.9279	1.393	.7178	.34
.65	.2275	.4770	.6142	1.288	1.755	.5698	−.2196	.3704	.4078	1.058	.9449	1.363	.7338	.35
.64	.2304	.4800	.6158	1.283	1.711	.5845	−.2095	.3741	.3725	1.039	.9623	1.333	.7500	.36
.63	.2331	.4828	.6174	1.279	1.669	.5993	−.1989	.3776	.3387	1.020	.9800	1.305	.7663	.37
.62	.2356	.4854	.6188	1.275	1.628	.6141	−.1876	.3808	.3061	1.002	.9980	1.277	.7829	.38
.61	.2379	.4877	.6200	1.271	1.590	.6290	−.1757	.3837	.2748	.9938	1.016	1.251	.7996	.39
.60	.2400	.4899	.6212	1.268	1.553	.6439	−.1631	.3863	.2447	.9659	1.035	1.225	.8165	.40
.59	.2419	.4918	.6223	1.265	1.518	.6589	−.1499	.3888	.2158	.9482	1.055	1.200	.8336	.41
.58	.2436	.4936	.6232	1.263	1.484	.6739	−.1361	.3909	.1879	.9307	1.074	1.175	.8510	.42
.57	.2451	.4951	.6240	1.260	1.451	.6891	−.1215	.3928	.1611	.9134	1.095	1.151	.8686	.43
.56	.2464	.4964	.6247	1.259	1.420	.7043	−.1063	.3944	.1353	.8964	1.116	1.128	.8864	.44
.55	.2475	.4975−	.6253	1.257	1.390	.7196	−.09043	.3958	.1105	.8796	1.137	1.106	.9045	.45
.54	.2484	.4984	.6258	1.256	1.360	.7351	−.07382	.3969	.0867	.8629	1.159	1.083	.9229	.46
.53	.2491	.4991	.6262	1.255	1.332	.7506	−.05650	.3978	.0637	.8464	1.181	1.062	.9417	.47
.52	.2496	.4996	.6264	1.254	1.305	.7662	−.03843	.3984	.0416	.8301	1.205	1.041	.9608	.48
.51	.2499	.4999	.6266	1.253	1.279	.7820	−.01960	.3988	.0204	.8139	1.229	1.020	.9802	.49
.50	.2500	.5000	.6267	1.253	1.253	.7979	−.00000	.3989	.0000	.7979	1.253	1.000	1.0000	.50

TABLE H Conversion of a Pearson *r* into a Fisher *Z* coefficient*

r	*Z*	*r*	*Z*	*r*	*Z*	*r*	*Z*	*r*	*Z*
.000	.000	.200	.203	.400	.424	.600	.693	.800	1.099
.005	.005	.205	.208	.405	.430	.605	.701	.805	1.113
.010	.010	.210	.213	.410	.436	.610	.709	.810	1.127
.015	.015	.215	.218	.415	.442	.615	.717	.815	1.142
.020	.020	.220	.224	.420	.448	.620	.725	.820	1.157
.025	.025	.225	.229	.425	.454	.625	.733	.825	1.172
.030	.030	.230	.234	.430	.460	.630	.741	.830	1.188
.035	.035	.235	.239	.435	.466	.635	.750	.835	1.204
.040	.040	.240	.245	.440	.472	.640	.758	.840	1.221
.045	.045	.245	.250	.445	.478	.645	.767	.845	1.238
.050	.050	.250	.255	.450	.485	.650	.775	.850	1.256
.055	.055	.255	.261	.455	.491	.655	.784	.855	1.274
.060	.060	.260	.266	.460	.497	.660	.793	.860	1.293
.065	.065	.265	.271	.465	.504	.665	.802	.865	1.313
.070	.070	.270	.277	.470	.510	.670	.811	.870	1.333
.075	.075	.275	.282	.475	.517	.675	.820	.875	1.354
.080	.080	.280	.288	.480	.523	.680	.829	.880	1.376
.085	.085	.285	.293	.485	.530	.685	.838	.885	1.398
.090	.090	.290	.299	.490	.536	.690	.848	.890	1.422
.095	.095	.295	.304	.495	.543	.695	.858	.895	1.447
.100	.100	.300	.310	.500	.549	.700	.867	.900	1.472
.105	.105	.305	.315	.505	.556	.705	.877	.905	1.499
.110	.110	.310	.321	.510	.563	.710	.887	.910	1.528
.115	.116	.315	.326	.515	.570	.715	.897	.915	1.557
.120	.121	.320	.332	.520	.576	.720	.908	.920	1.589
.125	.126	.325	.337	.525	.583	.725	.918	.925	1.623
.130	.131	.330	.343	.530	.590	.730	.929	.930	1.658
.135	.136	.335	.348	.535	.597	.735	.940	.935	1.697
.140	.141	.340	.354	.540	.604	.740	.950	.940	1.738
.145	.146	.345	.360	.545	.611	.745	.962	.945	1.783
.150	.151	.350	.365	.550	.618	.750	.973	.950	1.832
.155	.156	.355	.371	.555	.626	.755	.984	.955	1.886
.160	.161	.360	.377	.560	.633	.760	.996	.960	1.946
.165	.167	.365	.383	.565	.640	.765	1.008	.965	2.014
.170	.172	.370	.388	.570	.648	.770	1.020	.970	2.092
.175	.177	.375	.394	.575	.655	.775	1.033	.975	2.185
.180	.182	.380	.400	.580	.662	.780	1.045	.980	2.298
.185	.187	.385	.406	.585	.670	.785	1.058	.985	2.443
.190	.192	.390	.412	.590	.678	.790	1.071	.990	2.647
.195	.198	.395	.418	.595	.685	.795	1.085	.995	2.994

* Adapted with permission from Table V.B. of Fisher's *Statistical methods for research workers*. Edinburgh: Oliver & Boyd, 1932. For a discussion of the use of this table, see pp. 145, 163, and 330.

TABLE I Trigonometric functions*

Angle	*Sin*	*Cos*	*Tan*	*Angle*	*Sin*	*Cos*	*Tan*
0°	.000	1.000	.000	**45°**	.707	.707	1.000
1°	.018	.999	.018	**46°**	.719	.695	1.036
2°	.035	.999	.035	**47°**	.731	.682	1.072
3°	.052	.998	.052	**48°**	.743	.669	1.111
4°	.070	.997	.070	**49°**	.755	.656	1.150
5°	.087	.996	.087	**50°**	.766	.643	1.192
6°	.105	.994	.105	**51°**	.777	.629	1.235
7°	.122	.992	.123	**52°**	.788	.616	1.280
8°	.139	.990	.141	**53°**	.799	.602	1.327
9°	.156	.988	.158	**54°**	.809	.588	1.376
10°	.174	.985	.176	**55°**	.819	.574	1.428
11°	.191	.982	.194	**56°**	.829	.559	1.483
12°	.208	.978	.213	**57°**	.839	.545	1.540
13°	.225	.974	.231	**58°**	.848	.530	1.600
14°	.242	.970	.249	**59°**	.857	.515	1.664
15°	.259	.966	.268	**60°**	.866	.500	1.732
16°	.276	.961	.287	**61°**	.875	.485	1.804
17°	.292	.956	.306	**62°**	.883	.469	1.881
18°	.309	.951	.325	**63°**	.891	.454	1.963
19°	.326	.946	.344	**64°**	.899	.438	2.050
20°	.342	.940	.364	**65°**	.906	.423	2.144
21°	.358	.934	.384	**66°**	.914	.407	2.246
22°	.375	.927	.404	**67°**	.921	.391	2.356
23°	.391	.921	.424	**68°**	.927	.375	2.475
24°	.407	.914	.445	**69°**	.934	.358	2.605
25°	.423	.906	.466	**70°**	.940	.342	2.747
26°	.438	.899	.488	**71°**	.946	.326	2.904
27°	.454	.891	.510	**72°**	.951	.309	3.078
28°	.469	.883	.532	**73°**	.956	.292	3.271
29°	.485	.875	.554	**74°**	.961	.276	3.487
30°	.500	.866	.577	**75°**	.966	.259	3.732
31°	.515	.857	.601	**76°**	.970	.242	4.011
32°	.530	.848	.625	**77°**	.974	.225	4.331
33°	.545	.839	.649	**78°**	.978	.208	4.705
34°	.559	.829	.675	**79°**	.982	.191	5.145
35°	.574	.819	.700	**80°**	.985	.174	5.671
36°	.588	.809	.727	**81°**	.988	.156	6.314
37°	.602	.799	.754	**82°**	.990	.139	7.115
38°	.616	.788	.781	**83°**	.992	.122	8.144
39°	.629	.777	.810	**84°**	.994	.105	9.514
40°	.643	.766	.839	**85°**	.996	.087	11.430
41°	.656	.755	.869	**86°**	.997	.070	14.300
42°	.669	.743	.900	**87°**	.998	.052	19.081
43°	.682	.731	.933	**88°**	.999	.035	28.636
44°	.695	.719	.966	**89°**	.999	.018	57.290

* From Smail, L. L. *College Algebra*. New York: McGraw-Hill, 1931.

TABLE J Four-place logarithms of numbers*

N	*0*	*1*	*2*	*3*	*4*	*5*	*6*	*7*	*8*	*9*
0		0000	3010	4771	6021	6990	7782	8451	9031	9542
1	0000	0414	0792	1139	1461	1761	2041	2304	2553	2788
2	3010	3222	3424	3617	3802	3979	4150	4314	4472	4624
3	4771	4914	5051	5185	5315	5441	5563	5682	5798	5911
4	6021	6128	6232	6335	6435	6532	6628	6721	6812	6902
5	6990	7076	7160	7243	7324	7404	7482	7559	7634	7709
6	7782	7853	7924	7993	8062	8129	8195	8261	8325	8388
7	8451	8513	8573	8633	8692	8751	8808	8865	8921	8976
8	9031	9085	9138	9191	9243	9294	9345	9395	9445	9494
9	9542	9590	9638	9685	9731	9777	9823	9868	9912	9956
10	0000	0043	0086	0128	0170	0212	0253	0294	0334	0374
11	0414	0453	0492	0531	0569	0607	0645	0682	0719	0755
12	0792	0828	0864	0899	0934	0969	1004	1038	1072	1106
13	1139	1173	1206	1239	1271	1303	1335	1367	1399	1430
14	1461	1492	1523	1553	1584	1614	1644	1673	1703	1732
15	1761	1790	1818	1847	1875	1903	1931	1959	1987	2014
16	2041	2068	2095	2122	2148	2175	2201	2227	2253	2279
17	2304	2330	2355	2380	2405	2430	2455	2480	2504	2529
18	2553	2577	2601	2625	2648	2672	2605	2718	2742	2765
19	2788	2810	2833	2856	2878	2900	2934	2945	2967	2989
20	3010	3032	3054	3075	3096	3118	3139	3160	3181	3201
21	3222	3243	3263	3284	3304	3324	3345	3365	3385	3404
22	3424	3444	3464	3483	3502	3522	3541	3560	3579	3598
23	3617	3636	3655	3674	3692	3711	3729	3747	3766	3784
24	3802	3820	3838	3856	3874	3892	3909	3927	3945	3962
25	3979	3997	4014	4031	4048	4065	4082	4099	4116	4133
26	4150	4166	4183	4200	4216	4232	4249	4265	4281	4298
27	4314	4330	4346	4362	4378	4393	4409	4425	4440	4456
28	4472	4487	4502	4518	4533	4548	4564	4579	4594	4609
29	4624	4639	4654	4669	4683	4698	4713	4728	4742	4757
30	4771	4786	4800	4814	4829	4843	4857	4871	4886	4900
31	4914	4928	4942	4955	4969	4983	4997	5011	5024	5038
32	5051	5065	5079	5092	5105	5119	5132	5145	5159	5172
33	5185	5198	5211	5224	5237	5250	5263	5276	5289	5302
34	5315	5328	5340	5353	5366	5378	5391	5403	5416	5428
35	5441	5453	5465	5478	5490	5502	5514	5527	5539	5551
36	5563	5575	5587	5599	5611	5623	5635	5647	5658	5670
37	5682	5694	5705	5717	5729	5740	5752	5763	5775	5786
38	5798	5809	5821	5832	5843	5855	5866	5877	5888	5899
39	5911	5922	5933	5944	5955	5966	5977	5988	5999	6010
40	6021	6031	6042	6053	6064	6075	6085	6096	6107	6117
41	6128	6138	6149	6160	6170	6180	6191	6201	6212	6222
42	6232	6243	6253	6263	6274	6284	6294	6304	6314	6325
43	6335	6345	6355	6365	6375	6385	6395	6405	6415	6425
44	6435	6444	6454	6464	6474	6484	6493	6503	6513	6522
45	6532	6542	6551	6561	6571	6580	6590	6599	6609	6618
46	6628	6637	6646	6656	6665	6675	6684	6693	6702	6712
47	6721	6730	6739	6749	6758	6767	6776	6785	6794	6803
48	6812	6821	6830	6839	6848	6857	6866	6875	6884	6893
49	6902	6911	6920	6928	6937	6946	6955	6964	6972	6981
50	6990	6998	7007	7016	7024	7033	7042	7050	7059	7067
N	*0*	*1*	*2*	*3*	*4*	*5*	*6*	*7*	*8*	*9*

Prop. parts

	22	**21**	**20**	**19**	**18**	**17**	**16**	**15**	**14**	**13**	**12**	**11**	**9**	**8**
1	2.2	2.1	2.0	1.9	1.8	1.7	1.6	1.5	1.4	1.3	1.2	1.1	0.9	0.8
2	4.4	4.2	4.0	3.8	3.6	3.4	3.2	3.0	2.8	2.6	2.4	2.2	1.8	1.6
3	6.6	6.3	6.0	5.7	5.4	5.1	4.8	4.5	4.2	3.9	3.6	3.3	2.7	2.4
4	8.8	8.4	8.0	7.6	7.2	6.8	6.4	6.0	5.6	5.2	4.8	4.4	3.6	3.2
5	11.0	10.5	10.0	9.5	9.0	8.5	8.0	7.5	7.0	6.5	6.0	5.5	4.5	4.0
6	13.2	12.6	12.0	11.4	10.8	10.2	9.6	9.0	8.4	7.8	7.2	6.6	5.4	4.8
7	15.4	14.7	14.0	13.3	12.6	11.9	11.2	10.5	9.8	9.1	8.4	7.7	6.3	5.6
8	17.6	16.8	16.0	15.2	14.4	13.6	12.8	12.0	11.2	10.4	9.6	8.8	7.2	6.4
9	19.8	18.9	18.0	17.1	16.2	15.3	14.4	13.5	12.6	11.7	10.8	9.9	8.1	7.2

* From Smail, L. L. *College Algebra*. New York: McGraw-Hill, 1931.

TABLE K
Significance levels for the Spearman rank-difference coefficient of correlation

	Level of Significance for two-tail test						
N	*.10*	*.02*	*.01*	*N*	*.10*	*.02*	*.01*
5	.900	1.000		16	.425	.601	.665
6	.829	.943	1.000	18	.399	.564	.625
7	.714	.893	.929	20	.377	.534	.591
8	.643	.833	.881	22	.359	.508	.562
9	.600	.783	.833	24	.343	.485	.537
10	.564	.746	.794	26	.329	.465	.515
12	.506	.712	.777	28	.317	.448	.496
14	.456	.645	.715	30	.306	.432	.478
	.05	*.01*	*.005*		*.05*	*.01*	*.005*
	Level of Significance for one-tail test						

* Adapted and reproduced by permission from Dixon, W. J., and Massey, F. J., Jr. *Introduction to statisticsl analysis.* New York: McGraw-Hill, 1951. Table 17-6, p. 261. This table had been derived from Olds, E. G. The 5 per cent significance levels of sums of squares of rank differences and a correction. *Annals of Mathematical Statistics,* 1949, **20,** 117–118.

TABLE L Values to facilitate the estimation of the cosine-pi coefficient of correlation, with two-place accuracy*

$\frac{ad}{bc}$	$r_{\text{cos-pi}}$	$\frac{ad}{bc}$	$r_{\text{cos-pi}}$	$\frac{ad}{bc}$	$r_{\text{cos-pi}}$	$\frac{ad}{bc}$	$r_{\text{cos-pi}}$
1.013	.005†	1.940	.255	4.067	.505	11.512	.755
1.039	.015	1.993	.265	4.205	.515	12.177	.765
1.066	.025	2.048	.275	4.351	.525	12.906	.775
1.093	.035	2.105	.285	4.503	.535	13.702	.785
1.122	.045	2.164	.295	4.662	.545	14.592	.795
1.150	.055	2.225	.305	4.830	.555	15.573	.805
1.180	.065	2.288	.315	5.007	.565	16.670	.815
1.211	.075	2.353	.325	5.192	.575	17.900	.825
1.242	.085	2.421	.335	5.388	.585	19.288	.835
1.275	.095	2.490	.345	5.595	.595	20.866	.845
1.308	.105	2.563	.355	5.813	.605	22.675	.855
1.342	.115	2.638	.365	6.043	.615	24.768	.865
1.377	.125	2.716	.375	6.288	.625	27.212	.875
1.413	.135	2.797	.385	6.547	.635	30.106	.885
1.450	.145	2.881	.395	6.822	.645	33.578	.895
1.488	.155	2.957	.405	7.115	.655	37.818	.905
1.528	.165	3.095	.415	7.428	.665	43.100	.915
1.568	.175	3.153	.425	7.761	.675	49.851	.925
1.610	.185	3.251	.435	8.117	.685	58.765	.935
1.653	.195	3.353	.445	8.499	.695	71.046	.945
1.697	.205	3.460	.455	8.910	.705	88.984	.955
1.743	.215	3.571	.465	9.351	.715	117.52	.965
1.790	.225	3.690	.475	9.828	.725	169.60	.975
1.838	.235	3.808	.485	10.344	.735	293.28	.985
1.888	.245	3.935	.495	10.903	.745	934.06	.995

* Based upon a more detailed tabulation of the same values by Perry, N. C., Kettner, N. W., Hertzka, A. F., and Bouvier, E. A. Estimating the tetrachoric correlation coefficient via a cosine-pi table. Technical Memorandum No. 2. Los Angeles: University of Southern California, 1953.

† Example: If an obtained ratio *ad*/*bc* equals 3.472, we find that this value lies between tabled values of 3.460 and 3.571. The cosine-pi coefficient is therefore between .455 and .465; that is to say, it is .46. If *bc* is greater than *ad*, find the ratio *bc*/*ad* and attach a negative sign to $r_{\text{cos-pi}}$.

TABLE M Cell frequencies required to achieve significant chi squares at the .05 point (lightface) and at the .01 point (boldface) levels when each is parallel to the smallest cell frequency in a fourfold table*

Smallest cell frequency

N_i	0	1	2	3	4	5	6	7	8	9	10	11	12	13	14	15	16	17	18	19	20	21	22	23	24	25
4	4	—	—																							
	—	**—**	**—**																							
5	4	5	—	—																						
	5	**—**	**—**																							
6	5	6	—	—																						
	6	**—**	**—**																							
7	5	6	7	—																						
	6	**7**	**—**	**—**																						
8	5	6	7	8	—																					
	6	**8**	**8**	**—**	**—**																					
9	5	6	8	8	9																					
	6	**8**	**9**	**9**	**—**																					
10	5	7	8	9	10	10																				
	7	**8**	**9**	**10**	**—**	**—**																				
11	5	7	8	9	10	11																				
	7	**8**	**9**	**10**	**11**	**—**																				
12	5	7	8	9	10	11	12																			
	7	**8**	**10**	**11**	**11**	**12**	**—**																			
13	5	7	8	9	10	11	12																			
	7	**9**	**10**	**11**	**12**	**13**	**13**																			
14	5	7	8	10	11	12	12	13																		
	7	**9**	**10**	**11**	**12**	**13**	**14**	**14**																		
15	5	7	9	10	11	12	13	14																		
	7	**9**	**10**	**11**	**12**	**13**	**14**	**15**																		
16	5	7	9	10	11	12	13	14	15																	
	7	**9**	**10**	**12**	**13**	**14**	**14**	**15**	**16**																	
17	5	7	9	10	11	12	13	14	15																	
	7	**9**	**11**	**12**	**13**	**14**	**15**	**16**	**16**																	
18	5	7	9	10	11	12	13	14	15	16																
	7	**9**	**11**	**12**	**13**	**14**	**15**	**16**	**17**	**17**																
19	5	7	9	10	11	12	14	14	15	16																
	7	**9**	**11**	**12**	**13**	**14**	**15**	**16**	**17**	**18**																
20	5	7	9	10	11	13	14	15	16	16	17															
	7	**9**	**11**	**12**	**13**	**15**	**16**	**16**	**17**	**18**	**19**															
30	6	8	9	11	12	13	15	16	17	18	19	20	21	22	23	24										
	8	**10**	**12**	**13**	**15**	**16**	**17**	**18**	**19**	**20**	**21**	**22**	**23**	**24**	**25**	**26**										
40	6	8	9	11	12	14	15	16	18	19	20	21	22	23	24	25	26	27	28	29	30					
	8	**10**	**12**	**14**	**15**	**17**	**18**	**19**	**20**	**22**	**23**	**24**	**25**	**26**	**27**	**28**	**29**	**30**	**31**	**32**	**32**					
50	6	8	10	11	13	14	15	17	18	19	20	22	23	24	25	26	27	28	29	30	31	32	33	34	35	36
	8	**10**	**12**	**14**	**15**	**17**	**18**	**20**	**21**	**22**	**24**	**25**	**26**	**27**	**28**	**29**	**30**	**31**	**32**	**33**	**34**	**35**	**36**	**37**	**38**	**39**
N_i	0	1	2	3	4	5	6	7	8	9	10	11	12	13	14	15	16	17	18	19	20	21	22	23	24	25

* Adapted by permission from Mainland, D., and Murray, I. M. Tables for use in fourfold contingency tables. *Science*, 1952, **116,** 591–594.

INSTRUCTIONS Table *M* was designed for use in comparing frequencies in two corresponding categories, for two groups of equal size (N_i cases in each group), by means of a chi-square test of significance. For example, suppose that 10 men and 10 women were asked whether or not they liked to watch wrestling on television. Of the men, 8 said "Yes" and 2 said "No." Of the women, 4 said "Yes" and 6 said "No."

N_i is 10, and so we use the row of Table M that has 10 at the left. The smallest of the four frequencies is 2, and so we use the column with 2 at the top. At the intersection of the row for 10 and the column for 2, we see number 8 (in lightface) and number 9 (in boldface). These numbers mean that with 2 out of 10 men saying "No," for a chi square to be significant at the .05 level we should need to find that 8 women also said "No." Only 6 women actually said "No," and so the chi square that would be computed from this fourfold table would fail to reach the value that would equal any of the highest .05 of all chance-generated chi squares when there is one degree of freedom. Therefore, we should not reject the null hypothesis.

Had there been as many as 9 women saying "No," we could have rejected the null hypothesis at the .01 level. Had there been no men who said "No," it would require only 5 "No's" from the women to indicate a significant χ^2 at the .05 level and only 7 "No's" from the women to indicate a significant χ^2 at the .01 level. The reader should verify the last two statements by referring to the appropriate cell (intersection of row 10 and column 0) in Table M.

For further information, see p. 203.

TABLE N Cumulative proportions from the tail categories of binomial distributions for $(\frac{1}{2} + \frac{1}{2})^N$, with N varying from 6 to 25

Categories (C)

C N	C0 N	C1 (N − 1)	C2 (N − 2)	C3 (N − 3)	C4 (N − 4)	C5 (N − 5)	C6 (N − 6)	C7 (N − 7)	C8 (N − 8)	C9 (N − 9)
6	.016	.109								
7	.008	.062	.227							
8	.004	.035	.145							
9	.002	.020	.090	.254						
10	.001	.011	.055	.172						
11		.006	.033	.113						
12		.003	.019	.073	.194					
13		.002	.011	.046	.133					
14		.001	.006	.029	.090					
15			.004	.018	.059	.151				
16			.002	.011	.038	.105				
17			.001	.006	.025	.072	.166			
18			.001	.004	.015	.048	.119			
19				.002	.010	.032	.084	.180		
20				.001	.006	.021	.058	.132		
21				.001	.004	.013	.039	.095	.192	
22					.002	.008	.026	.067	.143	
23					.001	.005	.017	.047	.105	
24					.001	.003	.011	.032	.076	.154
25						.002	.007	.022	.054	.115

COMMENT Each entry is the probability of an outcome as extreme as the last category (0 heads or N heads, as in coin tossing), the next-to-the-last category (1 head or $N - 1$ heads), and so on (i.e., the probabilities are cumulative). Each probability is for one tail only. For a two-tail test, double the probability given.

TABLE O *T* values at the .05, .02, and .01 levels for different numbers of ranked differences. *T* is the smaller sum of ranks associated with differences all of the same sign, in two-tail sign-rank tests*

N	$p = .05$	$p = .02$	$p = .01$	N	$p = .05$	$p = .02$	$p = .01$
6	0			31	148	129	116
7	2	0		32	159	139	126
8	4	2	0	33	171	150	136
9	6	3	2	34	183	161	147
10	8	5	3	35	195	173	158
11	11	7	5	36	208	185	169
12	14	10	7	37	222	197	181
13	17	13	10	38	235	210	193
14	21	16	13	39	249	223	205
15	25	20	16	40	264	237	218
16	30	24	20	41	279	251	231
17	35	28	23	42	295	265	245
18	40	33	28	43	311	280	259
19	46	38	32	44	327	295	274
20	52	43	38	45	344	311	289
21	59	49	43	46	361	327	304
22	66	56	49	47	379	344	320
23	73	62	55	48	397	361	336
24	81	69	61	49	416	379	353
25	89	77	68	50	435	397	370
26	98	84	74				
27	107	92	82				
28	117	101	90				
29	127	110	98				
30	137	119	107				

* Reproduced by permission from Wilcoxon, F. *Some rapid approximate statistical procedures*. Stamford, Conn.: American Cyanamid Co., 1949, with addition of cases with *N* 26–50. For information on the use of this table, see p. 216.

TABLE P
Significant R values at the .05, .02, and .01 levels for different numbers of N_i cases in two samples of equal size. R is the smaller sum of ranks, in two-tail components-rank tests*

N_i	$p = .05$	$p = .02$	$p = .01$
5	18	16	15
6	27	24	23
7	37	34	32
8	49	46	44
9	63	59	56
10	79	74	71
11	97	91	87
12	116	110	105
13	137	130	125
14	160	152	147
15	185	176	170
16	212	202	196
17	241	230	223
18	271	259	252
19	303	291	282
20	338	324	315

* Reproduced by permission from Wilcoxon, F. *Some rapid approximate statistical procedures*. Stamford, Conn.: American Cyanamid Co., 1949. For further information, see p. 218.

TABLE Q Significance levels for the Pearson product-moment coefficient of correlation*

	Level of significance for two-tail test			
df *(N − 2)*	*.10*	*.05*	*.02*	*.01*
1	.988	.997	.9995	.9999
2	.900	.950	.980	.990
3	.805	.878	.934	.959
4	.729	.811	.882	.917
5	.669	.754	.833	.874
6	.622	.707	.789	.834
7	.582	.666	.750	.798
8	.549	.632	.716	.765
9	.521	.602	.685	.735
10	.497	.576	.658	.708
11	.476	.553	.634	.684
12	.458	.532	.612	.661
13	.441	.514	.592	.641
14	.426	.497	.574	.623
15	.412	.482	.558	.606
16	.400	.468	.542	.590
17	.389	.456	.528	.575
18	.378	.444	.516	.561
19	.369	.433	.503	.549
20	.360	.423	.492	.537
21	.352	.413	.482	.526
22	.344	.404	.472	.515
23	.337	.396	.462	.505
24	.330	.388	.453	.496
25	.323	.381	.445	.487
26	.317	.374	.437	.479
27	.311	.367	.430	.471
28	.306	.361	.423	.463
29	.301	.355	.416	.456
30	.296	.349	.409	.449
35	.275	.325	.381	.418
40	.257	.304	.358	.393
45	.243	.288	.338	.372
50	.231	.273	.322	.354
60	.211	.250	.295	.325
df	*.05*	*.025*	*.01*	*.005*
	Level of significance for a one-tail test.			

TABLE Q Significance levels for the Pearson product-moment coefficient of correlation* (continued)

	Level of significance for two-tail test			
df *(N − 2)*	*.10*	*.05*	*.02*	*.01*
70	.195	.232	.274	.303
80	.183	.217	.256	.283
90	.173	.205	.242	.267
100	.164	.195	.230	.254
125	.147	.174	.206	.228
150	.134	.159	.189	.208
200	.116	.138	.164	.181
300	.095	.113	.134	.148
400	.082	.098	.116	.128
500	.073	.088	.104	.115
1,000	.052	.062	.073	.081
df	*.05*	*.025*	*.01*	*.005*
	Level of significance for one-tail test			

* Abridged, with some extensions, from Fisher, R. A., and F. Yates, *Statistical tables for biological, agricultural, and medical research.* Published by Longman Group Ltd., London (previously published by Oliver & Boyd, Edinborough). Used with permission of authors and publishers.

TABLE R Coefficients of multiple correlation significant at the .05 level (lightface type) and at the .01 level (boldface type) for various df ($N - m$) and number of variables (m)*

Degrees of freedom	*Number of variables* 3	4	5	6	7	9
1	.999	.999	.999	1.000	1.000	1.000
	1.000	**1.000**	**1.000**	**1.000**	**1.000**	**1.000**
2	.975	.983	.987	.990	.992	.994
	.995	**.997**	**.998**	**.998**	**.998**	**.999**
3	.930	.950	.961	.968	.973	.979
	.976	**.983**	**.987**	**.990**	**.991**	**.993**
4	.881	.912	.930	.942	.950	.961
	.949	**.962**	**.970**	**.975**	**.979**	**.984**
5	.836	.874	.898	.914	.925	.941
	.917	**.937**	**.949**	**.957**	**.963**	**.971**
6	.795	.839	.867	.886	.900	.920
	.886	**.911**	**.927**	**.938**	**.946**	**.957**
7	.758	.807	.838	.860	.876	.900
	.855	**.885**	**.904**	**.918**	**.928**	**.942**
8	.726	.777	.811	.835	.854	.880
	.827	**.860**	**.882**	**.898**	**.909**	**.926**
9	.697	.750	.786	.812	.832	.861
	.800	**.836**	**.861**	**.878**	**.891**	**.911**
10	.671	.726	.763	.790	.812	.843
	.776	**.814**	**.840**	**.859**	**.874**	**.895**
11	.648	.703	.741	.770	.792	.826
	.753	**.793**	**.821**	**.841**	**.857**	**.880**
12	.627	.683	.722	.751	.774	.809
	.732	**.773**	**.802**	**.824**	**.841**	**.866**
13	.608	.664	.703	.733	.757	.794
	.712	**.755**	**.785**	**.807**	**.825**	**.852**
14	.590	.646	.686	.717	.741	.779
	.694	**.737**	**.768**	**.792**	**.810**	**.838**
15	.574	.630	.670	.701	.726	.765
	.677	**.721**	**.752**	**.776**	**.796**	**.825**
16	.559	.615	.655	.686	.712	.751
	.662	**.706**	**.738**	**.762**	**.782**	**.813**
17	.545	.601	.641	.673	.698	.738
	.647	**.691**	**.724**	**.749**	**.769**	**.800**
18	.532	.587	.628	.660	.686	.726
	.633	**.678**	**.710**	**.736**	**.756**	**.789**
19	.520	.575	.615	.647	.674	.714
	.620	**.665**	**.698**	**.723**	**.744**	**.778**
20	.509	.563	.604	.636	.662	.703
	.608	**.652**	**.685**	**.712**	**.733**	**.767**
21	.498	.552	.592	.624	.651	.693
	.596	**.641**	**.674**	**.700**	**.722**	**.756**
22	.488	.542	.582	.614	.640	.682
	.585	**.630**	**.663**	**.690**	**.712**	**.746**
23	.479	.532	.572	.604	.630	.673
	.574	**.619**	**.652**	**.679**	**.701**	**.736**

Degrees of freedom	*Number of variables* 3	4	5	6	7	9
24	.470	.523	.562	.594	.621	.663
	.565	**.609**	**.642**	**.669**	**.692**	**.727**
25	.462	.514	.553	.585	.612	.654
	.555	**.600**	**.633**	**.660**	**.682**	**.718**
26	.454	.506	.545	.576	.603	.645
	.546	**.590**	**.624**	**.651**	**.673**	**.709**
27	.446	.498	.536	.568	.594	.637
	.538	**.582**	**.615**	**.642**	**.664**	**.701**
28	.439	.490	.529	.560	.586	.629
	.530	**.573**	**.606**	**.634**	**.656**	**.692**
29	.432	.482	.521	.552	.579	.621
	.522	**.565**	**.598**	**.625**	**.648**	**.685**
30	.426	.476	.514	.545	.571	.614
	.514	**.558**	**.591**	**.618**	**.640**	**.677**
35	.397	.445	.482	.512	.538	.580
	.481	**.523**	**.556**	**.582**	**.605**	**.642**
40	.373	.419	.455	.484	.509	.551
	.454	**.494**	**.526**	**.552**	**.575**	**.612**
45	.353	.397	.432	.460	.485	.526
	.430	**.470**	**.501**	**.527**	**.549**	**.586**
50	.336	.379	.412	.440	.464	.504
	.410	**.449**	**.479**	**.504**	**.526**	**.562**
60	.308	.348	.380	.406	.429	.467
	.377	**.414**	**.442**	**.466**	**.488**	**.523**
70	.286	.324	.354	.379	.401	.438
	.351	**.386**	**.413**	**.436**	**.456**	**.491**
80	.269	.304	.332	.356	.377	.413
	.330	**.362**	**.389**	**.411**	**.431**	**.464**
90	.254	.288	.315	.338	.358	.392
	.312	**.343**	**.368**	**.390**	**.409**	**.441**
100	.241	.274	.300	.322	.341	.374
	.297	**.327**	**.351**	**.372**	**.390**	**.421**
125	.216	.246	.269	.290	.307	.338
	.266	**.294**	**.316**	**.335**	**.352**	**.381**
150	.198	.225	.247	.266	.282	.310
	.244	**.270**	**.290**	**.308**	**.324**	**.351**
200	.172	.196	.215	.231	.246	.271
	.212	**.234**	**.253**	**.269**	**.283**	**.307**
300	.141	.160	.176	.190	.202	.223
	.174	**.192**	**.208**	**.221**	**.233**	**.253**
400	.122	.139	.153	.165	.176	.194
	.151	**.167**	**.180**	**.192**	**.202**	**.220**
500	.109	.124	.137	.148	.157	.174
	.135	**.150**	**.162**	**.172**	**.182**	**.198**
1,000	.077	.088	.097	.105	.112	.124
	.096	**.106**	**.115**	**.122**	**.129**	**.141**

* Adapted from Wallace, H. A. and Snedecor, G. W. *Correlation and machine calculation.* Ames, Iowa: Iowa State College, 1931. By courtesy of the authors. All R values pertain to two-tail tests.

TABLE S Critical values of the Mann-Whitney U for significance at $\alpha = .05$ in a one-tail test and $\alpha = .10$ in a two-tail test.* (N_i for the larger sample is found at the top of a column.)

N_i	9	10	11	12	13	14	15	16	17	18	19	20
1											0	0
2	1	1	1	2	2	2	3	3	3	4	4	4
3	3	4	5	5	6	7	7	8	9	9	10	11
4	6	7	8	9	10	11	12	14	15	16	17	18
5	9	11	12	13	15	16	18	19	20	22	23	25
6	12	14	16	17	19	21	23	25	26	28	30	32
7	15	17	19	21	24	26	28	30	33	35	37	39
8	18	20	23	26	28	31	33	36	39	41	44	47
9	21	24	27	30	33	36	39	42	45	48	51	54
10	24	27	31	34	37	41	44	48	51	55	58	62
11	27	31	34	38	42	46	50	54	57	61	65	69
12	30	34	38	42	47	51	55	60	64	68	72	77
13	33	37	42	47	51	56	61	65	70	75	80	84
14	36	41	46	51	56	61	66	71	77	82	87	92
15	39	44	50	55	61	66	72	77	83	88	94	100
16	42	48	54	60	65	71	77	83	89	95	101	107
17	45	51	57	64	70	77	83	89	96	102	109	115
18	48	55	61	68	75	82	88	95	102	109	116	123
19	51	58	65	72	80	87	94	101	109	116	123	130

Critical values of U for significance at $\alpha = .01$ in a one-tail test and $\alpha = .02$ in a two-tail test

N_i	9	10	11	12	13	14	15	16	17	18	19	20
2					0	0	0	0	0	0	1	1
3	1	1	1	2	2	2	3	3	4	4	4	5
4	3	3	4	5	5	6	7	7	8	9	9	10
5	5	6	7	8	9	10	11	12	13	14	15	16
6	7	8	9	11	12	13	15	16	18	19	20	22
7	9	11	12	14	16	17	19	21	23	24	26	28
8	11	13	15	17	20	22	24	26	28	30	32	34
9	14	16	18	21	23	26	28	31	33	36	38	40
10	16	19	22	24	27	30	33	36	38	41	44	47
11	18	22	25	28	31	34	37	41	44	47	50	53
12	21	24	28	31	35	38	42	46	49	53	56	60
13	23	27	31	35	39	43	47	51	55	59	63	67
14	26	30	34	38	43	47	51	56	60	65	69	73
15	28	33	37	42	47	51	56	61	66	70	75	80
16	31	36	41	46	51	56	61	66	71	76	82	87
17	33	38	44	49	55	60	66	71	77	82	88	93
18	36	41	47	53	59	65	70	76	82	88	94	100
19	38	44	50	56	63	69	75	82	88	94	101	107

* Adapted and abridged from Tables 1, 3, 5 and 7 of Auble, D. Extended tables for the Mann-Whitney statistic. *Bulletin of the Institute of Educational Research at Indiana University*, **1**, No. 2, 1953, with the kind permission of the author and the publisher. Applications of the Mann-Whitney test are discussed beginning on p. 218.

TABLE T Coefficients of orthogonal polynomials for equally spaced intervals*

k	*Comparison*	*1*	*2*	*3*	*4*	*5*	*6*	*7*	*8*	*9*	*10*	a_i^2
3	**Linear**	−1	0	1								2
	Quadratic	1	−2	1								6
	Linear	−3	−1	1	3							20
4	**Quadratic**	1	−1	−1	1							4
	Cubic	−1	3	−3	1							20
	Linear	−2	−1	0	1	2						10
5	**Quadratic**	2	−1	−2	−1	2						14
	Cubic	−1	2	0	−2	1						10
	Quartic	1	−4	6	−4	1						70
	Linear	−5	−3	−1	1	3	5					70
6	**Quadratic**	5	−1	−4	−4	−1	5					84
	Cubic	−5	7	4	−4	−7	5					180
	Quartic	1	−3	2	2	−3	1					28
	Linear	−3	−2	−1	0	1	2	3				28
7	**Quadratic**	5	0	−3	−4	−3	0	5				84
	Cubic	−1	1	1	0	−1	−1	1				6
	Quartic	3	−7	1	6	1	−7	3				154
	Linear	−7	−5	−3	−1	1	3	5	7			168
8	**Quadratic**	7	1	−3	−5	−5	−3	1	7			168
	Cubic	−7	5	7	3	−3	−7	−5	7			264
	Quartic	7	−13	−3	9	9	−3	−13	7			616
	Linear	−4	−3	−2	−1	0	1	2	3	4		60
9	**Quadratic**	28	7	−8	−17	−20	−17	−8	7	28		2772
	Cubic	−14	7	13	9	0	−9	−13	−7	14		990
	Quartic	14	−21	−11	9	18	9	−11	−21	14		2002
	Linear	−9	−7	−5	−3	−1	1	3	5	7	9	330
10	**Quadratic**	6	2	−1	−3	−4	−4	−3	−1	2	6	132
	Cubic	−42	14	35	31	12	−12	−31	−35	−14	42	8580
	Quartic	18	−22	−17	3	18	18	3	−17	−22	18	2860

* Information on use of this table appears on pp. 277–283.

Name Index

Subject Index